INDETERMINATE STRUCTURAL ANALYSIS

By the late J. STERLING KINNEY
Rensselaer Polytechnic Institute

This book is designed to provide an introduction to indeterminate structural Every attempt has been made to develo entation which is coherent, lucid, and in It is written for the student and for the structural engineer or architect.

Although the text is introductory it is not elementary in the usual sense because of the extended treatment of many of the subjects which are considered. It is believed that the theoretical discussions and the solutions of examples, as herein presented, are unusually comprehensive and detailed. The level of presentation is either fourth-year undergraduate or first-year graduate. The reader is assumed to be thoroughly familiar with determinate structural analysis but to have no knowledge of indeterminate structures.

A brief history of the evolvement of structural theory is presented in the first chapter. The next three chapters are used to develop the theoretical foundation upon which the remainder of the book is based. Chapters five through thirteen include rather detailed discussions and demonstrations of those methods which are most frequently used in the analysis of indeterminate structures. The general method of Maxwell as modified by Mohr and Müller-Breslau, the method of least work, and an introduction to the column analogy are presented in Chapters 5, 6, and 7, respectively. Chapters 8, 9, and 10 are devoted to a rather comprehensive and detailed discussion of the method of moment distribution. The slope-deflection method, for both prismatic and nonprismatic members, is developed in Chapter 11. Influence lines for indeterminate structures are considered in Chapter 12. The analysis of arches, by either the elastic or the deflection theories, is presented in Chapter 13. Finally, in Chapter 14, basic principles and methods of direct and indirect structural model analysis are discussed.

INDETERMINATE STRUCTURAL ANALYSIS

This book is in the
ADDISON-WESLEY SERIES IN CIVIL ENGINEERING
J. STERLING KINNEY AND ROBERT B. BANKS
Consulting Editors

INDETERMINATE STRUCTURAL ANALYSIS

by

J. STERLING KINNEY

Department of Civil Engineering
Rensselaer Polytechnic Institute

COMPLETE EDITION

ADDISON-WESLEY PUBLISHING COMPANY, INC.
READING, MASSACHUSETTS, U.S.A.

Copyright © 1957
ADDISON-WESLEY PUBLISHING COMPANY, INC.

Printed in the United States of America

ALL RIGHTS RESERVED. THIS BOOK, OR PARTS THERE-
OF, MAY NOT BE REPRODUCED IN ANY FORM WITH-
OUT WRITTEN PERMISSION OF THE PUBLISHERS.

Library of Congress Catalog Card No. 57-6521

To Mary Agnes

PREFACE

The author, in this volume, has attempted to present indeterminate structural analysis in a manner which is rigorous, coherent and interesting. It is an elementary presentation in the sense that the reader is assumed to be unfamiliar with the subject. The treatment is, however, more detailed and extensive than would normally be associated with an elementary text.

Structural theory is based upon certain fundamental principles which were discovered many years ago. A deep sense of appreciation and a consequent understanding of the true significance of that theory can hardly fail to spring from a knowledge of the history of its development. Consequently the first chapter is devoted to a brief outline of that history. Historical notes also appear throughout the text. The reader will not, of course, be able to read the first chapter with complete comprehension unless he is familiar with the various methods of indeterminate structural analysis. A second reading, after completion of a study of the text, is therefore recommended.

No specific comments are necessary regarding Chapters 2 through 13. The subjects considered in these chapters are to a large extent, although not entirely, similar to those included in other texts. The treatment, however, will usually be found to be quite different. It is hoped that this presentation will be relatively easy to understand.

The material included in Chapter 14 is based on the author's experience in developing the structural model laboratories at the Rensselaer Polytechnic Institute. In the early stages of this development it became apparent that information relative to the details of structural model design and testing was scattered and incomplete. In the case of direct model analysis it seemed to be generally unavailable. It was necessary, therefore, to prepare the material of Chapter 14 and it seemed advisable to include it in this volume.

The book is designed for use as a text or for reference. As a text it is suited for use in senior undergraduate or first-year graduate courses. It is suggested that the material of Chapters 1 through 8 or 9 might be included in a first course and that the remainder be covered in a second course.

The author is indebted to a longtime friend and engineer, Martin P. Korn, who suggested several changes in the first draft of Chapter 1 and who has been a constant source of enthusiastic encouragement throughout the writing of this volume. Appreciation is expressed for several helpful suggestions received from Associate Professor John T. Watkins of the faculty of the Structural Engineering Division of the Civil Engineering Department at the Rensselaer Polytechnic Institute, as well as from John T. Percy, who was formerly an Associate Professor in the Structural Engineering Division at the Institute. The author also wishes to express his appreciation to the Portland Cement Association for permission to reproduce the information included in the appendix.

J.S.K.

Troy, New York
September 1957

CONTENTS

CHAPTER 1. A BRIEF HISTORY OF STRUCTURAL THEORY 1
   1–1   Introduction 1
   1–2   Before the Greeks 1
   1–3   The Greeks and the Romans (600 B.C.–476 A.D.) 4
   1–4   The Medieval Period (477–1492) 6
   1–5   The Early Period (1493–1687) 8
   1–6   The Pre-Modern Period (1688–1857) 9
   1–7   Since 1857—The Modern Period 11
   1–8   Conclusion 15

CHAPTER 2. STABILITY AND DETERMINATENESS OF STRUCTURES . . . 20
   2–1   General 20
   2–2   Stability 21
   2–3   Articulated structures and continuous frames 21
   2–4   Determinateness 22
   2–5   External stability and determinateness 22
   2–6   Internal stability and determinateness 27
   2–7   Combined external and internal indeterminateness . . . 35

CHAPTER 3. BASIC CONCEPTS 42
   3–1   General 42
   3–2   Determinate *vs.* indeterminate structures 42
   3–3   Analysis-design procedure for indeterminate structures . . 43
   3–4   Notation for deflections 45
   3–5   Deflection condition equations 48
   3–6   Principle of superposition 52
   3–7   Elastic, plastic, and deflection theories 53
   3–8   The principle of virtual displacements 55
   3–9   The principle of virtual work 57
   3–10 Maxwell's theorem of reciprocal deflections 61
   3–11 The Maxwell-Betti reciprocal theorem 65

CHAPTER 4. METHODS FOR COMPUTING DEFLECTIONS 72
   4–1   General 72

Part 1. Internal Strain Energy

   4–2   Internal work and deflections 73
   4–3   Expressions for internal strain energy 74

Part 2. Deflections by Real Work

4-4   Demonstration of the method 80

Part 3. Castigliano's First Theorem

4-5   Development and demonstration 84

Part 4. The Method of Virtual Work

4-6   General 96
4-7   Deflections resulting from axial strains 96
4-8   Deflections resulting from flexural strains 104
4-9   Deflections resulting from flexure of nonprismatic members . 114
4-10   Deflections resulting from shearing strains. 118
4-11   Deflections resulting from torsional strains 119
4-12   Maxwell's theorem of reciprocal deflections 124

Part 5. Moment Areas and Elastic Weights

4-13   General 126
4-14   The moment-area method 127
4-15   The conjugate beam method 129
4-16   Relationships between the real beam and the conjugate beam . 132

Part 6. The Conjugate Structure

4-17   General 143
4-18   Development of the method 143
4-19   Demonstration of the method 150
4-20   Application to multi-span frames 157

Part 7. The Williot-Mohr Diagram

4-21   General 164
4-22   The Williot diagram 164
4-23   The Mohr rotation diagram 172

Part 8. Elastic Weights Applied to Articulated Structures

4-24   General 179
4-25   Angle changes in articulated structures 180

CHAPTER 5. THE GENERAL METHOD 184

5-1   General 184
5-2   Analysis of beams 185
5-3   Analysis of articulated structures 198
5-4   Analysis of continuous frames 213
5-5   The elastic center 240

CONTENTS xi

CHAPTER 6. THE METHOD OF LEAST WORK 254
  6-1   General 254
  6-2   Development of Castigliano's second theorem. 254
  6-3   Analysis of continuous beams and frames 258
  6-4   Analysis of articulated structures 263

CHAPTER 7. THE COLUMN ANALOGY 278
  7-1   General 278
  7-2   The column flexure formula 278
  7-3   Development of the method 281
  7-4   Units in the column analogy. 290
  7-5   Analysis of continuous frames 291

CHAPTER 8. INTRODUCTION TO MOMENT DISTRIBUTION 302
  8-1   General 302
  8-2   Sign convention for moments 302
  8-3   Absolute stiffness and distribution factor 303
  8-4   Relative stiffness of a member 305
  8-5   Carry-over factor 306
  8-6   Evaluation of absolute stiffness of prismatic members . . . 307
  8-7   Fixed-end moments induced by displaced supports of prismatic members 308
  8-8   Why moment distribution works 309
  8-9   Fixed-end moments for various loads. Prismatic members . . 313
  8-10 Continuous beams with prismatic members 314
  8-11 Symmetry and antisymmetry 322
  8-12 Check on results of moment distribution. Prismatic members . 324
  8-13 Frames with one degree of freedom of joint translation . . . 338

CHAPTER 9. ADDITIONAL APPLICATIONS OF MOMENT DISTRIBUTION . . 368
  9-1   General 368
  9-2   Frames with two degrees of freedom of joint translation. . . 368
  9-3   Frames with several degrees of freedom 420
  9-4   Secondary stresses by moment distribution 426
  9-5   Frames with restrained joints 431
  9-6   Comments on the method of moment distribution 443

CHAPTER 10. ANALYSIS OF FRAMES WITH NONPRISMATIC MEMBERS BY MOMENT DISTRIBUTION 445
  10-1 General 445
  10-2 Stiffness and carry-over factors by the column analogy . . . 445
  10-3 Curves for stiffness and carry-over factors 454
  10-4 Fixed-end moments, stiffness, and carry-over factors by the conjugate beam 464
  10-5 Stiffness of a nonprismatic member with far end pinned . . 470
  10-6 Fixed-end moments induced in a nonprismatic member by relative displacement of the member ends 471

Chapter 11. The Slope-Deflection Method 477

- 11–1 General 477
- 11–2 Development of the method 477
- 11–3 Analysis of continuous beams 481
- 11–4 Frames with one degree of freedom 486
- 11–5 Analysis of gabled frames 495
- 11–6 Frames with several degrees of freedom 501
- 11–7 Secondary stresses 502

Chapter 12. Influence Lines 507

- 12–1 General 507
- 12–2 The Müller-Breslau principle 507
- 12–3 Influence lines for continuous beams with prismatic members . 513
- 12–4 Continuous beams with nonprismatic members 522
- 12–5 Influence lines by moment distribution 526
- 12–6 Influence lines for articulated structures 532
- 12–7 Qualitative influence lines by the Müller-Breslau principle . . 532

Chapter 13. Elastic Arches 536

- 13–1 General 536
- 13–2 Types of arches 539
- 13–3 Curve of arch axis 540
- 13–4 Analysis of two-hinged articulated arch ribs 541
- 13–5 Two-hinged solid arch ribs 543
- 13–6 Two-hinged parabolic arch with a secant variation of the moment of inertia 559
- 13–7 Hingeless arches 561
- 13–8 Secondary stresses in arches 580

Chapter 14. Model Analysis of Structures 584

- 14–1 General 584
- 14–2 Structural similitude 585
- 14–3 Fundamentals of indirect model analysis 586
- 14–4 Model materials for indirect types of analysis 596
- 14–5 The spline method of indirect analysis 597
- 14–6 Errors resulting from changes in geometry 598
- 14–7 The Beggs deformeter 599
- 14–8 The Eney deformeter 603
- 14–9 The R.P.I. deformeter 605
- 14–10 The contact indicator 607
- 14–11 The moment deformeter 607
- 14–12 The brass spring model for articulated structures 608
- 14–13 Model design for indirect analysis 608
- 14–14 Direct model analysis 611
- 14–15 Dimensional analysis 616
- 14–16 Design of models for direct analysis 620

14–17	Fabrication and loading of models for direct analysis	627
14–18	Auxiliary equipment for direct model analysis	628
14–19	The moment indicator	628
14–20	The cellulose acetate spring balance	632
Appendix		643
Index		651

CHAPTER 1

A BRIEF HISTORY OF STRUCTURAL THEORY

1-1 Introduction. "If I have been able to see a little farther than some others, it was because I stood on the shoulders of giants." So spoke Sir Isaac Newton, author of the greatest scientific book of all time, his *Philosophiae Naturalis Principia Mathematica*, or more commonly, his *Principia*. Thus he acknowledged his debt to the great men of science who preceded him.

We of this age have inherited a scientific legacy infinitely greater than that inherited by Sir Isaac Newton. During the more than two centuries since his time, progress has constantly accelerated. In the beginning, intervals of decades, or even centuries, separated the discovery of fundamental truths and basic principles. Each generation, however, "standing on the shoulders" of those who had passed on, reached new heights and glimpsed new horizons. Each new discovery quickened the pace and cleared the way for greater things. Down through the centuries the implications of known truths were explored, new principles were added to the growing store of knowledge, and the whole was assembled into a general pattern which constitutes the science of our time.

Five thousand years were required to discover and to organize crudely those principles of structural mechanics which the college student of today learns in several weeks. Another two hundred years were necessary to refine and expand these principles to their present state. The unceasing effort and dedication of great minds down through the ages have given us our modern structural theory.

1-2 Before the Greeks. All things as they existed in the beginning are hidden in the mists of the prehistoric ages. Sometime, somehow, during the centuries preceding the beginning of history, man invented simple machines. The first glimpse of the ancients afforded by history shows them in possession of the inclined plane, with its two variations, the wedge and the screw, and the lever, with its two derivations, the pulley and the wheel and axle. By the time Aristotle wrote his book on machines in 350 B.C., all six of these had been known for centuries and their origins long forgotten. It has been said (10)* that during all history only one simple machine has

---
* Numbers in parentheses designate references to be found at the ends of chapters.

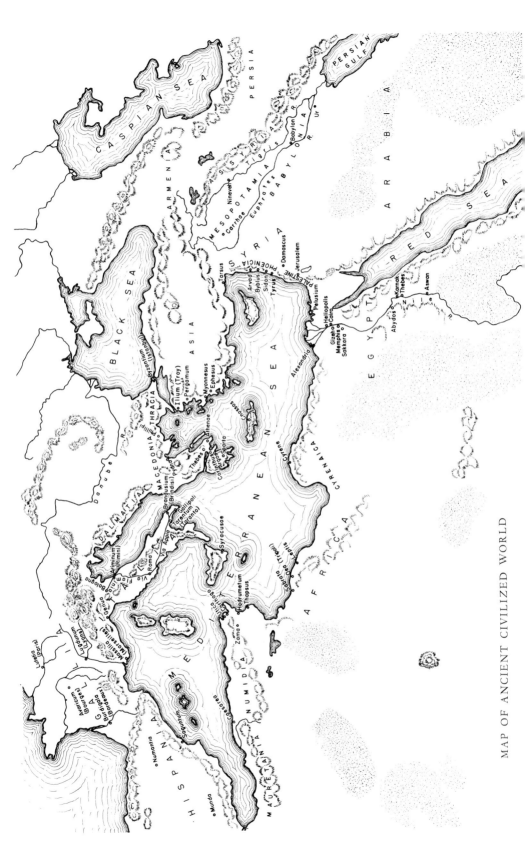

MAP OF ANCIENT CIVILIZED WORLD

been added to those known to prehistoric man. This one addition was Pascal's hydraulic press of 1620.

The beginning of history is indefinitely dated as perhaps 3400 B.C. and springs from records left by the peoples of the valleys of the Nile and the Euphrates and Tigris Rivers. In these valleys, to the best of our knowledge, man first organized societies on a large scale. Here writing was invented and, for the first time, records were consciously made for the use of those who were to follow.

The history of the time from 3400 B.C. to the birth of Christ is very obscure for certain periods. The "Old Kingdom" in Egypt endured from 3400 B.C. to 2431 B.C. and then collapsed. A period of anarchy and invasion by Asiatics followed, and during this time records were incomplete. Then, with the restoration of order in 2160 B.C., Egypt's "Middle Kingdom" was founded and existed until 1788 B.C. Invaders again came from the East and confusion reigned from 1788 to 1580 B.C., a period for which no reliable history is available. The invaders were driven out in 1580 B.C. and the "New Kingdom," or "Empire," was founded. For the centuries since 1580 B.C. Egyptian history is fairly complete.

In general, the early history of the peoples of the Euphrates and Tigris valleys is similar to that of Egypt, in that during certain periods, as a result of violent political upheavals or invasions, no reliable records were compiled.

Structural engineering is chronicled as existing at the time of the "Old Kingdom" in Egypt and a contemporary art existed in the valleys of the Euphrates and Tigris Rivers. It should be noted that structural engineering existed as an art, but not as a science, throughout antiquity. No record exists of any rational consideration, either as to the strength of structural members or as to the behavior of structural materials, until Galileo attempted to analyze the cantilever beam in 1638 A.D. The builders apparently were guided by rules of thumb, which were passed from generation to generation, guarded as secrets of the guild, and seldom supplemented by new knowledge. In spite of this fact, the structures erected during early historic periods are a constant source of amazement.

The first structural engineer of history seems to have been the Egyptian, Imhotep, one of only two commoners to be deified throughout the long history of Egypt. He is perhaps best known as the builder of the step pyramid of Sakkara about 3000 B.C., and his influence, because of other accomplishments, was great enough to initiate a new age of splendor in Egypt.

During the long centuries of ancient Egyptian history, many awe-inspiring structures were built, including the pyramids of Gizeh as well as numerous tombs and temples. It is interesting to note that practically the only structural elements used were the beam and the column. Only one true arch of ancient origin has been found throughout all Egypt, apparently built about 1500 B.C. The Egyptians did, at times, use corbeled arches as architectural rather than structural units.

During these same centuries civilizations in Assyria and Persia developed their own methods of construction. Wall cores were usually of sun-dried mud brick and were faced with kiln-burned brick glazed in different colors; stairs, beams, lintels, and columns were of stone. In their structures the Assyrians frequently used the corbeled arch, and perhaps the true arch, as well as the beam and column. Ancient Babylon, according to the Greek historian Herodotus, was an amazing city, with its two hundred and fifty towers and one hundred sets of bronze gates in its massive walls. The tallest tower of all is assumed to have been the Tower of Babel. The mud bricks could not resist the ravages of time as did some of the Egyptian construction, however, and in the words of Sir Banister Fletcher (9), "Babylon returned to the mud of which it was built, and only mounds now indicate its ancient site."

It is apparent from the preceding account that by 600 B.C. heavy construction had been practiced for centuries throughout the ancient civilized world, and yet no rational method of design had been evolved, even for a simple beam. Consequently, in the beginning it is not the story of structural analysis alone, but rather the story of the physical sciences in general, which is of interest. In this regard it is worth noting that the annual shifting of landmarks by the floodwaters of the Nile had caused the priests of Egypt to develop land surveying; and incidental to its development they had formulated some of the fundamentals of geometry. Also by 600 B.C., the Phoenicians and Mesopotamians had made some progress in astronomy.

1-3 The Greeks and the Romans (600 B.C.-476 A.D.). The age of the Greeks in science was initiated by the philosopher Thales, who lived about 600 B.C. and was a physicist best known for his knowledge of astronomy, which he derived to some extent from Mesopotamian sources. Wealthy, and able to devote much leisure time to study, he came to be called a scholar, or man of leisure (from the Greek *schole*, meaning "leisure"). While traveling in Egypt, Thales learned about surveying from the priests, and upon returning home he formulated the beginning of geometry, deriving the name from the Greek *ge* ("land") and *metron* ("measure"). It is interesting to note that one of Thales' students, Anaximander (611–547 B.C.), first proposed that the earth is poised in space, but he considered that its shape was cylindrical.

The Greek philosopher Pythagoras (born about 582 B.C.) founded his famous school, which was primarily a secret religious society, at Crotona in southern Italy. At this school he allowed neither textbooks nor recording of notes in lectures, on pain of death. He taught up to the age of ninety-five and is reported to have coined the word "mathematics," which means literally the "science of learning," and also the word "philosopher," meaning "one who loves wisdom." Pythagoras is best known to engineers for his theorem relative to the right triangle.

Among the great Greek scientists and philosophers, Democritus (460–370 B.C.) ranks second only to Archimedes and is included here because of his remarkable atomic hypothesis. He believed that the smallest particles of matter are indivisible and called them "atoms," from the Greek words *a* ("not") and *temno* ("cut" or "divide"). The following quotation from Simplicius' "De caelo" is most remarkable: "Leucippus and Democritus say . . . that the fundamental particles which they call 'atoms' or 'indivisibles,' are . . . indestructible, because they are solid and without pores . . . and separated from each other in the infinite void. As they encounter each other abruptly they come into collision. Democritus considers them so small that they escape our senses." Leucippus was a contemporary of Democritus but the latter is chiefly credited with formulating the atomic hypothesis. One can only guess at what this man might accomplish if he were alive today to "stand on the shoulders" of as many "giants" as do our modern scientists.

Aristotle (384–322 B.C.) must be mentioned in any history of structural analysis or mechanics. Dean of the Lyceum, a college just outside the eastern city gate of Athens, and a man of unquestioned ability, he is credited with having written in more than twenty-five different fields of knowledge (11). Probably no other man has ever recorded so many findings and opinions on so many subjects as Aristotle.

One of Aristotle's pupils was Alexander the Great (356–323 B.C.), who in 332 B.C. founded the city of Alexandria at the mouth of the Nile. After Alexander's death in 323 B.C., his most capable general became Pharaoh of Egypt as Ptolemy I and established a library at Alexandria, with the private library of Aristotle as a nucleus, that became the largest of the ancient world, containing some 700,000 scrolls. The first university (in the present-day significance of the word) was also founded at Alexandria by Ptolemy I and was the greatest of the ancient world, with a reported enrollment of 14,000 students. The first professor of geometry at the University of Alexandria was Euclid (315–250 B.C.).

The greatest of the Greeks was Archimedes (287–212 B.C.), who far surpassed Aristotle and, to some degree, Democritus. The greatest physicist of the ancient world and one of the greatest mathematicians of all time, his treatise "On Equilibrium" establishes Archimedes as the founder of statics. It was he who introduced the term "center of gravity," and it is believed that his work in geometry furnished Newton and Leibnitz with the information which led to their development of the calculus. He refused to write about such practical things as machines and thus his writings almost totally ignore his many famous mechanical inventions, including catapults, a spiral pump, and combinations of pulleys. In fact, it was his inventions which held the Roman armies at bay around Syracuse for three years. Archimedes was to have been spared when the city fell to the Romans, but was slain by an ignorant soldier who reportedly disobeyed orders. One version has it that he refused to appear before the Roman consul and conqueror of Syracuse, Marcellus, since to do so would have interrupted the solution of a problem, and consequently he was slain. With the passing of Archimedes the golden age of the Greek philosophers came to an end.

Science made much less progress under the Romans than under the Greeks. The Romans were of a more practical nature than the Greeks and thus were not as capable of abstract thinking, though excellent fighters and builders. Only one name from the history of the period of the Roman Empire will be added to the list of those great men of science already mentioned. This was Lucretius Carus (96?–55 B.C.), a poet and scientist of great ability who embraced the atomic theory of Democritus and who also surmised the existence of disease germs. Eighteen centuries before the French chemist Lavoisier attracted wide attention by his experimental derivation of the law of conservation of mass, Carus had arrived at the same law.

In many respects the Romans surpassed preceding and contemporary peoples in engineering accomplishments. As the empire expanded, the necessity for moving armies rapidly became ever greater and thus the Roman engineers developed their renowned ability to build bridges. A notable example was Caesar's bridge over the Rhine, a third of a mile in length and a pile type structure, which is reported to have been built in ten days. In addition to the pile-type bridge, the Romans developed the semicircular true masonry arch, which they used extensively in both bridges and aqueducts. Trajan's Bridge over the Danube near Turnu-Severin in Rumania, built in 104 A.D., was the longest bridge in the Roman Empire, with a total length of four thousand feet consisting of twenty timber arches.

As previously noted, Egyptian and Greek engineers and architects used stone columns and beams as principal structural elements. The Romans, in addition, made extensive use of masonry arches and, in the case of the Pantheon, the masonry dome, which was probably a Roman development of the Assyrian brick arch. The collapse of the Roman Empire in the West in 476 A.D., resulting from the invasion of Italy by the Teutonic tribes from the North, marked the end of the ancient period of world history.

1-4 The Medieval Period (477–1492). This period, also designated as the Middle Ages or the Dark Ages, was marked by a decline of civilization throughout Europe following the decadence and fall of the Western division of the Roman Empire. In Europe, it was an era of disruption of civil society by the incursions of new races and through general upheaval. The Eastern Roman Empire, however, was to continue in existence for centuries. The center of Greek life had, by this time, been transferred to Constantinople and its influence was for centuries exerted throughout Asia Minor and Egypt by Greek conquerors and traders. It was in Constantinople in 563 A.D. that the development of the structural dome reached its ultimate in the construction of the church of St. Sophia.

During the Dark Ages in Europe, the Arabians carried the torch of knowledge, but produced no scientist who might be classed with Democritus and Archimedes. One tremendously important development of this period, however, which came about through the efforts of the Arabians, was the invention of our system of numbers, credited to a group of mathematicians of Amjir, India, about 600 A.D. This system was adopted by the Arabian mathematicians and was subsequently transmitted through them to Europe and is therefore known to us as the system of "Arabic" numerals. The advantage of these Hindu numbers over both the Greek and Roman systems is very great, and it is quite unlikely that modern science could exist without them. (If in doubt, the reader should try to extract the square root of any number using Roman or Greek numerals.)

The first break in the scientific stagnation of the Dark Ages came with the establishment of Italian universities toward the end of the twelfth century. Progress was seriously retarded, however, because throughout Europe at this time scientists were suspected of heresy. Indeed, it was not until the eighteenth century that those pursuing scientific investigation were entirely free from suspicion and possible persecution.

Roger Bacon (1214–1294), a Franciscan Friar and spearhead of the attack against this intolerance, was persecuted, exiled from England, and imprisoned in Paris for assailing the ignorance of high officials. In his book, *Opus Majus*, he stressed the fundamental importance of experimentation in science and, in the optics section, predicted the discovery of the telescope and the microscope. Bacon was of great importance, not so much because of his specific contributions to science, but because of his influence on those who were to follow.

More than a century and a half separated Roger Bacon and the great Italian, Leonardo da Vinci (1452–1519), possibly the most versatile genius of all time. Da Vinci's every effort seems to have yielded valuable results: in music, sculpture, painting, canals and buildings, interpretation of fossils; and toward the development of the parachute, a proposed flying machine, a diving suit, and other contributions. He is of particular interest here because he was the first, in his statement of the law of the lever, to introduce the concept of the moment of a force. In his writings he also virtually set forth the principle which is now known as *Newton's third law of motion*.

1-5 The Early Period (1493-1687). Andrea Palladio (1518-1580), an Italian architect, is believed to have first used trusses, although his designs were not the result of rational analyses. Prior to this development all construction, except solid masonry, was composed of beams, columns, arches, and domes. It is interesting to note (12) that nearly two centuries were to pass before another man recognized the potentialities of the truss.

Simon Stevin (1548-1620), a noted Dutch engineer, is of interest in this discussion because of his book, published in 1586, dealing with statics and hydrostatics. Apparently he understood the composition and resolution of forces, for in his book he introduced the principle of the triangle of forces. Stevin investigated, and apparently solved (25), the problem of the loaded chord, a problem statically quite similar to that of the truss joint. In addition, he is regarded as a pioneer in the effort to discontinue the use of Latin in scientific writings.

It can be said that the Italian astronomer, Galileo Galilei (1564-1642), ushered in what might be called the age of reason in structural analysis. He apparently was the first to study the resistance of solids to rupture and may be said to have originated mechanics of materials. In his last publication, *Two New Sciences*, he discussed the problem of a cantilever beam loaded with its own or with an applied weight. Since his time, this has been known as *Galileo's problem*, and was not correctly and completely solved until 1855. Galileo considered that the entire beam was rigid except at the section of failure, and that compression was concentrated at the lower edge of this section, with uniform tension over the remainder. His results, of course, were erroneous but his contribution was very great, for he called attention to the existence and importance of what is now known as mechanics of materials.

One of the most outstanding of the Oxford men of science, whose association resulted in the formation of the Royal Society in 1660, was Robert Hooke (1635-1703), Professor of Geometry at Gresham College and Surveyor to the City of London. Of extraordinary ability and range of interest, he was unfortunately of poor health. As the laboratory assistant of the wealthy Robert Boyle (10) he did not receive proper credit for many of his accomplishments, and similar experiences with others led to his becoming bitter and suspicious, apparently with good cause. In the course of his study of elasticity, as the result of which he invented the spiral spring to replace the pendulum for timekeepers, he arrived at the law now bearing his name. Although this was in 1660, because of concern for the patent rights to his invention, he did not publish his law until 1676, and then only in the form of an anagram—"ceiiinossstuu"* (a solution to

---

*At the time Hooke wrote this anagram the two symbols u and v were employed interchangeably to denote either the vowel u or the consonant v. The general tendency, however, was to write v when it appeared as the initial letter in a word and to write u in other positions, without regard to phonetic considerations.

which he gave (28) in 1678)—containing the letters of the law in Latin: *Ut tensio sic vis*. Hooke did not apply his law to engineering problems, but in 1680 E. Mariotte made an independent announcement of the same law and applied it to the fibers of a beam. As the result of tests (31) which he conducted in 1680, Mariotte observed that some of the fibers of a beam are stretched and some are shortened, which led him to place the boundary between lengthening and shortening at the mid-depth of the cross section.

Sir Isaac Newton (1642–1727), born on Christmas day in the year of Galileo's death, was Professor of Mathematics at Cambridge University and for a quarter of a century president of the Royal Society. Newton was the author of the greatest book in the literature of science and yet was so retiring and modest that it was only through the insistence of one of his former students, Dr. Edmund Halley, that the manuscript of his *Principia* was brought forth from a dusty trunk and finally published in 1687. In it, among other things, he stated his laws of motion, the law of universal gravitation, and the infinitesimal calculus. Newton was notable for his willingness to give credit to his predecessors and to his contemporaries, as evidenced by the opening quotation of this chapter.

1–6 The Pre-Modern Period (1688–1857). James Bernoulli (1654–1705) of Basel, Switzerland, was the first member of his distinguished family to attain a wide reputation as a mathematician. He became interested in Galileo's problem and is credited with being the first to have assumed that a plane section of a beam remains plane during bending, but for some inexplicable reason decided that the position of the neutral surface was unimportant, and he failed to arrive at a satisfactory solution. Johann Bernoulli (1667–1748), brother of James and gifted with similar ability, is of interest because of his announcement in 1717 of the principle of virtual velocities. This principle, as will presently be demonstrated, is the basis for the most generally applicable of available methods for determining elastic deflections of structures. Daniel Bernoulli (1700–1782), son of Johann Bernoulli and equally able as a mathematician, became interested in the problem of determining the elastic curve of bent bars and the vibration of beams and rods and succeeded (about 1735) in obtaining a differential equation for the transverse vibrations of a bar. He interested his friend and fellow mathematician of Basel, Leonhard Euler (1707–1783), in the problem of determining the elastic curves of beams and columns, suggesting to Euler that the true elastic curve might be that which would cause the total internal work to be a minimum. Thus Euler employed the method of least work in a crude form and contributed his extremely valuable discussion of the buckling of columns.

It is interesting to note that during the time of Bernoulli and Euler (1758) a fellow Swiss, Ulric Grubenmann, realized the value of the trusses used two hundred years before by the Italian, Palladio, and bridged the

Rhine with a 170-foot timber span. Other timber bridges followed, although they combined the truss and the arch in a single structure, as did the ordinary covered timber bridges of the middle nineteenth century built in the United States. Finally, Ulric Grubenmann was joined by his brother Jean to build a great timber bridge with a span of 390 feet. Like the trusses of Palladio, however, these bridges were built by rule of thumb rather than by design based on rational analysis.

Charles Augustin Coulomb (1736–1806) began his career as a French military engineer but became a renowned physicist. He and Navier, who followed him by several decades, are considered to have founded the science of mechanics of materials. In 1776, Coulomb published (5) the first correct analysis of the fiber stresses in a flexed beam with rectangular cross section. He assumed that Hooke's law applied for the fibers, logically placed the neutral surface in the correct position, developed the equilibrium of forces on the cross section with external forces, and then correctly evaluated the stresses. Apparently recognizing the possible existence of a plastic stage, he indicated that at rupture, under certain conditions, the neutral surface might move to a different position. At the same time he presented his theory of earth pressure against a retaining wall and in a later paper (1784) set forth his theory for the torsion of shafts.

In 1816, Sir David Brewster, a Scottish physicist, published (1) a report of his discovery that deformations produce double refraction in glass. He had obtained patterns of colored fringes by passing polarized light through glass plates subjected to load and suggested that this method might be used to study arches and other structures, thus pioneering in the development of photoelasticity as a method for determining stresses.

Louis Marie Henri Navier (1785–1836), distinguished French engineer, mathematician, and professor, published in 1826 (23) the first edition of his *Leçons*, the first great textbook in mechanics of engineering. Not only did he present a sound treatment of the strength and deflection of beams of any cross section, but also he considered arches, columns under eccentric loads, suspension bridges, and other technical problems. To Navier belongs the signal honor of developing the first general theory of elastic solids as well as the first systematic treatment of the theory of structures. Even though Coulomb's work in beam analysis had been correct, publication of his results did not prevent subsequent muddled thinking and attempts to solve the beam problem. Navier's *Leçons*, however, after a few years seemed to bring order and reason to mechanics of materials and structural analysis.

It is worth noting that no clear division existed between the theory of elasticity and the theory of structures until about the middle of the nineteenth century. Theory of structures could not exist as such until the basic principles governing the behavior of materials had been established. Consequently, leading engineers entered into the study of the theory of elasticity. Coulomb and Navier would today be classed as professional structural engineers but, for the reason just stated, they entered into studies which resulted in the founding of mechanics of materials.

Three other structural engineers who pioneered in developing the theory of elasticity were Lamé, Clapeyron, and de Saint-Venant. Lamé (1795–1870) and Clapeyron (1799–1864) collaborated to publish in 1833 a notable paper on elasticity which presented important work on stresses in hollow cylinders and spheres and which introduced the idea of the stress ellipsoid, as well as the important principle of the equality of external and internal work of a strained structure. In 1852, Lamé published the first book on elasticity (13) and in it credited Clapeyron with the theorem of equality of external and internal work, a theorem of great importance in the analysis of indeterminate structures. In 1857, Clapeyron presented his "theorem of three moments" for the analysis of continuous beams (4).

The first contribution to the theory of structures from the United States came in 1847 when Squire Whipple (1804–1888) published his remarkable treatise, "Bridge Building," (32) in which he presented the first rational analysis of the jointed truss. Worthy of note is the fact that prior to 1850 the jointed truss was used almost exclusively in America. This type of construction developed steadily after the Revolutionary War and bridge spans up to 300 feet were built, but not from any rational design. Spans constructed in England and Europe were usually combinations of arches and trusses, or were otherwise rendered internally indeterminate. It was Whipple's contribution which made scientific design of jointed trusses a possibility.

The engineer Barre de Saint-Venant (1797–1886) was perhaps the greatest of elasticians. In the words of Southwell (28), "... he combined with high mathematical ability an essentially practical outlook which gave direction to all his work." In 1855 he presented famous memoirs dealing with torsion and in 1856, in memoirs dealing with flexure, shear stresses on cross sections of beams were correctly considered for the first time and problems of impact and vibration were discussed. He was also interested in the nature of molecular action and a theory of plasticity.

Saint-Venant's memoirs of 1855 and 1856 closed what this author considers to have been the Pre-Modern Period in the evolution of a theory of structures. During this period a method had been developed for determining the stresses in simple trusses, simple beam action had been correctly and completely analyzed, and there had been a beginning of a theory for column action. Thus the basic principles had been established upon which an adequate theory of structures could be developed.

1-7 Since 1857—The Modern Period. The evolution of a comprehensive theory of structures proceeded at an astonishing rate once the basic and requisite principles had been determined. Those who participated in this development were many and, consequently, only the most important can be mentioned in this discussion.

The French engineer, Bresse, published in 1859 a noteworthy treatise presenting very thorough and practical methods for analyzing curved beams and arches, and in 1867 the influence line was introduced by the German, E. Winkler. A year before Bresse's treatise appeared, the first great textbook in English dealing with mechanics of engineering and materials had been published by William John M. Rankine (1820–1872). The first edition of this book, the *Manual of Applied Mechanics*, appeared in 1858 and the twentieth edition in 1919. Rankine credited the empirical column formula bearing his name to Tredgold (1788–1829) as originator and to Gordon for having revived it. In Germany, in 1863, Professor August Ritter published his "Method of Sections" (26), now in wide use, and demonstrated how stresses in an articulated structure can be computed by the principle of moments.

James Clerk Maxwell (1830–1879), Cavendish Professor of Experimental Physics at Cambridge University, originated and published in 1864 (15) the first significant contribution to the development of a theory for indeterminate structural analysis. This was an analysis of a redundant framework by a method based on the equality of the internal strain energy of a loaded structure and the external work of the applied loads, an equality which had been previously established by Clapeyron. Maxwell expressed the necessary conditions for geometrical coherence by a set of equations in which the variables were the redundant stresses. This was the first method developed for the systematic analysis of indeterminate structures. In the course of his analysis Maxwell set forth his theorem of reciprocal deflections, but because his treatment was brief and was not illustrated by practical applications, the great importance of the principle was not appreciated for some time. The Italian, E. Betti, published a generalized form of Maxwell's theorem in 1872 and, consequently, it is often known as the *Maxwell-Betti reciprocal theorem*. In another paper published in 1864 (16), Maxwell presented his stress diagram for trusses, which combines all individual force polygons into a single figure. The method was subsequently extended by Cremona and the resulting diagram is now often called the *Maxwell-Cremona diagram*. As early as 1853, at the age of twenty-three, Maxwell had published a paper of considerable merit on elasticity, a paper which included a discussion of a series of photoelastic experiments and the interpretation of the fringes resulting from the use of circularly polarized light.

A remarkable book (8) on structural mechanics was published in 1866 by Professor Carl Culmann (1821–1881) of the Polytechnikum in Zurich, Switzerland, and it was largely as a result of this book that graphic statics, as an effective means of structural analysis, came into being. Varignon had studied the polygon of equilibrium for a loaded string but Culmann, discarding the idea of a material string, used the string polygon as a tool of analysis.

The contributions of the German, Otto Mohr (1835–1918), to the theory of structures are extremely significant. In 1868, he presented an outstanding paper (17) which discussed the representation of the elastic curve as a string polygon or, in other words, the method of elastic weights. On the same occasion he developed the closely allied method for computing deflections as bending moments in a fictitious beam loaded with elastic weights, that is, the conjugate beam method. In 1873, Professor Charles E. Greene of the University of Michigan presented a related principle of moments of elastic areas by which the deflection of any point of a beam, from a tangent to the elastic curve at another point, may be computed; this is usually called the moment-area method.

In 1874 (18), apparently without any knowledge of the work done previously by Maxwell, Mohr presented a simpler and more extensive derivation of the same general method for the analysis of indeterminate structures by the simultaneous solution of condition equations expressing the geometric coherence of the structure. Mohr's development, however, utilized the principles of virtual work, and illustrative examples of various applications of the method, including the effects of temperature variation, were appended. In 1887, he published (19) his correction diagram for use with the semi-graphical method for determining deflections of articulated structures, announced by Williot in 1877. Without Mohr's correction diagram, the Williot diagram would be of very limited usefulness, but the combined Williot-Mohr diagram is an extremely valuable method for determining deflections of articulated structures. Finally, in a paper (20) published in 1892, Mohr used a slope and deflection method of analysis in connection with a secondary stress problem.

An important basic principle of indeterminate structural analysis issued from the efforts of Alberto Castigliano (1847–1884), an engineer of the Italian railways, who first presented his theorem of least work as a thesis for an engineering diploma at Turin in 1873. A paper (2) published by him in 1876 presented the theorem of least work,* known as *Castigliano's second theorem*, as a corollary of his first theorem, his method for finding deflections. In 1879, Castigliano published a book (3) in Paris which was remarkable for its originality and, in addition, was much more comprehensive than the work done previously by Maxwell and Mohr. The importance of this book, in the development of indeterminate structural analysis, was very great.

---

* It is interesting to note that the Italian general, L. F. Menabrea, had stated the principle of least work in a paper published in 1858, but his proof was unsatisfactory.

In the same year that Castigliano published his book (1879), Manderla presented his analysis for secondary stresses in a truss with rigid joints, using the tangential angles at the member ends as the unknowns rather than the secondary stresses or moments.

Professor Heinrich Müller-Breslau (1851–1925), for many years a distinguished member of the faculty of Technische Hochschule in Berlin, is one of the most outstanding principals in the annals of the theory of structures. In 1886, he published a basic method (21) for the analysis of indeterminate structures, although this is essentially a variation of the previous work of Maxwell and Mohr. The condition equations for geometrical coherence of a structure, as written by Müller-Breslau, are obtained by superposition of displacements as caused by individual redundant stresses and reactions. The coefficients of the redundant stresses and reactions are the displacements due to unit stresses and reactions and these displacements may be found by any method desired. Of particular note is the fact that he recognized and pointed out the great value of Maxwell's theorem of reciprocal deflections in the evaluation of these displacements. While working on this problem, Müller-Breslau discovered that the influence line for a reaction or internal stress in a structure is, to some scale, the curve of the structure when deflected by an action similar to the reaction or stress. Known as *Müller-Breslau's principle*, this is the basis for various indirect methods for the analysis of structures, either determinate or indeterminate, by the use of models.

Müller-Breslau extended the applications of Castigliano's method of least work and also made significant contributions to the theory of space frames, both determinate and indeterminate. He made use of, and apparently originated, the method of tension coefficients, a method discussed in his book (22) published in 1924, as well as in earlier editions.

Claxton Fidler in 1887 introduced the method of "characteristic points" for the analysis of continuous beams, a method extended in 1905 and 1908 by the Dane, A. Ostenfeld, to include either linear or angular yield of supports.

A notable contribution to the development of photoelasticity as a method for determining stresses was the publication in 1891 by C. A. Carus Wilson of McGill University in Montreal of a study by photoelasticity of stresses in beams under concentrated loads. As the result of Wilson's work, the potentialities of photoelasticity became apparent.

In 1915, G. A. Maney of the University of Minnesota published (14) his independent development of the slope-deflection method. Mohr, as previously noted, had made use of a similar method in connection with the problem of secondary stresses as early as 1892, but his work in this connection apparently attracted little attention. Maney developed and presented slope deflection as a powerful method for the analysis of continuous frames.

The condition equations necessary to effect an analysis of a frame by slope deflection use deformations, that is, joint rotations and translations, as the unknown quantities, whereas in the general method of Maxwell and Mohr the unknowns are stresses and reactions. The general principles involved in methods of analysis of this kind, where deformations are the unknowns, were admirably set forth in a book (24) written by A. Ostenfeld and published in 1926.

Hardy Cross (1885–) began teaching his method of moment distribution at the University of Illinois in 1924. The method was published (6) in 1930 and has, to a large degree, revolutionized the analysis of continuous frames. Without question it is one of the greatest contributions of all time to indeterminate structural analysis. The moment distribution method is related, in some degree, to the deformation methods, because individual joints of the frame under consideration are allowed to rotate in successive steps until a condition of geometrical coherence results, although the attainment of this coherence is not directly indicated by the results of the analysis. The general concepts of the method of moment distribution were extended by Cross in 1936 (7) to the problem of flow in networks of pipes and conduits.

Professor R. C. Southwell of Oxford University developed the "relaxation method" of analysis, which is a method of calculation by successive approximations and which has varied applications. He published an initial paper (27) on the subject in 1935 and followed this with two books, one in 1940 (29) and the other in 1946 (30). Southwell was not, in the beginning, aware of moment distribution as previously developed by Cross, although the method is one of relaxation.

1–8 Conclusion. Art is the doing of things; science, the knowing of things. Science is knowledge reduced to law and organized in system. Thus it is apparent that structural engineering, beginning in the prehistoric ages, existed only as an art down through the ancient and medieval periods. It is interesting to find that those who practiced structural engineering were not, even as late as the beginning of the eighteenth century, known as structural engineers, or even as civil engineers. As a matter of fact, up to this time there was no professional distinction between the civil engineer and the architect. Now, in retrospect, it is possible to classify a builder in the centuries before the eighteenth as an engineer or as an architect only on the basis of his principal inclinations and accomplishments.

The emergence of the separate professions of the structural engineer and the architect was a slow process. The period of this adjustment can be considered to have been inaugurated early in the seventeenth century with Galileo's attempted analysis of the cantilever beam. His work pointed the way, and by the middle of the eighteenth century some principles of statics had been formulated and the physical properties of several important building materials had been investigated. Quite logically, therefore, at this time there was a beginning of scientific structural analysis and design. Those builders who first applied the new science were the pioneer structural engineers.

One of the first noteworthy applications of this new science was in the analysis of the dome of St. Peter's in 1742. The object was to determine the cause of cracks and to suggest corrective measures. It is interesting to find that the engineers who wrote the report were most apologetic for having used a scientific approach to the problem. By the middle of the nineteenth century, however, scientific analysis of structural problems was quite generally accepted. It will be recalled that by this time the Bernoullis, Euler, Coulomb, Navier, Clapeyron, Whipple, and de Saint-Venant, together with lesser contributors, had organized the beginning of a systematic and comprehensive theory of structures. Simple beams and trusses could then be correctly analyzed and there was a beginning of a theory for column action.

The Modern Period, designated herein as having begun in 1857, has seen the development of the remainder of the theory of determinate structures as well as most all of indeterminate structural theory. A list of the most important contributors to indeterminate structural theory would undoubtedly include Clapeyron, Maxwell, Mohr, Castigliano, Maney, Müller-Breslau, Cross, Ostenfeld, and Southwell.

In the absence of the facts, one might quite naturally assume that the theory of indeterminate structures is a fairly recent and logical extension of a previously developed complete theory of determinate structures. Obviously this is not the case. In actuality, the sequence of solution of the basic structural problems was in no way related to their theoretical complexity. Euler, for example, had developed his treatment of long columns and Navier had analyzed arches, suspension bridges, and eccentrically loaded columns before a method of analysis for simple trusses had been evolved. Many difficult problems relative to stresses in plates and cylinders, indeterminate to a high degree, were solved before the continuous girder was satisfactorily analyzed. It should be noted, however, that many of the early methods were very involved and that new and easier methods have since been introduced.

This brief discussion of the evolution of modern structural theory has been presented in the belief that it will broaden the interests, enhance the vision, and engender professional pride in the student. Perhaps he has found interest, even fascination, in these historical facts. If this is so, it is hoped that he will pursue the subject further and benefit greatly thereby. Understanding is most perfect when based on a knowledge of things as they were in the beginning. In the words of Aristotle, "He who considers things in their first growth and origin ... will obtain the clearest view of them."

Specific References

1. Brewster, D., *Trans. Roy. Soc.* (London), pp. 156–178, 1816.
2. Castigliano, A., "Nuova teoria intorno all' equilibrio dei sistemi elastici," *Trans. Acad. Sci.* (Turin), **11**, 127–286, 1876.
3. Castigliano, A., *Theorème de l'équilibre des systèmes élastiques et ses applications.* Paris, 1879. (*Note:* An English translation by E. S. Andrews, *Elastic Stresses in Structures*, was published in 1919 by Scott, Greenwood and Son, London.)
4. Clapeyron, B. P. E., *Comptes Rendus*, **45**, 1076, 1857.
5. Coulomb, C. A., "Essai sur une application des règles de maximis et minimis à quelques problèmes de statique, relatifs à l'architecture," *Memoirs de Mathématique et de Physique*, pp. 343–382, 1776.
6. Cross, H., "Analysis of Continuous Frames by Distributing Fixed End Moments," *Proc. Am. Soc. Civ. Engrs.*, May, 1930.
7. Cross, H., "Analysis of Flow in Networks of Conduits or Conductors," *Bull. Univ. Ill. Eng. Exp. Sta.*, No. 286, 1936.
8. Culmann, C., *Die Graphische Statik.* Zürich, 1866.
9. Fletcher, B., *A History of Architecture on the Comparative Method.* London: Scribner's, 1948.
10. Fraser, C. G., *Half Hours with the Great Scientists.* New York: Reinhold, 1948.
11. Girvin, H. F., *A Historical Appraisal of Mechanics.* Scranton: International, 1948.
12. Grinter, L. E., *Theory of Modern Steel Structures.* Vol. 1. New York: Macmillan, 1949.
13. Lamé, G., "Leçons sur la théorie mathématique de l'élasticité des corps solides." 1852.
14. Maney, G. A., *Studies in Engineering—No. 1.* University of Minnesota, 1915.
15. Maxwell, J. C., "On the Calculations of the Equilibrium and Stiffness of Frames," *Phil. Mag.* (4), **27**, 294, 1864.
16. Maxwell, J. C., "On Reciprocal Figures and Diagrams of Forces," *Phil. Mag.* (4), **27**, 250, 1864.
17. Mohr, O., "Beitrag zur Theorie der Holz- und Eisen Konstruktionen," *Zeitschrift des Architekten und Ingenieur Vereines zu Hannover*, 1868.
18. Mohr, O., "Beitrag zur Theorie des Fachwerks," *Zeitschrift des Architektenund Ingenieur Vereines zu Hannover*, 1874–5.
19. Mohr, O., "Über Geschwindigkeitspläne und Beschleunigungs pläne," *Zivilingineur*, 1887.
20. Mohr, O., "Die Berechnung der Fachwerke mit starren Knotenverbingungen," *Zivilingineur*, 1892.
21. Müller-Breslau, H. F. B., *Die Neuren Methoden der Festigkeitslehre und der Statik der Baukonstruktionen.* Berlin: 1886.
22. Müller-Breslau, H. F. B., *Die Neuren Methoden der Festigkeitslehre.* Leipzig, 1924.

23. NAVIER, L. M. H., *Résumé des leçons données à l'ecole des Ponts et Chaussées sur l'application de la méchanique à l'établissement des constructions et des machines.* 1826.
24. OSTENFELD, A., *Die Deformationsmethode.* Berlin: Springer, 1926.
25. PARCEL, J. I. and MANEY, G. A., *Statically Indeterminate Stresses.* New York: Wiley, 1936.
26. RITTER, A., *Elementare Theorie und Berechnung eisener Dach- und Brückenkonstruktionen.* Hannover, 1863.
27. SOUTHWELL, R. V., "Stress Calculation in Frameworks by the Method of Systematic Relaxation of Constraints," *Proc. Roy. Soc.*, **A151**, 56–95, 1935.
28. SOUTHWELL, R. V., *An Introduction to the Theory of Elasticity.* London: Oxford University Press, 1936.
29. SOUTHWELL, R. V., *Relaxation Methods in Engineering Science—A Treatise on Approximate Computation.* Oxford: Clarendon Press, 1940.
30. SOUTHWELL, R. V., *Relaxation Methods in Theoretical Physics.* London: Oxford University Press, 1946.
31. WESTERGAARD, H. M., "One Hundred Fifty Years Advance in Structural Analysis," *Trans. Am. Soc. Civ. Engrs.*, **94**, 226–240, 1930.
32. WHIPPLE, S., *Elementary and Practical Treatise on Bridge Building.* New York: 1847.

GENERAL REFERENCES

33. BREASTED, J. H., *A History of the Ancient Egyptians.* New York: Scribner's, 1908.
34. BREASTED, J. H., *The Conquest of Civilization.* New York: Harper, 1937.
35. BRIGGS, M. S., *The Architect in History.* Oxford: Clarendon Press, 1927.
36. CHIERA, E., *They Wrote on Clay.* Chicago: University of Chicago Press, 1938.
37. FLEMING, A. P. M. and BROCKLEHURST, H. J., *A History of Engineering.* London: Block, 1925.
38. FORBES, R. J., *Man the Maker—A History of Technology and Engineering.* New York: Schuman, 1950.
39. GRANGER, F., *Vitruvius on Architecture.* Cambridge: Cambridge University Press, 1945.
40. HART, J. D., *The Mechanical Investigations of Leonardo daVinci.* London: Chapman and Hall, 1925.
41. HOSKINS, C. H., *Studies on the History of Medieval Science.* Cambridge: Harvard University Press, 1924.
42. JEANS, SIR JAMES, *The Growth of Physical Science.* New York: Macmillan, 1948.
43. KIRBY, R. S., WITHINGTON, S., DARLING, A. B., and KILGOUR, F. G., *Engineering in History.* New York: McGraw-Hill, 1956.
44. LAYSON, J. F., *Famous Engineers in the Nineteenth Century.* London: Scott, 1885.
45. MOLITOR, D. A., *Kinetic Theory of Engineering Structures.* New York: McGraw-Hill, 1911.

46. SEDGWICK, W. T., and TYLER, E. W., *A Short History of Science*. New York: Macmillan, 1939.

47. STRAUB, H., *A History of Civil Engineering*. London: Hill, 1952.

48. TIMOSHENKO, S. P., *History of Strength of Materials*. New York: McGraw-Hill, 1953.

49. WESTERGAARD, H. M., *Theory of Elasticity and Plasticity*. Cambridge: Harvard University Press, 1952.

50. WOLF, A., *A History of Science, Technology and Philosophy in the XVIth and XVIIth Centuries*. 2nd ed. London: Allen and Unwin, 1950.

51. WOLF, A., *A History of Science, Technology and Philosophy in the XVIIIth Century*. 2nd ed. London: Allen and Unwin, 1950.

CHAPTER 2

STABILITY AND DETERMINATENESS OF STRUCTURES

2-1 General. Squire Whipple, as indicated in Chapter 1, published his treatise, *Bridge Building*, in 1847 and as a consequence engineers were able for the first time to apply a rational analysis to jointed trusses. All types of structures had previously been built with little, if any, regard as to whether or not they were determinate, and indeed many of these structures, particularly bridges, were highly indeterminate. This is not surprising in view of the fact that comparatively little existed in the way of a rational theory prior to 1847.

Whipple's analysis was one of many notable contributions to structural theory published during the last half of the nineteenth century. As a result, an aversion to indeterminateness in structures appears to have developed in the United States as structural engineers became aware of their incompetence in indeterminate structural analysis and of the dangerous inadequacy of the old methods of design. Apparently this antipathy had become quite general by the end of the nineteenth century.

The fact that indeterminate structures are now often used in the United States, when they represent the best structural solution to a given problem, can be attributed in large degree to four developments. The first of these was the introduction of reinforced concrete in the period from 1900 to 1910. Here was a building material, with potential advantages which could not be ignored, that naturally produced a continuous, and therefore indeterminate, structure. Methods of analysis had, of necessity, to be devised. The second development came quite logically in 1915 with the publication of Professor Maney's presentation of the slope-deflection method. For certain types of structures, this was superior to alternate available methods of analysis and served to stimulate considerable interest in indeterminate structures. With the introduction of arc welding of structural steel, beginning about 1920, came the third development. By 1935 welding was recognized not only as an excellent method for making connections, but also as permitting a new and inspiring flexibility in the design of steel structures. Here again was a new type of construction which naturally resulted in continuity. Finally, the method of moment distribution, developed by Professor Cross, was published in 1930. This was received with enthusiasm by the majority of the profession and, since it was easy to understand, did much to remove what little remained of the former general distrust of the indeterminate structure.

In briefly tracing the evolution of the classification of structures as determinate or indeterminate, it is interesting to observe that such a classification was probably impossible before the latter part of the nineteenth century because there was apparently no general comprehension as to exactly what structural characteristics constitute determinateness or even theoretical stability. Since an understanding of the principles involved is important, they will be discussed in the following sections of this chapter.

2-2 Stability. *A stable structure will support any conceivable system of applied loads, resisting these loads elastically and immediately upon their application, the strength of all members and the capacity of all supports being considered infinite.* In other words, the stability of a structure depends on the number and arrangement of the reaction components and component parts, rather than on the strength of the supports and parts of the structure. Even though a structure may be stable for a particular load or system of loads, unless it is also stable for any other conceivable system of loads, it is classified as unstable. Quite often an unstable structure will be stable under a particular system of applied loads; when it is in this condition it is said to be in a state of *unstable equilibrium*. Since a structure, to be classified as stable, must be stable under any conceivable system of loads, it is advisable to omit all loads when considering the question of stability and determinateness. Therefore no loads will be shown on the illustrations in sections which follow.

2-3 Articulated structures and continuous frames. *A truss, or an articulated structure, is composed of links or bars, assumed to be connected by frictionless pins at the joints, and arranged so that the area enclosed within the boundaries of the structure is subdivided by the bars into geometrical figures which are usually triangles.* Since the pins at the joints are assumed to be frictionless, the bars of an articulated structure are considered, in the majority of cases, to be subjected to axial stresses only. These are called the *primary stresses*. Actually, of course, joints are bolted, riveted, or welded. Consequently, the members at a joint are not free to rotate relative to one another, as they tend to do, when the structure deflects under load. As a result, bending stresses are induced in the bars. These are known as the *secondary stresses* (to be discussed in a later section) and, in some cases, are important.

A continuous frame is a structure which is dependent, in part, for its stability and load-carrying capacity upon the ability of one or more of its joints to resist moment. In other words, one or more joints are more or less rigid. The members of a continuous frame are usually subjected to axial loads, shear, and moment.

A structure, whether a continuous frame or articulated, is either stable or unstable and either determinate or indeterminate depending upon the number and arrangement of internal component parts and external reaction components.

2-4 Determinateness. As to determinateness, structures are conveniently divided into two groups, the two-dimensional and the three-dimensional or space frame. This discussion will be confined to the two-dimensional problem. *An indeterminate structure may be defined as one for which the reaction components and internal stresses cannot be completely determined by the application of the three condition equations for static equilibrium.* These equations, of course, are $\Sigma H = 0$, $\Sigma V = 0$, and $\Sigma M = 0$. Indeterminate structures differ in the extent or degree of indeterminateness. *The degree of indeterminateness for a given structure is the number of unknowns over and above the number of condition equations available for solution.*

It is convenient to consider stability and determinateness as follows:

(a) With respect to reactions; that is, external stability and determinateness.

(b) With respect to members; that is, internal stability and determinateness.

(c) A combination of external and internal conditions; that is, total stability and determinateness.

2-5 External stability and determinateness. *Three reaction components are necessary, but not always sufficient, for external stability of two-dimensional structures.* A fixed support will provide three reaction components, consisting of a moment component and two force components, as shown in Fig. 2-1(a). These force components are usually taken as parallel to horizontal and vertical axes. A knife-edge or pin support, as in Fig. 2-1(b), will provide two force components but no moment component. As indicated

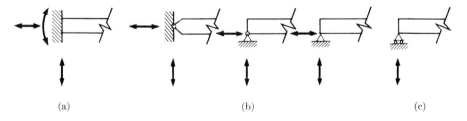

FIGURE 2-1

in Fig. 2–1(c), only one force component, normal to the plane of action of the rollers, can exist at a roller support. The rollers are considered to be capable of providing a reaction acting with either sense in a direction normal to the plane.

If three reaction components act on a two-dimensional structure, the arrangement of these components is important. When, for example, the lines of action of the three components are concurrent, the structure is externally unstable because, even though complete collapse probably will not occur, a small initial rotation about the point of concurrency will be possible before elastic restraint is developed. The structure will also be unstable if the three reaction components have parallel lines of action.

If the number of reaction components is less than the number of independent condition equations for equilibrium of the structure, then the structure will be externally unstable. If the number of reaction components exceeds the number of independent condition equations for the equilibrium of the structure, then the structure is externally indeterminate to the degree by which the number of external reaction components exceeds the number of condition equations. These condition equations are $\Sigma H = 0$, $\Sigma V = 0$, and $\Sigma M = 0$ applied to the structure as a whole, and, in addition, any other equations provided by special features of construction, such as internal pins or links. (A link consists of a short bar with a pin at each end.)

Whenever a structure is externally determinate and stable, the number of reaction components will be equal to the number of different types of movement which would be possible if the reaction components were removed. These movements, which may be either of the entire body or relative movements of the various parts, must be possible without the inducement of internal elastic strains. One condition equation, for the evaluation of reaction components, may be written for each type of movement prevented. Thus each pin inserted in a structure in such a position as to make possible the relative rotation of parts of the structure will provide one additional condition equation for the evaluation of reaction components.

FIGURE 2–2

Consider the case of the propped cantilever in Fig. 2–2. There are five reaction components, and since three condition equations are available from statics for external equilibrium, the beam is indeterminate externally to the second degree.

In Fig. 2–3 there is a beam with pin ends but each end is mounted on rollers. The number of independent reaction components is one less than the three required for stability and the beam is unstable.

FIGURE 2–3

The continuous frame shown in Fig. 2–4 has six reaction components and consequently is externally indeterminate to the third degree.

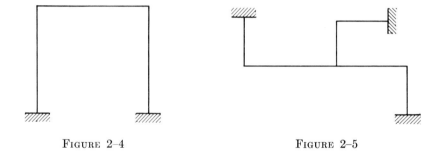

FIGURE 2–4　　　　　　　　　　FIGURE 2–5

In Fig. 2–5, a continuous frame is indicated which is externally indeterminate to the sixth degree.

Figure 2–6 shows a fixed arch which is externally indeterminate to the third degree.

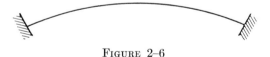

FIGURE 2–6

Figure 2–7 indicates a three-hinged arch with four reaction components that is externally determinate, since a fourth condition equation for equilibrium is available because of the center hinge. The insertion of the hinge at the crown makes possible the rotation of either part of the arch with respect to the other, unless this rotation is prevented by the reaction

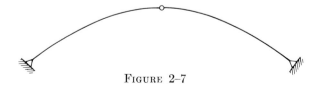

FIGURE 2–7

components; hence the fourth condition equation expressing the fact that the moments about the hinge must be zero for the portion of the arch on either side of the hinge.

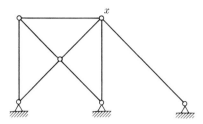

FIGURE 2–8

Consider the articulated structure in Fig. 2–8. In this case there are six reaction components. In addition to the three condition equations expressing the equilibrium of the articulated structure as a whole, the pin at joint x makes it possible to write a fourth condition equation indicating that the moments of all loads and reactions acting on the portion of the structure on either side of the pin must be zero. Consequently the structure is externally indeterminate to the second degree.

The beam of Fig. 2–9 is externally determinate. There are four reaction components, but a fourth condition equation is provided by the fact that ΣM must equal zero about the pin.

FIGURE 2–9

The structure indicated in Fig. 2–10 has five reaction components. In this case, two internal pins exist, one at each end of the link. The addition of each pin makes one additional type of movement possible if reaction components are removed. Two condition equations are therefore provided by the link. Hence a total of five condition equations are available and the structure is externally stable and determinate. If rollers were inserted at either end support, then only four reaction components would exist and the structure would be unstable.

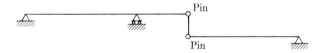

FIGURE 2–10

Actually, as previously indicated, each condition equation expresses a condition which must be fulfilled in order to prevent a particular type of movement of a part or of the whole structure, such movement occurring without elastic resistance to the applied loads. Thus, in Fig. 2–10, the satisfaction of the equations $\Sigma H = 0$, $\Sigma V = 0$, and $\Sigma M = 0$ for the structure as a whole will respectively ensure that the structure in its entirety will neither move horizontally or vertically, nor rotate. In addition, the satisfaction of the equation $\Sigma M = 0$ for the portion of the structure on either side of the pins will guarantee that these portions will not rotate about the pins. Finally, the satisfaction of $\Sigma H = 0$ for the portion of the structure on either side of the pins will prevent the horizontal movement of one portion of the structure relative to the other. Thus five independent conditions are set up which must be satisfied simultaneously by the reaction components in order to prevent the five possible movements of the structure. Five reaction components are necessary to do this. Any fewer than five reaction components will result in instability and any more than five will result in indeterminateness, since in this case two or more reaction components will cooperate in preventing one of the possible movements of the structure. With the condition equations available, it will be impossible to determine to what extent each of the cooperating reaction components prevents this movement.

If the number of reaction components is equal to the number of condition equations, the structure will usually, but not always, be externally determinate and stable. It is possible for the above equality between reaction components and condition equations to exist and still result in a structure which is unstable.

Consider, for example, Fig. 2–11. Four reaction components exist and four condition equations are available, these four equations consisting of the three equations of static equilibrium applied to the structure as a whole and $\Sigma M = 0$, for either part of the structure, about the center pin.

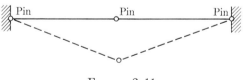

FIGURE 2–11

When a load is applied, however, a small initial displacement will occur which will not be resisted elastically by the structure. Consequently, the definition for a stable structure, which specifies that immediate elastic restraint will oppose an applied load, is not satisfied. Very often when this occurs the structure will collapse. In this case, collapse will not occur, the structure coming to rest in some position as shown by the dotted lines. This particular type of instability, then, is classed as *geometric instability*, a term which has significance in view of the fact that the instability actually exists because of the geometric arrangement of the members. If the two bars of Fig. 2–11 had originally been placed in the dotted position, the structure would have been stable. (The reader is referred to Fife and Wilbur (1) for a profound discussion of this type of instability.)

2–6 Internal stability and determinateness. There is some question as to the advisability of considering that a condition of internal determinateness or indeterminateness may exist in a structure. Actually, the fact that a given structure is internally determinate or indeterminate is largely of academic interest. In the analysis of an indeterminate structure, the question of external or internal indeterminateness is usually incidental; only the degree of total indeterminateness (that is, the summation of the degrees of external and internal indeterminateness) is of primary importance. Nevertheless, it is felt that a discussion of internal indeterminateness is desirable as an approach to a consideration of total indeterminateness. Internal stability or instability, however, and the skill to determine which condition exists in a given case, is of considerable practical importance. Articulated structures will be considered first in this section, followed by a discussion of continuous frames.

The decision as to whether a given articulated structure is internally indeterminate or determinate, and the degree of internal indeterminateness, will depend upon what is considered to constitute internal determinateness. Therefore it is necessary that a clear concept be established as to what is meant herein by an internally determinate structure. *A structure is internally determinate if, with all reaction components necessary for external stability known and acting on the structure, it is possible to determine all internal stresses by applications of the three condition equations for static equilibrium.*

First, consider the Pratt truss shown in Fig. 2–12. Each joint in the truss yields two condition equations, $\Sigma H = 0$ and $\Sigma V = 0$, for the determination of all the unknown forces acting at the joints, including bar stresses and reaction components. This results in the equation

$$2j = b + r,$$

where j is the number of joints and b is the number of bars. The value of r, regardless of the number of reaction components actually made available by the various supports of the structure, must be the number of reaction components required for external stability and determinateness. In other words, since an investigation into internal conditions cannot be completely

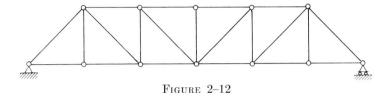

FIGURE 2-12

divorced from a consideration of reaction components, these must be assumed consistent with external stability and determinateness since, by so doing, any indication of instability or indeterminateness will clearly be due to internal conditions. The proper value for r, as indicated in Section 2-5, will be the number of condition equations available for the evaluation of reaction components, since this is also the number of reaction components necessary for external stability and determinateness. The left-hand side of the equation, $2j$, represents the number of simultaneous condition equations available for the solution of the unknown bar stresses and reactions, $b + r$.

The above equation expressing the relationship between the number of condition equations available for the solution and the number of bar stresses and reactions is usually written as

$$b = 2j - r.$$

Particular attention is called to the fact that *the satisfaction of this equation is a necessary condition for internal determinateness and stability of an articulated structure, but it is not sufficient.* To understand the significance of this statement, consider the truss of Fig. 2-12. The number of joints is 12, the number of bars 21; and, by substituting in the above equation,

$$21 = 24 - 3 = 21$$

and the truss is stable and determinate, according to this equation.

Suppose, however, that this truss is changed as shown in Fig. 2-13. The number of joints is now 13 and the number of bars 23. Again the condition equation is satisfied. It is obvious from an inspection of the structure, however, remembering that all joints are considered to be pinned, that

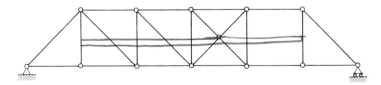

FIGURE 2-13

collapse will result from the rotation of the top and bottom chord members in the panel where no diagonal exists. It is apparent, therefore, that satisfaction of the above equation is not a sufficient condition for internal stability of an articulated structure.

Moreover, it is impossible to determine the stresses in the two crossed diagonals of Fig. 2–13 by use of the three condition equations for static equilibrium. It should be apparent that any vertical shear existing in the panel having the two crossed diagonals will be divided (not necessarily equally) between the two diagonals. In other words, two possible stress paths are provided for the stability of the panel, whereas only one is necessary. It is impossible to determine in what proportions this vertical shear will be divided between the two diagonals by use of the three condition equations for static equilibrium. Hence the structure is internally indeterminate to the first degree, in addition to being internally unstable. Consequently, the given condition equation is sufficient neither for internal determinateness nor for internal stability.

The reader will find after some experience that *in most cases the question of stability and internal indeterminateness of an articulated structure can most easily be settled by inspection and by a consideration of stress paths, and that very little attention will have to be given to equations.*

If one desires to use the above condition equation (its use is sometimes a necessity), then it should be remembered that satisfaction of this equation means that the structure in question *may* be, but not necessarily is, internally determinate and stable. A final decision that the structure is both determinate and stable should be based upon common sense and upon a consideration of stress paths. If the condition equation is not satisfied, the structure is either internally unstable or internally indeterminate, or possibly both. If b is less than $2j - r$, the structure is internally unstable. If b is larger than $2j - r$, the structure is internally indeterminate, usually to the degree indicated by the excess of b over $2j - r$. The application of the condition equation will be demonstrated with several problems.

Consider the small structure of Fig. 2–14. By inspection it is externally indeterminate to the first degree. For use in the equation, $b = 7$, $j = 5$, and $r = 3$, since it is possible to remove either horizontal reaction component (that is, connect either support to rollers operating so that a reaction component either up or down will be available, but no horizontal reaction component can exist) and still have stability. Therefore in $b = 2j - r$, the result is

$$7 = 2 \times 5 - 3 = 7$$

and the necessary condition for internal stability and determinateness is satisfied. Inspection of the structure shows that it is stable. Therefore,

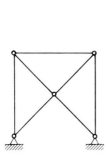

FIGURE 2–14   FIGURE 2–15

if the reaction components (all of which may now be considered to act) are known, it is possible to determine all bar stresses by applying the three equations for static equilibrium and this, it will be remembered, in this discussion constitutes internal determinateness.

In the case of the tower of Fig. 2–15, the first tendency is to class it as externally indeterminate to the first degree, since four reaction components exist. Actually, however, in addition to the three condition equations for static equilibrium, applied to the structure as a whole, a fourth condition equation is provided by the fact that ΣM must equal zero for AB about the pin at B. The reaction component H_1 will exist when an action, other than a vertical force, is applied to AB between A and B. For the purpose of the criterion, or condition equation, for internal indeterminateness and stability, $b = 8$, $j = 6$, and $r = 4$. Attention is again called to the fact that, when investigating internal indeterminateness and stability, r must be the number of reaction components necessary for external stability and determinateness. In the given case, the reaction H_1 is absolutely necessary for stability and hence $r = 4$. Substitution in $b = 2j - r$ gives

$$8 = 2 \times 6 - 4 = 8$$

and the necessary condition equation for internal stability and determinateness is satisfied. Inspection of the structure will establish internal stability and determinateness as a fact.

The propped cantilever truss of Fig. 2–16 is externally indeterminate to the second degree. In this case, $b = 18$ (without the dotted member AB), $j = 11$, and $r = 4$, since either H_1 and V_1 or H_2 and H_3 can be removed and the structure will still be stable. Substitution in $b = 2j - r$ will give

$$18 = 22 - 4 = 18.$$

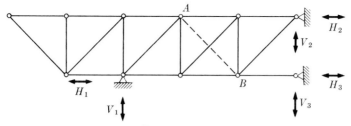

FIGURE 2-16

This indication of internal stability and determinateness may be verified by inspection. Addition of the member AB (indicated by dashed lines) will increase b to 21 and j to 12, and therefore

$$21 \neq 24 - 4 = 20$$

and the structure is internally indeterminate to the first degree.

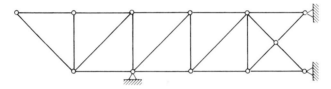

FIGURE 2-17

If the structure of Fig. 2-16 is modified to that shown in Fig. 2-17, the external indeterminateness is increased to the third degree. A count gives $b = 21$, $j = 12$, and $r = 3$. Substitution in the condition equation will result in

$$21 = 2 \times 12 - 3 = 21$$

and the indication is internal determinateness and stability, which is verified by inspection.

As an additional example consider Fig. 2-18, a case where the number of joints is 6, the number of bars 8, and the number of reaction components (necessary for external stability and determinateness) is 4. Possible ar-

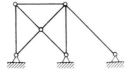

FIGURE 2-18

rangements of these 4 reaction components are shown in Fig. 2–19. Substitution in the condition equation results in

$$8 = 2 \times 6 - 4 = 8,$$

which indicates that the structure is internally determinate and stable, and this is confirmed by inspection.

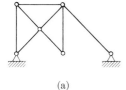 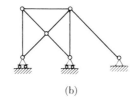 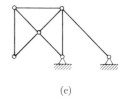

(a) (b) (c)

FIGURE 2–19

In the case of the tower of Fig. 2–20, $b = 22$, $j = 12$, and $r = 4$, and hence

$$22 \neq 2 \times 12 - 4 = 20.$$

Thus the indication is that the structure is internally indeterminate to the second degree, which inspection will corroborate.

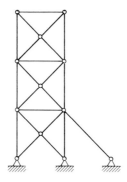

FIGURE 2–20

In the case of the continuous truss shown in Fig. 2–21, j is 28, b is 53, and r is 3, and hence

$$53 = 2 \times 28 - 3 = 53.$$

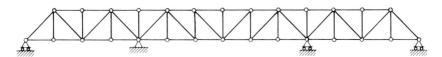

FIGURE 2–21

Internal stability and determinateness are indicated, which is, in fact, the case.

As a final example, a three-hinged structure is shown in Fig. 2–22, where j is 23, b is 42, and r is 4. Therefore,

$$42 = 2 \times 23 - 4 = 42$$

and the indication is that the structure is internally determinate and stable. Inspection will show that this is the case.

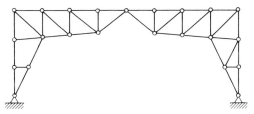

FIGURE 2–22

Any discussion of internal indeterminateness of a continuous frame is academic and usually without practical significance. Such a problem can be investigated by means of a formula, as with articulated structures, but the situation is quite readily analyzed by inspection. If a formula is preferred, it might be written as

$$D = 3n,$$

where D is the degree of internal indeterminateness and the value of n is the number of segments of area, within the outline of the frame, which are completely enclosed by the frame members. Thus segments adjacent to foundations are not counted.

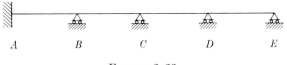

FIGURE 2–23

Consider the continuous beam in Fig. 2–23. Assuming that all reaction components have been determined, it is possible, if the loads on each span are known, to determine the internal moments and shears successively at B, C, and D merely by the application of the three equations of static equilibrium to each individual span. The continuous beam is therefore internally determinate, a fact which agrees with the equation, for n therein would be zero.

34   STABILITY AND DETERMINATENESS OF STRUCTURES   [CHAP. 2

Consider next the frame of Fig. 2–24. Assuming that the reaction components, that is, moment, thrust, and shear, have been determined at the column bases, then by treating the first-story length of each column, that is, AE, BF, CG, and DH, as a free body, the internal stresses in these members at E, F, G, and H can be determined by the three equations for static equilibrium. At joint E then, when considered as a free body, there are still six unknowns consisting of moment, thrust, and shear in EI and EF. The three condition equations for static equilibrium might arbitrarily be assigned to the solution of the internal stresses in EI, but there are still three unknowns at the joint in the stub of the member EF, and thus three degrees of internal indeterminateness are contributed by the member EF.

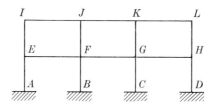

FIGURE 2–24

Once the internal stresses EF (a designation which will be taken to mean the moment, thrust, and shear at end E of the member EF, whereas the internal stresses FE would have the same significance for end F) are determined, the internal stresses FE can be easily found by statics applied to EF as a free body. The internal stresses FB having been previously determined, there are again six unknowns at F. Again, if we assign the condition equations for static equilibrium to the determination of the internal stresses FJ, it is apparent that the member FG contributes three degrees of internal indeterminateness. Proceeding to joint G, the internal stresses GF and GC may be determined as in the discussion of joint F and the member GH contributes three degrees of indeterminateness. If we consider joint H, the internal stresses HG and HD may be determined as before, which leaves the internal stresses HL as unknowns. Since the three condition equations for equilibrium apply to joint H, no indeterminateness exists at H. Proceeding to the top level, and using similar reasoning, we find that no indeterminateness exists at I or L. This means that the internal stresses JI and KL may be found by statics, which is also true for the internal stresses JF and KG, and therefore there are only three unknowns, and hence no indeterminateness, at joints J and K. This will be found to be true for the top level of any frame. Addition of the degrees of internal indeterminateness enumerated above will give a total of nine;

a check by the equation, with $n = 3$, also indicates internal indeterminateness of the ninth degree.

As a final example, consider the frame of Fig. 2–25. A count will give n as 22 and the frame is therefore internally indeterminate to the sixty-sixth degree.

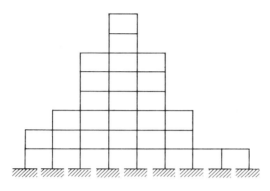

Figure 2–25

2–7 Combined external and internal indeterminateness. The reader will presently understand that the question of external or internal indeterminateness of a structure, when either is considered without respect to the other, is quite academic. This is true because the external and internal conditions cannot be divorced from each other in an analysis. In other words, if a structure is externally determinate and internally indeterminate, the reaction components will usually be evaluated before the bar stresses are determined; on the other hand, if a structure is internally determinate but externally indeterminate, no solution for the reactions is possible that is independent of the bar stresses. Obviously, then, if a structure is both externally and internally indeterminate, both reactions and bar stresses enter into the solution.

It will be found that when an indeterminate structure is being analyzed by a method which necessitates the solution of simultaneous equations, one equation is required for each degree of indeterminateness, regardless of whether the determinateness is external or internal. It is therefore apparent that, in the final analysis, it is the total indeterminateness of the structure which is of interest. For this purpose, the equations previously considered must be revised either as to interpretation or as to use. In the case of articulated structures, the criterion equation was

$$2j = b + r.$$

The only revision necessary to utilize this equation in checking for total indeterminateness is in the significance of r, which should now be entered as the total number of reaction components. Thus, for the continuous truss shown in Fig. 2–21, j is 28, b is 53, and r is 5. Substitution yields

$$2 \times 28 \neq 53 + 5;$$
$$56 \neq 58,$$

and the structure is totally indeterminate to the second degree.

The equation for checking internal indeterminateness of a continuous structure has been given as

$$D = 3n.$$

This should be revised to read

$$D = 3n + r - 3,$$

where r is the total number of reaction components and 3 represents the condition equations supplied by the requirements for static equilibrium of the structure as a whole. Consequently, for the continuous beam shown in Fig. 2–23, r is 7, n is 0, and

$$D = 7 - 3 = 4.$$

In Fig. 2–24, r is 12, n is 3, and

$$D = 3 \times 3 + 12 - 3 = 18.$$

In Fig. 2–25, r is 27, n is 22, and

$$D = 3 \times 22 + 27 - 3 = 90.$$

Problems

Determine the external and internal conditions of stability and determinateness for the following structures:

Answers

	External	*Internal*
2–1.	Unstable	Stable Determinate
2–2.	Stable Determinate	
2–3.	Stable Determinate	Stable Indeterminate Third degree
2–4.	Stable Indeterminate Sixth degree	
2–5.	Stable Indeterminate Third degree	Stable Indeterminate Third degree

2–1.

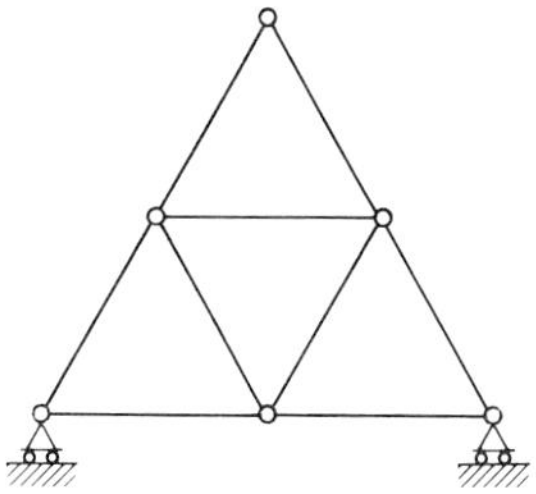

2–2.

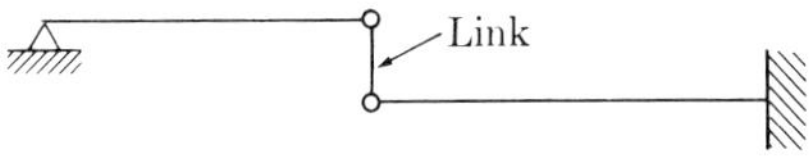

2–3.

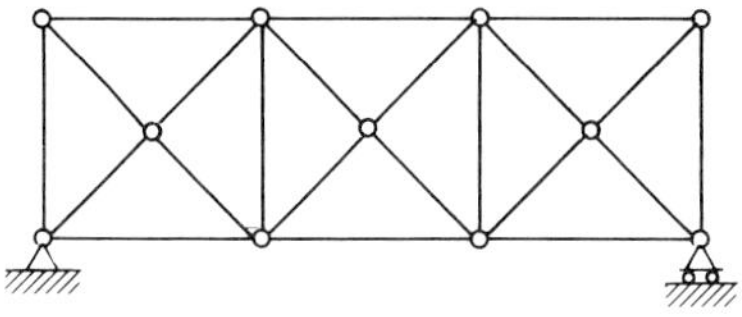

2–4.

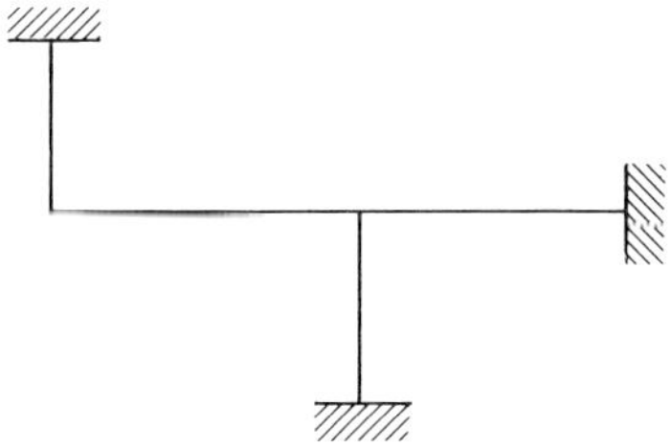

2–5.

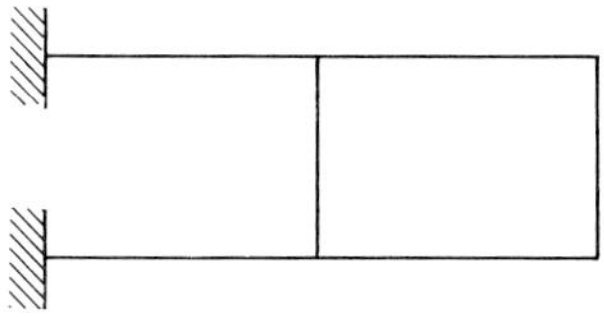

38   STABILITY AND DETERMINATENESS OF STRUCTURES   [CHAP. 2

ANSWERS

	External	Internal
2–6.	Stable Indeterminate First degree	Stable Indeterminate Second degree
2–7.	Stable Determinate	Stable Indeterminate First degree
2–8.	Stable Determinate	Stable Indeterminate Third degree
2–9.	Stable Indeterminate Ninth degree	Stable Indeterminate Twelfth degree

PROBLEMS 39

	ANSWERS	
	External	*Internal*

2–10.

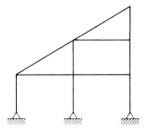

Stable
Indeterminate
Third degree

Stable
Indeterminate
Ninth degree

2–11.

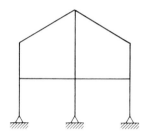

Stable
Indeterminate
Third degree

Stable
Indeterminate
Third degree

2–12.

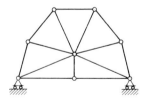

Unstable

Stable
Indeterminate
First degree

2–13.

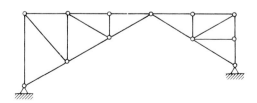

Stable
Determinate

Stable
Determinate

40 STABILITY AND DETERMINATENESS OF STRUCTURES [CHAP. 2

ANSWERS

	External	Internal
2–14.	Stable Indeterminate First degree	Stable Determinate

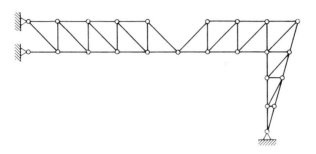

| 2–15. | Stable
Determinate | Stable
Determinate |

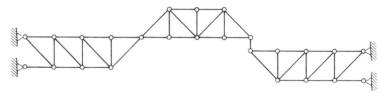

2–16.

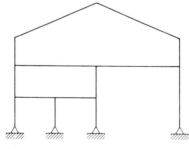

| | Stable
Indeterminate
Fifth degree | Stable
Indeterminate
Sixth degree |

2–17.

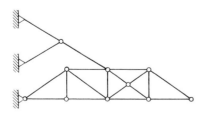

| | Stable
Determinate | Stable
Indeterminate
First degree |

PROBLEMS

ANSWERS

	External	Internal
2–18.	Geometrically unstable	Stable
	Indeterminate First degree	Determinate

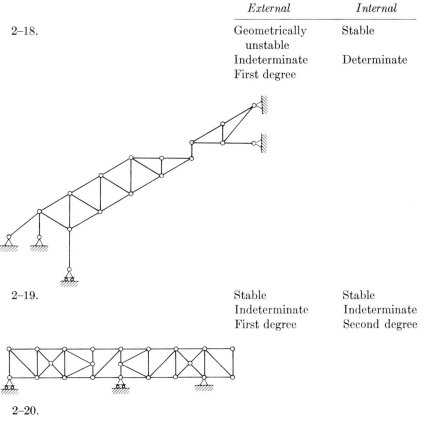

2–19.	Stable	Stable
	Indeterminate	Indeterminate
	First degree	Second degree

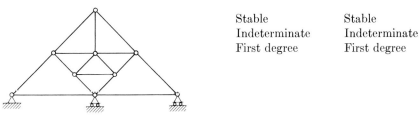

2–20.

	Stable	Stable
	Indeterminate	Indeterminate
	First degree	First degree

REFERENCES

1. FIFE, W. M. and WILBUR, J. B., *Theory of Indeterminate Structures*. New York: McGraw-Hill, 1937.

CHAPTER 3

BASIC CONCEPTS

3-1 General. Any treatment of indeterminate structural analysis should include some brief discussion of the advantages and disadvantages which are, in general, attributed to indeterminate structures. Moreover, the reader should understand that indeterminateness in a structure will require certain departures, aside from the theory involved, from the analysis-design sequence as usually applied to determinate structures. In addition, several theoretical principles may be regarded as fundamental to, and should therefore be presented before, the development of methods of analysis.

The object of this chapter, therefore, is to discuss those facts and to develop those theoretical principles which, in the opinion of the author, are prerequisite to a proper and appreciative understanding of indeterminate structural analysis.

3-2 Determinate *vs*. indeterminate structures. Probably the first, and often the most important basis for the comparison of two or more competitive designs for a given structure will be that of economy. No general statement can be made as to the relative economy of determinate and indeterminate designs. Since each structure usually constitutes a new problem, the aspect of economy will often be decided by the particular site, fabrication, or service conditions. Depending upon these conditions, either type may be found to be more economical than the other.

It is probable that an indeterminate structure will be more beautiful than an alternate determinate structure. The usual graceful lines of an arch, a suspension bridge, or a rigid frame can seldom, if ever, be approached in a determinate structure. One disadvantage of the indeterminate structure, however, arises from the fact that, since relative positions of members and supports cannot be varied without causing stresses in all or part of the structure, the effects of temperature changes, or settling supports, or errors in fabrication or erection may be significant. Consequently, foundation conditions must be good and fabrication and erection should be carefully controlled.

The analysis of an indeterminate structure is, of course, more involved than that of a corresponding determinate structure. This cannot be considered to be a serious disadvantage, however, since analysis usually represents a small percentage of the total cost.

3-3 Analysis-design procedure for indeterminate structures. The engineering activity on any project may be considered to proceed through four different stages: planning, analysis, design, and construction. These four stages will be briefly discussed before considering the analysis-design procedure as applied to indeterminate structures.

Planning may be thought of as including all activity dealing with the development of an initial idea into a general plan. Different types of structures often have to be considered as possible optimum solutions. Erection problems peculiar to each type must, if they exist, be recognized. Cost estimates will be required, and consequently some preliminary analysis and design will usually be necessary before a definite structure type can be selected and before general plans can be prepared. These plans would, at the least, show over-all dimensions and general arrangement of the project.

The analysis of a structure would normally be considered to include all work dealing with the evaluation of axial stresses, shearing stresses, or bending moments as caused by any action the structure may be required to resist. A method of construction or erection may have to be defined during this stage because erection stresses, if important, must be computed. Sometimes certain parts of a structure are stressed higher during erection than at any other time. In structural analysis, "stress" is often considered to mean total load rather than load per unit area. Thus the expressions "axial stress" and "axial stress component" are taken to mean the total axial load in a member; "shearing stress" and "shearing stress component" are considered to mean the total shear in a member; and "flexural stress component" may be used to designate the total bending moment at a given section. Load per unit area is identified by "unit stress." The word "load" is thus reserved to designate actions on the structure such as dead load, live load, wind load, or snow load. (The amount of load or stress is usually expressed in kilopounds (kips), designated by k; thus, 1 k indicates 1000 lb.)

The design stage includes all work concerned with the sizing of the component parts of the structure. All parts must be adequate, with maximum economy, to resist the stresses as determined by the analysis. The component parts of an indeterminate structure are usually designed in accordance with the same specifications as apply to determinate structures.

As previously indicated, because indeterminate structures are sensitive to errors in fabrication, errors in location of supports, or settlement of foundations, fabrication and field construction should be carefully controlled.

The analysis and design stages, in the case of an indeterminate structure, are more closely related and interdependent than for a determinate structure. Stress components at all points of a structure must, of course, be known before it can be designed. If the structure is determinate, these stress components, as caused by applied loads, are independent of the sizes of the component parts of the structure. For example, if a simple beam on a span of 20 ft is loaded at the center with a 5 k concentration, the center moment, caused by the concentration, is 25 ft·k, and this moment is independent of the size of the beam. Dead load moments will, of course, be different for different sized beams. The point to be emphasized, however, is that one analysis of a determinate structure is sufficient to determine the stresses if the loads are known.

If, now, the structure is indeterminate, the internal stress components, as caused by a given load, will change when the relative resistances of the component parts to elastic strains are changed. Consequently, since the parts of the structure cannot be designed until the stress components are known, and since the stress components cannot be determined until the relative strengths or sizes of the various structural parts are determined, it is usually necessary to proceed with the analysis and design of indeterminate structures simultaneously.

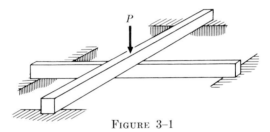

FIGURE 3-1

An example will serve to illustrate the point. Consider the two simple beams of Fig. 3–1. These two beams have unequal spans, they cross at 90°, and the top beam is just in contact with the bottom beam at the centers of the spans. A concentrated vertical load P acts down at the center of the top beam. This arrangement is indeterminate to the first degree. One condition equation, in addition to those available from statics, is required and is written to express the fact that the two beams must deflect equally at their centers. Let P_T and P_B represent the parts of P taken by the top and bottom beams, respectively. L_T and L_B will indicate the span lengths, and I_T and I_B the moments of inertia, of top and bottom beams. Using the expression for center line deflection from mechanics, it is possible to write

$$\frac{P_T L_T^3}{48EI_T} = \frac{P_B L_B^3}{48EI_B},$$

from which

$$\frac{P_T}{P_B} = \frac{I_T L_B^3}{I_B L_T^3}. \tag{3-1}$$

Since in any design problem the span lengths will normally be fixed, it is apparent that the load P will be divided between the two beams in proportion to their moments of inertia, that is, in proportion to their ability to resist flexural elastic strain. Consequently, if, during the design procedure, the ratio of I_T to I_B is changed, the part of P taken by each beam will change.

One design procedure, therefore, is to assume the sizes of the two beams, perhaps with equal moments of inertia. The part of the load taken by each beam is then determined as indicated above and the beams are designed, each to support its part of the load. The division of the load is again computed, on the basis of the new beam sizes, and the cycle is repeated as many times as necessary.

The procedure just described is the *cyclic method of design*. The initial assumptions as to the sizes of component parts of any structure may be based on experience, the parts may be designed by neglecting continuity, or the analyst may simply guess at the sizes. In any case, the end result will be the same. A poor guess will simply require additional work before a satisfactory design is achieved; usually three or four cycles will be sufficient.

Occasionally an alternate method of design may be used. To illustrate, Eq. (3–1) indicates that if I_T and I_B are varied in the same proportion, then the part of the load P taken by each beam will be unchanged. Therefore it is possible, when starting a design, to arbitrarily select the relative strengths or stiffnesses of the component parts of an indeterminate structure. An analysis is made on the basis of these relative stiffnesses and the component parts are then designed so that each part is at least adequate to resist the stresses obtained from the first analysis, and so that the relative stiffnesses are unchanged. Since the relative stiffnesses are not changed, the stresses obtained from the first analysis will still be valid. The weakness in this procedure is that it is very unlikely that the variation in relative stiffnesses of the component parts, as first selected, will be consistent with the variation in stress among these parts. Consequently, certain parts of the structure are likely to be overdesigned.

3–4 Notation for deflections. In the analysis of the crossed beams of Fig. 3–1, one condition equation was required in addition to those available from statics. This additional equation was written to express the fact that the vertical centerline deflections of the two beams are equal. In other words, it expresses a condition which the deformations of the structure must satisfy in order to be consistent with the structural constraints. This type of equation, which the author prefers to call a *deflection condition equation*, is essential in the application of one of the most useful methods for indeterminate structural analysis—Maxwell's "general method" and its variations. If the method is to be applied effectively, it is necessary to develop a notation for deflections which will be adequate to express any kind of deflection component, of any point on a structure, as caused by any type of action on that structure. The following notation will suffice.

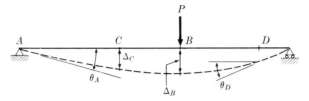

FIGURE 3-2

Consider the beam of Fig. 3-2 loaded with a real load P at any point B. Linear deflection components of any point on a structure loaded with a system of real loads will always be represented by Δ (the Greek capital *delta*); rotational deflection components will always be represented by θ (the Greek lower-case *theta*). A subscript added to the Δ or to the θ will designate the particular point on the structure to which the Δ or the θ pertains. Thus, in Fig. 3-2, Δ_C and Δ_B indicate vertical linear deflections of the points C and B, respectively, while θ_A and θ_D represent rotational deflections of points A and D. Since, in this case, only vertical deflections are possible (deflections due to change in length of a member resulting from bending are negligible), only one subscript is necessary to specify the linear deflection component represented by a given Δ. In Fig. 3-3, however, this is not true. In this case, points A and B have three deflection components each, two linear and one angular or rotational. Consequently, two subscripts are necessary for each Δ. The first designates the point of the linear deflection component and the second indicates which component is intended. One subscript is still sufficient for the θ designations, but for

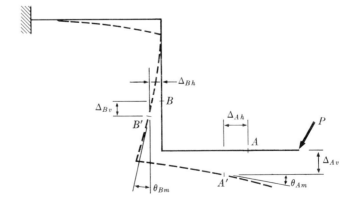

FIGURE 3-3

reasons which will be discussed later the author prefers to add the m, although it may appear at this point to be superfluous.

Usually the great majority of the deflection components which must be evaluated in the solution of any problem are those resulting from the application of a unit force or a unit couple (usually 1 k or 1 ft·k). Linear deflections resulting from the action of a 1 k force are represented by δ (lower-case Greek *delta*); angular or rotational deflections from a 1 k force are represented by α' (lower-case Greek *alpha* prime). Linear deflections caused by a 1 ft·k couple are represented by δ' and angular deflections by α. Double subscripts are often required to represent deflections resulting from the action of a unit load or a unit couple. When two subscripts are used, the first always indicates the point for which the deflection component is designated, and the second the point of application of the unit load or unit couple causing the deflection. A study of Fig. 3–4 and Fig. 3–5 should clarify this system of notation.

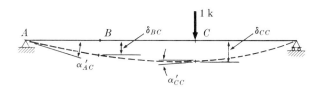

FIGURE 3–4

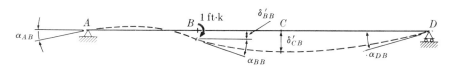

FIGURE 3–5

Very often two subscripts are insufficient to identify properly a given deflection component. In Fig. 3–6 application of a unit horizontal force at B causes the frame to deflect as shown. Point A moves to A' and B to B'. The various deflection components of these points are properly labeled in the figure. The vertical deflection component at A, for example, as caused by a unit horizontal force at B, is designated by δ_{AvBh}. In every case the first capital letter of the subscript represents the point of the displacement and the adjacent lower-case letter (v or h) represents the direction of the designated deflection component (either vertical or horizontal). The second capital letter indicates the point of application of the unit force or couple, and, if it is a unit force, the last lower-case letter (v or h) shows the direction of the unit force (either vertical or horizontal). Note

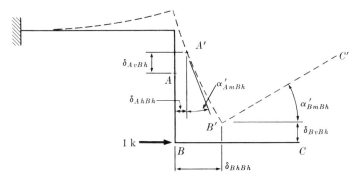

FIGURE 3-6

that no lower-case letter is necessary in connection with a capital letter to designate the direction of a unit couple or the direction of a rotational deflection component, but in order to be consistent in notation the lower-case letter m is often used. For example, the rotational deflection of point A due to a unit horizontal force at B is designated as α'_{AmBh}. (The reader will soon become accustomed to this system of notation and should have no difficulty with it.)

3–5 Deflection condition equations. Three examples are included in this section to illustrate, first, how the above notation is used in writing deflection condition equations, and second, how these equations are utilized in indeterminate structural analysis. These examples should also clearly indicate why a thorough knowledge of the various methods for computing elastic deflections of structures is important.

As a beginning, consider the propped cantilever of Fig. 3–7. In this case, as in many cases where deflections are utilized directly to effect a solution, the structure is first "cut back" so that it is statically determinate and stable. The exact manner of reducing an indeterminate structure to one which is determinate is immaterial. Any combination of redundant reactions and internal stresses may be removed temporarily to render the structure determinate. Their evaluation becomes the immediate object of the analysis.

In the given case, the vertical reaction at B is taken as the redundant

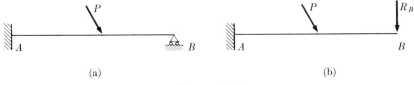

FIGURE 3-7

and removed. This having been done, the end B is free to deflect under the action of the load, as shown in Fig. 3–8. The load P is now removed and a unit vertical load is applied at the point and along the line of action

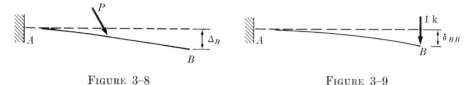

FIGURE 3–8   FIGURE 3–9

of the redundant reaction, as shown in Fig. 3–9. It is necessary to apply a unit load at B because the value of R_B is unknown. Since the principle of superposition applies, however, the deflection caused by R_B will be R_B times the deflection caused by the 1 k load. Now, since the total deflection at B must be zero, the deflection condition equation for this solution is

$$\Delta_B + R_B \cdot \delta_{BB} = 0, \qquad (3\text{–}2)$$

which is readily solved after the two deflections Δ_B and δ_{BB} are known.

Attention is called to the fact that the 1 k load in Fig. 3–9, and therefore R_B itself, was assumed to act down, although structural sense would indicate that it should act up. Actually, it is immaterial whether this unit load is assumed in the correct sense; that is, whether it is assumed to act up or down. If it is assumed incorrectly, then the sign of R_B, as determined from Eq. (3–2), will be negative.

Referring again to the propped cantilever of Fig. 3–7, we see that the beam could have been "cut back" (as shown in Fig. 3–10) to give an

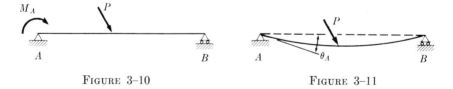

FIGURE 3–10   FIGURE 3–11

alternate solution. Under the action of P (in Fig. 3–10), Fig. 3–11 results, with θ_A as the angular deflection caused by the load P. Applying a unit couple at A results in Fig. 3–12. Since in the original structure the total rotation at A must be zero (unless the support rotates), the required deflection condition equation is

$$\theta_A + M_A \cdot \alpha_{AA} = 0$$

and the value of M_A may be determined.

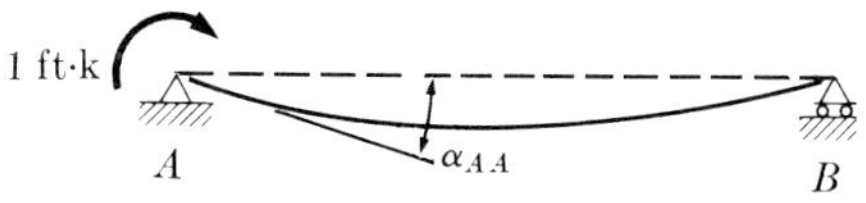

FIGURE 3–12

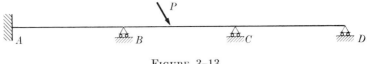

Figure 3–13

Figure 3–14

As a second example, consider Fig. 3–13. This beam will be "cut back" as in Fig. 3–14, although any other combination of redundant reactions may be removed to reduce it to determinateness. Under the action of P, the result is as shown in Fig. 3–15. With 1 k acting down at B, the beam deflects as in Fig. 3–16; with 1 k acting down at C, Fig. 3–17 results. Finally, with a unit couple acting at A, the beam deflects as shown in Fig. 3–18.

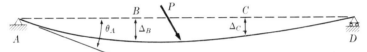

Figure 3–15

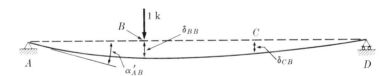

Figure 3–16

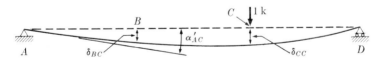

Figure 3–17

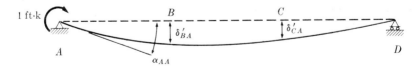

Figure 3–18

In the original structure the total rotational deflection at A is zero and the total vertical deflection at B and C is zero. By adding the deflections at these points from Figs. 3–15 through 3–18, we obtain the following deflection condition equations:

$$\theta_A + R_B \cdot \alpha'_{AB} + R_C \cdot \alpha'_{AC} + M_A \cdot \alpha_{AA} = 0,$$

$$\Delta_B + R_B \cdot \delta_{BB} + R_C \cdot \delta_{BC} + M_A \cdot \delta'_{BA} = 0, \qquad (3\text{-}3)$$

$$\Delta_C + R_B \cdot \delta_{CB} + R_C \cdot \delta_{CC} + M_A \cdot \delta'_{CA} = 0.$$

These equations may be solved simultaneously for the required redundant reactions after the various deflection components are evaluated.

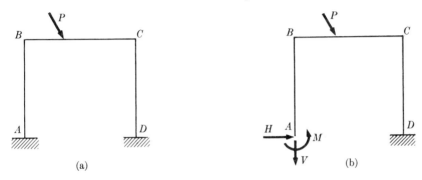

FIGURE 3–19

As a final example, consider the rigid frame of Fig. 3–19. The structure is "cut back" by removing the support at A. Under the action of P, the frame will deflect as in Fig. 3–20, and with a unit horizontal force acting

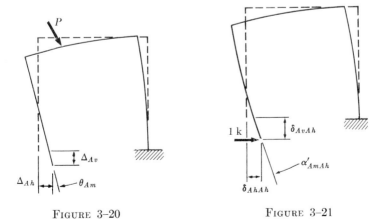

FIGURE 3–20   FIGURE 3–21

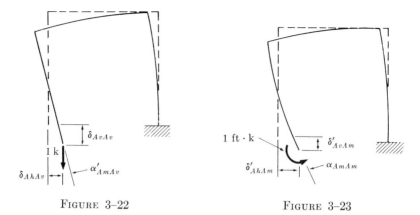

FIGURE 3-22   FIGURE 3-23

at A, Fig. 3–21 results. Figure 3–22 is obtained by applying a unit vertical force at A. Finally, a unit couple applied at A results in Fig. 3–23. In Figs. 3–19 through 3–23, the use of M in the rotational deflection designations is, to some extent, superfluous. It does, however, maintain the double subscript system and is useful in checking the final equations.

The required deflection condition equations express the fact that the three deflection components of point A will be zero. These equations are

$$\Delta_{Ah} + H \cdot \delta_{AhAh} + V \cdot \delta_{AhAv} + M \cdot \delta'_{AhAm} = 0,$$

$$\Delta_{Av} + H \cdot \delta_{AvAh} + V \cdot \delta_{AvAv} + M \cdot \delta'_{AvAm} = 0, \qquad (3\text{--}4)$$

$$\theta_{Am} + H \cdot \alpha'_{AmAh} + V \cdot \alpha'_{AmAv} + M \cdot \alpha_{AmAm} = 0.$$

The simultaneous solution of these equations will yield the desired redundant reaction components.

The three foregoing examples, which are actually illustrations of Maxwell's "general method" for the analysis of indeterminate structures, as modified by Mohr and Müller-Breslau, should have clearly demonstrated that it is necessary for the reader to become thoroughly familiar with the various methods for computing deflections of loaded structures.

3–6 Principle of superposition. A brief discussion of the advantages of indeterminate structures and a few remarks relative to their analysis and design have been presented in the preceding sections of this chapter. Several theoretical principles fundamental to indeterminate structural analysis will now be introduced.

Quite logically, since it is the basis of all structural analysis by the elastic theory, the first of these is the *principle of superposition*. This may be stated as follows:

If the displacements of, and stresses at, all points of a structure are proportional to the loads causing them, then the total displacements and stresses resulting from the application of several loads will be the sum of the displacements and stresses caused by these loads when applied separately. In other words, if the principle of superposition is to apply, a linear relationship must exist, or be assumed to exist, among loads, stresses, and deflections (strains). The loads may be either forces, or couples, or both.

A nonlinear relationship will exist under either of two conditions. The first of these will occur when the strains in the structural material are not proportional to the stresses; that is, when the material does not follow Hooke's law. The second will hold when the geometry of the structure changes significantly during the application of loads.

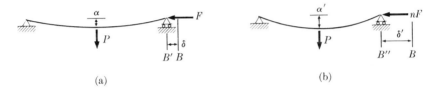

FIGURE 3–24

As an illustration of this second case, consider Fig. 3–24. A deflected beam column is shown in (a) under the action of loads F and P. Centerline deflection is represented by α and the horizontal movement of the right end B is represented by δ. If the axial load is gradually increased to nF, where nF may be equal to or less than the critical axial load, the final deflections will be α' and δ'. In the final position, however, $\alpha' \neq n\alpha$ and $\delta' \neq n\delta$. Moreover, the fiber stresses in (b) will not be n times the fiber stresses in (a). The reason is that, as the beam column deflects, the eccentricity of the axial load at any given section of the member will increase. Consequently, the increments of moment, caused by succeeding constant increments of axial load, will increase as the total axial load and resultant deflections increase. Therefore, because of the changing geometry, a linear relationship cannot exist between the axial load and deflections or stresses.

3–7 Elastic, plastic, and deflection theories. It is apparent from the discussion in Section 3–6 that two basic assumptions must be either actually or practically correct in order to make the principle of superposition applicable to a given structure. The first of these is the assumption that a linear relationship exists between stresses and strains throughout the range of working stresses. The second assumption is that the change in the shape of the structure as loads are applied may be neglected. There are many structures for which one or the other of these assumptions is invalid;

consequently, different methods of analysis are necessary, depending upon whether either one, or both, of the above basic assumptions are correct.

The various theorems and methods for the analysis of indeterminate structures which are dependent for their validity upon the applicability of the principle of superposition constitute what has come to be known as the *elastic theory*. This means that stresses and deformations are computed on the basis of the original dimensions and shape of the unloaded structure, and it is assumed that the various consequences of any change in geometry due to applied loads are negligible.

When the first basic assumption of proportionality between stresses and strains is invalid, because stresses have entered the plastic range, the structure must be analyzed by the *plastic theory*. At the time of this writing much has been accomplished in developing plastic design in structural steel.

The *deflection theory* includes those methods of analysis that are applicable to structures in which significant changes in internal stresses and/or reactions result from changes in geometry as loads are applied. (Stresses and strains are still assumed to be proportional.) The essential difference between the elastic theory and the deflection theory is that in the former the stresses and reactions are computed on the basis of the unloaded position of the structure, whereas in the latter they are computed from the final deflected position. In both theories the structure is considered to be elastic. (It has been suggested that a better terminology for "elastic theory" and "deflection theory" would be "theory of rigid systems" and "theory of flexible systems.")

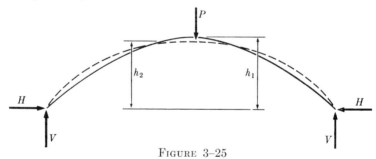

FIGURE 3–25

Considerable work has been done with the deflection theory in the analysis of suspension bridges and long-span arches, and, indeed, an arch is a good illustration of a structure in which stresses may be materially affected by deflections. In Fig. 3–25 the solid curve represents the axis of an undeflected arch rib, and the dashed-line curve the axis of the deflected rib. If the elastic theory is used in the analysis, the positive moment at the crown would be considered to be reduced by the horizontal reaction H acting with a lever arm h_1. Actually, however, because of the deflection, the true lever arm for H should be h_2, somewhat smaller than h_1, and the deflection theory would use h_2 in the analysis. In the case of arches, an analysis by the elastic theory will give stresses in the arch rib which will be lower than those actually existing. In the case of a 950-ft two-hinged trial arch designed as a preliminary study for the Rainbow Arch Bridge at Niagara Falls, it was found (1) that deflection of the rib increased the quarter point moment by 64%, increased the quarter point stress by 29%, and decreased the factor of safety from 1.87 to 1.30. In the hingeless arch which was actually constructed, the analysis by the deflection theory resulted in quarter point moments 18% and quarter point stresses 7% in excess of those given by the elastic theory. (A bibliography which appears at the end of this chapter includes a number of references pertaining to the plastic and deflection theories.)

3-8 The principle of virtual displacements. This principle, originated and first used by Johann Bernoulli in 1717, is the basis for the method of virtual work, the most versatile of the methods available for computing the deflections of structures. The word "virtual" means "being in essence or effect, but not in fact." A *virtual displacement* denotes a hypothetical displacement, either finite or infinitesimal, of a point or system of points on a rigid body in equilibrium such that the equations of equilibrium of the body are not violated. *Virtual velocity* is the term used to designate the projection of the virtual displacement of the point of application of any particular force or couple on the line of action of that particular force or in the plane of rotation of that particular couple. Virtual velocity, therefore, is actually a component of linear displacement or a rotational displacement. It is always considered that virtual displacements are caused by some action or actions other than the loads which hold the rigid body in equilibrium. Thus the loads of the original system move at their full value while the virtual displacement takes place. The product of each load and its virtual velocity will give the *virtual work* done by that load during the displacement.

Figure 3–26 is used to demonstrate the principle of virtual displacements. This demonstration, as well as that of the next section, is similar in many respects to that of Wilbur and Norris (2). The *rigid body* shown, within which there can be no relative movement of its parts, is acted upon and held in equilibrium by the P system of forces and couples. The various forces may be broken into components parallel to the x- and y-axes, as shown for P_2. Since static equilibrium exists,

$$\Sigma P_x = 0, \tag{3-5}$$

$$\Sigma P_y = 0, \tag{3-6}$$

$$\Sigma M + \Sigma P_x \cdot y + \Sigma P_y \cdot x = 0. \tag{3-7}$$

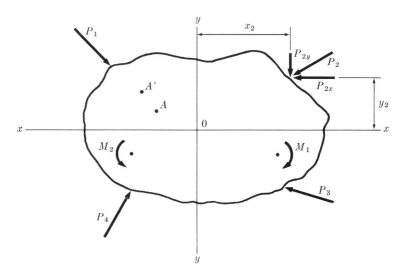

Figure 3-26

Assume that the whole body is displaced a small amount, such as $\overline{AA'}$, without any rotation. The two components of this displacement parallel to the x- and y-axes are designated as δ_x and δ_y. Since the translation is very small, it is assumed that the forces are not changed in direction or magnitude and that the body remains in equilibrium at all times. The work done by the P forces during the translation can be written as

$$\Sigma P_x \cdot \delta_x + \Sigma P_y \cdot \delta_y.$$

Since δ_x and δ_y are constant for all points, this becomes

$$\delta_x \Sigma P_x + \delta_y \Sigma P_y,$$

which, by Eqs. (3-5) and (3-6), is zero.

If, now, it is assumed that the rigid body, with P forces acting, is rotated a small angle α about the origin O (which could be any point), the component of displacement of any other point parallel to the x-axis will be $y\alpha$ and parallel to the y-axis will be $x\alpha$. The total work done by the components of the P forces and the couples will be

$$\Sigma M \cdot \alpha + \Sigma P_x \cdot y\alpha + \Sigma P_y \cdot x\alpha = \alpha(\Sigma M + \Sigma P_x \cdot y + \Sigma P_y \cdot x).$$

This, by Eq. (3-7), is zero.

Since any small displacement of the rigid body can be represented as the sum of a translation and a rotation about some point, and since the total work of the P system of forces and couples in the case of either translation or rotation has been shown to be zero, *Johann Bernoulli's principle of virtual displacements* can be stated as follows:

Given a rigid body held in equilibrium by a system of forces and/or couples, the total virtual work done by this system of forces and/or couples during a virtual displacement is zero.

A simple example will serve to illustrate one application of the principle.

EXAMPLE 3–1. Using the principle of virtual displacements, find the magnitude of R_B for the indicated beam of Fig. 3–27.

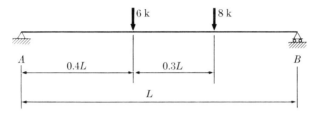

FIGURE 3–27

A virtual displacement is impressed by causing end B to be raised a distance Δ, and Fig. 3–28 is the result. Applying the principle of virtual displacements, and recognizing that both the 6 k and the 8 k loads will do negative virtual work, we find the result to be

$$R_B \cdot \Delta - 6(0.4\Delta) - 8(0.7\Delta) = 0,$$

from which

$$R_B = +8.0 \text{ k}.$$

The positive sign of the answer indicates that the upward sense assumed in Fig. 3–28 for R_B is correct.

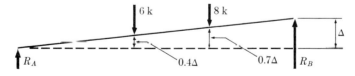

FIGURE 3–28

3–9 The principle of virtual work. The principle of virtual displacements will now be used to develop the principle of virtual work. Consider the deformable body shown in Fig. 3–29 assumed, for convenience, to be a thin slice. This body is in equilibrium under the action of the single

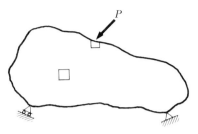

FIGURE 3-29

force P and the reactions. Any number of P loads (either forces or couples) could have been shown as acting on the body without complicating the demonstration. The two segments indicated in Fig. 3-29 are shown as free bodies in Fig. 3-30. One of these, as shown in (a), is an internal segment and is stressed on all sides by the reactions of adjacent segments. The other is on the boundary at the point of application of P, and is therefore subjected to the action of P on one side and to intersegmental stresses on the other three sides. Internal stresses also exist in each segment, and both segments are in equilibrium.

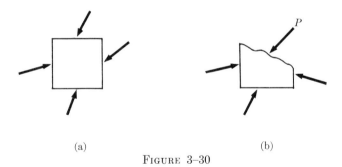

(a)   (b)

FIGURE 3-30

Now assume a virtual action on the body which will result in a virtual deformation of the entire body and all its segments. This action is entirely apart from the P system of loads. All intersegmental stresses, all stresses within the segments, and the external load P will move through their respective virtual displacements with their values unchanged, and thus virtual work will result. Let dw_e represent the total virtual work done by the actions on the boundaries of one segment. In the case of segment (a), these actions are the intersegmental stresses acting on the four sides. In the case of segment (b), these actions are the intersegmental stresses acting on three sides and the load P on the fourth side. Actually, then, dw_e represents the virtual work of the external forces acting on a segment.

In the general case there will be virtual translation, virtual rotation, and virtual deformation (virtual strain) of the segment. Therefore the external virtual work or energy of the forces applied to the segment will be dissipated in two forms as follows: (1) the virtual work of rotation and translation of the segment treated as a rigid body, represented by dw_{rt}, and (2) the virtual work of deformation of the segment, that is, the internal virtual strain energy of the segment, designated by dw_i. Therefore

$$dw_e = dw_i + dw_{rt},$$

but, by the principle of virtual displacements,

$$dw_{rt} = 0,$$

and therefore

$$dw_e = dw_i.$$

An integration over the whole body will result in

$$W_e = W_i, \tag{3-8}$$

where W_e represents the total virtual work of the load P and all the intersegmental forces, and where W_i represents the internal virtual strain energy of the entire body.

Some additional explanation is necessary for a complete understanding of the significance of W_e. For every internal surface of a differential segment with intersegmental stresses or forces acting upon it, there is an opposing surface of an adjacent segment with equal but opposite stresses acting upon it (see Fig. 3-31).

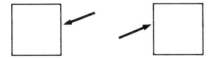

FIGURE 3-31

Thus it is seen that any positive or negative virtual work resulting from the intersegmental stress acting on a surface of a given differential segment is canceled in the final summation by an equal negative or positive virtual work due to the same intersegmental stress on the opposing surface of the adjacent differential segment. Consequently, the total virtual work resulting from the intersegmental stresses is zero, and the actual value of W_e is the virtual work due to the external P load acting on the structure.

In view of the above discussion the *principle of virtual work* can be stated as follows: *If a deformable structure, in equilibrium and sustaining a given* load or system of loads, is subjected to a virtual deformation as the result of some additional action, the external virtual work of the given load or system of loads is equal to the internal virtual work of the stresses caused by the given load or system of loads. In other words, the external virtual work is equal in magnitude to the internal virtual strain energy. An alternate and sometimes preferable statement is simply that the algebraic summation of all virtual work done, both external and internal, must be zero. (Total external virtual work will be opposite in sign to total internal virtual work.)

It will be recalled from an earlier discussion that virtual displacements may be either finite or infinitesimal. In either case, however, they must satisfy two conditions. First, the virtual displacements must be such that the applied loads will remain, or can be considered to remain, constant during the displacement. Second, the virtual displacements must be consistent with the geometry of the structure and its constraints. Actually, there are two forms of the principle of virtual work and there are certain differences between them. If the virtual displacements and/or deformations are infinitesimal, then the principle is used in the infinitesimal form. If, on the other hand, displacements and deformations are finite (as will be the case in all applications in this book), the principle is applied in the finite form and with certain conditions imposed. A discussion of these two different forms follows.

It can be considered that there are three types of forces acting on, and in, a body in equilibrium under a system of loads. These are the external loads, the reactions, and the internal stresses, all of which may do virtual work during a virtual deformation. Let the external loads be represented by P, the reactions by R, the internal stresses by F; and let the infinitesimal virtual velocity of each load P be represented by δp; of each reaction R, by δr (in most cases the reactions do not move and therefore do no work); and of each internal stress F, by δf. Then, by the principle of virtual work,

$$\Sigma P \cdot \delta p + \Sigma F \cdot \delta f + \Sigma R \cdot \delta r = 0, \tag{3-9}$$

and this is the infinitesimal form of the principle. Attention is called to the fact that no conditions of proportionality have been imposed.

Assume now that the principle of superposition applies to p, f, and r, where p, f, and r are finite virtual velocities. If the initial conditions are arbitrarily defined by $p = f = r = 0$, then for any subsequent condition it is possible to write $p = Af = Br$, in which A and B are constants of proportionality. Consequently, $\delta p = A \cdot \delta f = B \cdot \delta r$, and, for any specific condition of the structure defined by $p = p_1, f = f_1$, and $r = r_1$, it follows that

$$\frac{\delta p}{p_1} = \frac{\delta f}{f_1} = \frac{\delta r}{r_1} = \text{constant}. \tag{3-10}$$

Substituting Eq. (3–10) in (3–9) yields

$$\Sigma P \cdot p_1 + \Sigma F \cdot f_1 + \Sigma R \cdot r_1 = 0, \qquad (3\text{–}11)$$

where p_1, f_1, and r_1 are finite virtual velocities of the loads, stresses, and reactions subject to the condition that they be consistent with the geometry of a structure to which the principle of superposition will apply. Equation (3–11) is the finite form of the principle of virtual work and in Chapter 4 will be developed into the most versatile method available for computing the elastic deflections of structures.

3–10 Maxwell's theorem of reciprocal deflections. It will be recalled from Chapter 1 that in 1864 Clerk Maxwell set forth the *theorem of reciprocal deflections* in connection with his development of the first systematic method for the analysis of indeterminate structures. Because his treatment was very brief, however, and without illustrative practical applications, the significance of the theorem was not appreciated for some time. In fact it was not until 1886, when Müller-Breslau published his very much improved version of the Maxwell-Mohr general method for indeterminate structural analysis and in which he called attention to the great value of the reciprocal deflection theorem, that Maxwell's work received proper attention. This theorem will now be developed.

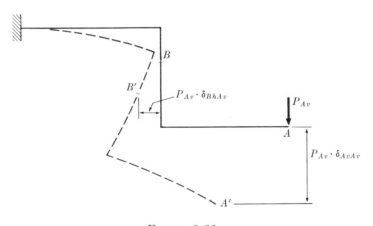

FIGURE 3–32

In Fig. 3–32, a frame is shown which is loaded with the vertical force P_{Av} applied at A. The resulting deflected structure is indicated by the dashed lines. In this figure, δ_{BhAv} and δ_{AvAv} are components of deflections which would result if a unit vertical load were to be applied at A. Since the principle of superposition applies, there is a linear relationship between

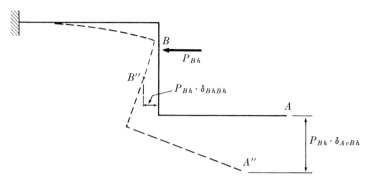

FIGURE 3-33

loads and deflections, and thus the final deflections caused by P_{Av} will be P_{Av} times the deflections caused by a unit vertical load at A.

In Fig. 3-33, the same frame is loaded with a horizontal force P_{Bh}, with the resulting deflections indicated. Now, if P_{Av} and P_{Bh} are slowly and simultaneously applied, the following equation will be the expression for the total external work of the applied loads:

$$W = \tfrac{1}{2}P_{Av}(P_{Av} \cdot \delta_{AvAv} + P_{Bh} \cdot \delta_{AvBh}) \\ + \tfrac{1}{2}P_{Bh}(P_{Av} \cdot \delta_{BhAv} + P_{Bh} \cdot \delta_{BhBh}). \tag{3-12}$$

Consider that both loads are now removed and that P_{Av} is individually and slowly applied to the frame. The total work done when P_{Av} comes to rest will be

$$\tfrac{1}{2}P_{Av}^2 \cdot \delta_{AvAv}.$$

If P_{Bh} is slowly added to the frame, with P_{Av} in position, point A, with P_{Av} "riding along," will deflect to give an additional vertical deflection component equal to $P_{Bh} \cdot \delta_{AvBh}$ and point B will deflect with an additional horizontal deflection component equal to $P_{Bh} \cdot \delta_{BhBh}$. The additional external work will be

$$P_{Av}P_{Bh} \cdot \delta_{AvBh} + \tfrac{1}{2}P_{Bh}^2 \cdot \delta_{BhBh}.$$

The total external work resulting from this second loading sequence is therefore

$$W = \tfrac{1}{2}P_{Av}^2 \cdot \delta_{AvAv} + P_{Av}P_{Bh} \cdot \delta_{AvBh} + \tfrac{1}{2}P_{Bh}^2 \cdot \delta_{BhBh}. \tag{3-13}$$

The sequence of loading, however, will not affect the total external work done and therefore Eqs. (3–12) and (3–13) may be equated. This results in

$$\tfrac{1}{2}P_{Av}P_{Bh} \cdot \delta_{AvBh} = \tfrac{1}{2}P_{Bh}P_{Av} \cdot \delta_{BhAv},$$

from which

$$\delta_{AvBh} = \delta_{BhAv}. \qquad (3\text{–}14)$$

Note that A and B could have been any two points on the frame and that any two directions, other than vertical and horizontal, could have been used for the lines of action of the forces and the deflection components. Therefore, Maxwell's theorem of reciprocal deflections for the case of two forces applied separately to a structure (either articulated or a rigid frame) may be stated as follows:

PROPOSITION 1. *Any linear deflection component of any point A, resulting from the application of a unit force at any other point B, is equal in magnitude to the linear deflection component of B (in the direction of the first applied force at B) resulting from the application of a unit force at A applied in the direction of the original deflection component of A.*

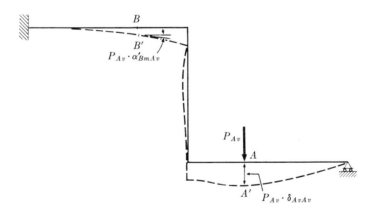

FIGURE 3–34

The reciprocal deflection theorem is valid for rotational deflections as well as for combinations of linear and rotational deflections. The latter case will now be demonstrated. In Fig. 3–34, α'_{BmAv} is the rotation of B in radians, and δ_{AvAv} is the vertical deflection component at A, as caused by a *unit vertical load* at A. The total rotation at B and vertical deflection component at A will therefore be as indicated in Fig. 3–34.

In Fig. 3–35 the same frame is loaded with a couple M_B at B. In this case, α_{BmBm} is the rotational deflection at B in radians and δ'_{AvBm} is the vertical deflection at A, both being caused by a unit couple at B. The total

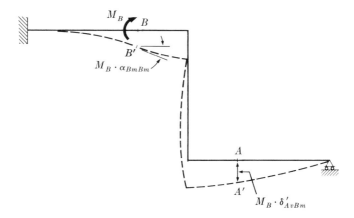

FIGURE 3–35

rotation at B and vertical deflection component at A will be as shown in Fig. 3–35.

It is now assumed that P_{Av} and M_B are slowly and simultaneously applied. The total external work done by P_{Av} and M_B will be

$$\tfrac{1}{2}M_B(P_{Av} \cdot \alpha'_{BmAv} + M_B \cdot \alpha_{BmBm}) + \tfrac{1}{2}P_{Av}(P_{Av} \cdot \delta_{AvAv} + M_B \cdot \delta'_{AvBm}). \tag{3-15}$$

Assume that P_{Av} and M_B are removed. With P_{Av} acting alone as in Fig. 3–34, the total work done by P_{Av} as it is slowly applied and comes to rest will be

$$\tfrac{1}{2}P_{Av}^2 \cdot \delta_{AvAv}.$$

The couple M_B is now slowly applied to the frame with P_{Av} remaining in action. Point A deflects vertically an additional amount $M_B \cdot \delta'_{AvBm}$ and point B rotates $M_B \cdot \alpha_{BmBm}$ radians. The additional external work will be

$$P_{Av}M_B \cdot \delta'_{AvBm} + \tfrac{1}{2}M_B^2 \cdot \alpha_{BmBm}$$

and the total external work is

$$\tfrac{1}{2}P_{Av}^2 \cdot \delta_{AvAv} + P_{Av}M_B \cdot \delta'_{AvBm} + \tfrac{1}{2}M_B^2 \cdot \alpha_{BmBm}. \tag{3-16}$$

Equations (3–15) and (3–16) must be equal, since the sequence of loading will not alter the total external work done. If these are equated, the result is

$$\tfrac{1}{2}M_B P_{Av} \cdot \alpha'_{BmAv} = \tfrac{1}{2}P_{Av}M_B \cdot \delta'_{AvBm},$$

and consequently,
$$\alpha'_{BmAv} = \delta'_{AvBm}. \qquad (3\text{-}17)$$

Maxwell's theorem of reciprocal deflections for the case of a force and a couple applied separately to any structure may therefore be stated as follows:

PROPOSITION 2. *Any linear deflection component of any point A, as caused by a unit couple applied at any other point B, is equal in magnitude to the rotational deflection of B in radians resulting from the application of a unit force at A in the direction of the original linear deflection component of A.*

A similar demonstration could be given for the case of two couples applied separately at two different points on a structure. The statement of the theorem for this case is:

PROPOSITION 3. *The rotational deflection of any point A on a structure, as caused by a unit couple acting at any other point B, is equal in magnitude to the rotational deflection of B as caused by a unit couple acting at A.*

The great value of the theorem of reciprocal deflections will not be apparent from the above demonstrations. It will be found in succeeding chapters, however, that the use of this theorem will result in a considerable saving of labor in the solution of problems by the general method. In addition, it is the basis for certain types of model analysis.

3–11 The Maxwell-Betti reciprocal theorem. In 1872 the Italian, E. Betti, published a generalized form of the reciprocal theorem. This will be used a number of times in the following chapters to derive or to demonstrate various principles, and because it is of great importance, the reader should become thoroughly familiar with it.

Consider Fig. 3–36(a), which shows a system of forces and a couple, designated "System 1," acting on a structure. Figure 3–36(b) indicates an alternate system of forces and a couple, designated "System 2." The forces and couple and resulting deflections of System 2 are capped with a bar to distinguish them from System 1. Note that the deflection components which are designated on the illustrations and are to be used in this discussion are, for each system, "parallel" to the force or couple acting at the same point in the other system. The expression "parallel to a couple" simply means that a rotational deflection caused by one system is designated for each point on the structure at which a couple acts in the other system. Components of deflections caused by one system, "parallel" to the loads of the other system, are said to be "corresponding deflection components." Since only corresponding deflection components are used in this discussion, the lower-case letters ordinarily used to indicate directions of loads and deflection components are omitted from all subscripts.

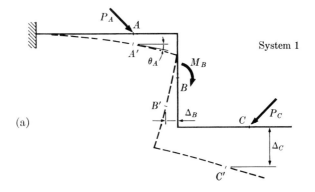

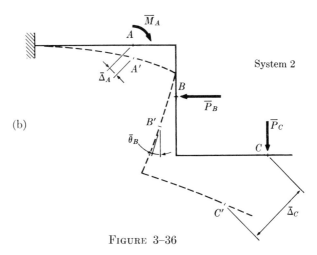

FIGURE 3–36

The various deflection components resulting from the action of System 1 are as follows:

$$\theta_A = P_A \cdot \alpha'_{AA} + M_B \cdot \alpha_{AB} + P_C \cdot \alpha'_{AC},$$
$$\Delta_B = P_A \cdot \delta_{BA} + M_B \cdot \delta'_{BB} + P_C \cdot \delta_{BC}, \quad (3\text{–}18)$$
$$\Delta_C = P_A \cdot \delta_{CA} + M_B \cdot \delta'_{CB} + P_C \cdot \delta_{CC}.$$

The deflection components caused by System 2 are

$$\bar{\Delta}_A = \overline{M}_A \cdot \delta'_{AA} + \overline{P}_B \cdot \delta_{AB} + \overline{P}_C \cdot \delta_{AC},$$
$$\bar{\theta}_B = \overline{M}_A \cdot \alpha_{BA} + \overline{P}_B \cdot \alpha'_{BB} + \overline{P}_C \cdot \alpha'_{BC}, \quad (3\text{–}19)$$
$$\bar{\Delta}_C = \overline{M}_A \cdot \delta'_{CA} + \overline{P}_B \cdot \delta_{CB} + \overline{P}_C \cdot \delta_{CC}.$$

3–11] THE MAXWELL-BETTI RECIPROCAL THEOREM

Multiplication of the forces and couples of System 1 by the corresponding deflection components caused by System 2 (arbitrarily used as virtual velocities for System 1) will give

$$P_A[\overset{(1)}{\overline{M}_A \cdot \delta'_{AA}} + \overset{(2)}{\overline{P}_B \cdot \delta_{AB}} + \overset{(3)}{\overline{P}_C \cdot \delta_{AC}}]$$

$$+ M_B[\overset{(4)}{\overline{M}_A \cdot \alpha_{BA}} + \overset{(5)}{\overline{P}_B \cdot \alpha'_{BB}} + \overset{(6)}{\overline{P}_C \cdot \alpha'_{BC}}]$$

$$+ P_C[\overset{(7)}{\overline{M}_A \cdot \delta'_{CA}} + \overset{(8)}{\overline{P}_B \cdot \delta_{CB}} + \overset{(9)}{\overline{P}_C \cdot \delta_{CC}}]. \quad (3\text{--}20)$$

Multiplication of the forces and couples of System 2 by the corresponding deflection components caused by System 1 (arbitrarily used as virtual velocities for System 2) will result in

$$\overline{M}_A[\overset{(1)}{P_A \cdot \alpha'_{AA}} + \overset{(4)}{M_B \cdot \alpha_{AB}} + \overset{(7)}{P_C \cdot \alpha'_{AC}}]$$

$$+ \overline{P}_B[\overset{(2)}{P_A \cdot \delta_{BA}} + \overset{(5)}{M_B \cdot \delta'_{BB}} + \overset{(8)}{P_C \cdot \delta_{BC}}]$$

$$+ \overline{P}_C[\overset{(3)}{P_A \cdot \delta_{CA}} + \overset{(6)}{M_B \cdot \delta'_{CB}} + \overset{(9)}{P_C \cdot \delta_{CC}}]. \quad (3\text{--}21)$$

The various terms of these equations carry numbers from (1) to (9) above them. Applying the reciprocal deflection theorem to like-numbered terms in these two equations will show them to be equal after the multiplications indicated by the brackets are performed.

From the two equations above, it is important to note that it is not required that all points of load application be loaded in both systems. As a matter of fact, any one or more of the points can be loaded in either system, and the above equality will exist. For example, if P_A is zero, then terms (1), (2), and (3) reduce to zero in both equations.

A statement of the *Maxwell-Betti reciprocal theorem* is as follows: *Given any stable structure with a linear load-deformation relationship on which arbitrary points have been chosen, at any or all of which points in either of two different loading systems forces and/or couples are considered to act, the virtual work done by the forces and/or couples of the first system acting through the corresponding displacements as caused by the second system will be equal to the virtual work done by the forces and/or couples of the second system acting through the corresponding displacements as caused by the first system.*

Reaction components can be included in the system if desired, but if the supports are unyielding, whether or not they are included is immaterial. If, however, a support is considered to yield under the action of either system, then the reaction components (forces or moment) corresponding to the deflection components of the yielding support must be considered in the alternate system. Internal stress components (thrust, shear, or moment) at a given section of a structure may be included in either system by considering that the structure is cut (the cut structure must be stable) at the given section and that the type of restraint imposed by the stress component (only one at a given time) is removed from the structure and replaced by a pair of corresponding actions (thrust, shear, or moment) on the cut ends.

Two examples will be presented to illustrate possible applications of the Maxwell-Betti reciprocal theorem. Unfortunately, to do this, it is necessary to introduce several concepts prematurely. It is believed, however, that the examples will be of considerable assistance in comprehending the significance of the theorem, even though the reader may not be fully prepared to understand the significance of the examples themselves. This understanding must of necessity come later.

EXAMPLE 3–2. Consider the two propped cantilever beams shown in Fig. 3–37. Prove that the moment induced at the fixed end B of Beam 1 by an impressed unit rotation of end A will be equal to the moment induced at the fixed end A of Beam 2 by an impressed unit rotation of end B. The two beams are identical except for the arrangement of the supports.

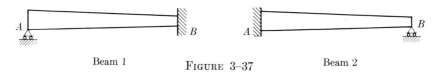

Beam 1   FIGURE 3–37   Beam 2

Before proceeding with the solution of the problem, it is necessary to explain what is meant by *absolute stiffness* and by *carry-over factor*. The *absolute stiffness* of a flexural member is arbitrarily defined as the magnitude of the moment necessary to rotate a simply supported end of the flexural member through one radian. The opposite end of the member may be simply supported, restrained, or fixed. This absolute stiffness is commonly designated by K.

When the moment K is applied to the simply supported end of a member, causing it to rotate, a restraining moment is induced at the opposite end if this end is fixed or restrained. If the opposite end is fixed, as in the present case, the induced moment will be given by CK, where C is the *carry-over factor*.

Returning to the problem, Beam 1 is loaded at end A with the moment K_A and Beam 2 is loaded at end B with the moment K_B. The two systems

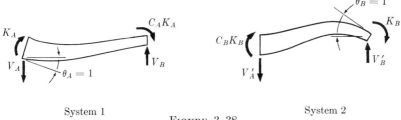

System 1   FIGURE 3-38   System 2

of forces and couples and the resulting distortions are shown in Fig. 3-38. Applying the Maxwell-Betti reciprocal theorem,

$$V_A(0) + V_B(0) + K_A(0) + C_A K_A(1)$$
$$= V'_A(0) + V'_B(0) + K_B(0) + C_B K_B(1),$$

from which
$$C_A K_A = C_B K_B. \qquad (3\text{-}22)$$

EXAMPLE 3-3. Find an expression for the horizontal reaction component H_A as caused by the load P acting on the frame of Fig. 3-39.

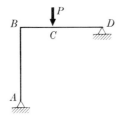

FIGURE 3-39

System 1 is obtained directly from the loaded structure as shown in Fig. 3-40. System 2 consists of four forces, one vertical and one horizontal at each of the frame ends, at A and C. These cause A to move a horizontal distance Δ'_{AhAh}. By the Maxwell-Betti reciprocal theorem,

$$H_A(\Delta'_{AhAh}) + V_A(0) - P(\Delta'_{CvAh}) + H_D(0) + V_D(0)$$
$$= H'_A(0) + V'_A(0) + H'_D(0) + V'_D(0),$$

from which
$$H_A = \frac{\Delta'_{CvAh}}{\Delta'_{AhAh}} \cdot P. \qquad (3\text{-}23)$$

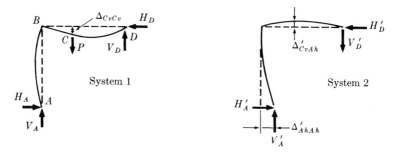

FIGURE 3-40

The coefficient of P in the above equation is the ratio of two deflections which are independent of P. These deflections are caused by the forces of System 2, but the forces do not appear in the expression. Consequently, any convenient value may be assigned to H'_A and the remaining forces of the system determined by statics. The resulting deflections of C and A are then computed, by methods which will be discussed in Chapter 4, and H_A is evaluated.

An alternate procedure for determining the values of the deflections of A and C would be to impress any convenient horizontal deflection at A and to measure the resulting vertical deflection of C. It would not be practicable, of course, to do this on the structure itself; instead, a model can be made and a horizontal deflection impressed on A. The deflection of C is then measured on the model.

The point C can be anywhere between B and D. The value of H_A is therefore seen to be proportional to the deflection of any point of application of the load P as caused by the impressed horizontal deflection of A. In other words, the deflected shape of member BD is, to some scale, an influence line for H_A for vertical loads on BD. Similarly, it can be shown that the deflected curve of AB is an influence line for H_A for horizontal loads acting on AB. The above problem is actually a simple demonstration of the Müller-Breslau principle, mentioned in Chapter 1.

Specific References

1. "Rainbow Arch Bridge over Niagara Gorge—A Symposium," *Trans. Am. Soc. Civ. Engrs.*, **110**, 1–178, 1945.
2. WILBUR, J. B. and NORRIS, C. H., *Elementary Structural Analysis*. New York: McGraw-Hill, 1948.

General References

3. BAKER, J. F., "A Review of Recent Investigations into the Behaviour of Steel Frames in the Plastic Range," *J. Inst. Civ. Eng.*, **31**, 188, 1949.
4. BAKER, J. F., HORNE, M. R., and HEYMAN, J., *The Steel Skeleton—Volume 2—Plastic Behaviour and Design*. Cambridge: University Press, 1956.
5. BEEDLE, L. S., THURLIMANN, B., and KETTER, R. L., "Plastic Design in Structural Steel," Lecture Notes, Lehigh University and Am. Inst. Steel Const., 1955.
6. CROSS, H., "The Relation of Analysis to Structural Design," *Trans. Am. Soc. Civ. Engrs.*, **101**, 1363–74, 1951.
7. FOULKES, J., "Minimum Weight Design and the Theory of Plastic Collapse," *Quart. Appl. Math.*, **10**, 347, 1953.
8. GREENBERG, H. J. and PRAGER, W., "On Limit Design of Beams and Frames," *Proc. Am. Soc. Civ. Engrs.*, Vol. 77, 1951. (Also Separate No. 59.)
9. HEYMAN, J., "Plastic Design of Beams and Plane Frames for Minimum Material Consumption," *Quart. Appl. Math.*," Vol. 8, 1951.
10. HILL, R., "On the State of Stress in a Plastic-Rigid Body at the Yield Limit," *Phil. Mag.*, Ser. 7, **42**, 868, 1951.
11. KIST, N. C., "Die Zähigkeit des Materials als Grundlage für die Berechnung von Brücken, Hochbauten, und ähnlichen Konstruktionen aus Flusseisen," *Der Eisenbau*, **11**, 425, 1929.
12. NEWMARK, N. M., "Numerical Procedure for Computing Deflections, Moments and Buckling Loads," *Trans. Am. Soc. Civ. Engrs.*, **108**, 1161–88, 1943.
13. PHILLIPS, A., *Introduction to Plasticity*. New York: Ronald, 1956.
14. SOUTHWELL, R. V., *Relaxation Methods in Engineering Science—A Treatise on Approximate Computation*. Oxford: Clarendon Press, 1940.
15. SYMONDS, P. and NEAL, B. G., "Recent Progress in the Plastic Methods of Structural Analysis," *J. Franklin Inst.*, Vol. 252, 1951. Part I, p. 383; Part II, p. 469.
16. TIMOSHENKO, S., *Theory of Elastic Stability*. New York: McGraw-Hill, 1936.
17. VANDENBROECK, J. A., *Theory of Limit Design*. New York: Wiley, 1948.
18. WESTERGAARD, H. M., "Buckling of Elastic Structures," *Trans. Am. Soc. Civ. Engrs.*, **85**, 576–654, 1922.
19. WESTERGAARD, H. M., "On the Method of Complementary Energy," *Trans. Am. Soc. Civ. Engrs.*, **107**, 765–793, 1942.
20. WHITNEY, C. S., "Plastic Theory of Reinforced Concrete Design," *Trans. Am. Soc. Civ. Engrs.*, **107**, 251–282, 1942.

CHAPTER 4

METHODS FOR COMPUTING DEFLECTIONS

4-1 General. Several examples were presented in Chapter 3 to illustrate how elastic deflections are used in condition equations for the analysis of indeterminate structures. This particular application in itself warrants a rather detailed treatment of the various methods for computing deflections. There are, however, other applications.

Bridge trusses, for example, should always be cambered so that their decks, when supporting full dead load plus part live load, will not deflect below a straight line joining the deck ends. Long-span roof trusses should be cambered so that the roof deck will not sag under load. In these cases, truss deflections must be computed to determine the amount of camber to be built into the erected trusses.

Observed deflections must constantly be checked against computed deflections during the erection of many structures. This is especially true when the cantilever method of erection is used, as, for example, in the case of an arch spanning a deep chasm. It would be quite embarrassing, to say the least, after cantilevering each half of the arch out from opposite sides to the span center, to find the end of one half lower than the end of the other. To prevent such an occurrence the theoretical elevations of the ends of the two cantilevers must be computed for all stages of erection. Weights and positions of crawler cranes must also be known for each stage, and elastic strains in bents and backstays must be considered. Any appreciable difference between observed and computed elevations for any given point is cause for immediate investigation and correction.

The erection of the towers of short-span suspension bridges, say with main spans up to 400 ft, is another case where computed deflections are essential for the proper erection of a structure. The two columns of each tower may be wide-flange sections, their webs being parallel to the bridge centerline, with a portal connecting the tops of the two columns. The main cables, composed of wire strands, will probably be clamped to the saddles at the top of the tower columns. The bases of these columns will likely be pinned so that the tops of the towers are free to move parallel to the bridge centerline, as required by load and temperature changes and consequent changes in the length of the backstays, without causing bending stresses in the tower columns. When the towers are erected, they must be tilted back, each toward its near anchorage. The strands of the cables are then placed and clamped to the saddles, and as the stiffening truss and deck are added, and as the backstays elongate because of increased tensions, the towers gradually rotate and come to rest in a vertical position. In this case the amount of backward tilt of the towers must be computed.

As indicated by the preceding discussion, the ability to evaluate deflections is of great importance in the analysis, design, and erection of structures. Deflections resulting from different actions on different types of structures must be determined. No one method of computation is the best for all problems; consequently, a number of different methods are discussed in this chapter. These are preceded by a development of expressions for the different types of internal strain energy.

Part 1. Internal Strain Energy

4-2 Internal work and deflections. It will be recalled that in 1833 Clapeyron established the equality between the external work done by the loads deflecting a structure and the internal strain energy. This is simply written as

$$\text{External work} = \text{Internal work}. \qquad (4\text{--}1)$$

That this equality should exist seems quite obvious, for when loads are applied to a structure they will do work as their points of application are displaced. If the reactions of the structure do not move, then all the external work of the applied loads must be dissipated in straining the structural material. If the elastic limit is not exceeded, all the external work is converted into elastic strain energy, which is recoverable and which returns the structure to its original position when the loads are removed. If, on the other hand, the elastic limit is exceeded, some permanent deformation occurs. As previously indicated, all structures dealt with herein are considered to be subjected to elastic strains.

Equation (4-1) is the basis of one method for finding deflections. As an illustration, consider the truss of Fig. 4-1. As the load P is gradually applied, the point of application deflects and comes to rest with a total

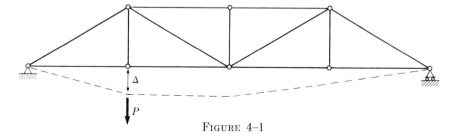

FIGURE 4-1

deflection Δ. Since the principle of superposition applies, there is a linear relationship between load and deflection. The average value of the load acting during the deflection, therefore, is $P/2$, and, consequently, the external work is $P\Delta/2$. Substituting in Eq. (4-1), we obtain

$$\frac{P\Delta}{2} = \text{Internal work.} \qquad (4\text{-}2)$$

If an expression can be found for the internal work (strain energy) in terms of P and the properties of the various members, the deflection of the point of application of P can be found.

In Fig. 4-1 internal strain energy is the result of axial stresses and elastic axial strains. Energy will also be stored internally when a material is subjected to elastic flexural, shearing or torsional strains. In the case of Fig. 4-2, for example, the deflection Δ is the result of bending and shearing strains.

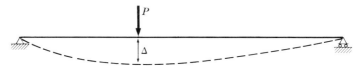

FIGURE 4-2

The beam of Fig. 4-3 is loaded with a couple which causes a translation and a rotation θ of the point of application of the couple. In this case, Eq. (4-1) becomes

$$\frac{M\theta}{2} = \text{Internal work.} \qquad (4\text{-}3)$$

FIGURE 4-3

It is obvious from the foregoing that expressions are required for the internal strain energy (internal work) resulting from, or necessary to cause, the various types of elastic strain. These expressions will be developed in the next section.

4-3 Expressions for internal strain energy.

(a) *Axial loads.* Consider the bar of Fig. 4-4 with a constant cross-sectional area A and length L. A gradual application of the force S will

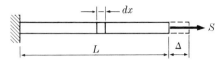

FIGURE 4-4

result in a final deflection Δ. The internal work in a length dx will be equal to the average load times the strain in the length dx. This will be

$$dW_i = \frac{S}{2} \cdot \frac{\Delta}{L} \cdot dx = \frac{S}{2} \cdot \frac{SL}{AE} \cdot \frac{dx}{L} = \frac{S^2\,dx}{2AE},$$

and the total internal work for the entire bar will be

$$W_i = \int_0^L \frac{S^2\,dx}{2AE} = \frac{S^2 L}{2AE}. \tag{4-4}$$

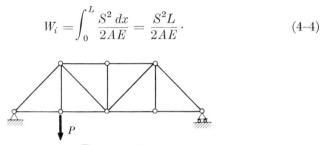

FIGURE 4-5

In the case of the truss indicated in Fig. 4-5, if S is the total stress in each bar of the truss (the value of S may be different for the various bars) due to the load P, then the internal work in each bar is, as in Eq. (4-4),

$$\frac{S^2 L}{2AE}.$$

The total internal work for the entire structure will be

$$W_i = \sum \frac{S^2 L}{2AE}, \tag{4-5}$$

and this is the required expression for internal strain energy resulting from elastic axial strain in the members of an articulated structure.

(b) *Flexure.* The beam of Fig. 4-6 is considered to be subjected to some external action which will cause an internal moment M_x at any section x along the beam. As this external action is gradually applied, the entire member (except at the ends and points of contraflexure, if they exist) is caused to flex. As this takes place, the internal work in a length dx of

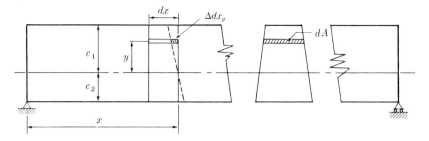

FIGURE 4-6

a fiber will be equal to the average force acting on the fiber times the total strain in the length dx. The strain in a fiber y distance from the neutral axis in the length dx, represented in Fig. 4-6 by Δdx_y, will be

$$\Delta dx_y = \frac{M_x \cdot y}{I} \cdot \frac{dx}{E}.$$

The final unit stress in this same fiber will be

$$\frac{M_x \cdot y}{I}.$$

Since the external action is gradually applied, the value of the average load acting on the area dA is

$$\frac{1}{2} \cdot \frac{M_x \cdot y \cdot dA}{I}.$$

The internal strain energy in the length dx of the fiber in question is therefore

$$\frac{1}{2}\left(\frac{M_x \cdot y}{I}\right)^2 \frac{dA \cdot dx}{E},$$

and the total internal work for all fibers in the length dx will be

$$dW_i = \frac{1}{2}\int_{-c_2}^{c_1}\left(\frac{M_x \cdot y}{I}\right)^2 \frac{dA \cdot dx}{E}$$

$$= \frac{1}{2} \cdot \frac{M_x^2\, dx}{EI^2}\int_{-c_1}^{c_2} y^2\, dA = \frac{1}{2} \cdot \frac{M_x^2\, dx}{EI}.$$

In the length L the total internal strain energy will be (omitting the subscript x)

$$W_i = \int_0^L \frac{M^2\,dx}{2EI}. \tag{4-6}$$

Attention is called to the fact that Eq. (4–6) can be applied correctly only to members which are initially straight; it is approximately correct for those with some initial curvature. However, when a member has a large initial curvature, say with a ratio of radius of centerline curvature to depth of section less than 10, application of the ordinary theory of flexure, which was used in deriving Eq. (4–6), may result in a substantial error. In these cases a more exact treatment, due to Winkler, is required. [This is extremely well presented by Pippard and Baker (4).]

(c) *Shear.* The expression for internal shearing strain energy will now be developed for the simple beam of Fig. 4–7(a), a beam which will be assumed to have a rectangular cross section which is constant throughout its length. It is also assumed to be subjected to some external action which causes a total shear V_x on any section x distance from the left end. This total shear V_x is not uniformly distributed over the cross section and thus the intensity of the shear on any fiber, at a distance y from the neutral axis, will be represented by v_{xy}.

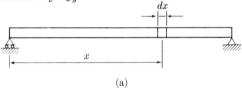

(a)

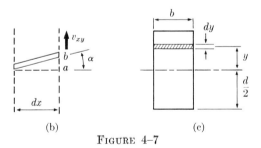

FIGURE 4–7

In Fig. 4–7(b) the shearing distortion of the fiber y distance from the neutral axis is indicated. The angle of this shearing distortion is very small and is designated by α. Consequently, $\sin\alpha = \tan\alpha = \alpha$. The expression for the external work done by v_{xy} while the segment dx is being distorted, and consequently an expression for the internal shearing strain energy of the fiber, will be

$$\partial w_i = \tfrac{1}{2}(v_{xy}\,dA)\overline{ab} = \tfrac{1}{2}(v_{xy}\,dA)\alpha\,dx, \tag{4-7}$$

where $dA = b\,dy$, and $\alpha = v_{xy}/G$, where G is the modulus of rigidity. When these substitutions are made, the above reduces to

$$\partial w_i = \frac{bv_{xy}^2}{2G}\,dx\,dy. \tag{4-8}$$

The shearing intensity is given by

$$v_{xy} = \frac{V_x Q_y}{Ib}, \tag{4-9}$$

where Q_y is the first moment, with respect to the neutral axis, of that portion of the beam which is outside the fiber y distance from the neutral axis. The moment of inertia of the cross section, referred to the neutral axis, is represented by I, and the value of Q_y is given by

$$Q_y = b\left(\frac{d}{2} - y\right)\left[y + \left(\frac{d/2 - y}{2}\right)\right] = \frac{b}{8}(d^2 - 4y^2). \tag{4-10}$$

Substituting (4-10) in (4-9), and substituting the result in (4-8), we obtain

$$\partial w_i = \frac{bV_x^2(d^2 - 4y^2)^2}{128 I^2 G}\,dx\,dy. \tag{4-11}$$

This is the internal shearing strain energy in the fiber y distance from the neutral axis in the length dx. To obtain the value for the entire segment, Eq. (4-11) is integrated over the depth of the beam as follows:

$$dw_i = \frac{bV_x^2\,dx}{128 I^2 G}\int_{-d/2}^{d/2}(d^2 - 4y^2)^2\,dy = \frac{bd^5 V_x^2}{240 I^2 G}\,dx.$$

Since the beam cross section is rectangular, $bd^3/12$ may be substituted for I, with the result that

$$dW_i = \frac{1.2 V_x^2\,dx}{2AG}.$$

The expression for the internal shearing strain energy for the entire beam is obtained by integrating the above equation over the length of the beam, with the result that

$$W_i = \int_0^L \frac{1.2 V_x^2\,dx}{2AG}. \tag{4-12}$$

This expression applies only to a member with a constant rectangular cross section. A more general form is

$$W_i = K \int_0^L \frac{V_x^2\, dx}{2AG}, \qquad (4\text{-}13)$$

where the value of the factor K depends on the shape of the cross section, and, as just demonstrated, for rectangular cross sections with two faces parallel to the force plane, its value is 1.2. In a similar manner it can be shown that K is 10/9 for circular sections and is (almost) 1 for steel WF or I beams. In this last case, A is the area of the web.

(d) *Torsion.* The cantilever round shaft of Fig. 4–8 is subjected to a twisting action which results in a torque T_x at any section x along the

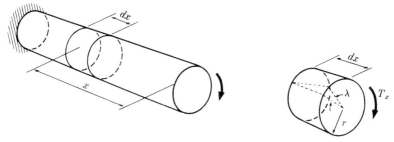

Figure 4–8

shaft. A segment dx in length is shown enlarged. The twisting action being gradually applied, the torsional strain energy for the segment will be

$$\tfrac{1}{2} T_x \cdot \lambda,$$

where λ is the torsional strain in radians of one face of the segment relative to the other face. The value of λ will be

$$\frac{T_x r}{JG} \cdot \frac{dx}{r},$$

where J is the polar moment of inertia of the cross section and G is the modulus of rigidity of the material. Consequently, the torsional strain energy for the segment will be

$$dW_i = \frac{T_x^2\, dx}{2JG},$$

and, for the entire member of length L, the total strain energy (omitting the subscript x) will be given by

$$W_i = \int_0^L \frac{T^2\,dx}{2JG}. \tag{4-14}$$

Attention is called to the fact that if the shaft is of any section other than circular, J is not the polar moment of inertia. The proper value to use for J in these instances can be calculated (1). It has been established (6) that for a rectangular section, with a long-side dimension h and a short-side dimension b, the value to be used in place of J is given by the expression Cb^3h. The values of C, for various ratios of h/b, are shown in the accompanying table.

h/b	C	h/b	C	h/b	C
1	0.141	2.5	0.249	6	0.299
1.2	0.166	3	0.263	10	0.312
1.5	0.196	4	0.281	∞	0.333
2	0.229	5	0.291		

If a formula is desired, the following will apply (5):

$$C = \frac{hb^3}{16}\left[\frac{16}{3} - 3.36\frac{b}{h}\left(1 - \frac{b^4}{12h^4}\right)\right]. \tag{4-15}$$

Part 2. Deflections by Real Work

4–4 Demonstration of the method. If linear deflection components are to be computed by real work, Eq. (4–2) applies:

$$\frac{P\Delta}{2} = \text{Internal work.}$$

If rotational deflection components are desired, Eq. (4–3) is used:

$$\frac{M\theta}{2} = \text{Internal work.}$$

The method is quite limited in application. The unknown in any problem will be Δ or θ, depending on whether linear or rotational deflection components are required. If more than one P or more than one M are applied simultaneously to a structure, then (usually) more than one unknown Δ or θ will appear on the left side of the equation and a solution is impossible. In general, therefore, the method can only be used to find the linear deflection component of the point of application of a force in the direction of that force, or to find the rotation of the point of application of a couple in the plane of that couple. In most cases the force or the couple must act alone on the structure. If, however, two equal forces or two equal couples act symmetrically on a symmetrical structure, the equation can be solved since, by reason of the symmetry, the two deflection components are known to be equal.

EXAMPLE 4–1. Using the method of real work, find the deflection under the load if the beam of Fig. 4–9 is a steel section with a moment of inertia of 200 in^4. Use $E = 30{,}000$ k/in^2 and neglect shearing strains.

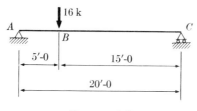

FIGURE 4–9

By inspection, $R_A = 12$ k and $R_C = 4$ k, and

$$\frac{P\Delta}{2} = \int \frac{M^2\,dx}{2EI}.$$

Therefore,

$$\Delta = \frac{1}{P}\int \frac{M^2\,dx}{EI}.$$

TABLE 4–1.

Section	$x = 0$ at	x increasing from	M (ft·k)
AB	A	A to B	$12x$
CB	C	C to B	$4x$

$$\int M^2\,dx = \int_0^5 (12x)^2\,dx + \int_0^{15} (4x)^2\,dx$$

$$= \frac{144x^3}{3}\Big]_0^5 + \frac{16x^3}{3}\Big]_0^{15} = 24{,}000.$$

Therefore,

$$\Delta = \frac{24{,}000 \times 1728}{16 \times 30{,}000 \times 200} = 0.43 \text{ in.}$$

Attention is called to the multiplication of the numerator of the answer by the number 1728. This, of course, is to change ft^3 to in^3 so that the units of length will be consistent in both numerator and denominator.

EXAMPLE 4–2. Using the method of real work, find the vertical deflection component of the point of application of the 10 k load of Fig. 4–10. The truss is steel and the sectional areas of the various members in square inches are circled on the illustration. $E = 30,000$ k/in^2.

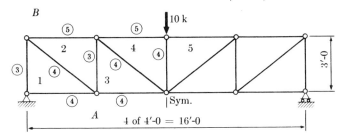

FIGURE 4–10

$$\frac{P\Delta}{2} = \sum \frac{S^2 L}{2AE}$$

TABLE 4–2.

Member	S (k)	L (ft)	A (in^2)	$\dfrac{S^2 L}{A}$
B1	−5.0	3	3.0	+25.0
B2	−6.67	4	5.0	+35.6
B4	−13.33	4	5.0	+142.4
A1	0	4	4.0	0
A3	+6.67	4	4.0	+44.4
12	+8.33	5	4.0	+86.7
23	−5.0	3	3.0	+25.0
34	+8.33	5	4.0	+86.7
½(45)	−10.0	1.5	4.0	+37.5
				+483.3

$$\Delta = \frac{2 \times 483.3 \times 12}{10 \times 30,000} = 0.039 \text{ in.}$$

Note that since the truss is symmetrical, it is only necessary to tabulate the members and evaluate $S^2 L/A$ for half the truss. It is for this reason

PROBLEMS

4-3. Given a simple beam with a span of L ft and with a concentrated vertical load of P k acting at the center. The beam is a steel wide-flange section with web area of A in^2. The modulus of rigidity is G k/in^2. Using the method of real work, derive the expression for the vertical deflection of the point of application of P as caused by the shearing strain. [*Ans.*: $3PL/AG$ in.]

4-4. Find the vertical deflection of the center of the simple beam shown in Fig. 4-11 as caused by (a) flexural strains, and (b) shearing strains. The beam is a 12WF27 with a moment of inertia of 204 in^4, web area = 2.85 in^2, E = 30,000 k/in^2, and G = 12,000 k/in^2. [*Ans.*: (a) 0.47 in., (b) 0.02 in.]

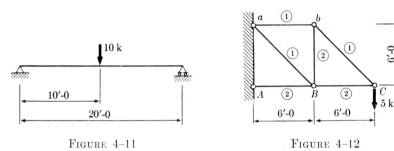

FIGURE 4-11   FIGURE 4-12

4-5. Find the vertical deflection component of point C (see Fig. 4-12) as caused by the 5 k load. Sectional areas of members (in square inches) are circled on the illustration. E = 30,000 k/in^2. [*Ans.*: 0.12 in.]

4-6. For the cantilever shown in Fig. 4-13, find:
(a) The vertical deflection component of point A resulting from the action of the 3 k vertical load acting alone.
(b) The rotational deflection component of point A as caused by the 2 ft·k couple acting alone.
Use E = 30,000 k/in^2. Neglect shearing strains. [*Ans.*: (a) 2.54 in., (b) 0.0015 rad.]

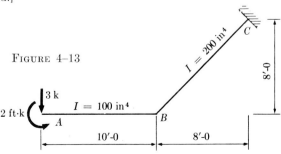

FIGURE 4-13

Part 3. Castigliano's First Theorem

4–5 Development and demonstration. In 1876, Alberto Castigliano published a notable paper in which his two important theorems were presented, the second theorem (to be considered in detail in another chapter) appearing as a corollary of the first. Castigliano's first theorem is of immediate interest since it provides one of the most important methods for determining elastic deflections of structures. It is a method which enables the analyst to find any deflection component of any point on a structure. The deflection component may be either rotational or linear, in any direction, and caused by any system of applied loads.

The first theorem of Castigliano can be stated as follows: *The deflection component of the point of application of an action on a structure, in the direction of that action, will be obtained by evaluating the first partial derivative of the total internal strain energy of the structure with respect to the applied action.*

In this statement an action is taken to mean either a force or a couple. Thus a deflection component in the direction of a force will be along the line of action of the force, and a deflection component in the direction of a couple will be a rotation in the plane of the couple.

The following demonstration of the first theorem is similar to that presented by Church (2), but has been somewhat revised. Let W represent the total internal strain energy in the deflected beam of Fig. 4–14. The

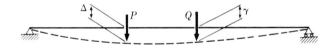

Figure 4–14

deflected position is represented by the dashed line. P and Q have been gradually and simultaneously applied and a linear relationship exists between these loads and the resulting deflections. Consequently, the average value of each load during the deflection of the beam will be one-half its final value. It has been previously demonstrated that the internal strain energy is equal to the external work of the applied loads; therefore it is possible to write

$$W = \frac{P\Delta}{2} + \frac{Q\gamma}{2}. \qquad (4\text{–}16)$$

A small additional load dP is now added to P, causing an additional deflection of the beam to the position indicated by the dashed line of Fig. 4–15. The resulting increment of internal strain energy will be equal to the sum of the products obtained by multiplying the additional

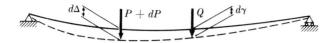

FIGURE 4-15

deflection of each point of load by the average load acting through this additional deflection. This is written as

$$dW = \left[\frac{P + (P + dP)}{2}\right] d\Delta + Q\, d\gamma.$$

Neglecting the product of two differentials, the above becomes

$$dW = P\, d\Delta + Q\, d\gamma. \tag{4-17}$$

It is possible to express dW in another way. Apply $P + dP$ and Q gradually and simultaneously, and let W' represent the total internal strain energy in the beam in the final deflected position with all loads acting. It is now possible to write

$$W' = \left(\frac{P + dP}{2}\right)(\Delta + d\Delta) + \frac{Q}{2}(\gamma + d\gamma),$$

and, by neglecting the product of two differentials,

$$W' = \frac{P\Delta}{2} + \frac{Q\gamma}{2} + \frac{dP\,\Delta}{2} + \frac{P\,d\Delta}{2} + \frac{Q\,d\gamma}{2}. \tag{4-18}$$

But, $dW = W' - W$ and, therefore, from Eqs. (4-18) and (4-16),

$$dW = \frac{dP\,\Delta}{2} + \frac{P\,d\Delta}{2} + \frac{Q\,d\gamma}{2}.$$

If Eq. (4-17) is substituted in this expression, the result is

$$dW = \frac{dP\,\Delta}{2} + \frac{dW}{2},$$

and therefore

$$dW = dP \cdot \Delta,$$

from which

$$\Delta = \frac{dW}{dP}, \tag{4-19}$$

which was to be demonstrated.

When solving for a deflection by Castigliano's first theorem, it should be noted that, if the sign of the answer is negative, the actual deflection is opposite to the sense of the action with respect to which the derivative is taken. Furthermore, if a deflection component is required for a point where no action is applied, or if an action exists at the point but not in the direction of the desired deflection component, then an imaginary action is applied at the point and in the desired direction until the partial derivative of the total internal strain energy has been found. The imaginary action is then reduced to zero. These points will be illustrated in subsequent examples.

A considerable amount of work may be saved in the computations by differentiating under the integral. The general expression for deflections by Castigliano's first theorem should be written as a partial derivative, since more than one action will usually be applied to the structure. For linear deflection components, this expression will be

$$\Delta = \frac{\partial W}{\partial P}. \qquad (4\text{--}20)$$

If the internal work results from bending, the above expression becomes

$$\Delta = \frac{\partial}{\partial P} \int \frac{M^2 \, dx}{2EI}.$$

If the indicated operation is performed, it is necessary to square the various expressions for M (which are often quite lengthy), integrate, and then evaluate the partial derivative. It is much easier to differentiate under the integral sign, which results in

$$\Delta = \int M \left(\frac{\partial M}{\partial P}\right) \frac{dx}{EI}. \qquad (4\text{--}21)$$

Similarly, when finding a linear deflection of an articulated structure,

$$\Delta = \sum S \left(\frac{\partial S}{\partial P}\right) \frac{L}{AE}. \qquad (4\text{--}22)$$

Expressions of similar form apply for deflections resulting from shearing or torsional strains. For rotational deflections the left side of the above equations will be θ, and the partial derivative will be taken with respect to a moment acting at the point of the desired rotational deflection.

EXAMPLE 4–7. Using Castigliano's first theorem, find the vertical deflection component of the point of application of the 16 k load (see

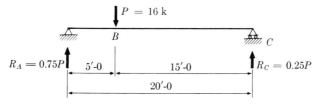

FIGURE 4-16

Fig. 4-16). The steel beam has a moment of inertia of 200 in^4. Use $E = 30{,}000$ k/in^2 and neglect shearing strains.

TABLE 4-3.

Section	$x = 0$ at	x increasing	M (ft·k)	$\dfrac{dM}{dP}$
AB	A	A to B	$0.75Px$	$0.75x$
CB	C	C to B	$0.25Px$	$0.25x$

$$\Delta = \int M\left(\frac{\partial M}{\partial P}\right)\frac{dx}{EI},$$

$$\int M\left(\frac{\partial M}{\partial P}\right)dx = \int_0^5 0.75Px(0.75x)\,dx + \int_0^{15} 0.25Px(0.25x)\,dx$$

$$= \frac{0.563Px^3}{3}\bigg]_0^5 + \frac{0.063Px^3}{3}\bigg]_0^{15} = 1508,$$

$$\Delta = \frac{1508 \times 1728}{30{,}000 \times 200} = 0.43 \text{ in.}$$

A brief explanation of the multiplication by the number 1728 in the evaluation of the linear deflection component of the above example, and a similar multiplication by 144 in the case of a rotational deflection component, is desirable. The expression for a linear deflection component is

$$\Delta = \int M\left(\frac{\partial M}{\partial P}\right)\frac{dx}{EI}.$$

If the proper dimensions are substituted, the result is

$$\Delta = (\text{ft·k})\left(\frac{\text{ft·k}}{\text{k}}\right)\left(\frac{\text{ft}}{\text{k/in}^2}\right)\left(\frac{1}{\text{in}^4}\right) = \frac{\text{ft}^3}{\text{in}^2},$$

and it is apparent that a multiplication by 1728 is necessary to obtain an answer in inches; obviously a multiplication by 144 would give a deflection in feet.

In the case of a rotational deflection component, the correct expression would be

$$\theta = \int M \left(\frac{\partial M}{\partial M_A}\right) \frac{dx}{EI}.$$

Substituting the correct dimensions, we find that

$$\theta = (\text{ft·k}) \left(\frac{\text{ft·k}}{\text{k}}\right) \left(\frac{\text{ft}}{\text{k/in}^2}\right) \left(\frac{1}{\text{in}^4}\right) = \frac{\text{ft}^2}{\text{in}^2},$$

and the necessity for a multiplication by 144 is apparent, since the answer must be in radians and units must cancel.

EXAMPLE 4–8. Find the vertical deflection component of section B of the steel beam shown in Fig. 4–17. The moment of inertia is 300 in^4. Use $E = 30{,}000$ k/in^2.

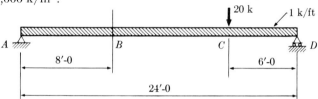

FIGURE 4–17

No concentrated force acts at section B, as is required to obtain the partial derivative. It is necessary, therefore, to assume an imaginary force P acting at B, as shown in Fig. 4–18. Note that the reactions, also indicated in this figure, must be expressed in terms of P. Even

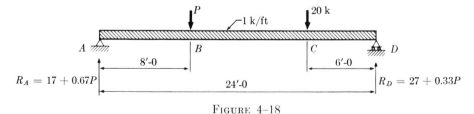

FIGURE 4–18

if the original loading had included a concentrated force at B, this force would still be represented by a letter until the partial derivative were obtained. The suggested tabular arrangement for moments and partial derivatives is as follows:

TABLE 4-4.

Section	$x = 0$ at	x increasing	M (ft·k)	$\dfrac{\partial M}{\partial P}$
AB	A	A to B	$(17 + 0.67P)x - \dfrac{x^2}{2}$	$+0.67x$
BC	B	B to C	$(17 + 0.67P)(8 + x) - Px - \dfrac{(8+x)^2}{2}$	$-0.33x + \dfrac{16}{3}$
DC	D	D to C	$(27 + 0.33P)x - \dfrac{x^2}{2}$	$+0.33x$

Substituting these values in the expression

$$\Delta = \int M \left(\frac{\partial M}{\partial P} \right) \frac{dx}{EI}$$

and reducing P to zero (the actual value of P in this case), we obtain

$$EI\,\Delta = \int_0^8 \left(17x - \frac{x^2}{2} \right)(+0.67x)\, dx$$

$$+ \int_0^{10} \left(9x + 104 - \frac{x^2}{2} \right)\left(-0.33x + \frac{16}{3} \right) dx$$

$$+ \int_0^6 \left(27x - \frac{x^2}{2} \right)(+0.33x)\, dx.$$

If we perform the indicated operations, it is possible to solve for Δ. Note that an adjustment must be made for units, specifically the multiplication by 1728, as in Example 4-7.

EXAMPLE 4-9. Using Castigliano's first theorem, find the vertical deflection component of L_1 in the indicated truss (see Fig. 4-19) when loaded as shown. The areas of the various members are given in Table 4-5. Use $E = 30,000$ k/in^2.

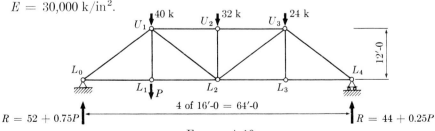

FIGURE 4-19

TABLE 4–5.

Member	$\dfrac{L}{\text{ft}}$	$\dfrac{A}{\text{in}^2}$	$\dfrac{L}{A}$	S Real loads (k)	P	$\dfrac{\partial S}{\partial P}$	$\dfrac{SL}{A}$	$\dfrac{SL}{A}\cdot\dfrac{\partial S}{\partial P}$
L_0U_1	20	8	2.50	−86.7	−1.25P	−1.25	−217.0	+271
U_1U_2	16	7	2.28	−85.4	−0.667P	−0.667	−194.7	+130
U_2U_3	16	7	2.28	−85.4	−0.667P	−0.667	−194.7	+130
U_3L_4	20	8	2.50	−73.4	−0.417P	−0.417	−183.7	+76
L_0L_2	32	5	6.40	+69.3	+1.0P	+1.000	+443.0	+443
L_2L_4	32	5	6.40	+58.6	+0.333P	+0.333	+375.0	+125
U_1L_1	12	2	6.00	0	+1.0P	+1.000	0	0
U_1L_2	20	3	6.67	+20.0	−0.417P	−0.417	+133.0	−55
U_2L_2	12	4	3.00	−32.0	0	0	−96.0	0
L_2U_3	20	3	6.67	+33.3	+0.417P	+0.417	+222.0	+93
U_3L_3	12	2	6.00	0	0	0	0	0
							Total	+1213

Therefore,

$$\Delta = \dfrac{+1213 \times 12}{30{,}000} = +0.48 \text{ in.}$$

Note that strict attention must be given to the signs of the stresses in the above table. The author prefers to designate a tensile stress with a positive sign and a compressive stress with a negative sign. A negative sign for the value in the last column for any member indicates that the elastic strain in that particular member tends to cause a deflection of the point in question with a sense opposite to that of the force P. Obviously, then, a positive answer signifies a deflection having the same sense as the force P.

EXAMPLE 4–10. Using Castigliano's first theorem, find the horizontal, vertical, and rotational deflection components of point A (see Fig. 4–20) as caused by the 10 k load. Use $E = 30{,}000$ k/in^2 and consider any moment causing tension on the outside of the frame as positive.

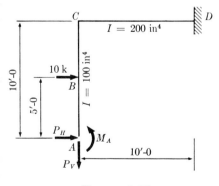

FIGURE 4–20

TABLE 4-6.

Sec-tion	$x=0$ at	x increasing	\multicolumn{4}{c}{M (ft·k)}	$\dfrac{\partial M}{\partial P_H}$	$\dfrac{\partial M}{\partial P_V}$	$\dfrac{\partial M}{\partial M_A}$
			10 k	P_H	P_V	M_A			
AB	A	A to B	0	$+P_H \cdot x$	0	$+M_A$	$+x$	0	$+1$
BC	B	B to C	$+10x$	$+5P_H + P_H \cdot x$	0	$+M_A$	$+5+x$	0	$+1$
CD	C	C to D	$+50$	$+10P_H$	$+P_V x$	$+M_A$	$+10$	$+x$	$+1$

$$E\Delta_{Ah} = \int M\left(\frac{\partial M}{\partial P_H}\right)\frac{dx}{I}$$

$$= \int_0^5 (10x)(5+x)\frac{dx}{100} + \int_0^{10} (50)(10)\frac{dx}{200} = +35.42,$$

$$\Delta_{Ah} = \frac{+35.42 \times 1728}{30,000} = +2.04 \text{ in.},$$

$$E\Delta_{Av} = \int M\left(\frac{\partial M}{\partial P_V}\right)\frac{dx}{I} = \int_0^{10} (50) \times \frac{dx}{200} = +12.5,$$

$$\Delta_{Av} = \frac{+12.5 \times 1728}{30,000} = +0.72 \text{ in.},$$

$$E\theta_A = \int M\left(\frac{\partial M}{\partial M_A}\right)\frac{dx}{I}$$

$$= \int_0^5 (10x)(1)\frac{dx}{100} + \int_0^{10} (50)(1)\frac{dx}{200} = +3.75,$$

$$\theta_A = \frac{+3.75 \times 144}{30,000} = +0.018 \text{ rad.}$$

Note in the above that, since all real moments and all partial derivatives are positive, the positive signs have been omitted for the various substitutions behind the integral signs. It is extremely important that any negative signs be carried throughout computations of this sort, since the sign of the answer indicates the sense of the required deflection component relative to the force or couple with respect to which the partial derivative is taken. As indicated previously, a positive answer signifies a deflection component of the same sense as this force or couple.

EXAMPLE 4–11. Given a 4-in. φ standard pipe bracket (see Fig. 4–21) having a 90° angle at B and located in a horizontal plane. Find the vertical deflection component of point A and the rotational deflection component of A about the axis of AB, as caused by the 1 k vertical load at A. The plane moment of inertia of the pipe is 7.23 in^4 and the polar moment of inertia is 14.46 in^4. $E = 30{,}000$ k/in^2 and $G = 12{,}000$ k/in^2.

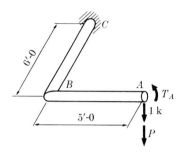

FIGURE 4–21

TABLE 4–7.

Section	$x = 0$ at	x increasing	M (ft·k)	T (ft·k)	$\dfrac{\partial M}{\partial P}$	$\dfrac{\partial M}{\partial T_A}$	$\dfrac{\partial T}{\partial P}$	$\dfrac{\partial T}{\partial T_A}$
AB	A	A to B	$+x + Px$	$+T_A$	$+x$	0	0	$+1$
BC	B	B to C	$+x + Px + T_A$	$+5 + 5P$	$+x$	$+1$	$+5$	0

The expression for the vertical deflection of A, from Castigliano's first theorem, is

$$\Delta_{Av} = \int M \left(\frac{\partial M}{\partial P}\right) \frac{dx}{EI} + \int T \left(\frac{\partial T}{\partial P}\right) \frac{dx}{GJ}.$$

The temporary load P and moment T_A are reduced to their actual value of zero in the various terms in Table 4–7 before substituting in the above expression. Therefore,

$$\Delta_{Av} = \int_0^5 \frac{(+x)(+x)\, dx}{30{,}000 \times 7.23} + \int_0^6 \frac{(+x)(+x)\, dx}{30{,}000 \times 7.23} + \int_0^6 \frac{(+5)(+5)\, dx}{12{,}000 \times 14.46}$$

$$= +2.40 \text{ in}.$$

Similarly,

$$\theta_A = \int M\left(\frac{\partial M}{\partial T_A}\right)\frac{dx}{EI} + \int T\left(\frac{\partial T}{\partial T_A}\right)\frac{dx}{GJ}$$

$$= \int_0^6 \frac{(+x)(+1)\,dx}{30{,}000 \times 7.23}$$

$$= +0.012 \text{ rad.}$$

Problems

4–12. Using Castigliano's first theorem, find (see Fig. 4–22):

(a) Vertical deflection component of section C.
(b) Rotational deflection component of section B.

I of beam is 200 in^4 and E is 30,000 k/in^2. Consider flexural strains only. [*Ans.:* (a) 3.64 in., (b) 0.026 rad.]

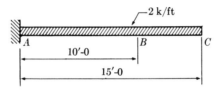

Figure 4–22

4–13. Find:

(a) The horizontal deflection component of point A due to the 20 k load shown in Fig. 4–23.

(b) The horizontal deflection component of point A due to a force of 1 k applied horizontally at A acting toward the left.

Consider flexural strains only. $E = 30{,}000$ k/in^2. [*Ans.:* (a) 0.88 in. to left, (b) 1.59 in. to left.]

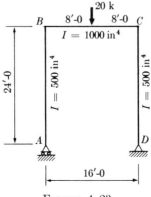

Figure 4–23

4-14. Find the rotational deflection component of point A on the frame shown in Fig. 4-24. Consider flexural strains only. $E = 30,000$ k/in^2. [*Ans.:* 0.0096 rad.]

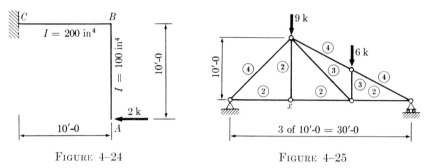

FIGURE 4-24

FIGURE 4-25

4-15. Find the vertical deflection component of point x (see Fig. 4-25). Use $E = 10,000$ k/in^2. Cross-sectional areas of members in square inches are indicated in the circles on the illustration. [*Ans.:* 0.24 in.]

4-16. Given a 4-in. ϕ standard pipe bracket (see Fig. 4-26) having a 90° angle at B and C and located in a horizontal plane. Find:

(a) The vertical deflection component of D.
(b) The rotational deflection component of D in the plane normal to the axis of CD.

Plane moment of inertia $= 7.23$ in^4, $G = 12,000$ k/in^2, $E = 30,000$ k/in^2. [*Ans.:* (a) 4.61 in., (b) 0.022 rad.]

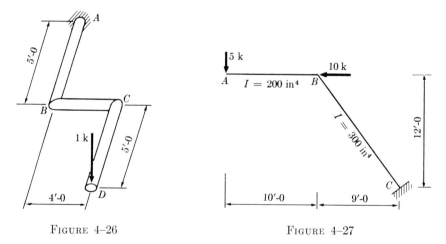

FIGURE 4-26

FIGURE 4-27

4-17. Find the three deflection components of point A as caused by the loads shown in Fig. 4-27. $E = 30,000$ k/in^2. [*Ans.:* horizontal $= 2.76$ in., vertical $= 6.36$ in., $\theta = 0.038$ rad.]

4-18. The pipe bracket shown in Fig. 4-28 is constructed of 5-in. ϕ standard pipe for which the plane moment of inertia is 15.16 in^4. Considering flexure and torsion, find the horizontal deflection component of point E in a direction parallel to the member BC. Members AB, BC, and CD are in a horizontal plane, and DE is vertical. $E = 30,000$ k/in^2 and $G = 12,000$ k/in^2. [*Ans.:* 4.04 in.]

4-19. Find the three deflection components of point A as caused by the 10 k load shown in Fig. 4-29. $E = 30,000$ k/in^2. [*Ans.:* horizontal = 0.73 in., vertical = 18.9 in., θ = 0.065 rad.]

4-20. A steel frame supporting a sign 20 ft high is subjected to the wind loads

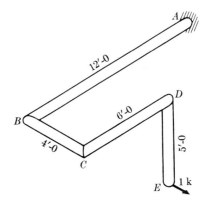

FIGURE 4-28

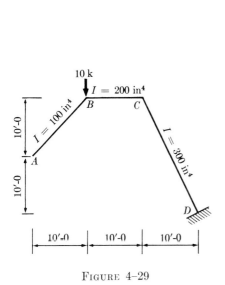

FIGURE 4-29

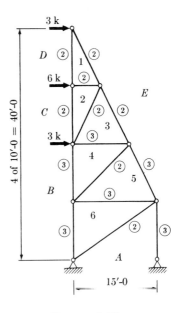

FIGURE 4-30

indicated in Fig. 4-30. Find the horizontal deflection of the top of the frame resulting from these loads. $E = 30,000$ k/in^2 and cross-sectional areas of members are indicated in square inches in the circles on the illustration. [*Ans.:* 0.43 in.]

Part 4. The Method of Virtual Work

4-6 General. The principle of virtual work, based on the principle of virtual velocities which Johann Bernoulli presented in 1717, constitutes the most versatile method available for evaluating elastic deflections of structures. Not only is it possible to determine deflections resulting from loads of any type, causing any kind of strain in a structure, but it is also possible to compute deflections resulting from temperature changes, errors in fabrication, or shrinkage of the structural material. These deflections may be linear or angular and in any direction. The only restriction is that when the principle of virtual work is used in its finite form, as will be the case herein, the principle of superposition must apply to the structures considered.

The principle of virtual work was previously stated as follows: *If a deformable structure, in equilibrium and sustaining a given load or system of loads, is subjected to a virtual deformation as the result of some additional action, the external virtual work of the given load or system of loads is equal to the internal virtual work of the stresses caused by the given load or system of loads.*

As a consequence, the basic relation for the method of virtual work is

$$\text{External virtual work} = \text{Internal virtual work.} \qquad (4\text{-}23)$$

It is only necessary to express the two sides of this equation in a manner consistent with the type of the desired deflection component and the internal strains, and then to solve the resulting equation for the deflection component. In the pages immediately following, the proper expressions for the left side of this equation will be developed for linear and rotational deflection components. Expressions for the right side will be derived or written for axial, flexural, torsional, and shearing strains.

4-7 Deflections resulting from axial strains. Assume that it is required to find the vertical deflection component Δ of point A as caused by the given loads on the truss of Fig. 4-31. First, consider that the given loads are removed from the truss and that a unit force is applied at point A

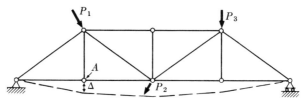

FIGURE 4-31

acting in the direction of the required deflection component. This unit force will be 1 k if the real loads are expressed in kilopounds, or 1 lb if the real loads are expressed in pounds; and it is, of course, imaginary. (Some writers refer to it as a "dummy" unit force; the author prefers to call it a "fictitious" force.) The structure is in equilibrium under the action of this fictitious unit force and therefore it constitutes the "given load or system of loads" in the preceding statement of the principle of virtual work. Figure 4–32 shows this fictitious force in position, with the bottom chord of the truss, as deflected by the fictitious force, indicated by the dashed line.

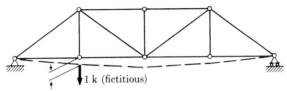

FIGURE 4–32

The truss is now considered to be subjected to virtual displacements which are identical to the deflections resulting from the action of the real loads. The obvious procedure is to assume that the real loads are superposed on the truss of Fig. 4–32. To repeat, *the resulting deflections are considered to be virtual displacements.* Consequently, point A is caused to deflect with a virtual vertical component Δ. The fictitious 1 k will move through the distance Δ with its value unchanged, and, therefore, the external virtual work will be $1 \times \Delta$, which is the proper form for the left side of the basic virtual work equation when a linear deflection component is required.

An expression for the internal virtual work must now be written. The fictitious bar stresses in the truss, resulting from the action of the fictitious unit force, will be designated by u. When the real loads are added, the fictitious stress u, acting on each member, will move through a distance equal to the elastic strain in that member as caused by the real loads. This elastic strain is considered to be a virtual displacement of one end of the member relative to the other. If the total stress or load in each member caused by the real loads is designated by S, the sectional area by A, the length by L, and the modulus of elasticity by E, then the elastic strain in each member will be SL/AE. For example, consider one panel length of the bottom chord as a free body. The unit fictitious force having been applied to the truss, the ends of the member are acted upon by stresses u as shown in Fig. 4–33(a). When the real loads are superposed, one end of the member will move away from the other end a distance SL/AE, as in Fig. 4–33(b). The stress u "rides along" at its full value. Consequently,

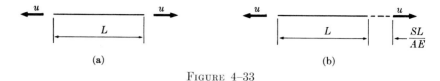

FIGURE 4-33

the expression for the internal virtual work for one member will be uSL/AE and the total internal virtual work for the entire structure will be given by $\Sigma uSL/AE$. Therefore, by the principle of virtual work,

$$1 \times \Delta = \sum \frac{uSL}{AE}. \qquad (4\text{-}24)$$

Attention is again called to the fact that Δ is not the total deflection of point A, but is, instead, the linear component of this total deflection along the line of action of the unit fictitious force.

To supplement the preceding discussion, several observations will be helpful in understanding the application of the method of virtual work to articulated structures.

(a) If a linear deflection component of a point is desired, a unit fictitious force must be applied at the point and along the line of the desired deflection component. The sense along this line, as assumed for this fictitious force, is immaterial. If the sign of the answer is positive (in this case, if the sign of $\Sigma uSL/AE$ is positive), then the actual deflection is of the same sense as the unit fictitious force. If the sign is negative, then the actual deflection will be opposite. It is apparent, therefore, that it is extremely important that the proper sign be placed before every value of u and S. It is immaterial whether a positive sign is used to indicate tension or compression, but throughout this book a positive sign before a stress will designate tension. This seems to be more logical than the alternate negative designation, since a tensile stress will cause a positive increment, or increase, in the length of a member. The final sign of uSL/AE for each member is determined by the signs of S and u and by the usual rules of algebra.

(b) An interesting alternate concept as to the physical significance of the fictitious stress u is that of its being a deflection coefficient. When multiplied by the change in length (due to any cause whatsoever) of the corresponding member, this coefficient u will give the effect (or contribution) of the change in length of that member on (or to) the required deflection component.

(c) The placing of the unit fictitious loads for various deflections may be better understood by reference to the figures which follow. Consider Fig. 4-34. If, for any system of applied loads, it is required to find the vertical deflection component of any panel point, say B, then place a unit fictitious vertical force at B. If the horizontal deflection component of B is required, then place a unit fictitious horizontal force at B.

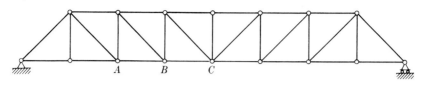

FIGURE 4-34

If the rotation of the member AB is desired, then a unit fictitious couple is applied as in Fig. 4-35. Note the value of the forces of the couple. In this

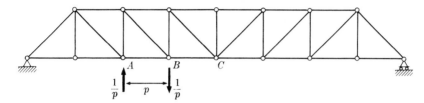

FIGURE 4-35

case the left side of Eq. (4-24) is written as $1 \times \theta$, where θ represents the rotation of member AB in radians. The author also prefers to change the u in the right side of Eq. (4-24) to u_α, the subscript α indicating that the several stresses represented by u_α are caused by a unit couple rather than by a unit force. If it is necessary to find the rotation of member BC, the unit fictitious couple is applied to BC. However, if the angle between AB and BC is desired, the individual rotations of AB and BC can be determined as above and the answers combined for the final result, although this will involve two complete solutions and considerable labor. A much easier method would be to place both unit fictitious couples on the truss at one time, as shown in Fig. 4-36. In this case, only six members need be considered for the solution instead of working through the entire structure twice.

It is desirable that the reader compare each method for finding deflections with alternate methods so that he will know which can be applied most effectively to a given problem. In this connection it is important to realize that the value of u, in the method of virtual work, is exactly the same as the value of $\partial S/\partial P$ in Castigliano's first theorem, the latter being

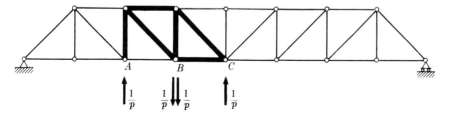

FIGURE 4-36

merely the rate of change of S with respect to P, or the change per unit value of P. Obviously this is the stress caused by a unit force and is the same as u. The equality of the two methods is expressed by

$$\sum S\left(\frac{\partial S}{\partial P}\right)\frac{L}{AE} = \sum \frac{SuL}{AE}. \tag{4-25}$$

The reader should consider the relative amounts of work necessary to evaluate $\partial S/\partial P$ and u in any given problem (it will usually be found that there is slightly less labor involved in the solution by virtual work), for therein is the only difference between the two methods. Several examples are presented below to demonstrate the application of the method.

EXAMPLE 4-21. Using virtual work, find the vertical deflection component of point Y in the truss of Fig. 4-37. The area of each member is 10 in². Use $E = 30{,}000 \text{ k/in}^2$.

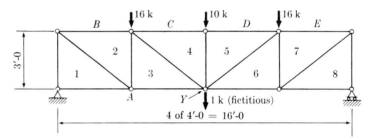

FIGURE 4-37

TABLE 4-8.

Member	S (k)	u (k)	L (ft)	SuL
B2	−28	−0.667	4	+75
C4	−35	−1.333	4	+190
A1	0	0	4	0
A3	+28	+0.667	4	+75
B1	−21	−0.5	3	+31
12	+35	+0.833	5	+150
23	−21	−0.5	3	+31
34	+8	+0.833	5	+30
½(45)	−10	0	1.5	0
				+582

$$\Delta_Y = \frac{+582 \times 2 \times 12}{10 \times 30{,}000} = +0.047 \text{ in.}$$

EXAMPLE 4–22. A six-panel highway bridge truss having the dimensions indicated in Fig. 4–38 is constructed with sidewalks outside the trusses so that the bottom chords are shaded. What will be the vertical deflection component of the bottom chord at the center of the bridge when the temperature of the bottom chord is 40°F below that of the top chord, endposts, and webs? (Temperature coefficient of expansion of steel is 0.0000065 per degree F.)

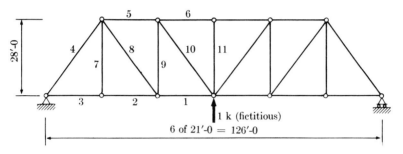

FIGURE 4–38

The expression for deflection may now be written as

$$1 \times \Delta = \Sigma u \cdot dL,$$

where dL is the temperature change in the length of each member.

TABLE 4–9.

Member	L (ft)	40 × 0.0000065L = dL	u (k)	u dL
4	35	+0.00910	+0.625	+0.00568
5	21	+0.00546	+0.75	+0.00409
6	21	+0.00546	+1.13	+0.00616
7	0	0	0	0
8	35	+0.00910	−0.625	−0.00568
9	28	+0.00728	+0.5	+0.00364
10	35	+0.00910	−0.625	−0.00568
11	0	0	0	0
				+0.00821

$$\Delta = +2 \times 0.00821 \times 12 = +0.20 \text{ in. (up)}.$$

EXAMPLE 4-23. It is desired to provide 3 in. of camber at the center of the truss shown in Fig. 4-39 by fabricating the endposts and top chord members additionally long. How much should the length of each endpost and each panel of the top chord be increased?

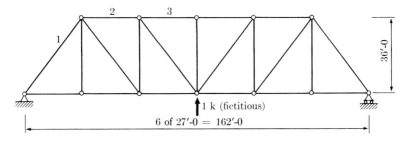

FIGURE 4-39

Assume that each endpost and each section of top chord is increased 0.1 in.

TABLE 4-10.

Member	u	dL	$u\,dL$
1	+0.625	+0.1	+0.0625
2	+0.750	+0.1	+0.0750
3	+1.125	+0.1	+0.1125
			+0.2500

$$= 2 \times 0.250 = +0.50 \text{ in.}$$

The required increase of length for each section will be

$$\frac{3.0}{0.50} \times 0.1 = 0.60 \text{ in.}$$

If we use the practical value of 0.625 in., the theoretical camber will be

$$6.25 \times 0.50 = 3.125 \text{ in.}$$

PROBLEMS

4-24. Find the vertical deflection component of point E (see Fig. 4-40). Sectional areas of members are indicated. $E = 30,000$ k/in². [Ans.: 0.082 in.]

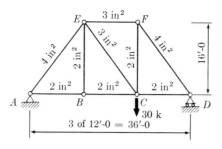

FIGURE 4-40

4-25. Find the rotation, in seconds, of the lower chord member 2-3 resulting from the dead load stresses shown in Fig. 4-41. Sectional areas are indicated. $E = 30{,}000$ k/in^2. [*Ans.:* 79 sec.]

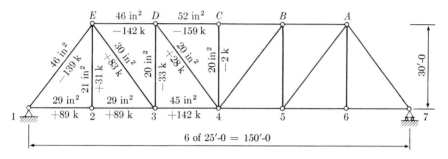

FIGURE 4-41

4-26. Determine the vertical deflection component of point B (see Fig. 4-42) as caused by the 100 k load at B. Sectional areas of the members in square inches are shown on the illustration. $E = 30{,}000$ k/in^2. [*Ans.:* 0.17 in.]

4-27. Find the horizontal deflection component of the right end of the structure in Fig. 4-43. All members have sectional areas of 3 in^2. $E = 30{,}000$ k/in^2. [*Ans.:* 1.48 in. to right.]

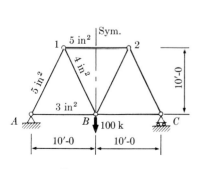

FIGURE 4-42

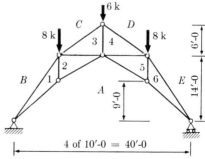

FIGURE 4-43

4–28. It is required to build a single-lane bridge over a canyon for a road to a new mine. Conditions at the site, equipment, and materials available lead to a solution as indicated in Fig. 4–44(a). Members AB and BC are 1-in. diameter wire strands with a sectional area of 0.577 in. and a modulus of elasticity of 24,000 k/in^2. Two sets of strands are to be used, one in the vertical plane of each side of the roadway, and two adjustable members BD, one for each set of strands, support the ends of the simple trusses at the span center. Ends C of the two strands are connected to the top of a steel tower. Each strand works against a vertical truss having dimensions and member sectional areas as shown in auxiliary Fig. 4–44(b). E for the towers is 30,000 k/in^2. The maximum load which will be permitted over the bridges will increase the stress in each member BD by 20 k. Neglecting the elastic strain in BD, what is the theoretical amount that D should be placed above E and F so that D will not deflect below their level during the passage of the maximum load? What is the maximum horizontal deflection component of B? [*Ans.:* vertical = 7.91 in., horizontal = 1.62 in.]

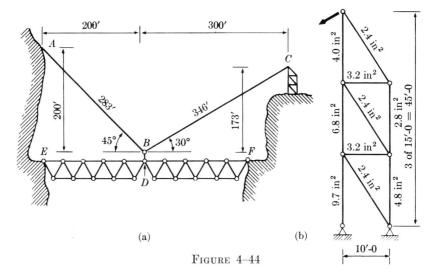

FIGURE 4–44

4–8 Deflections resulting from flexural strains. It has been established that whenever a linear deflection component is to be computed by the method of virtual work, a unit fictitious force is applied at the point, and in the direction, of the desired deflection. In this case the external virtual work is expressed as $1 \times \Delta$. If a rotational deflection component is required, a unit fictitious couple is applied at the point, and in the plane, of the required deflection and, in this second case, the external virtual work is written as $1 \times \theta$.

Actually, of course, the type of internal strain in a structure will not affect the form of the expression for the external virtual work. Consequently, the above expressions for external virtual work will apply in all cases, regardless of whether the internal strain is axial, flexural, torsional, or shearing. Therefore, in order to extend the method of virtual work to compute deflections resulting from strains other than axial, it is only necessary to develop the correct expressions for internal virtual work for these other types of strains. The proper expression for internal virtual work resulting from flexural strains will now be developed.

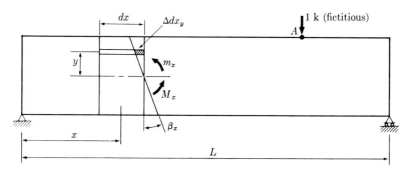

FIGURE 4–45

Consider the simple beam shown in Fig. 4–45. Assume that as the result of any conceivable system of real loads (the real loads are not shown in the figure), each section along the beam x distance from the left end is subjected to an internal moment M_x. For the purpose of this discussion assume that it is desired to find the vertical deflection component of point A as caused by the real loads. Since a linear deflection is desired, a unit fictitious vertical force is applied at A as shown.

For the time being, assume that the real loads are removed from the beam and that the 1 k fictitious force acts alone. Each section x along the beam is subjected to an internal fictitious moment m_x caused by this unit fictitious force. Virtual displacements are now impressed on the beam by replacing the real loads, and the internal fictitious moment on one face of the differential segment is caused to rotate through some virtual angle β_x relative to the other face, because of the flexural strain resulting from the application of the real loads. The internal virtual work for a length dx is then

$$m_x \cdot \beta_x.$$

It is necessary to evaluate β_x in terms of the real loads and the properties of the beam. Under the action of these real loads the unit stress in the fiber y distance from the neutral axis is

$$f_y = \frac{M_x \cdot y}{I},$$

and the unit strain in this fiber will be

$$\frac{\text{Unit stress}}{E} = \frac{f_y}{E} = \frac{M_x \cdot y}{EI}.$$

Consequently, the total strain (represented by $\Delta\, dx_y$), or increment of length, of this fiber in a length dx is then

$$\Delta dx_y = \frac{M_x \cdot y}{EI} \cdot dx.$$

Since the angle is small, the rotation of the right face of the segment relative to the left face will be

$$\beta_x = \tan \beta_x = \frac{\Delta dx_y}{y},$$

and therefore

$$\beta_x = \frac{M_x \cdot y}{EI} \frac{dx}{y} = \frac{M_x\, dx}{EI}. \tag{4-26}$$

(The derivation of this important expression for the flexural strain in a length dx of a member as caused by a moment M_x should be remembered, for it will be used in future derivations.)

If the value of β_x from Eq. (4–26) is substituted in the previously derived expression for the internal virtual work in a segment of length dx, the result is

$$m_x \cdot \frac{M_x\, dx}{EI}.$$

To complete the derivation of the expression for the internal virtual work resulting from flexural strains in the total length L, we integrate the above and obtain

$$\int_0^L \frac{m\, M\, dx}{EI}.$$

(The subscripts x are omitted from M and m in this final expression, the significance of these terms being apparent without them.)

If a linear deflection component resulting from flexure is desired, the basic equation is therefore

$$1 \times \Delta = \int \frac{m\, M\, dx}{EI}. \tag{4-27}$$

If a rotational deflection component is required, the equation is

$$1 \times \theta = \int \frac{m_\alpha M\, dx}{EI}, \tag{4-28}$$

where M is the internal moment at section x as caused by the real loads, m is the internal moment at section x as caused by a unit fictitious force, and m_α is the internal moment at section x resulting from a unit fictitious couple. It is suggested that the subscript α be used whenever m is caused by a unit fictitious couple, since, in certain cases, this procedure will serve to avoid considerable confusion.

EXAMPLE 4–29. Find the vertical deflection of the center of the beam of Fig. 4–46, using the method of virtual work. Use $E = 30{,}000$ k/in^2.

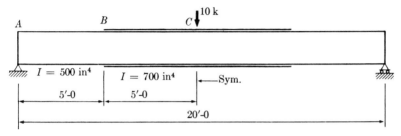

FIGURE 4–46

TABLE 4–11.

Section	$x = 0$ at	x increasing	M (ft·k)	m (ft·k)
AB	A	A to B	$+5x$	$+0.5x$
BC	B	B to C	$+25 + 5x$	$+2.5 + 0.5x$

$$E\Delta = \int \frac{m\,M\,dx}{I}$$

$$= \frac{2}{500}\int_0^5 (+0.5x)(+5x)\,dx + \frac{2}{700}\int_0^5 (+2.5 + 0.5x)(+25 + 5x)\,dx$$

$$= \frac{2}{500}\int_0^5 +2.5x^2\,dx + \frac{2}{700}\int_0^5 +62.5\,dx + \frac{2}{700}\int_0^5 +25x\,dx$$

$$+ \frac{2}{700}\int_0^5 2.5x^2\,dx,$$

$E\Delta = 2.50$,

$$\Delta = \frac{2.50 \times 1728}{30{,}000} = 0.14 \text{ in.}$$

EXAMPLE 4–30. Using the method of virtual work, find the horizontal, vertical, and rotational deflection components of point A (see Fig. 4–47) as caused by the 10 k load. Consider that any internal moment resulting in tension on the outside of the frame is positive and assume the necessary unit fictitious forces and the unit fictitious couple to act in the directions indicated. (Note that these fictitious loads have all been assumed with senses opposite to the obvious deflection components and, therefore, all answers should be negative.)

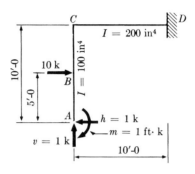

FIGURE 4–47

TABLE 4–12.

Section	$x = 0$ at	x increasing	M (ft·k)	m_h (ft·k)	m_v (ft·k)	m_α (ft·k)
AB	A	A to B	0	$-x$	0	-1
BC	B	B to C	$+10x$	$-5-x$	0	-1
CD	C	C to D	$+50$	-10	$-x$	-1

$$\Delta_{Ah} = \int \frac{m_h M\, dx}{EI},$$

$$E\,\Delta_{Ah} = \int_0^5 \frac{(-5-x)(+10x)\,dx}{100} + \int_0^{10} \frac{(-10)(+50)\,dx}{200} = -35.42,$$

$$\Delta_{Ah} = \frac{-35.42 \times 1728}{30{,}000} = -2.04 \text{ in.,}$$

$$\Delta_{Av} = \int \frac{m_v M\, dx}{EI},$$

$$E\Delta_{Av} = \int \frac{(-x)(+50)\,dx}{200} = -12.5,$$

$$\Delta_{Av} = \frac{-12.5 \times 1728}{30{,}000} = -0.72 \text{ in.},$$

$$\theta_A = \int \frac{m_\alpha M\,dx}{EI},$$

$$E\theta_A = \int_0^5 \frac{(-1)(+10x)\,dx}{100} + \int_0^{10} \frac{(-1)(+50)\,dx}{200} = -3.75,$$

$$\theta_A = \frac{-3.75 \times 144}{30{,}000} = -0.018 \text{ rad.}$$

The negative answers in the above equations indicate that the deflection components are opposite in sense to the assumed fictitious loads. Note that this is an alternate solution for Example 4–10. The two solutions should be carefully compared.

EXAMPLE 4–31. Using virtual work, find the horizontal, vertical, and rotational deflection components of point A (see Fig. 4–48). Consider that tension on the outside of the frame is positive and assume fictitious actions as indicated. $E = 30{,}000$ k/in^2.

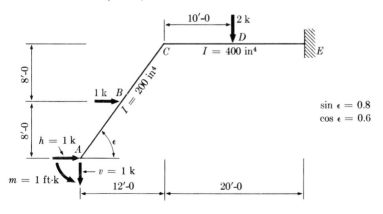

FIGURE 4–48

TABLE 4–13.

Section	$x=0$ at	x increasing	M (ft·k)	m_h (ft·k)	m_v (ft·k)	m_α (ft·k)
AB	A	A to B	0	$+0.8x$	$+0.6x$	$+1$
BC	B	B to C	$+0.8x$	$+8+0.8x$	$+6+0.6x$	$+1$
CD	C	C to D	$+8$	$+16$	$+12+x$	$+1$
DE	D	D to E	$+8+2x$	$+16$	$+22+x$	$+1$

$$E\,\Delta_{Ah} = \int \frac{m_h M\,dx}{I}$$

$$= \int_0^{10} \frac{(+8+0.8x)(+0.8x)\,dx}{200} + \int_0^{10} \frac{(+16)(+8)\,dx}{400}$$

$$+ \int_0^{10} \frac{(+16)(+8+2x)\,dx}{400} = +13.07,$$

$$\Delta_{Ah} = \frac{+13.07 \times 1728}{30{,}000} = +0.75 \text{ in.,}$$

$$E\,\Delta_{Av} = \int \frac{m_v M\,dx}{I}$$

$$= \int_0^{10} \frac{(+6+0.6x)(+0.8x)\,dx}{200} + \int_0^{10} \frac{(+12+x)(+8)\,dx}{400}$$

$$+ \int_0^{10} \frac{(+22+x)(+8+2x)\,dx}{400} = +17.97,$$

$$\Delta_{Av} = \frac{+17.97 \times 1728}{30{,}000} = +1.03 \text{ in.,}$$

$$E\theta_A = \int \frac{m_a M \, dx}{I}$$

$$= \int_0^{10} \frac{(+1)(+0.8x) \, dx}{200} + \int_0^{10} \frac{(+1)(+8) \, dx}{400}$$

$$+ \int_0^{10} \frac{(+1)(+8 + 2x) \, dx}{400} = +0.85,$$

$$\theta_A = \frac{+0.85 \times 144}{30{,}000} = +0.0041 \text{ rad.}$$

PROBLEMS

4-32. Find the vertical, horizontal, and rotational deflection components of point A (see Fig. 4-49). Consider flexural strains only. $E = 30{,}000$ k/in². [Ans.: $\Delta_{Av} = 0.96$ in., $\Delta_{Ah} = 0.29$ in., $\theta_A = 0.0096$ rad, clockwise.]

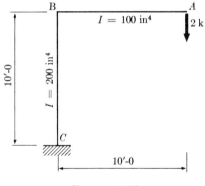

FIGURE 4-49

4-33. A 21WF62 beam supports the loads indicated in Fig. 4-50. If the beam is cut at the center support and considered as two simple spans, what will be the angular opening between the ends of the two spans at the center support? I of the beam is 1327 in⁴ and E is 30,000 k/in². [Ans.: 0.0030 rad.]

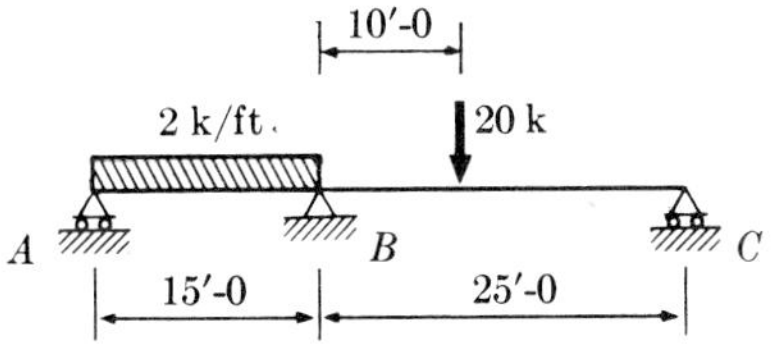

FIGURE 4-50

4-34. Find (from Fig. 4-51):

(a) The rotational deflection component of B due to a 1 k force acting down at A.

(b) The vertical deflection component of A due to a counterclockwise couple of 1 ft·k acting at B.

(c) The horizontal deflection component of C due to a 1 k force acting down at A.

Consider flexural strains only. $E = 30,000$ k/in^2. [*Ans.:* (a) 1.44×10^{-3} rad, (b) 1.44×10^{-3} ft, (c) 0.024 ft.]

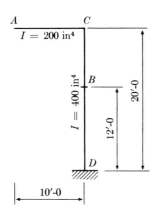

Figure 4-51

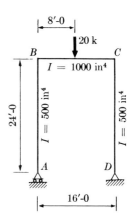

Figure 4-52

4-35. Find (from Fig. 4-52):

(a) The horizontal deflection component of point A due to the load shown.

(b) The horizontal deflection component of point A due to 1 k applied horizontally at A.

Consider flexural strains. $E = 30,000$ k/in^2. [*Ans.:* (a) 0.88 in., (b) 1.59 in.]

4-36. Find the vertical, horizontal, and rotational deflection components of point A (see Fig. 4-53) as caused by the 10 k load shown. Also, find the horizontal deflection at A as caused by a 1 k load acting to the right at A. [*Ans.:* $\Delta_{Av} = 1.56$ in. down, $\Delta_{Ah} = 0.14$ in. right, $\theta_A = 0.013$ rad, $\delta_{AhAh} = 0.31$ in. right.]

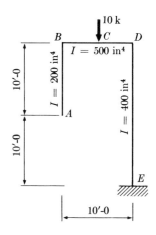

Figure 4-53

4-37. Find the vertical, horizontal, and rotational deflection components of point A (see Fig. 4-54) as caused by flexural strains. $E = 30,000$ k/in^2. I is 100 in^4. [Ans.: $\Delta_{Av} = 0.29$ in. up, $\Delta_{Ah} = 1.01$ in. right, $\theta_A = 0.0036$ rad, counterclockwise.]

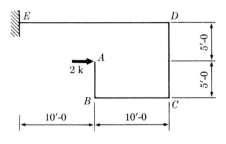

FIGURE 4-54

4-38. Find the vertical, horizontal, and rotational deflection components of point A (see Fig. 4-55). Consider flexural strains only. $E = 30,000$ k/in^2. I is 200 in^4. [Ans.: $\Delta_{Av} = 1.16$ in. down, $\Delta_{Ah} = 0.53$ in. right, $\theta_A = 0.0049$ rad, counterclockwise.]

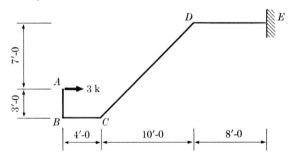

FIGURE 4-55

4-39. Given a prestressed concrete beam with a constant moment of inertia I throughout its length. The compressor elements are placed in a parabolic curve with an eccentricity at the beam ends of y.

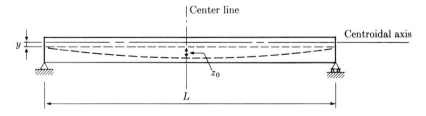

FIGURE 4-56

The equation of the parabola, referred to the left end of the cable, is

$$z = \frac{4z_0}{L^2}(Lx - x^2).$$

Find the expression for the upward deflection at the center of the beam span as caused by the prestressing. Assume that the horizontal component of the compressor elements P is constant throughout the length of the beam. The cable eccentricity at the center line is $y + z_0$. [*Ans.:* $(PL^2/8EI)$ $(\tfrac{5}{6}z_0 + y)$].

4–9 Deflections resulting from flexure of nonprismatic members. The flexural deflection problems considered thus far, with the exception of Example 4–29, have dealt with members, or frames composed of members, with constant moments of inertia. Members having a constant moment of inertia are known as *prismatic members*. The reinforced beam of Example 4–29, although nonprismatic, does have constant moments of inertia throughout definite segments of its length. Any flexural deflection, either linear or rotational, of any point on the beam may be found by methods similar to those illustrated in that example.

Prismatic members will be the only type found or designed in many structures, although, in some cases, nonprismatic members are preferable to prismatic, from the standpoint of either economy or appearance or both. Modern methods of analysis of indeterminate structures, which enable the designer to analyze frames composed of nonprismatic members with accuracy and speed, and modern methods of fabrication, have had the effect of greatly increasing the use of nonprismatic members.

The summation of the effects of bending throughout the length of a nonprismatic member may quite often be managed in two ways. The first, when possible, is by direct integration. The second way consists of dividing the member into a number of segments and considering that the two variables, moment of inertia and bending moment, will have the same values throughout the length of each segment as they have at the center of that segment. Example 4–40, which follows, will illustrate both methods. A preliminary discussion of one point in that example will be of value. The statement of Example 4–40 specifies that the uncracked section shall be used for determining the various moments of inertia. This is customary in the analysis of reinforced concrete structures and simply means that the effect of the reinforcing steel is neglected and the concrete is not considered to crack on the tension side.

EXAMPLE 4–40. The reinforced concrete cantilever beam shown in Fig. 4–57 is of 3000 lb concrete. The width is 1 ft and the depths vary as indicated. $E = 3000 \text{ k/in}^2$. Using the method of virtual work, find the deflection of point B as caused by the 10 k load. Use the uncracked section for I.

4-9] FLEXURE OF NONPRISMATIC MEMBERS 115

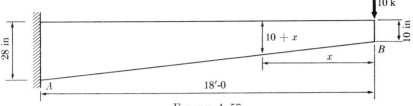

FIGURE 4-57

Solve in the following ways:
  (a) Direct integration.
  (b) Using segments 2 ft long.
  (c) Using segments 3 ft long.
  (d) Using segments 4.5 ft long.
  (e) Using segments 6 ft long.

(a) The dimensions of the beam are such that if x is expressed in feet, the depth at any section x is $10 + x$ inches. Therefore,

$$\Delta = \int_0^{18} \frac{mM\,dx}{EI} = \frac{12}{12E}\int_0^{18} \frac{(+x)(+10x)\,dx}{(10+x)^3} = \frac{10}{E}\int_0^{18} \frac{x^2\,dx}{(10+x)^3},$$

$\int \frac{x^2\,dx}{(10+x)^3}$ is in the form $\int \frac{x^n\,dx}{(a+x)^m}$.

Let $y = a + x$, $n = 2$. Then $x = y - a$, $m = 3$; $dx = dy$, $a = 10$. Therefore,

$$\int_0^{18} \frac{x^2\,dx}{(10+x)^3} = \int_0^{18} \frac{(y-a)^2\,dy}{y^3} = \int_0^{18}(y^2 - 2ay + a^2)\frac{dy}{y^3}$$

$$= \int_0^{18} \frac{dy}{y} - 2a\int_0^{18} y^{-2}\,dy + a^2\int y^{-3}\,dy$$

$$= \left[\log_e y + \frac{2a}{y} - \frac{a^2}{2y^2}\right]_0^{18}$$

$$= \left[\log_e (a+x) + \frac{2a}{a+x} - \frac{a^2}{2(a+x)^2}\right]_0^{18}$$

$$= \left[\log_e 28 + \frac{20}{28} - \frac{100}{1568}\right] - \left[\log_e 10 + \frac{20}{10} - \frac{100}{200}\right]$$

$$= [3.333 + 0.714 - 0.064] - [2.303 + 2.000 - 0.500]$$
$$= 0.180,$$
$$\Delta = \frac{10 \times 0.180 \times 1728}{3000} = 1.038 \text{ in.}$$

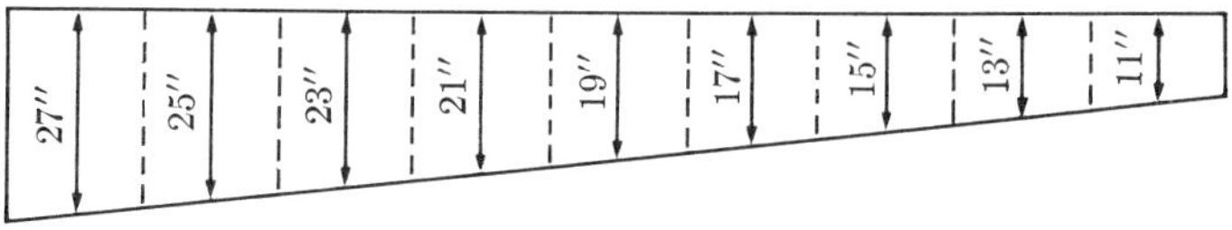

FIGURE 4-58

(b) $\quad \Delta = \sum \frac{mM \, \Delta x}{EI} = \sum \frac{(x)(10x)2}{E \cdot 1 \cdot d^3/12} = \frac{240}{E} \sum \frac{x^2}{d^3}.$

TABLE 4-14.

Segment	x (ft)	x^2	$d = 10 + x$ (in.)	d^3	$\dfrac{x^2}{d^3}$
1	1	1	11	1,331	0.00075
2	3	9	13	2,197	0.00411
3	5	25	15	3,375	0.00742
4	7	49	17	4,913	0.00997
5	9	81	19	6,859	0.01182
6	11	121	21	9,261	0.01307
7	13	169	23	12,167	0.01391
8	15	225	25	15,625	0.01441
9	17	289	27	19,683	0.01470
					0.09016

$$\Delta = \frac{240 \times 0.09016 \times 144}{3000} = 1.038 \text{ in.}$$

Parts (c), (d), and (e) are solved in the manner illustrated for part (b). The results of the various solutions are shown in Table 4-15.

TABLE 4-15.

Integration	9 Segments	6	4	3
1.038	1.038	1.039	1.041	1.048

It is apparent for the above member and loading that satisfactory deflections may be obtained with a relatively small number of segments. This would not necessarily be true for other shapes of members and for other types of loads. However, if the segments are taken relatively short, the deflections thus obtained will be in close agreement with those which result from direct integration. Deflections of nonprismatic members are often obtained by the segment method rather than by direct integration, since, in many cases, it is impossible to integrate exactly the expressions for deflection.

Problems

4-41. A reinforced concrete-slab highway bridge is to have a longitudinal section (see Fig. 4-59) which will hold throughout the entire 24 ft of roadway width. What will be the upward deflection of points C and C' as caused by a line load of 4 k per ft of width of roadway at the center of the span? E is 3000 k/in^2. Determine the deflection by:
(a) Integration.
(b) Segments 3 ft long in center span. [*Ans.:* (a) 0.55 in., (b) 0.55 in.]

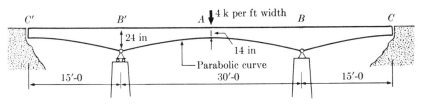

Figure 4-59

4-42. A simple prestressed concrete beam is rectangular in section and 1 ft wide, and the elevation is as indicated in Fig. 4-60. The tension in the compressor elements is 350 k. E is 6000 k/in^2. Determine:

(a) The vertical deflection at beam center due to the weight of the beam. Use the segment method with segments 5 ft long. Assume the concrete to weigh 150 lb/ft^3.

(b) The vertical deflection at beam center resulting from prestressing, neglecting dead load. Use direct integration. [*Ans.:* (a) 1.19 in. down, (b) 1.19 in. up.]

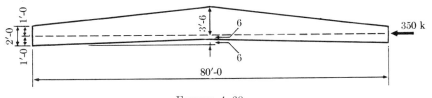

Figure 4-60

4-10 Deflections resulting from shearing strains. In a manner similar to that already demonstrated for articulated structures and for flexural deflections, it can be shown that the internal virtual work of shear in a member of length L is

$$K \int_0^L \frac{vV\,dx}{AG},$$

where V is the shear resulting from the real load system, v is the shear resulting from the fictitious load, and A is the area of the cross section of the member, except in the case of WF or I steel beams, where A is the area of the web. As previously explained in the discussion of shearing internal work, K is a form factor.

Attention is called to the fact that in the case of most beams encountered in practice, the deflection due to shearing strains is relatively small compared to that resulting from flexural strains. Table 4-16 shows the ratio of shearing strain deflections to flexural strain deflections as computed for a 12WF27 steel beam.

TABLE 4-16.

Span to depth ratio	Concentrated load at center	Uniform load
5	0.60	0.48
10	0.15	0.12
15	0.07	0.05
20	0.04	0.03

PROBLEMS

4-43. Using virtual work, find the vertical deflection of the free end of the cantilever (see Fig. 4-61) resulting from (a) flexure, and (b) shear. $E = 30{,}000$ k/in^2, $G = 12{,}000$ k/in^2, $I_x = 597$ in^4, and web area $= 5.27$ in^2. [*Ans.:* (a) 1.08 in., (b) 0.03 in.]

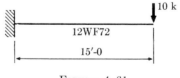

FIGURE 4-61

4-44. Using virtual work, find the vertical deflection at the center of the 10WF45 beam resulting from (a) flexure, and (b) shear. Cross-sectional area of web $= 3.5$ in^2, $I = 249$ in^4, $E = 30{,}000$ k/in^2, and $G = 12{,}000$ k/in^2. [*Ans.:* (a) 0.38 in., (b) 0.01 in.]

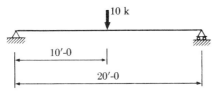

FIGURE 4-62

4-11 Deflections resulting from torsional strains. The expression for the internal virtual work resulting from torsional strains will now be derived. Assume that the unit fictitious torque t is first applied to the free end of the cantilever in Fig. 4–63. All internal sections will be subjected

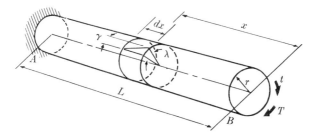

FIGURE 4-63

to this torque. The real torque T is now superimposed on the free end of the cantilever and, as a result, the shaft is strained throughout its length. Consider the segment dx in length. The face of the segment nearer the free end will, as the result of the action of the torque T, rotate through λ radians relative to the face nearer the fixed end. The fictitious torque t already acting on the near face will "ride" through λ radians at its full value and, consequently, the internal virtual work in the segment will be $t\lambda$.

It is now necessary to evaluate λ. Under the action of T, the unit shearing stress on the surface of the shaft is

$$\tau_r = \frac{Tr}{J},$$

where J is the polar moment of inertia of the shaft cross section. The torsional shearing strain, represented by γ, on the surface of the shaft in a length dx is

$$\gamma = \frac{\tau_r}{G} \cdot dx = \frac{Tr}{JG} \cdot dx$$

and the value of λ becomes

$$\lambda = \frac{Tr}{JG}\frac{dx}{r} = \frac{T\,dx}{JG}.$$

The internal virtual work in a length dx is therefore

$$t\lambda = t \cdot \frac{T\,dx}{JG},$$

and in length L is

$$\int_0^L \frac{tT\,dx}{JG}.$$

For cross sections other than circular the remarks of Section 4–3(d) relative to the values of J apply.

EXAMPLE 4–45. Given a 4-in. standard pipe bracket (see Fig. 4–64) having a 90° angle at B and C, and located in a horizontal plane. Find the vertical deflection component of A and the rotation of end A of segment AB about the axis of AB, as caused by a vertical load of 2 k applied at A. The plane moment of inertia of the pipe is 7.23 in^4 and the polar moment of inertia is 14.46 in^4. $G = 12{,}000$ k/in^2 and $E = 30{,}000$ k/in^2. Use the method of virtual work.

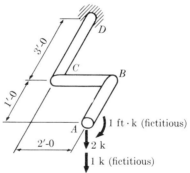

FIGURE 4–64

TABLE 4–17.

Segment	$x = 0$ at	x increasing	M (ft·k)	T (ft·k)	m (ft·k)	m_α (ft·k)	t (ft·k)	t_α (ft·k)
AB	A	A to B	$+2x$	0	$+x$	0	0	$+1$
BC	B	B to C	$+2x$	$+2$	$+x$	$+1$	$+1$	0
CD	C	C to D	$+2 + 2x$	$+4$	$+1 + x$	0	$+2$	$+1$

In Table 4–17, for any section x in the respective segments:

(i) M is the plane bending moment caused by the 2 k load.
(ii) T is the torque, or polar moment, resulting from the 2 k load.
(iii) m is the plane bending moment due to the 1 k fictitious force.
(iv) m_α is the plane bending moment resulting from the 1 ft·k fictitious couple.
(v) t is the torque due to the 1 k fictitious force.
(vi) t_α is the torque caused by the 1 ft·k fictitious couple.

The expression for vertical deflection is

$$1 \times \Delta = \int \frac{mM\,dx}{EI} + \int \frac{tT\,dx}{GJ},$$

$$1 \times \Delta = \frac{1}{EI}\left[\int_0^1 (+x)(+2x)\,dx + \int_0^2 (+x)(+2x)\,dx \right.$$
$$\left. + \int_0^3 (+1+x)(+2+2x)\,dx\right]$$
$$+ \frac{1}{GJ}\left[\int_0^2 (+1)(+2)\,dx + \int_0^3 (+2)(+4)\,dx\right],$$

$$\Delta = \frac{48 \times 1728}{30{,}000 \times 7.23} + \frac{28 \times 1728}{12{,}000 \times 14.46} = +0.66 \text{ in.}$$

For rotation,

$$1 \times \theta = \int \frac{m_\alpha M\,dx}{EI} + \int \frac{t_\alpha T\,dx}{GJ}$$

$$= \int_0^2 \frac{(+1)(+2x)\,dx}{EI} + \int_0^3 \frac{(+1)(+4)\,dx}{GJ}$$

$$= \frac{4 \times 144}{30{,}000 \times 7.23} + \frac{12 \times 144}{12{,}000 \times 14.46} = +0.013 \text{ rad.}$$

PROBLEMS

4–46. Using virtual work, find the vertical deflection component of point D (see Fig. 4–65) as caused by the 0.2 k vertical load at D. Points A, B, and C are in one vertical plane; points B, C, and D are in one horizontal plane. Consider flexure and torsion. The bracket is of 3-in. diameter standard pipe with a plane moment of inertia of 3 in^4 and area of 2.23 in^2. $E = 30,000$ k/in^2 and $G = 12,000$ k/in^2. [*Ans.:* 0.92 in. down.]

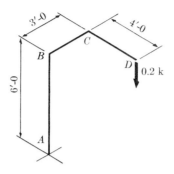

FIGURE 4-65

REVIEW PROBLEMS

4-47. Find the horizontal deflection component of A (see Fig. 4-66) due to the 2 k load at A, by using the methods of:
(a) Real work.
(b) Castigliano's first theorem.
(c) Virtual work.
Consider flexural strains only. $E = 30,000$ k/in^2. [*Ans.:* 0.19 in.]

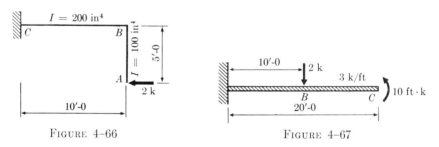

FIGURE 4-66 FIGURE 4-67

4-48. Find the vertical and rotational deflection components of point C on the cantilever of Fig. 4-67 by:
(a) Castigliano's first theorem.
(b) Virtual work.
Consider flexural strains only. $I = 5000$ in^4, $E = 30,000$ k/in^2. [*Ans.:* $\Delta_{Cv} = 0.69$ in., $\theta_C = 0.0038$ rad, clockwise.]

4-49. Find the vertical deflection component of point D (see Fig. 4-68) by:
(a) Real work.
(b) Virtual work.
(c) Castigliano's first theorem.
Consider flexural strains only. $E = 30,000$ k/in^2. [*Ans.:* 2.88 in.]

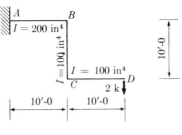

FIGURE 4-68

4-50. Find the vertical deflection component of point A (see Fig. 4-69) by virtual work. Neglect shearing strains in the beam. $E = 30{,}000$ k/in^2. [*Ans.:* 1.21 in.]

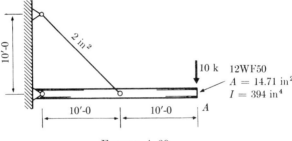

FIGURE 4-69

4-51. The support for a monorail hoist is cantilevered as shown in Fig. 4-70 and suitable lateral bracing is provided. Find the vertical deflection component of point A by virtual work. Consider that the splice to the left of B will not transmit moment. The cross-sectional areas of the various members are noted on the illustration. $E = 30{,}000$ k/in^2. [*Ans.:* 1.22 in.]

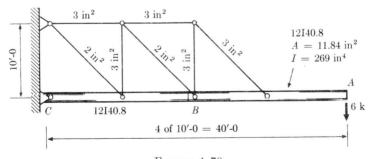

FIGURE 4-70

4-52. Find the vertical deflection component of points A and A' (see Fig. 4-71) when these points are simultaneously loaded with vertical downward forces of 2 k each. The member CDE is a 10WF21 steel beam with a moment of inertia of 106 in^4 and A of 6.19 in^2. The members ABC and $A'B'C'$ are 5-in. ϕ standard weight pipe with a plane moment of inertia of 15.16 in^4. Sectional areas for the truss members, in square inches, are shown on the illustration. Use a value for E of 30,000 k/in^2, and for G of 12,000 k/in^2. Consider axial, flexural, and torsional strains. [*Ans.:* 6.24 in.]

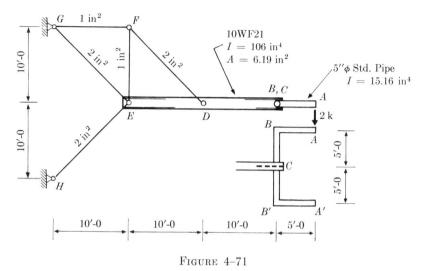

FIGURE 4–71

4–12 Maxwell's theorem of reciprocal deflections. Maxwell's reciprocal deflection theorem was developed in the form of three propositions in Section 3–10. Interesting and easily understood alternate demonstrations are possible, however, by the method of virtual work and it is advisable that these be considered at this point.

Proposition 1, for the case of two unit forces applied separately to a structure, has been stated as follows: *Any linear deflection component of any point A, resulting from the application of a unit force at any other point B, is equal in magnitude to the linear deflection component of B (in the direction of the first applied force at B) resulting from the application of a unit force at A applied in the direction of the original deflection component of A.* The alternate demonstration will be developed in connection with Fig. 4–72, where A and B may be any two points.

Proposition 1, for the two structures shown in Fig. 4–72(a) and (b), can be expressed as

$$\delta_{AvBh} = \delta_{BhAv}.$$

This may be proved as follows:

$$\delta_{AvBh} = \int \frac{m\,M\,dx}{EI} = \int \frac{m_{Av} \cdot m_{Bh} \cdot dx}{EI},$$

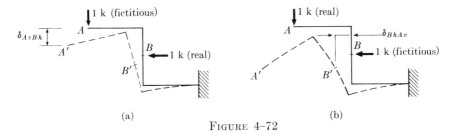

Figure 4-72

where m_{Bh} is the moment at any section caused by the 1 k real force at B in Fig. 4-72(a), and m_{Av} is due to the 1 k fictitious force at A. Also,

$$\delta_{BhAv} = \int \frac{m_{Bh} \cdot m_{Av} \cdot dx}{EI},$$

where m_{Av} now represents the moment at any section due to the 1 k real vertical force at A in Fig. 4-72(b), and m_{Bh} is the moment resulting from the 1 k fictitious force at B. Comparison of the two integrals shows that

$$\delta_{AvBh} = \delta_{BhAv}.$$

Proposition 2, covering the case of a force and a couple applied separately at any two different points on a structure, has been stated as follows: *Any linear deflection component of any point A, as caused by a unit couple applied at any other point B, is equal in magnitude to the rotational deflection of B in radians resulting from the application of a unit force at A in the direction of the original linear deflection component of A.* Assuming for the purpose of the demonstration that a vertical deflection component is to be considered at A, this proposition is expressed as

$$\delta'_{AvBm} = \alpha'_{BmAv}.$$

(As previously indicated, the lower-case m following B in the subscript is redundant, but the author often prefers to use it in order to maintain the four-letter subscript system.) Imagine that a unit real couple is applied at B, in Fig. 4-72 (a), in place of the unit real load shown. Then the expression for δ'_{AvBm} is

$$\delta'_{AvBm} = \int \frac{m \, M \, dx}{EI} = \int \frac{m_{Av} m_{\alpha Bm} \, dx}{EI},$$

where $m_{\alpha Bm}$ is the moment at any section due to the unit real couple at B, and m_{Av} is the moment at any section as caused by the unit fictitious vertical load at A. Now, in Fig. 4–72(b), consider that the unit fictitious horizontal force at B is replaced by a unit fictitious couple. The value of α'_{BmAv} is given by

$$\alpha'_{BmAv} = \int \frac{m\,M\,dx}{EI} = \int \frac{m_{\alpha Bm} m_{Av}\,dx}{EI},$$

and comparison of the above expression with that for δ'_{AvBm} indicates that

$$\delta'_{AvBm} = \alpha'_{BmAv}.$$

For the case of the two unit couples applied separately to any two points of a structure, Proposition 3 has been previously stated as follows: *The rotational deflection of any point A on a structure, as caused by a unit couple acting at any other point B, is equal in magnitude to the rotational deflection of B as caused by a unit couple acting at A.* The above proposition is expressed by

$$\alpha_{AmBm} = \alpha_{BmAm},$$

and that this relationship is true can easily be proved by virtual work, using methods similar to those used in verifying the first two propositions.

Attention is called to the fact that, although a frame composed of members which are primarily flexural was used in the above demonstrations, the resulting relationships could have been proved just as easily with an articulated structure or for deflections resulting from shearing or torsional strains.

These propositions are extremely important, for not only do they serve to reduce the amount of work involved in analyzing many indeterminate structures, but also they are the basis for various methods of model analysis. Since they will be used repeatedly in subsequent problems, the reader should be certain that he thoroughly understands them.

Part 5. Moment Areas and Elastic Weights

4–13 General. The moment-area method and the method of elastic weights, both useful for finding slopes and deflections of flexural members, are rather closely related. Each method comprises two principles or propositions. The two propositions considered first frame the moment-area method and are due to Professor Charles E. Greene of the University of Michigan, who introduced them in 1873. The last two propositions constitute the method of elastic weights, more commonly known as the *conjugate beam method*, and were announced in an outstanding paper (3) presented by Otto Mohr in 1868. It was in this same paper that Mohr discussed the representation of the elastic curve as a string polygon.

In general, the conjugate beam method is of much greater practical importance than the moment-area method. Occasionally, however, a problem or demonstration will be encountered in which moment areas can be applied to better advantage.

4-14 The moment-area method. PROPOSITION 1: *The difference in slope between any two sections of a loaded flexural member is equal to the area of the M/EI diagram between these two sections.*

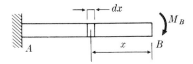

FIGURE 4-73

As a first, and quite simple, demonstration of the above proposition, consider the cantilever beam of Fig. 4-73. Assume that the whole beam is rigid, except for the differential slice dx. Under the action of M_B, the beam will deflect as in Fig. 4-74.

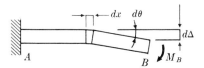

FIGURE 4-74

It has been previously demonstrated (Eq. 4-26) that

$$d\theta = \frac{M\,dx}{EI},$$

where M is the internal moment in the differential slice as caused by any system of applied loads. The bending moment diagram will be of the shape indicated in Fig. 4-75. If each ordinate is made M_B/EI, it is ob-

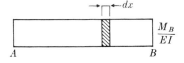

FIGURE 4-75

vious that $d\theta$ will equal the area of the indicated differential slice. Actually the whole beam is elastic, and thus each differential slice will contribute a $d\theta$. If we wish to determine the difference in slope between any two points, as A and B, the value of θ will be

$$\theta = \int_A^B \frac{M_B\, dx}{EI}.$$

This is the area of the M/EI diagram between points A and B.

PROPOSITION 2: *The tangential deviation of point B on a loaded flexural member, from a tangent to the deflection curve at point A, is equal to the moment of the area of the M/EI diagram between A and B about B.*

For a demonstration of this proposition we refer again to Fig. 4-74. It is apparent that the flexural strain in the differential slice dx will result in a small deflection $d\Delta$ at B, and this deflection will be

$$d\Delta = x \cdot d\theta = \frac{M_B\, dx}{EI} \cdot x.$$

If, then, it is required to find the deflection of B relative to A, it can be found by summing all the values of $d\Delta$ as contributed by all the differential slices. This would obviously be the summation of the moments of all the differential slices of the M/EI diagram about B, or, in other words, the moment of the entire M/EI diagram between A and B about B.

EXAMPLE 4-53. The steel beam shown in Fig. 4-76 has a moment of inertia of 200 in^4 and an E of 30,000 k/in^2. Using the moment-area method, find:

(a) The difference in slope between X and Y.
(b) The deflection of X from a tangent at Y.

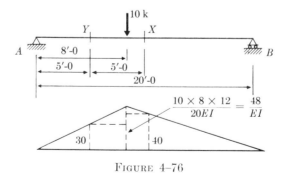

FIGURE 4-76

(a) The area of the M/EI diagram between X and Y is

$$\frac{1}{EI}\left[\left(\frac{30+48}{2}\right)3 + \left(\frac{40+48}{2}\right)2\right] = \frac{205}{EI}.$$

Change in slope between X and Y is

$$\frac{205 \times 144}{30{,}000 \times 200} = 0.00492 \text{ rad}.$$

(b) To obtain the moment of the M/EI diagram between Y and X about X, it is convenient to divide the diagram into rectangles and triangles as shown in Fig. 4–76.

$$\Delta_X = \frac{1}{EI}\left[30 \times 3 \times 3.5 + \frac{18}{2} \times 3 \times 3 + 40 \times 2 \times 1 + \frac{8}{2} \times 2 \times \frac{4}{3}\right]$$

$$= \frac{486.7}{EI}$$

$$= \frac{486.7 \times 1728}{30{,}000 \times 200} = 0.14 \text{ in}.$$

PROBLEMS

4–54. Using the moment-area method, find (from Fig. 4–77):
(a) The difference in slope between X and Y.
(b) The deflection of X from a tangent at Y.
$E = 30{,}000$ k/in^2 and $I = 200$ in^4. [*Ans.:* (a) 0.0057 rad, (b) 0.27 in.]

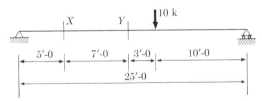

FIGURE 4–77

4–15 The conjugate beam method. Propositions 3 and 4, upon which the *conjugate beam method* is based, will now be developed. For a demonstration of Proposition 3, consider the simple beam of Fig. 4–78, loaded so as to produce the bending moment diagram indicated. If the entire beam, with the exception of the segment dx, is considered to be rigid, then the flexural strain in the segment dx will result in the deflected beam shown in (b).

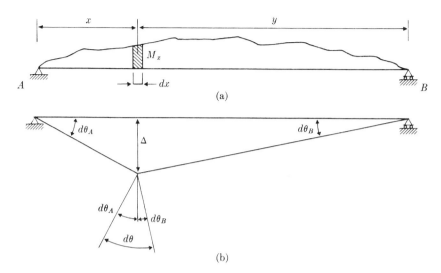

Figure 4–78

Now,
$$d\theta_A + d\theta_B = d\theta = \frac{M_x\,dx}{EI}, \tag{4-29}$$

and
$$x \cdot d\theta_A = y \cdot d\theta_B, \qquad d\theta_A = \frac{y}{x} d\theta_B. \tag{4-30}$$

Substituting Eq. (4–30) in (4–29),
$$\frac{y}{x} d\theta_B + d\theta_B = \frac{M_x\,dx}{EI},$$

$$d\theta_B = \left(\frac{x}{x+y}\right)\frac{M_x\,dx}{EI} = \frac{x}{L} \cdot \frac{M_x\,dx}{EI}, \tag{4-31}$$

and
$$d\theta_A = \frac{y}{L} \cdot \frac{M_x\,dx}{EI}. \tag{4-32}$$

But Eq. (4–31) would be the right-hand reaction of a simple beam (called the *conjugate beam*) loaded with the elastic weight $M_x\,dx/EI$ and Eq. (4–32) would be the left-hand reaction. If the entire beam is considered to be elastic, then the resulting slope at the right end would be

$$\theta_B = \int_0^L \frac{x}{L} \cdot \frac{M_x\, dx}{EI} \qquad (4\text{-}33)$$

and, at the left end,

$$\theta_A = \int_0^L \frac{L - x}{L} \cdot \frac{M_x\, dx}{EI}. \qquad (4\text{-}34)$$

But Eqs. (4-33) and (4-34) are, respectively, the right and left reactions for a simple conjugate beam loaded with the entire M/EI diagram. Therefore the slopes at the ends of the real beam are the reactions (that is, the shears) at the corresponding ends of the conjugate beam. Moreover, the slope at any other section of the real beam will be the end slope minus the flexural strain from that end to the section in question. By Proposition 1, however, this flexural strain is given by the area of the M/EI diagram, which is the load on the conjugate beam, between these two points. That is, the slope at any section of the real beam is equal to either end reaction of the conjugate beam minus the elastic load between that end and the section in question. This, obviously, would be the shear in the conjugate beam at the section. Therefore, Proposition 3 is stated as follows:

The slope at any section of a loaded beam, relative to the original axis of the beam, is equal to the shear in the conjugate beam at the corresponding section.

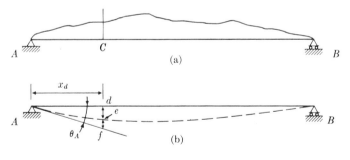

FIGURE 4-79

The beam of Fig. 4-79, which will be used to develop Proposition 4, is loaded with a system of loads causing the bending moment diagram indicated in (a). The deflected beam is shown in (b), and we wish to find the deflection de. From the figure,

$$de = df - ef,$$

which is equal to $x_d \cdot \theta_A$ minus the moment of the M/EI diagram from A to C about C. But, since θ_A is the left reaction of the conjugate beam, the above expression for de is obviously the internal moment at section C of the conjugate beam. Therefore, Proposition 4 is stated as:

The deflection at a given section of a loaded beam, relative to its original position, is equal to the bending moment at the corresponding section of the conjugate beam.

4-16 Relationships between the real beam and the conjugate beam. Propositions 3 and 4 having been demonstrated, it is now possible to establish certain definite relationships between the conjugate beam and the real beam. Several of these relationships, which are obvious from the statements of the propositions, are as follows:

(a) The span of the conjugate beam is equal to the span of the real beam.

(b) The load of the conjugate beam is the M/EI diagram of the real beam.

(c) The shear at any section of the conjugate beam is equal to the slope of the corresponding section of the real beam.

(d) The moment at any section of the conjugate beam is equal to the deflection at the corresponding section of the real beam.

In addition to the above, the conjugate beam must be supported in a manner consistent with the constraints of the real beam. The proper supports for the conjugate beam can be easily deduced from relationships (c) and (d) as stated above. For example, if the real beam has a fixed support (which means no rotation or deflection of the beam at the support), the corresponding section of the conjugate beam cannot have any shear or moment acting on it. Consequently, this end of the conjugate beam is entirely free and unsupported. On the other hand, if one end of the real beam is free, as, for example, the free end of a cantilever, it will have both slope and vertical deflection when the beam is loaded. The corresponding end of the conjugate beam will therefore be fixed, since it must have both shear and moment acting on it. In other words, if the real beam is a cantilever, the conjugate beam is also a cantilever, but with the fixed support on the opposite end. If the support at the end of the real beam is simple (that is, no moment restraint is provided), then this end of the real beam will rotate but will not deflect under load. The support at the corresponding section of the conjugate beam must provide shear, but no moment, and consequently it will also be a simple support.

The foregoing discussion of types of support for the conjugate beam at points corresponding to the *ends* of the real beam (the real beam may be either a single span or continuous and of two or more spans) can best be summarized as in Table 4-18.

TABLE 4–18.

Real beam support at end	Conjugate beam support at corresponding end
Free end of cantilever	Fixed
Simple	Simple
Fixed	Free end of cantilever

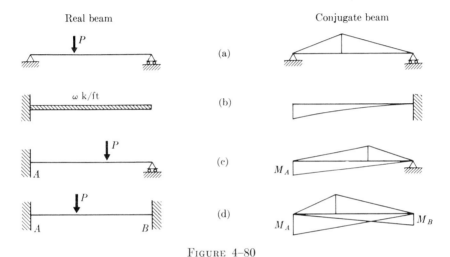

FIGURE 4–80

In Fig. 4–80, the loaded real beams, and the corresponding conjugate beams, will serve to illustrate the above discussion. It is important to note here that the M/EI diagrams have been drawn on the side of the conjugate beam corresponding to the compression side of the real beam. The M/EI load is always considered to push against the conjugate beam, whether it be up or down. If these rules are followed, then the real beam, at any section, will deflect toward the tension side of the conjugate beam at the corresponding section.

If the real beam is continuous, as in Fig. 4–81, then the conjugate beam

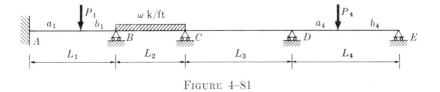

FIGURE 4–81

can be represented by Fig. 4–82. The left end of the real beam is fixed so no support exists at this end of the conjugate beam. The support at the right end of the conjugate beam must be simple, since the simple support at this end of the real beam permits slope but no deflection. At the interior supports of the real beam there is no deflection, but there is a common slope for the ends of adjoining spans. Consequently, a pin must be considered to exist in the conjugate beam at this point, since this is the only arrangement which will ensure zero moment and equal shears on the ends of adjoining spans.

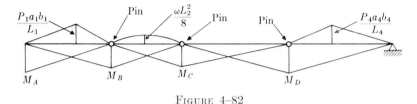

FIGURE 4–82

If the real beam consists of two cantilevers with a suspended span between, as in Fig. 4–83, then the conjugate beam is an overhanging beam,

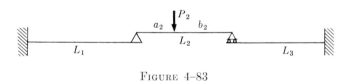

FIGURE 4–83

as shown in Fig. 4–84. Note that the conjugate beam must be determinate, for an indeterminate conjugate beam would require an unstable real beam.

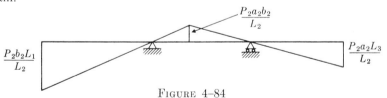

FIGURE 4–84

EXAMPLE 4–55. Using the conjugate beam method, find (from Fig. 4–85):
(a) The slope at A.
(b) The deflection of point C.
(c) The section of maximum deflection.
(d) The value of maximum deflection.

$E = 30{,}000 \text{ k/in}^2$ and $I = 200 \text{ in}^4$.

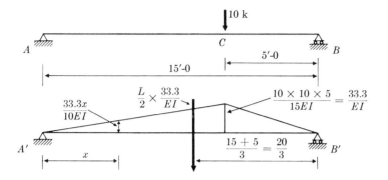

FIGURE 4–85

(a) $$\Sigma M_{B'} = 0,$$

$$R'_A \times L - \frac{L}{2} \times \frac{33.3}{EI} \times \frac{20}{3} = 0,$$

from which,

$$R'_A = \frac{111}{EI}.$$

Therefore,

$$\theta_A = \frac{111 \times 144}{30{,}000 \times 200} = 0.00264 \text{ rad.}$$

(b) $$\Delta_C = \frac{111 \times 10}{EI} - \frac{33.3}{2EI} \times 10 \times \frac{10}{3}$$

$$= \frac{1110}{EI} - \frac{555}{EI} = \frac{555 \times 1728}{30{,}000 \times 200} = 0.16 \text{ in.}$$

(c) Maximum deflection in the real beam will occur at the section of maximum moment in the conjugate beam, which, in turn, will be at the section of zero shear in the conjugate beam. Accordingly, we write the equation for shear at any section x in the segment of the beam to the left of the load, and equate to zero as follows:

$$\text{Shear in conjugate beam} = \frac{111}{EI} - \frac{3.33x^2}{2EI} = 0,$$

$$x = 8.16 \text{ ft.}$$

(d) $$\Delta_{max} = \frac{111}{EI} \times 8.16 - \frac{3.33 \times 8.16^2}{2EI} \times \frac{8.16}{3}$$

$$= \frac{603 \times 1728}{30{,}000 \times 200} = 0.17 \text{ in.}$$

EXAMPLE 4-56. Using the conjugate beam method, find (see Fig. 4-86):
(a) The deflection at B.
(b) The section of maximum deflection in the span.
$E = 4000$ k/in^2.

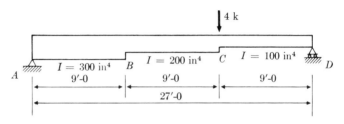

FIGURE 4-86

The loaded conjugate beam is represented by Fig. 4-87.

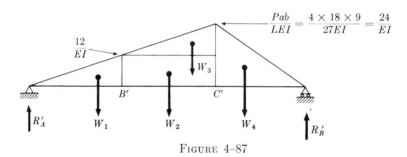

FIGURE 4-87

(a) Assuming a basic moment of inertia of 100 in^4, the relative moments of inertia of the three segments of the beam, from left to right, are 3, 2, and 1. The M/EI diagram is subdivided into triangles and a rectangle, as indicated in Fig. 4-87, and the values of W_1 through W_4 are computed below:

$$W_1 = 12 \times \tfrac{9}{2} \times \tfrac{1}{3} = 18, \qquad W_2 = 12 \times 9 \times \tfrac{1}{2} = 54,$$

$$W_3 = 12 \times \tfrac{9}{2} \times \tfrac{1}{2} = 27, \qquad W_4 = 24 \times \tfrac{9}{2} = 108.$$

These values act through the centroids of their respective M/EI triangle or rectangle. Each reaction of the conjugate beam is found by taking moments about the other reaction. The results are

$$R'_A = 77, \qquad R'_D = 130.$$

The deflection of B on the real beam is found by computing the moment at B' in the conjugate beam:

$$\Delta_B = (77 \times 9 - 18 \times 3)\frac{1728}{4000 \times 100} = 2.76 \text{ in.}$$

(b) The section of maximum deflection in the real beam will correspond to the section of maximum moment in the conjugate beam, and this, of course, will be at the section of zero shear in the conjugate beam. The shear at B' will be $77-18 = 59$. By inspection, therefore, the section of zero shear will be between B' and C'. The load diagram for the conjugate beam between these points, adjusted for the relative moment of inertia of the section, is shown in Fig. 4–88.

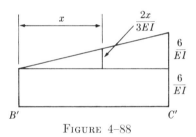

FIGURE 4–88

The area of the M/EI diagram from B' to any point x to the right of B' (between B' and C') is

$$\frac{x^2}{3} + 6x.$$

Since the shear at B' is 59, the required equation for locating the section of zero shear is

$$\frac{x^2}{3} + 6x - 59 = 0,$$

and

$$x = 7.07 \text{ ft.}$$

The section of maximum deflection is 16.07 ft to the right of A.

EXAMPLE 4-57. Using the conjugate beam method, find (see Fig. 4-89):

(a) The deflection of F.
(b) The deflection of B.
(c) The slope at B.
(d) The slope at C.

$E = 30,000$ k/in^2.

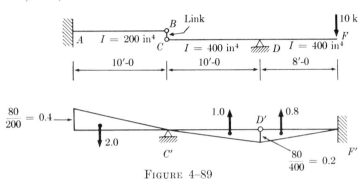

FIGURE 4-89

The conjugate beam, with ordinates adjusted for the different moments of inertia, is shown in Fig. 4-89. Note that E has been omitted in these ordinates for convenience, but that it is introduced in the final evaluation of each answer. When they are constant, it is most convenient to temporarily omit both E and I in this same manner.

(a) $E\Delta_F = \left[2.0 \times \dfrac{20}{3} + 1.0 \times \dfrac{20}{3}\right] \dfrac{8}{10} + 0.8 \times \dfrac{16}{3} = 20.26,$

$\Delta_F = \dfrac{20.26 \times 1728}{30,000} = 1.17$ in.

(b) $E\Delta_B = 2.0 \times \dfrac{20}{3} = \dfrac{40}{3},$

$\Delta_B = \dfrac{40 \times 1728}{3 \times 30,000} = 0.77$ in.

(c) $E\theta_B = 2.0,$

$\theta_B = \dfrac{2.0 \times 144}{30,000} = 0.0096$ rad.

(d) The shear on the pin at D' is

$$\left[2.0 \times \frac{20}{3} + 1.0 \times \frac{20}{3}\right]\frac{1}{10} = 2.0$$

and the shear to the right of C' will be

$$2.0 - 1.0 = 1.0,$$

$$\theta_C = \frac{1.0 \times 144}{30,000} = 0.0048 \text{ rad.}$$

The conjugate beam method provides a means for evaluating redundant moments in continuous beams. Usually, however, it is not a preferred method of analysis for this type of problem. Nevertheless, since it occasionally may be found useful when a continuous beam is nonprismatic, this kind of application will be demonstrated in Example 4–58. A preliminary discussion of one point is in order, however, before proceeding with the example. When a continuous beam is to be analyzed by the conjugate beam method, the internal moments at supports are always selected as the redundants. Each redundant moment necessitates the writing of one condition equation. Particular attention is called to the fact that each condition equation, although written to directly express a particular requirement for equilibrium of the whole or a part of the conjugate beam, at the same time indirectly expresses a requirement for the geometrical coherence of the real beam.

EXAMPLE 4–58. Using the conjugate beam method, find (see Fig. 4–90):
(a) The moments at A and B.
(b) The section of maximum deflection in BC.
(c) The maximum deflection in BC.

$I = 200 \text{ in}^4$ and $E = 30,000 \text{ k/in}^2$.

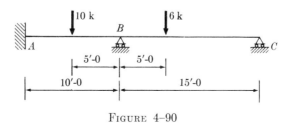

FIGURE 4–90

The beam under discussion is indeterminate to the second degree and the redundant moments are M_A and M_B. It is suggested that redundant moments always be assumed as positive; that is, that they be assumed with senses to cause compression on the top side of the beam. This has been done in loading the conjugate beam of Fig. 4–91. Note that since E and I are constant, they have been omitted from the loading.

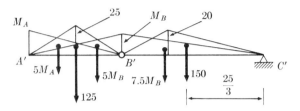

FIGURE 4–91

(a) Since two unknowns exist, two equations are required, each expressing a necessary condition for the equilibrium of a part, or of all, of the conjugate beam. Perhaps the most obvious of these in the beam of Fig. 4–91 is the requirement that the sum of the first moments of all the elastic loads on the span $A'B'$ about B' must be zero. In other words, $\Sigma M_{B'} = 0$. This, from the conjugate beam of Fig. 4–91, is written as

$$5M_A \times \tfrac{20}{3} + 5M_B \times \tfrac{10}{3} + 125 \times \tfrac{15}{3} = 0,$$

from which

$$2M_A + M_B + 37.5 = 0. \qquad (4\text{–}35)$$

The second condition equation is, in the author's opinion, most conveniently written to express the requirement that the pin at B', when considered as a free body, must be in equilibrium. That is, $\Sigma V = 0$ for the pin. This is expressed as

$$5M_A + 5M_B + 125 + \frac{150 \times 25/3 + 7.5M_B \times 10}{15} = 0,$$

from which

$$M_A + 2M_B + 41.7 = 0. \qquad (4\text{–}36)$$

The simultaneous solution of Eqs. (4–35) and (4–36) will result in

$$M_A = -11.1 \text{ ft·k} \quad \text{and} \quad M_B = -15.3 \text{ ft·k}.$$

The complete load on the conjugate beam, omitting EI, can now be represented as shown in Fig. 4–92.

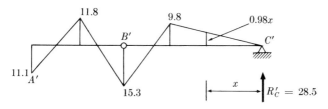

FIGURE 4–92

(b) The section of maximum deflection in the span BC of the real beam will coincide with the section of zero shear in $B'C'$. In order to locate this section of zero shear, it is first necessary to solve for R'_C. This is most conveniently done by considering M_B (in Fig. 4–91) to have its correct sign and magnitude and then to take moments about B'. When this is done, the resulting equation, assuming R'_C is up, is

$$15R'_C + 7.5 \times 15.3 \times 5 - 150 \times \tfrac{20}{3} = 0,$$

from which

$$R'_C = +28.5.$$

This reaction is shown in Fig. 4–92, as well as the value of the ordinate of the M diagram at any section x from C'. The expression for the shear at any section distance x from C' is then written, and this, when equated to zero and solved for x, will locate the section of zero shear in the span $B'C'$. This expression is

$$28.50 - 0.98 \frac{x^2}{2} = 0,$$

from which

$$x = 7.65 \text{ ft.}$$

(c) In accordance with the above, the section of maximum deflection in span BC is 7.65 ft to the left of C. If we take moments about the corresponding section of the conjugate beam, the result is

$$EI\Delta_{\max} = 28.50 \times 7.65 - 0.98 \times \frac{7.65^2}{2} \times \frac{7.65}{3} = 145,$$

$$\Delta_{\max} = \frac{145 \times 1728}{30{,}000 \times 200} = 0.04 \text{ in.}$$

Problems

4-59. Using the conjugate beam method, find (see Fig. 4-93):

(a) The deflection at B.
(b) The section of maximum deflection in the span.

$E = 30,000$ k/in^2. [*Ans.:* (a) 0.46 in., (b) 16.98 ft to right of A.]

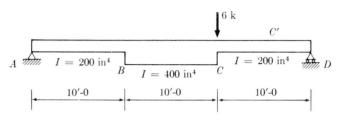

FIGURE 4-93

4-60. Using the conjugate beam method, find (see Fig. 4-94):

(a) The deflection of B and C.
(b) The deflection of E.
(c) The slope at B.
(d) The slope at D.

$E = 30,000$ k/in^2. [*Ans.:* (a) 0.48 in., (b) 0.15 in., (c) 0.006 rad, (d) 0.007 rad.]

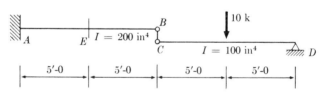

FIGURE 4-94

4-61. Using the conjugate beam method, find (see Fig. 4-95):

(a) The moment at B.
(b) The distance from A to the section of maximum deflection.
(c) The value of the maximum deflection.

$E = 30,000$ k/in^2 and $I = 172.8$ in^4. [*Ans.:* (a) 74.8 ft·k, (b) 11.1 ft, (c) 0.51 in.]

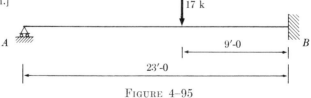

FIGURE 4-95

4-62. Find the moment at A, in the prismatic fixed end beam shown in Fig. 4-96, by the conjugate beam method. [$Ans.$: $-Pab^2/L^2$.]

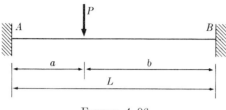

FIGURE 4-96

4-63. Using the conjugate beam method as required, find (see Fig. 4-97):

(a) The moments at A and B.
(b) The distance from C to the point of maximum deflection in BC.
(c) The maximum deflection in span BC.
(d) The slope at B.

$I = 500$ in^4 and $E = 30,000$ k/in^2. [$Ans.$: (a) $M_A = 18.75$ ft·k, $M_B = -37.5$ ft·k, (b) 14.14 ft, (c) 0.41 in., (d) 0.0018 rad.]

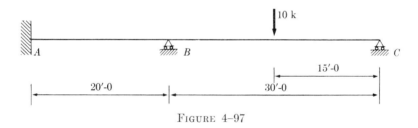

FIGURE 4-97

Part 6. The Conjugate Structure

4-17 General. The conjugate structure is an extension of the method of elastic weights into two dimensions. The method, as hereafter set forth, was developed independently by the author in 1946. (The principles are so simple, however, that others have probably originated similar applications.) In the evaluation of deflections of single-story rigid frames, either single or multi-span, the method is extremely useful.

4-18 Development of the method. Consider the frame of Fig. 4-98. If all of the frame is considered to be entirely rigid, except the segment dx,

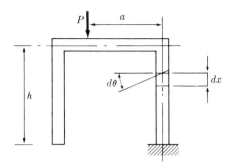

FIGURE 4-98

then, under the action of the load P, the flexural strain in the segment will be

$$d\theta = \frac{M\,dx}{EI},$$

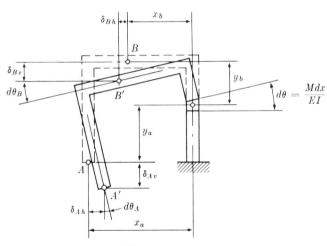

FIGURE 4-99

where M is the moment on the segment caused by P. The frame will then assume the shape shown in Fig. 4-99. It is apparent that

$$\delta_{Av} = x_a\,d\theta = x_a \cdot \frac{M\,dx}{EI}, \qquad \delta_{Ah} = y_a\,d\theta = y_a \cdot \frac{M\,dx}{EI},$$

$$d\theta_A = d\theta = \frac{M\,dx}{EI},$$

and that

$$\delta_{Bv} = x_b\, d\theta = x_b \cdot \frac{M\, dx}{EI}, \qquad \delta_{Bh} = y_b\, d\theta = y_b \cdot \frac{M\, dx}{EI},$$

$$d\theta_B = d\theta = \frac{M\, dx}{EI}.$$

Now consider a structure which is identical with the given frame as to the lengths of its members and their relative positions. This alternate structure, hereafter called the *conjugate structure*, is located in a horizontal plane and the end corresponding to the point A is fixed. Figure 4-100

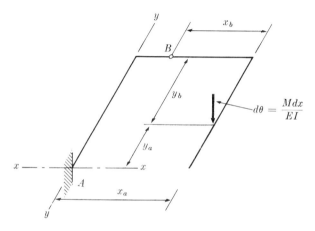

FIGURE 4-100

results. If the flexural strain in the segment is considered as a vertical load acting on the conjugate structure at a point corresponding to the position of the segment in the original structure (as shown in Fig. 4-100), then, taking moments at A,

$$M_y = x_a\, d\theta = x_a \cdot \frac{M\, dx}{EI} = \delta_{Av},$$

$$M_x = y_a\, d\theta = y_a \cdot \frac{M\, dx}{EI} = \delta_{Ah}.$$

Also,

$$\text{Shear at } A = d\theta = \frac{M\, dx}{EI} = d\theta_A.$$

Taking moments about B,

$$M_y = x_b \cdot d\theta = \delta_{Bv}, \qquad M_x = y_b \cdot d\theta = \delta_{Bh}.$$

Also,

$$\text{Shear at } B = d\theta = d\theta_B.$$

Now, if the frame is considered to be elastic throughout, the conjugate structure will be as shown in Fig. 4-101. Note that when the M/EI diagram is of a shape such that the centroid is known, the whole load (area of M/EI diagram) can be considered to act through this centroid when taking moments.

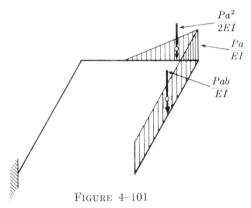

FIGURE 4-101

Having demonstrated how the conjugate structure is proportioned and used, we will now formulate a group of governing principles. Principles (9), (10), and (11) will be found to be somewhat involved, but fortunately, since the senses of deflections and rotations can usually be determined by inspection, it is seldom necessary to use these. The principles are:

(1) The conjugate structure, for a given real structure, is identical to the real structure with regard to the lengths of the members and their relative position.

(2) The conjugate structure is positioned in a horizontal plane.

(3) Two different concepts as to the exact manner of loading the conjugate structure are tenable. The first of these conceives of a conjugate structure composed of members represented by lines and loaded with an M/EI diagram in exactly the same manner as the conjugate beam. Thus, in a length dx along a member, the load is $M\,dx/EI$. The alternate concept considers that the members of the conjugate structure have a definite

width. This width, at any section, is equal to the value of $1/EI$ for the real structure at the corresponding section. Consequently, in a length dx along a member, the area is given by dx/EI. This is called an *elastic area* and it represents the elastic flexural strain at the corresponding section of the real structure, in a length dx, as caused by a unit moment. The intensity of load on this elastic area is the value of the bending moment at this same section of the real structure. Thus, as before, the load on the conjugate structure in a length dx along a member is given by $M\,dx/EI$. This is recognized as the elastic flexural strain at the corresponding section of the real structure in the length dx. Consequently, the load on the conjugate structure (or the conjugate beam) is often designated as an *elastic load*. It is apparent that the conjugate structure (or the conjugate beam) is actually loaded with the flexural strains of the real structure.

(4) If the flexural strain at a given section of the real structure is such as to cause tension on the outside fibers, then this flexural strain is represented as a downward load on the conjugate structure. If compression exists on the outside fibers of the real structure, the load on the corresponding section of the conjugate structure is up.

(5) The conjugate structure, under the action of the flexural strains of the real structure as loads, and the consequent reactions of the conjugate structure, must satisfy three equilibrium condition equations:

$$\Sigma M_x = 0, \qquad \Sigma M_y = 0, \qquad \Sigma V = 0.$$

The x- and y-axes are as shown in Fig. 4–100.

(6) The shear at any section of the conjugate structure is the slope of the corresponding section of the real structure.

(7) The internal moment on any section of the conjugate structure is the deflection of the corresponding section of the real structure in a direction perpendicular to the lever arm used to find any particular moment.

(8) The end of the conjugate structure corresponding to the end of the real structure that deflects always has a fixed support.

(9) If a section be passed through any point of the conjugate structure and if the portion of the conjugate structure to the right of the section tends to move down with respect to the part to the left of the section, then the rotation of the corresponding point of the real structure is counterclockwise. Diagonal sections are passed from upper right to lower left through right-hand vertical members, and from upper left to lower right through left-hand vertical members.

(10) If the moment at any point on a horizontal or inclined member of the conjugate structure, about an axis through that point parallel to the

y-y-axis, results in tension in the member top fibers normal to this axis, then the vertical deflection of the corresponding point on the real structure is down.

(11) If the moment at any point on a vertical or inclined member of the conjugate structure, about an axis through that point parallel to the x-x-axis, results in tension in the member top fibers normal to this axis, and if the supported end of the member has an algebraically larger y-coordinate than the unsupported end, then the horizontal deflection component of the corresponding point of the real structure is toward the right, provided the fixed support of the conjugate structure is to the right, but toward the left if the fixed support of the conjugate structure is to the left. If, however, the supported end of the conjugate structure member has an algebraically smaller y-coordinate than the unsupported end, then tension in the member top fibers as described above will signify a horizontal deflection component of the corresponding point on the real structure to the left if the conjugate structure fixed support is to the right, and a deflection to the right when the conjugate structure fixed support is to the left.

This last principle will be clarified by reference to Fig. 4–102. Figure 4–102(a) shows a real structure and Fig. 4–102(b) shows the corresponding conjugate structure. The supported end of AB in the conjugate structure is at B, and the supported end of BC is at C, and therefore each member has a supported end with a y-coordinate algebraically larger than the

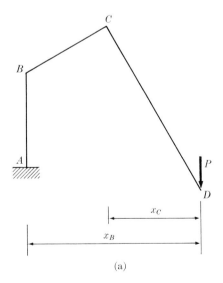

Figure 4–102(a)

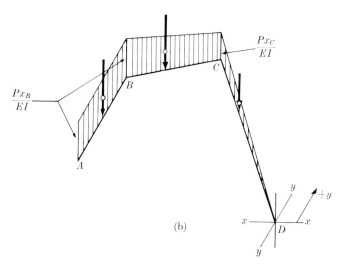

Figure 4–102(b)

unsupported end. Consequently, the tension on the top fibers normal to the x-axis, which under the given loading will exist throughout AB and BC, signifies a horizontal deflection component to the right for all corresponding points of AB and BC on the real structure. The member CD on the conjugate structure, however, with its supported end at D, falls into the opposite classification, for which tension on the top side indicates (in this case) a horizontal deflection component of corresponding points on the real structure toward the left. It is probable that some point along CD of the real structure will have zero horizontal deflection. The corresponding point on the conjugate structure will be the point at which the moment of the elastic loads, about an axis parallel to the x-x-axis and through the point, will be zero. All sections between this point and C will have compression on the top fibers of the conjugate structure, and corresponding sections of the real structure will deflect to the right; while all sections between this point and D will have tension on the top fibers, and corresponding sections of the real structure will deflect to the left.

Note particularly that the sign conventions outlined in Principles (9), (10), and (11) will apply only if the sign convention of (4) is adopted. Attention is again called to the fact that in most cases it will be unnecessary to use Principles (9), (10), and (11), since the senses of the various deflection components can usually be determined by inspection.

150   METHODS FOR COMPUTING DEFLECTIONS   [CHAP. 4

4-19 Demonstration of the method. Several examples will serve to demonstrate the application of the method.

EXAMPLE 4-64. Find the horizontal, vertical, and rotational deflection components of point A (see Fig. 4-103) as caused by the 10 k load. $E = 30,000$ k/in². (This is the same problem as in Examples 4-10 and 4-30. The three methods of solution should be carefully compared.)

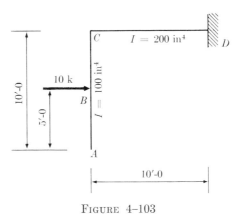

FIGURE 4-103

The conjugate structure would be represented by Fig. 4-104:

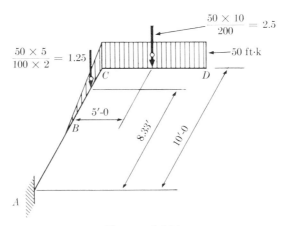

FIGURE 4-104

$$E\Delta_{Ah} = 1.25 \times 8.33 + 2.5 \times 10 = 35.41,$$

$$\Delta_{Ah} = \frac{35.41 \times 1728}{30{,}000} = 2.04 \text{ in.}$$

The direction of the above deflection is to the right, from Principle (11).

$$E\Delta_{Av} = 2.5 \times 5 = 12.5,$$

$$\Delta_{Av} = \frac{12.5 \times 1728}{30{,}000} = 0.72 \text{ in.},$$

and this deflection must be down, from Principle (10).

$$E\theta_A = 1.25 + 2.5 = 3.75,$$

$$\theta_A = \frac{3.75 \times 144}{30{,}000} = 0.018 \text{ rad.}$$

The above rotation must be counterclockwise, from Principle (9).

EXAMPLE 4-65. Using the conjugate structure, find the horizontal, vertical, and rotational deflection components of point A (see Fig. 4-105). $E = 30{,}000$ k/in^2. (This is an alternate solution for the problem of Example 4-31.)

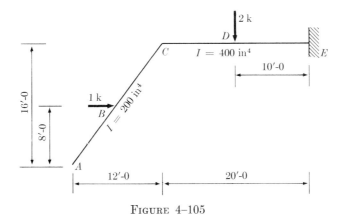

FIGURE 4-105

The conjugate structure is represented by Fig. 4-106:

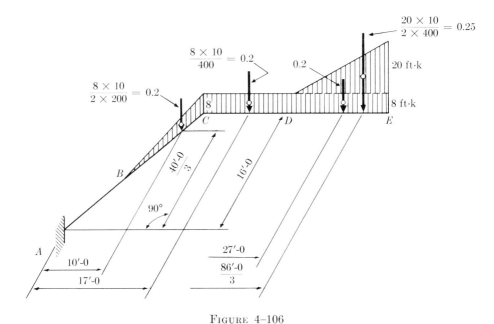

Figure 4-106

$$E\Delta_{Ah} = 0.2 \times \frac{40}{3} + 0.2 \times 16 + 0.2 \times 16 + 0.25 \times 16 = 13.07,$$

$$\Delta_{Ah} = \frac{13.07 \times 1728}{30{,}000} = 0.75 \text{ in. to the right [Principle (11)]},$$

$$E\Delta_{Av} = 0.2 \times 10 + 0.2 \times 17 + 0.2 \times 27 + 0.25 \times \frac{86}{3} = 17.97,$$

$$\Delta_{Av} = \frac{17.97 \times 1728}{30{,}000} = 1.03 \text{ in. down [Principle (10)]},$$

$$E\theta_A = 0.2 + 0.2 + 0.2 + 0.25 = 0.85,$$

$$\theta_A = \frac{0.85 \times 144}{30{,}000} = 0.0041 \text{ rad counterclockwise [Principle (9)]}.$$

EXAMPLE 4-66. Using the conjugate structure, find the vertical, horizontal, and rotational deflection components of point A (see Fig. 4-107) as caused by the 10 k load. Also, find the horizontal deflection component of point A as caused by a 1 k load acting to the right at A. $E = 30{,}000$ k/in^2.

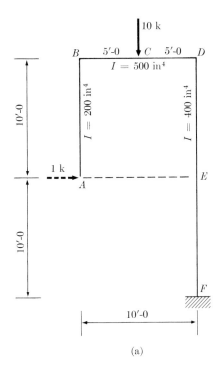

(a)

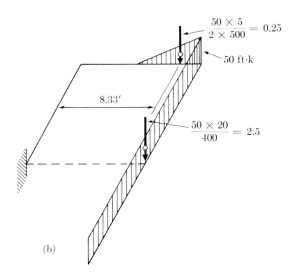

(b)

Figure 4–107

$$E\Delta_{Av} = 0.25 \times 8.33 + 2.5 \times 10 = 27.08,$$

$$\Delta_{Av} = \frac{27.08 \times 1728}{30{,}000} = 1.56 \text{ in. down},$$

$$E\Delta_{Ah} = 0.25 \times 10 = 2.5,$$

$$\Delta_{Ah} = \frac{2.5 \times 1728}{30{,}000} = 0.14 \text{ in. to the right},$$

$$E\theta_A = 2.5 + 0.25 = 2.75,$$

$$\theta_A = \frac{2.75 \times 144}{30{,}000} = 0.013 \text{ rad counterclockwise}.$$

For a 1 k load acting horizontally to the right at A, the conjugate structure would be represented by Fig. 4–108:

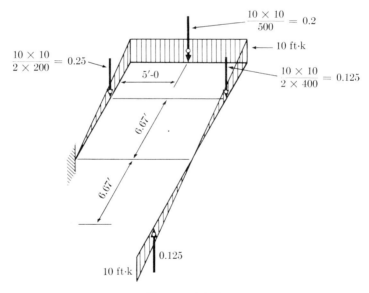

FIGURE 4–108

$$E\delta_{Ah} = 0.25 \times 6.67 + 0.2 \times 10 + 2 \times 0.125 \times 6.67 = 5.34,$$

$$\delta_{Ah} = 0.31 \text{ in. to the right}.$$

EXAMPLE 4-67. Using the conjugate structure, find (see Fig. 4-109):

(a) The horizontal deflection of point A caused by the 20 k load.

(b) The horizontal deflection of A caused by 1 k applied horizontally at A acting toward the left.

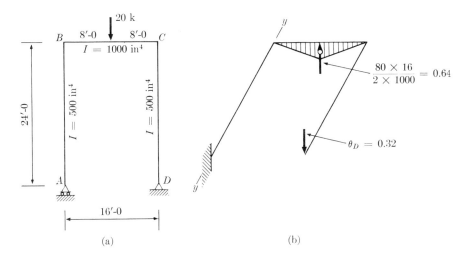

FIGURE 4-109

(a) In this case point A has no vertical deflection component, and thus, even though this end of the conjugate structure must be fixed, the moment about the y-axis through A must be zero. This can be accomplished only by applying an external load θ_D at the free end of the conjugate structure to give $\Sigma M_{A_{yy}} = 0$. Assuming θ_D as down on the conjugate structure (as it obviously would have to be), the equation is

$$\theta_D \times 16 - 0.64 \times 8 = 0.$$

Therefore,
$$\theta_D = 0.32,$$

and this value of θ_D, if multiplied by 144 and divided by $E = 30{,}000$ k/in^2, will give the slope of the right-hand leg. Note that this value of θ_D does not enter into the solution for Δ_{Ah}, since its y lever arm is zero. If, however, the frame had unequal legs, it would have appeared in the equation for Δ_{Ah}.

$$E\Delta_{Ah} = 0.64 \times 24 = 15.38,$$

$$\Delta_{Ah} = \frac{15.38 \times 1728}{30{,}000} = 0.88 \text{ in. to the left.}$$

156 METHODS FOR COMPUTING DEFLECTIONS [CHAP. 4

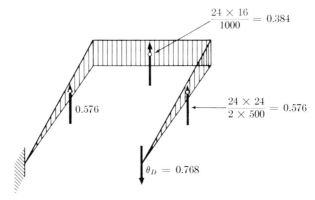

FIGURE 4–110

(b) $\qquad \theta_D \times 16 - 0.384 \times 8 - 0.576 \times 16 = 0,$

$$\theta_D = 0.768,$$

$$E\delta_{Ah} = 0.384 \times 24 + 2 \times 0.576 \times 16 = 27.63,$$

$$\delta_{Ah} = \frac{27.63 \times 1728}{30{,}000} = 1.59 \text{ in. to the left.}$$

EXAMPLE 4–68. For the concrete culvert section shown in Fig. 4–111, consider that a cut is made at A. Using the conjugate structure, and considering that end A of AD is fixed in the real structure after cutting, find the three deflection components of end A of AB. Consider a length of culvert of 1 ft perpendicular to the paper. $E_c = 3000 \text{ k/in}^2$.

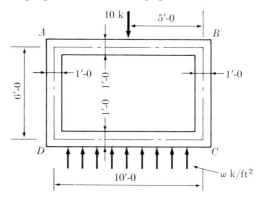

FIGURE 4–111

$$I = \frac{1 \times 1^3}{12} \times 12^4 = 1728 \text{ in}^4.$$

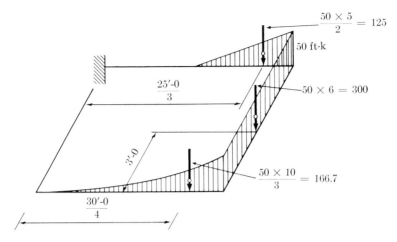

Figure 4–112

$$E\Delta_{Ah} = 300 \times 3 + 166.7 \times 6 = 1900,$$

$$\Delta_{Ah} = \frac{1900 \times 1728}{3000 \times 1728} = 0.63 \text{ in. to the left,}$$

$$E\Delta_{Av} = 125 \times \frac{25}{3} + 300 \times 10 + 166.7 \times \frac{30}{4} = 5291,$$

$$\Delta_{Av} = \frac{5291 \times 1728}{3000 \times 1728} = 1.76 \text{ in. down,}$$

$$E\theta_A = 125 + 300 + 166.7 = 591.7,$$

$$\theta_A = \frac{591.7 \times 144}{3000 \times 1728} = 0.016 \text{ rad counterclockwise.}$$

4–20 Application to multi span frames. As previously indicated, the conjugate structure is useful in finding deflections of multi-span single-story rigid frames after they have been reduced to statically determinate and stable structures by removal of redundant reaction components. Suppose, for example, that we wish to find the stresses in the frame of Fig. 4–113. If the solution is to be effected by the general method (discussed in Section 3–5 and to be further discussed in detail in Chapter 5), the first step is to reduce the structure to determinateness by removing reaction components. As previously explained, this can be done by re-

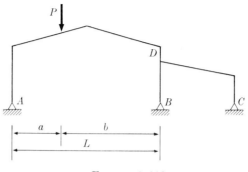

FIGURE 4-113

moving any combination of reaction components desired, so long as the final structure is determinate and stable. Assume that in the given case both reaction components are removed at C and the horizontal component is removed at B. The result is shown in Fig. 4-114, with the dashed lines indicating how the frame will deflect under the action of the load P. A number of other deflection components of B and C under certain other loads are necessary for a complete solution of the problem by the general method, but the immediate object of this discussion is to explain how the conjugate structure can be used to find the deflection components indicated in Fig. 4-114.

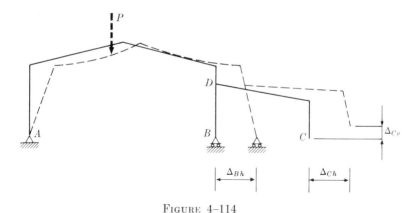

FIGURE 4-114

As a beginning, the conjugate structure for the span AB is as shown in Fig. 4-115. The horizontal deflection components of B and D, and the rotational deflection component of D, may be found from this structure.

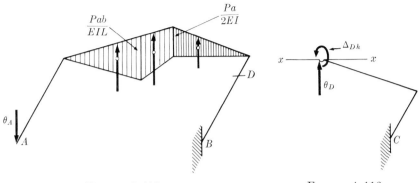

FIGURE 4–115   FIGURE 4–116

The conjugate structure for span BC is shown in Fig. 4–116. This is loaded at its unsupported end D with the rotation and horizontal deflection of D, as determined from the conjugate structure in Fig. 4–115. The rotation of D is properly represented as a concentrated load, and the horizontal displacement of D as a couple about an axis through D parallel to the x-x axis. The required vertical and horizontal deflection components of C may be readily found from Fig. 4–116. In a similar manner, deflection components resulting from any system of applied loads on a frame with any number of spans may be determined. The application of the method to a two-span rigid frame is demonstrated in the example which follows.

EXAMPLE 4–69. In order to analyze a two-span rigid frame by the general method, reaction components have been removed to reduce the actual structure to that shown in Fig. 4–117. One step in the solution requires that a 1 k horizontal load be applied at A and that the resulting deflection components, vertical and horizontal at A, and horizontal at F,

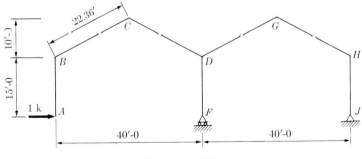

FIGURE 4–117

be determined. Assuming that I is 400 in^4 and that E is 30,000 k/in^2, evaluate these deflection components. The conjugate structure for the span FJ, with elastic loads, is shown in Fig. 4–118. The upward reaction

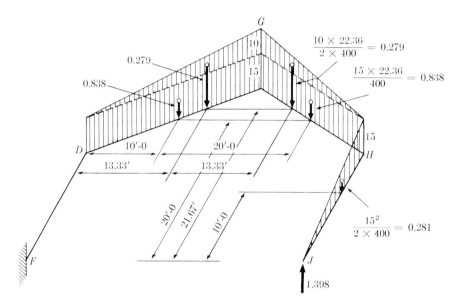

FIGURE 4–118

at J is found by taking moments about a y-axis through F. This having been evaluated as 1.398, then we find $E \cdot \delta_{DhAh}$ by taking moments about an x-axis through D as follows:

$$E \cdot \delta_{DhAh} = 1.398 \times 15 + 2 \times 0.279 \times 6.67$$
$$+ 2 \times 0.838 \times 5 - 0.281 \times 5$$
$$= 31.7 \text{ (relative value)},$$

$$\delta_{DhAh} = \frac{31.7 \times 1728}{30,000} = 1.83 \text{ in.,}$$

$$E \cdot \theta_D = \text{shear at } D = 1.117 \text{ (relative value)}.$$

Taking moments about an x-axis through F,

$$E \cdot \delta_{FhAh} = 2 \times 0.279 \times 21.67 + 2 \times 0.838 \times 20 + 0.281 \times 10$$

$$= 48.5 \text{ (relative value)}$$

$$= \frac{48.5 \times 1728}{30,000} = 2.79 \text{ in.}$$

The conjugate structure for the span AF is shown in Fig. 4–119.

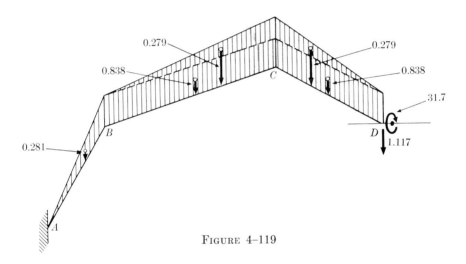

FIGURE 4–119

Moments about an x-axis through A will give

$$E \cdot \delta_{AhAh} = 48.5 + 1.117 \times 15 + 31.7$$

$$= 97.0 \text{ (relative value)},$$

$$\delta_{AhAh} = \frac{97.0 \times 1728}{30,000} = 5.58 \text{ in.}$$

Moments about a y-axis through A will result in

$$E \cdot \delta_{AvAh} = 1.117 \times 40 + 20(2 \times 0.279 + 2 \times 0.838)$$

$$= 89.4 \text{ (relative value)},$$

$$\delta_{AvAh} = \frac{89.4 \times 1728}{30,000} = 5.15 \text{ in.}$$

Attention is called to the fact that in a complete analysis by the general method, there would be no particular advantage in using absolute values for deflections. Relative values are ordinarily used in the condition equations, which are solved simultaneously to evaluate the redundant reactions.

In the above example it was possible to consider each portion of the elastic load, that is, each triangle or rectangle, as a single elastic force acting through its centroid. This, however, cannot be done if the members of the frame are nonprismatic. In this latter case, it is necessary to divide the frame into segments, assuming that the values for M and I which exist at the center of each segment will hold throughout the entire length of that segment.

Problems

4-70. By the conjugate structure, find (see Fig. 4-120):

(a) The horizontal deflection component of A.
(b) The rotation of D.
(c) The rotational deflection component of A.
(d) The horizontal deflection component of C.

$E = 30,000$ k/in^2. [*Ans.:* (a) 4.35 in., (b) 0.0086 rad, (c) 0.0095 rad, (d) 2.07 in.]

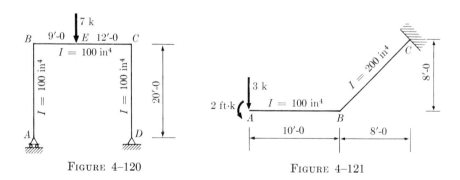

Figure 4-120   Figure 4-121

4-71. For the cantilever shown in Fig. 4-121 find:

(a) The vertical deflection component of point A as caused by the 3 k vertical load acting alone.

(b) The rotational deflection component of point A resulting from the action of the 2 ft·k couple acting alone.

Use $E = 30,000$ k/in^2. [*Ans.:* (a) 2.54 in., (b) 0.0015 rad.]

4-72. Find the three deflection components of point A (see Fig. 4-122) using the conjugate structure. $E = 30,000$ k/in^2. [Ans.: $\Delta_{Ah} = 9.3$ in., $\Delta_{Av} = 36.4$ in., $\theta_A = 0.089$ rad.]

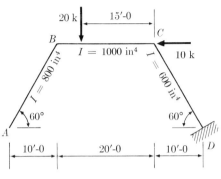

FIGURE 4-122

4-73. Find the horizontal deflection component of point A (see Fig. 4-123) as caused by the 10 k concentrated load at span center. The various moments of inertia are indicated on the members of the frame. $E = 30,000$ k/in^2. [Ans.: 29.1 in.]

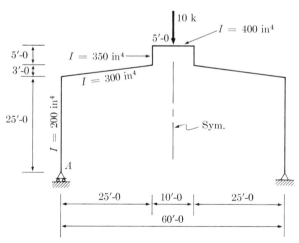

FIGURE 4-123

4-74. Find the vertical and horizontal deflection components of point A and the horizontal deflection component of point F of the frame of Fig. 4-117, as caused by a vertical load of 1 k acting down at A. [Ans.: $\Delta_{AvAv} = 6.85$ in., $\Delta_{AhAv} = 5.14$ in., $\Delta_{FhAv} = 2.57$ in.]

Part 7. The Williot-Mohr Diagram

4-21 General. The Williot-Mohr diagram provides an excellent method for determining the absolute displacements of the joints of articulated structures. The French engineer Williot developed the original idea in 1877; however, when used alone, his diagram is of limited value. The deflections obtained by its use will be absolute only if the graphical construction is referred to a structural member which remains fixed in direction with at least one end fixed in location during deflection. When all members of a structure move, that is, either translate or rotate or both, then the Williot diagram will give relative deflections.

In 1887 Professor Otto Mohr, of Dresden, having realized the potential value of the Williot diagram, as well as its particular shortcomings, published his rotation diagram. This, combined with the work of Williot, provides a method which is unequaled as a means for determining the absolute displacements of the joints of articulated structures.

4-22 The Williot diagram. As a beginning, consider the small cantilever truss shown full-size in Fig. 4–124(a). Assume that the structural material will behave elastically in spite of very large strains, and consider that each member, except bc, is strained one unit, one unit being equal to one-tenth the length of the panel. The strain for each member is written in units on the sketch of the truss, a positive sign indicating a strain causing an increase in length of the member.

We wish to find the deflections of all panel points resulting from the indicated strains. In the case of the small truss under consideration, this can easily be done as shown in Fig. 4–124(b). Starting with the points a' and d', drawn in their true positions, the new location of e, that is, e', must be determined first. This is because it is the only joint tied directly to the fixed points a and d by two members, ae (2) and de (3). The member ae is first drawn from a' in its original direction and length, and the positive strain of one unit is then added to the original length of the member (this is shown in Fig. 4–124(b) and labeled "2" to identify the strain with its member). The new location of e (that is, e') must be on the arc through which the end e' of the strained member ae, which might now be designated $a'e'$, will pass when the member is rotated about a' as a center. A similar operation is then performed with the member de. From d' the member de is drawn in its original direction and length, and the length is then decreased by the amount of the compressive strain, labeled "3" in this case. The strained member $d'e'$ is then rotated about d' as a center so that e' strikes an arc. Obviously, the correct location of e' will be at the intersection of the two arcs.

4–22] THE WILLIOT DIAGRAM

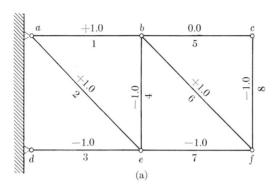

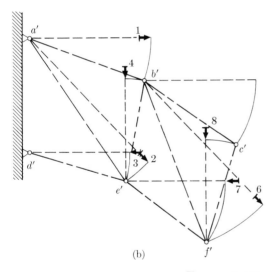

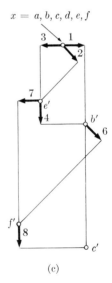

Figure 4–124

Having located e', we can now locate b', inasmuch as this point is tied to a and e by the two members ab (1) and be (4). We proceed as before, the member ab being drawn from a' in its original direction and length, the strain 1 added, and an arc swung. Then the member eb is drawn from e' in its original direction and length, the compressive strain 4 is subtracted from the length, and an arc is swung to intersect the arc previously formed by rotating $a'b'$. The intersection of the two arcs is the location of b'. Similarly, f', and then c', are located. If the unstrained truss of Fig. 4–124(a) is superimposed on the deflected structure of 4–124 (b), with a on a' and d on d', the absolute displacement of each joint can be measured directly.

The above construction is perfectly satisfactory for the assumed small truss. Obviously, however, a full-scale construction as demonstrated above would be impractical for an actual structure. The strains in an actual structure are very small, and angles of rotation are so small that the angles and their sines and tangents are equal. Hence a full-scale construction would be of little value.

Williot perceived that if exactly the same operations as previously described are performed in the same sequence, but on the basis of the original lengths of all members being zero, then exactly the same information may be obtained from the resulting diagram. In addition, since no consideration need be given to the actual size of the structure, the analyst is free to choose the scale of the graphical construction so as to magnify the strains, with resulting increase in accuracy.

For an illustration of the Williot construction, refer again to Fig. 4–124(a). Consider that, with point a held in position, the lengths of all members are reduced to zero. This brings all panel points, including d, into coincidence with a, and this point of coincidence, labeled x in Fig. 4–124(c), is the starting point of the Williot diagram. Each deflected panel point must be located by means of the strains in two members intersecting at that panel point and having their other ends at points which are already located. Joints a and d are fixed in position, and therefore the deflected position of e (that is, e') must first be determined, followed by b', f', and c', in that order.

In order to locate e' in Fig. 4–124(c), draw the strain labeled "2" from the starting point in the proper amount (note that the scale of the strains has been doubled in 4–124(c) to give a larger diagram) and parallel to member 2. Since a tensile strain is indicated for this member, e in the structure will move away from a, and therefore strain 2 is drawn down and to the right. Member 3 has a compression strain, and consequently e moves to the left relative to d. Strain 3 is therefore drawn to the left from the starting point.

Reference to Fig. 4–124(b) shows that the strains 2 and 3 appear in the same relative positions therein as in 4–124(c). In Fig. 4–124(b), after strains 2 and 3 had been drawn, the next step was to swing arcs with the ends of the strained members 2 and 3, and the intersection of these arcs located e' in that figure. In the Williot construction, however, these arcs are assumed to be straight lines which are perpendicular to the original directions of members 2 and 3. Obviously this is a poor assumption for the truss under discussion, because of the very large strains, but in the usual practical problem it is entirely valid. Consequently, in Fig. 4–124(c) the arcs of 4–124(b) are replaced by perpendicular lines drawn at the ends of strains 2 and 3, and the intersection of these perpendiculars locates e'. Note that the strains 2 and 3 and the arcs of 4–124(b) are in the same relative positions as the strains 2 and 3 and the perpendiculars of 4–124(c). In (c), however, the scale has been chosen to magnify the strains in order to give greater accuracy. In 4–124(b) such a magnification is impossible without a corresponding increase in the drawn lengths of the members.

The position of e' having been determined in Fig. 4–124 (c), the next step is to draw the strains 1 and 4 from a and e' in the proper magnitude and direction. For example, because of the compressive strain in member 4, the joint b will move down with respect to e, and therefore strain 4 is drawn down from e'. The strain in member 1 is tensile and b will move to the right relative to a. Consequently, strain 1 is drawn to the right from point a in 4–124 (c). The intersection of perpendiculars drawn through the ends of these strains in 4–124(c) will locate b'.

Points f' and c' are similarly found. Note that in locating c', since no strain exists in member 5, no strain 5 is drawn from b' in Fig. 4–124(c). In spite of this, however, a line is drawn from the end of this zero strain (actually b') in a direction perpendicular to member 5. This, of course, corresponds to the arc which was swung with the end of the unstrained member 5 in 4–124(b).

The deflections obtained from the Williot diagram are always relative to an assumed fixed point which is located at one end of a member assumed fixed in direction. If the point and the member are in fact fixed in location and direction, then the deflections are relative to an actual fixed point and direction, and are absolute.

In most cases, when the deflections obtained by the Williot diagram are relative, they are of little practical value. There are certain cases, however, in which relative deflections are useful. For example, if the deflections of the joints of a symmetrical truss that is loaded symmetrically are required, and if the Williot diagram is referred to a center vertical member (if such does not exist, one may be assumed), then the resulting deflections, even though relative, will give the desired information. Relative deflections are also used in secondary stress problems.

168   METHODS FOR COMPUTING DEFLECTIONS   [CHAP. 4

EXAMPLE 4–75. Find the vertical deflection of panel points c and d (see Fig. 4–125) by the Williot diagram. Strains are marked on the truss members, with a positive sign indicating a tensile strain. Consider that panel point d is the fixed point. The member de will not rotate during deflection of the truss.

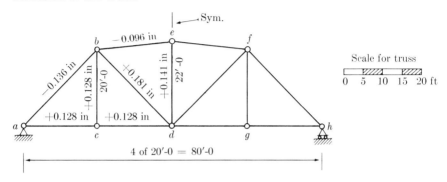

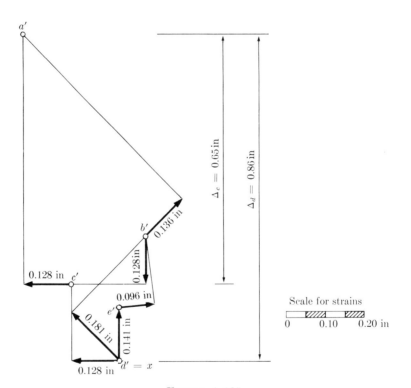

FIGURE 4–125

EXAMPLE 4-76. A camber of 6 in. is required at the center of the truss of Fig. 4-126. What increase in each top chord panel length and each end-post will be necessary to obtain this camber?

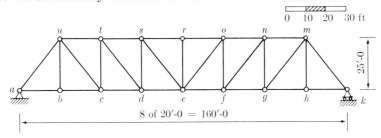

FIGURE 4-126

A solution will first be completed to find the camber resulting from an increase of one-quarter inch in the length of each top chord member and each endpost. Then, by direct proportion, the required increase for a 6 in. camber will be determined. The member er will not rotate and point e is taken as the fixed point.

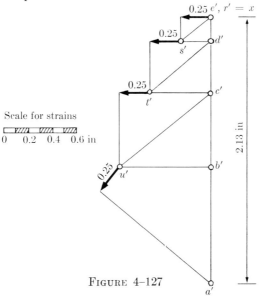

FIGURE 4-127

$$\Delta L = 6 \times \frac{0.25}{2.13} = 0.71 \text{ in. (use 0.75 in.)},$$

$$\text{Actual camber} = \frac{0.75}{0.71} \times 6 = 6.36 \text{ in.}$$

170 METHODS FOR COMPUTING DEFLECTIONS [CHAP. 4

EXAMPLE 4–77. Find the vertical and horizontal deflection components of joint a in the truss of Fig. 4–128, as caused by the application of the indicated loads. $E = 29{,}000$ k/in^2.

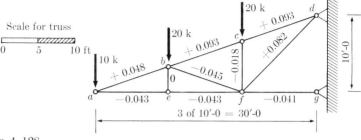

FIGURE 4–128

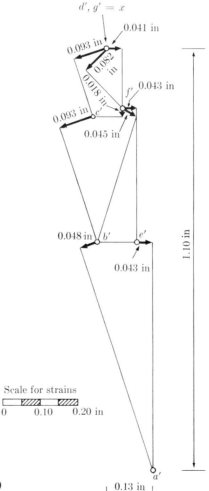

FIGURE 4–129

All stresses having been found, the strains in the truss members are computed by SL/AE. The value of A is 3 in^2 for all members. The resulting strains are marked on the members, a positive sign indicating a tensile strain. The Williot diagram is drawn as shown in Fig. 4–129. The line "dg" is taken as the line fixed in direction (as it actually is) and joint d is taken as the fixed point.

Problems

4–78. What would be the sag at the center of the truss shown in Fig. 4–130 if all members of the lower chord were fabricated 0.25 in. too long? Assume point e as fixed in location and member er as fixed in direction. Suggested scales: truss, 1 in. = 30 ft; strains, 1 in. = 0.50 in. [*Ans.:* 1.40 in.]

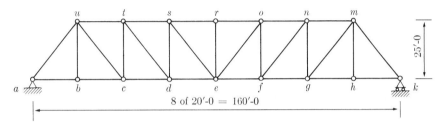

Figure 4–130

4–79. Find the horizontal and vertical deflection components of panel point 4 for the dead load strains indicated in Fig. 4–131. Strains are symmetrical. Assume joint 4 as fixed in location and member $C4$ as fixed in direction. Suggested scales: truss, 1 in. = 30 ft; strains, 1 in. = 0.10 in. [*Ans.:* vertical = 0.41 in., horizontal = 0.09 in.]

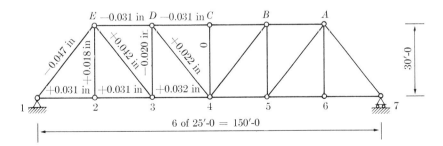

Figure 4–131

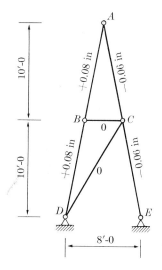

4-80. An aerial conveyor tower is strained as indicated in Fig. 4-132. Find the horizontal and vertical deflection components of point A. The line between D and E is fixed in direction and length. Suggested scales: tower, 1 in. = 5 ft; strains, 1 in. = 0.2 in. [*Ans.*: horizontal = 0.74 in., vertical = 0.03 in.]

FIGURE 4-132

4-23 The Mohr rotation diagram. As indicated in Section 4-22, if the construction of the Williot diagram can be referred to a nontranslating end of a nonrotating member in a deflecting articulated structure, then the deflections obtained from the Williot diagram will be absolute. In many deflecting articulated structures, however, no such joint and member will exist. Consequently, the joint displacements obtained from the Williot construction will be relative to some joint and member which are not, in fact, fixed in position and direction. In most cases these relative displacements, with the exceptions noted in Section 4-22, are of little value.

Actually, of course, the Williot diagram can be constructed with respect to any assumed fixed joint located at one end of an assumed nonrotating member. For example, consider the solid-line truss of Fig. 4-133(a). This is the same small cantilever truss, drawn full-scale, as shown in Fig. 4-124(a). In this case, however, the displacements of the truss joints are to be determined on the assumption that joint c will remain fixed in position and that member cf will not rotate. The distorted truss, indicated by long and short dashed lines in Fig. 4-133(b), is developed by following exactly the same procedure as previously explained in connection with Fig. 4-124(b), except that the construction is now started with the joint marked c'. Starting with c', the sequence in which the joints are located in Fig. 4-133(b) is f', b', e', a', and d'. If the unstrained truss is superimposed on the strained truss, with a on a' and d on d', then the distance from each double-primed to the primed letter is the magnitude of the deflection of that joint. The unstrained truss is indicated by dot-and-dash lines in Fig. 4-133(b). The direction of each displacement, however,

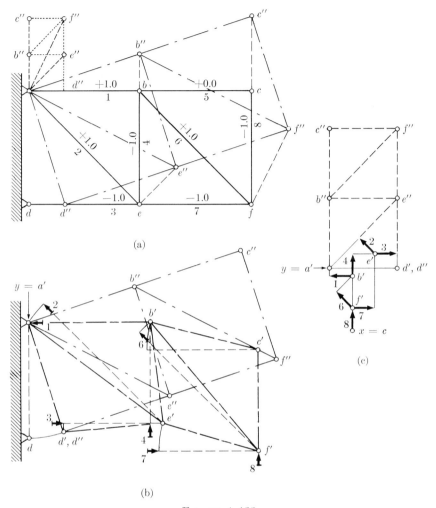

FIGURE 4-133

is in error by the angle between $a'd'$ and the vertical. In other words, the correct direction of each displacement would be obtained by rotating the entire figure in a clockwise direction until $a'd'$ were vertical. This error, although very considerable in the construction used for this demonstration, becomes insignificant if the same operation is performed with an actual structure. The strains in the truss members have been taken to be extremely large, about 0.10, so as to demonstrate satisfactorily the various steps in the graphical construction, and these very large strains result in the above error. In an actual steel truss the strains would probably not exceed 0.0007.

A full-scale construction such as shown in Fig. 4–133(b) is practically impossible for an actual structure. The same information regarding the displacements of all joints (relative to c as fixed in location and member cf as fixed in direction) may be readily obtained, however, by the Williot diagram. The solid lines in Fig. 4–133(c) show this diagram [the scale has been doubled as compared to 4–133(b)] but the information which may be obtained from it is quite useless. The difficulty is due to the fact that the construction has been based on false assumptions, and consequently the indicated joint displacements are not consistent with the constraints. The Williot diagram cannot be rotated; obviously, then, the only alternative is to rotate the unstrained structure so that its constraints (reactions) are consistent with the indicated joint displacements of the Williot diagram. This, of course, was done in Fig. 4–133(b) in the full-scale construction. The Mohr rotation diagram accomplishes the same thing.

To understand why the Mohr construction will give the displacement of joints as caused by rotation of the structure, consider Fig. 4–133(a). The dot-dash truss has been rotated about a to the same position as in Fig. 4–133(b). Note, however, that b'' and c'' are shown directly above b and c, respectively, instead of being on arcs passing through b and c with a as a center. This is because in the case of an actual structure the angle of rotation is extremely small, and, for all practical purposes, there is no horizontal displacement of b and c during the rotation. Similarly, there is no vertical displacement of d.

All joint displacements are proportional to the length of the radial line from the joint to the center of rotation. Since the angle of rotation is very small, each joint displacement will occur in a direction perpendicular to the radial line to the joint. If, now, in Fig. 4–133(a) the lengths of all members are reduced to zero, all joints of the unrotated truss become coincident with a. In addition, all lines of joint displacement, such as dd'', ee'', cc'', etc., are pulled in to originate at the joint and center of rotation a. These lines are shown in Fig. 4–133(a), all drawn from a. If the lines $c''f''$, $b''f''$, and $b''e''$ are drawn (shown in 4–133(a) as short dashed lines), the result will be a scale drawing of the original truss, but rotated 90° from its original position. This small-scale drawing is the *Mohr rotation diagram*.

In order to be of practical value the rotation diagram must be superimposed on the Williot diagram. It must show the effects of a rotation of the unstrained structure about a pinned reaction joint, and this joint is designated as the center of rotation y in the Mohr diagram. This rotation brings the other reaction joint into a position so that the indicated absolute displacement of this other reaction joint (the distance from the double-primed to the single-primed letter for that joint) will be consistent with the constraint imposed by that reaction. In the case of the cantilever truss under discussion, the structure is rotated about the center of rotation $y = a'$ in Fig. 4–133(b) so as to bring joint d into coincidence with d'. Therefore, a'' and d'' of the Mohr diagram coincide with a' and d' in Fig. 4–133(c). The remainder of the rotation diagram is readily constructed by drawing each member of the Mohr truss perpendicular to the corresponding member of the original structure. The absolute deflection of any joint (except for the errors due to large strains, as previously discussed) will be the distance from any double-primed letter to the corresponding single-primed letter.

The above demonstration illustrates the fundamentals of the rotation diagram. It does not, however, indicate that two necessary principles are available for locating the position of the rotated reaction so as to make it consistent with the Williot diagram. It has been established that in the Mohr diagram the distance from the center of rotation y to any double-primed joint designation is the displacement of that joint due to rotation. This displacement will always be perpendicular to the direction of the radial line, in the original structure, from the center of rotation to the joint. Consequently, the first principle is as follows:

The rotated constraint (reaction) will always be located in the Mohr diagram on a line drawn through the center of rotation y of the Mohr diagram in a direction perpendicular to the straight line joining the two exterior constraints (reactions) of the original structure.

In addition to the above, the fact has also been established that after the Mohr rotation diagram has been superimposed on the Williot diagram, the line from any double-primed joint designation (in the Mohr diagram) to the corresponding single-primed joint designation (in the Williot diagram) is, in both magnitude and direction, the absolute displacement of the joint. Consequently, a line between the double-primed and single-primed designation for the joint at the rotated reaction must be in the direction of the constrained motion (if motion exists) of that joint as the structure deflects. If, for example, one end of a truss is on rollers, then this end will move along the plane of the rollers as the truss deflects under load. Therefore the statement of the second principle is as follows:

The double-primed designation (in the Mohr diagram) for the joint of the structure at which the rotated reaction acts will always be located on a line drawn through the single-primed designation (in the Williot diagram) of the same joint, this line being parallel to the constrained motion of this joint as the structure deflects. The joints d' and d'' coincide in the Williot-Mohr diagram of Fig. 4–133(c) because in this case there is no actual movement of joint d during the deflection of the structure. In other words, the line $d'd''$ is of zero length.

The application of these two principles will be demonstrated in the examples which follow.

EXAMPLE 4-81. Find the vertical and horizontal deflection components of joints b and d in the truss indicated in Fig. 4-134. Strains in inches are marked on the members. Assume that joint a is fixed in location and that member ab is fixed in direction.

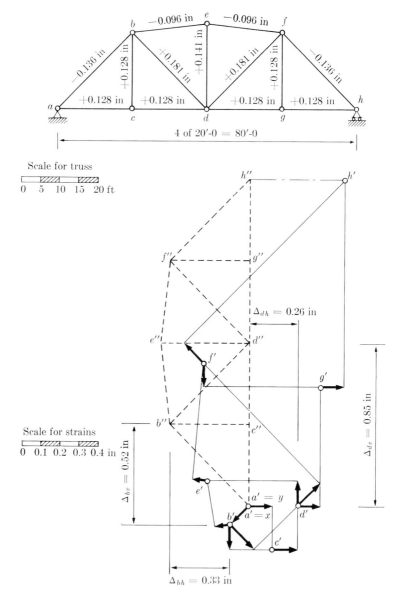

FIGURE 4-134

Note that h'' is located as follows: a vertical line is drawn through $a' = y$, in accordance with the first principle previously discussed. A horizontal line is drawn through h', in accordance with the second principle. The intersection of these two lines locates h''.

EXAMPLE 4–82. Find the vertical and horizontal deflection components of joint b (see Fig. 4–135) and the deflection of joint a resulting from the indicated strains. Assume that joint a is fixed in location and that member ab is fixed in direction.

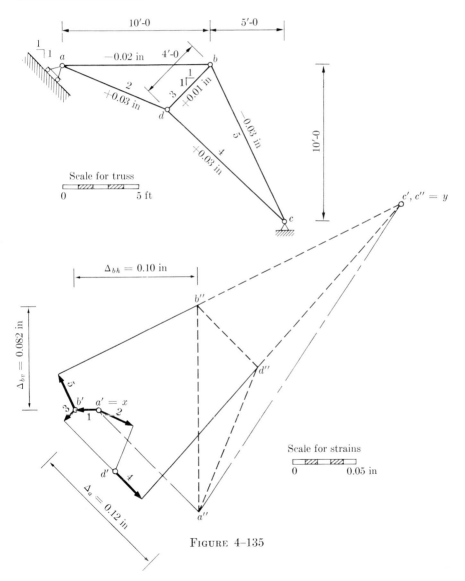

FIGURE 4–135

Problems

4-83. Find the horizontal and vertical deflection components of point B in Fig. 4-136. Strains are indicated on the members. Consider point A as fixed in location and member AB as fixed in direction. [Ans.: $\Delta_{Bh} = 0.04$ in., $\Delta_{Bv} = 0.17$ in.]

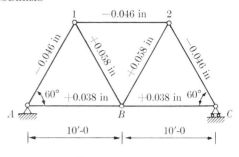

Figure 4-136

4-84. Determine reactions and stresses graphically for the truss shown in Fig. 4-137 and compute strains in all members, using $E = 30{,}000$ k/in². Construct the Williot-Mohr diagram. Use point a as the fixed point and ab as a line fixed in direction. Determine the absolute displacement of c along the plane of rollers and also the vertical and horizontal components of the deflection of point b. [Ans.: $\Delta_{bh} = 0.004$ in., $\Delta_{bv} = 0.031$ in., $\Delta_C = 0.022$ in.]

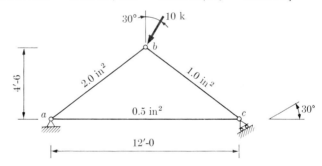

Figure 4-137

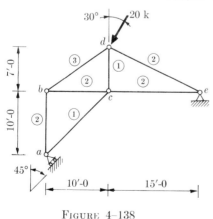

Figure 4-138

4-85. Determine reactions and stresses for the truss of Fig. 4-138. Compute the strains in all members, using $E = 30{,}000$ k/in². The cross-sectional areas of the several truss members (in in²) are circled on the sketch. Use point a as the fixed point and ab as the member fixed in direction. Determine the displacement of joint a and the horizontal and vertical deflection components of joint c. [Ans.: $\Delta_a = 0.54$ in., $\Delta_{ch} = 0.11$ in., $\Delta_{cv} = 0.54$ in.]

Part 8. Elastic Weights Applied to Articulated Structures

4-24 General. The concept of elastic weights was introduced in connection with discussions of the conjugate beam and the conjugate structure. The structures considered therein were composed entirely of flexural members. Elastic weights can also be used for computing deflections of articulated structures.

It will be recalled that the angle change (flexural strain) in a flexural member for a length ds is given by $M\,ds/EI$. It will also be recalled that it is often convenient to divide a structure into segments having a finite length ΔS and to compute the angle change for each segment as $M\,\Delta S/EI$. This angle change for each segment is then assumed to be concentrated at the center of the segment and to act as a concentrated load at that point on the conjugate beam or the conjugate structure.

In the case of articulated structures, which are analyzed as though pinned at the joints, the angle changes actually are concentrated at the joints. Consider, for example, the truss of Fig. 4-139(a). When the truss is loaded, the various members are strained. Consequently, the internal angles of the various triangles in the structure will change. Suppose we wish to find the resulting vertical deflection components of the bottom chord. It is first necessary to compute the total angle change at each bottom chord panel point, and these total angle changes are then applied as concentrations on a conjugate beam, as shown in Fig. 4-139(b). The bending moment at any section along this loaded conjugate beam will be the vertical deflection component of the corresponding point on the truss bottom chord. Obviously, the only new feature in the problem under consideration is the method for evaluating the angle changes at the bottom chord panel points. This method will be developed in the section which follows.

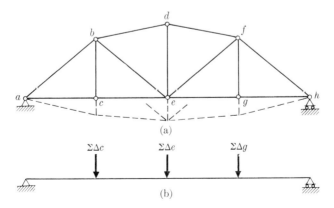

Figure 4-139

4–25 Angle changes in articulated structures. Consider the triangle 1-2-3 in Fig. 4–140. It is required to find the changes in the three internal angles resulting from a change in length of the three sides. Initial values of the internal angles are designated by A, B, and C, and the original lengths of the opposing sides by a, b, and c.

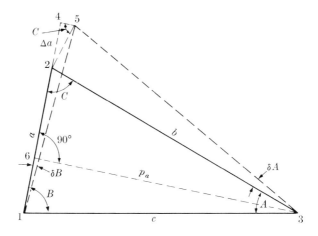

Figure 4–140

Assume that the side a is increased in length by the increment Δa, shown in Fig. 4–140 as the length 2–4, an extension of the side a (1–2). The new position of joint 2, shown as point 5 in the figure, is obtained by drawing 4–5 perpendicular to 1–4, and 2–5 perpendicular to 2–3. The new position of joint 2 must be at the intersection of these two perpendiculars. The angle changes resulting from Δa are easily computed from Fig. 4–140 as follows: Since Δa is very small compared to a,

$$\delta B = -\frac{4\text{-}5}{1\text{-}2} = -\frac{\Delta a \cot C}{a} = -\epsilon_a \cot C,$$

where ϵ_a is the unit strain in the member a. Also,

$$\delta A = +\frac{2\text{-}5}{2\text{-}3} = +\frac{\Delta a}{b \sin C} = +\frac{\epsilon_a \cdot a}{b \sin C} = +\frac{\epsilon_a \cdot a}{p_a},$$

but $a = (2\text{-}6) + (6\text{-}1) = p_a (\cot C + \cot B)$ and thus

$$\delta A = +\epsilon_a (\cot C + \cot B).$$

Since the sum of the internal angles of the triangle must remain a constant value,
$$\delta C = -(\delta B + \delta A) = -\epsilon_a \cot B.$$

Similarly, the angle changes caused by an increase Δb in the length of b are found to be

$$\delta C = -\epsilon_b \cot A, \qquad \delta A = -\epsilon_b \cot C, \qquad \delta B = +\epsilon_b (\cot C + \cot A).$$

In like manner, for an increase Δc in the length of c, the resulting angle changes are:

$$\delta A = -\epsilon_c \cot B, \qquad \delta B = -\epsilon_c \cot A, \qquad \delta C = +\epsilon_c (\cot A + \cot B).$$

Adding the three changes for each angle, we find the total angle changes are

$$\Delta A = (\epsilon_a - \epsilon_b) \cot C + (\epsilon_a - \epsilon_c) \cot B,$$
$$\Delta B = (\epsilon_b - \epsilon_a) \cot C + (\epsilon_b - \epsilon_c) \cot A, \qquad (4\text{--}37)$$
$$\Delta C = (\epsilon_c - \epsilon_b) \cot A + (\epsilon_c - \epsilon_a) \cot B.$$

Note that each term in the above expressions consists of parentheses enclosing the unit strain for the side opposite the angle of the computed change minus the unit strain for an adjacent side, and that each enclosure is multiplied by the cotangent of the included angle.

EXAMPLE 4–86. Find the vertical deflection of points c and d of the truss of Fig. 4–141. Unit strains, multiplied by 10^{-3}, are indicated on the various members. (This is the same as Example 4–75.)

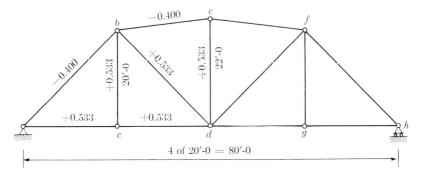

FIGURE 4–141

The angle changes are computed by Eq. (4-37):

$$\Delta acb = (\epsilon_{ab} - \epsilon_{ac}) \cot cab + (\epsilon_{ab} - \epsilon_{bc}) \cot abc$$
$$= (-0.400 - 0.533)\,1 + (-0.400 - 0.533)\,1$$
$$= 1.866 \times 10^{-3} \text{ rad},$$
$$\Delta dcb = (\epsilon_{bd} - \epsilon_{cd}) \cot bdc + (\epsilon_{bd} - \epsilon_{bc}) \cot dbc$$
$$= (+0.533 - 0.533)\,1 + (0.533 - 0.533)\,1 = 0,$$

and therefore

$$\Delta c = \Delta g = -1.866 \times 10^{-3} \text{ rad},$$
$$\Delta bde = (\epsilon_{be} - \epsilon_{bd}) \cot ebd + (\epsilon_{be} - \epsilon_{ed}) \cot bed$$
$$= (-0.400 - 0.533)\,0.816 + (-0.400 - 0.533)\,0.102$$
$$= -0.856 \times 10^{-3} \text{ rad}.$$
$$\Delta bdc = 0 \text{ (same substitutions as for } \Delta dcb),$$
$$\Delta fde = -0.856 \times 10^{-3} \text{ rad},$$

and therefore

$$\Delta d = -1.712 \times 10^{-3} \text{ rad}.$$

The angle changes at c, d, and g are applied as concentrated loads on a conjugate beam (see Fig. 4-142):

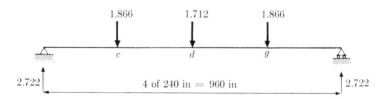

FIGURE 4-142

Deflection at $c = 2.722 \times 10^{-3} \times 240 = 0.65$ in.,
Deflection at $d = 2.722 \times 10^{-3} \times 480 - 1.866 \times 10^{-3} \times 240 = 0.85$ in.

This method is particularly valuable in determining the necessary increase in the fabricated length of the top chord members of a truss in order to obtain a desired camber. Deflection components of articulated structures in both the horizontal and vertical directions may easily be found by applying angle changes as loads on a conjugate structure.

Specific References

1. BAIRSTOW, L. and PIPPARD, A. J. S., "The Determination of the Stresses in a Shaft of any Cross Section," *Proc., Inst. Civ. Engs.*, Vol. 214, 1921–22.
2. CHURCH, I. P., *Mechanics of Internal Work*. New York: Wiley, 1910.
3. MOHR, O., "Beitrag zur Theorie der Holz—und Eisen Konstruktionen," *Zeitschrift des Architekten und Ingenieur Vereines zu Hannover*, 1868.
4. PIPPARD, A. J. S. and BAKER, J. F., *The Analysis of Engineering Structures*. London: Arnold, 1943.
5. ROARK, R. J., *Formulas for Stress and Strain*. New York: McGraw-Hill, 1943.
6. TIMOSHENKO, S., *Theory of Elasticity*. 2nd ed. New York: McGraw-Hill, 1943.

General References

7. ANDERSEN, P., *Statically Indeterminate Structures*. Chaps. 2 and 3. New York: Ronald, 1953.
8. GRINTER, L. E., *Theory of Modern Steel Structures*. Vol. 2, chap 3. New York: Macmillan, 1949.
9. HOFF, N. J., *The Analysis of Structures*. Part 1. New York: Wiley, 1956.
10. PARCEL, N. J. and MOORMAN, R. B. B., *Analysis of Statically Indeterminate Structures*. Chaps. 1 and 2. New York: Wiley, 1955.
11. SOUTHWELL, R. V., *An Introduction to the Theory of Elasticity*. Oxford: Clarendon Press, 1936.
12. WANG, C. K., *Statically Indeterminate Structures*. Chaps. 2 and 3. New York: McGraw-Hill, 1953.
13. SUTHERLAND, H. and BOWMAN, H. L., *Structural Theory*. 3rd ed., chap. 7. New York: Wiley, 1942.
14. WILLIAMS, C. D., *Analysis of Statically Indeterminate Structures*. Chaps. 1 and 2. Scranton: International, 1943.

CHAPTER 5

THE GENERAL METHOD

5-1 General. The "general method" for analyzing indeterminate structures, also known as the *method of consistent distortions or consistent displacements*, is credited to Clerk Maxwell, Otto Mohr, and Heinrich Müller-Breslau. It was Maxwell who, in 1864, first introduced the method and called it a "general method." His development was based on Clapeyron's statement of the equality between the external work of the loads applied to a structure and the resulting internal strain energy. At the same time he developed his *theorem of reciprocal deflections*. As previously indicated, however, his presentation was devoid of any illustrations and was so abstract that it attracted little attention.

In 1874, without any knowledge of Maxwell's previous work, Otto Mohr developed the same method. His derivation was different, however, for he used the concept of virtual work. Mohr also presented various examples illustrating the application of the method, including the effects of temperature variation.

In 1886, Heinrich Müller-Breslau published his variation of the previous work of Maxwell and Mohr. The condition equations for geometrical coherence of a structure, as written by Müller-Breslau, are obtained by superposition of displacements as caused by the applied loads and individual redundant stresses and reactions. The coefficients of these redundant stresses and reactions are the deflections due to unit stresses and reactions, and these deflections may be found by any method desired. The Müller-Breslau version of the general method will be applied to subsequent illustrative problems.

As previously explained in Section 3-5, the first step in the application of the general method, as herein described, is to remove the redundant stresses and/or reaction components and, by so doing, to reduce or "cut back" the structure to a condition of determinateness and stability. Any combination of redundant stresses and/or reaction components may be removed. Deflection condition equations are then written, one for each point of application of a redundant stress or reaction component. The left side of each equation is a summation of all deflection components (as caused by all real loads and redundant stresses and/or reaction components acting on the structure) of the point of application, and in the direction, of one of the redundant stresses or reaction components. The right side of the equation is the predetermined value for the sum of these deflection components, and is usually zero. When solved simultaneously, these equations will give the magnitudes and senses of the redundant stresses and/or reaction components. The method is quite easy to understand and can be most effectively demonstrated by a series of illustrative problems.

5-2 Analysis of beams. The general method is not usually selected as a method for analyzing continuous beams if the moment of inertia is constant throughout individual spans, since a solution by moment distribution is somewhat easier. If, however, the individual spans have varying moments of inertia, the general method is unsurpassed as a means of analysis, unless tabulated information is available regarding the stiffness and carry-over factors of the various spans. If this standard information is available, a solution by moment distribution is to be preferred.

If the general method is to be used, the various deflection components may be evaluated either by virtual work or by the conjugate beam. Often the beam is divided into segments, as demonstrated for the method of virtual work in Example 4–40.

EXAMPLE 5-1. Determine the reaction at B for the beam indicated in Fig. 5-1. Obtain the necessary deflection components by:

(a) Virtual work, assuming no settlement of the support at B.

(b) Conjugate beam, assuming a settlement in the support at B of 0.25 in. Assume that $I = 200$ in^4 and that $E = 30{,}000$ k/in^2.

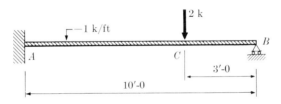

FIGURE 5-1

Since the reaction at B is required, it is taken as the redundant and the support at B is removed. The resulting statically determinate and stable beam is shown in its deflected position in Fig. 5-2. If the real loads are

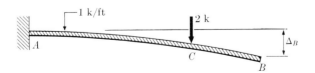

FIGURE 5-2

considered to be removed, and a single force of 1 k applied at B, the deflected beam of Fig. 5-3 results. This illustration of the deflected beam might well be called the R_B *deflection diagram*, because the deflection of

FIGURE 5-3

point B as caused by R_B will be $R_B \cdot \delta_{BB}$.

(a) Since no settlement of the support at B has been indicated in the example statement, the net deflection of B will be zero. Consequently, the deflection condition equation will be

$$\Delta_B + R_B \cdot \delta_{BB} = 0.$$

In order to find the values of Δ_B and δ_{BB} by virtual work, it is convenient to use a tabular arrangement for the moment expressions. In Table 5-1, m_B is the moment as caused by a unit load acting down at B. This unit load is used as both a real load and a fictitious load. M is the moment caused by the applied loads. A moment causing tension on the top fibers is considered to be positive.

TABLE 5-1.

Section	$x = 0$ at	x increasing	M (ft·k)	m_B (ft·k)
BC	B	B to C	$+\dfrac{x^2}{2}$	$+x$
CA	C	C to A	$+\dfrac{(3+x)^2}{2} + 2x$	$+3+x$

$$EI\Delta_B = \int m_B M \, dx = \int_0^3 (+x)\left(+\frac{x^2}{2}\right) dx$$

$$+ \int_0^7 [+3+x]\left[\frac{+(3+x)^2}{2} + 2x\right] dx = +1625,$$

$$EI\delta_{BB} = \int m_B^2 \, dx = \int_0^3 x^2 \, dx + \int_0^7 (3+x)^2 \, dx = +333.3.$$

The positive signs of these two deflections signify that both have the same sense as the unit load acting at B; that is, both are downward. The sense of any assumed unit load at any point must be considered to be the positive sense for deflections and redundant reaction components in the same line of action at that point.

Substituting in the deflection condition equation, after multiplying each term by EI, yields

$$1625 + 333.3\, R_B = 0,$$

from which

$$R_B = -4.88 \text{ k}.$$

The negative sign indicates that the true sense of R_B is opposite to that of the unit fictitious load assumed at B; that is, R_B acts with an upward sense.

(b) The conjugate beam for the simplified real beam, that is, with the support at B removed, is shown in Fig. 5–4(a) and (b).

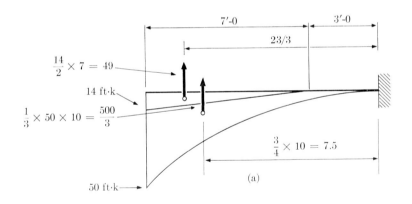

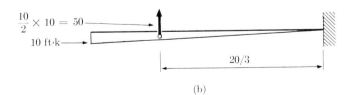

FIGURE 5-4

From Fig. 5-4(a),

$$EI\,\Delta_B = 49 \times \frac{23}{3} + \frac{500}{3} \times 7.5 = 1625,$$

$$\Delta_B = \frac{1625 \times 1728}{200 \times 30{,}000} = 0.47 \text{ in. down.}$$

From Fig. 5-4(b),

$$EI\,\delta_{BB} = 50 \times \frac{20}{3} = 333.3,$$

$$\delta_{BB} = \frac{333.3 \times 1728}{200 \times 30{,}000} = 0.096 \text{ in. down.}$$

These values agree with those previously found by virtual work. Since the example statement specified a settlement of 0.25 in. at B, the condition equation will be

$$0.47 + 0.096 R_B = 0.25,$$

$$R_B = -\frac{0.22}{0.096} = -2.3 \text{ k.}$$

This reaction, as before, will be up.

EXAMPLE 5-2. Determine the magnitude and direction of the reactions at B and C (see Fig. 5-5) by the general method. Use the conjugate beam to find the necessary deflection components. The moment of inertia is constant.

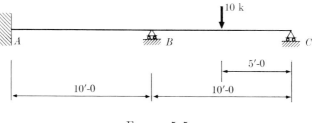

FIGURE 5-5

The reactions at B and C are taken as the redundants, and consequently the necessary deflection condition equations are

$$\Delta_B + R_B \delta_{BB} + R_C \delta_{BC} = 0,$$

$$\Delta_C + R_B \delta_{CB} + R_C \delta_{CC} = 0.$$

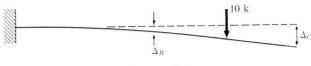

FIGURE 5-6

The supports at B and C having been removed, the deflected beam will be as shown in Fig. 5-6; the conjugate beam is shown in Fig. 5-7.

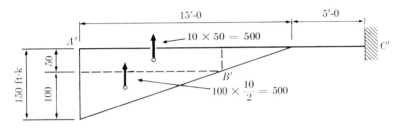

FIGURE 5-7

From this conjugate beam, if we take moments at B' and C', respectively,

$$EI\, \Delta_B = 500 \times \tfrac{20}{3} + 500 \times 5 = 5833,$$

$$EI\, \Delta_C = 150 \times \tfrac{15}{2} \times 15 = 16{,}900.$$

Since, with the conjugate beam sign convention used herein, the deflection of the real beam is always toward the tension side of the conjugate beam, both these deflections will be down.

The 10 k load is now considered to be removed from the simplified real beam of Fig. 5-6 and a load of 1 k is assumed to be applied at B acting along the line of action of R_B. Although it is not necessary to do so, the author prefers to assume this unit load to have the same sense as the deflection caused by the real load. The deflected structure is shown in Fig. 5-8, and the conjugate beam in Fig. 5-9.

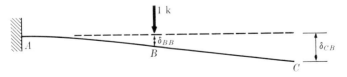

FIGURE 5-8

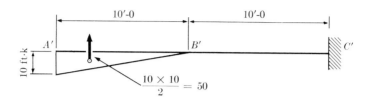

FIGURE 5-9

If we take moments at B' and C', respectively,

$$EI\ \delta_{BB} = 50 \times \tfrac{20}{3} = 333,$$
$$EI\ \delta_{CB} = 50 \times \tfrac{50}{3} = 833.$$

These deflections are down.

The unit load is next moved to C and the deflected beam is shown in Fig. 5-10.

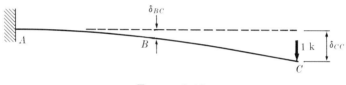

FIGURE 5-10

The conjugate beam is represented by Fig. 5-11.

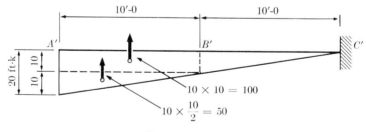

FIGURE 5-11

Moments at B' and C', respectively, will give

$$EI\ \delta_{BC} = 100 \times 5 + 50 \times \tfrac{20}{3} = 833,$$
$$EI\ \delta_{CC} = 20 \times \tfrac{20}{2} \times \tfrac{40}{3} = 2667.$$

Note that $\delta_{CB} = \delta_{BC}$, which is in accordance with Maxwell's theorem of reciprocal deflections. (This theorem should be used as a check on computed deflection components whenever possible.)

Attention is called to the fact that the various deflection components must be substituted in the deflection condition equations with strict regard to signs. In this example all unit forces have been assumed with a downward sense, and downward deflections must therefore be considered as positive.

Substitution of the various deflection components in the deflection condition equations will result in

$$5833 + 333R_B + 833R_C = 0,$$

$$16{,}900 + 833R_B + 2667R_C = 0.$$

It is important to note that in writing these equations the assumption has necessarily been made that the redundants R_B and R_C act with the same sense as the unit loads assumed to act at B and C. These unit loads were assumed to act down. Simultaneous solution of the above equations will result in

$$R_B = -7.53 \text{ k} \quad \text{and} \quad R_C = -3.99 \text{ k}.$$

The negative signs of these answers indicate that the true sense of the reactions is opposite to the sense of the unit loads assumed at B and C. In other words, R_B and R_C actually act up.

EXAMPLE 5-3. From Fig. 5-12, find M_A and R_A directly by the general method. Use the method of virtual work to find the necessary deflection components. The moment of inertia is constant.

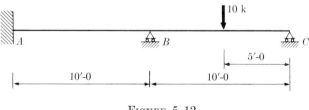

FIGURE 5-12

The example statement indicates that M_A and R_A are to be taken as the redundant reaction components. Accordingly, the support at A is removed. The "cut-back" structure is shown as a solid line in Fig. 5-13.

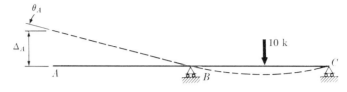

FIGURE 5–13

This arrangement appears to violate the rule that the cut-back structure must be stable, but actually, of course, the beam is stable for vertical loads. No indeterminateness exists so far as the horizontal reaction component at A is concerned. Therefore any inclined loads may be broken into vertical and horizontal components, and only the vertical components need be considered as loads on the cut-back beam of Fig. 5–13. Under the action of the 10 k load, the beam will deflect as shown by the dashed line in that figure. The 10 k load is then removed and a 1 k force is considered to act vertically at A. This force may be considered to act either up or down. The author indicated in the discussion of Example 5–2 that his preference, when assuming the sense of a unit force or couple, is to assume it to have the same sense as the corresponding deflection of its point of application as caused by the real load. In order to demonstrate that the opposite procedure can be used equally well, the 1 k force will be

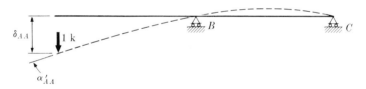

FIGURE 5–14

assumed to act down at A. The "R_A diagram" is shown in Fig. 5–14. If a counterclockwise unit couple is assumed to act alone at A, the "M_A diagram" will be as shown in Fig. 5–15.

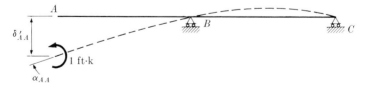

FIGURE 5–15

The deflection condition equations must express the fact that the net vertical deflection component and the net rotational deflection component of A must be zero. These equations will be

$$\Delta_A + R_A \delta_{AA} + M_A \delta'_{AA} = 0,$$

$$\theta_A + R_A \alpha'_{AA} + M_A \alpha_{AA} = 0.$$

It is necessary to evaluate the various deflection components in the above equations and the example statement has specified virtual work as the method to be used in this case. The cut-back beam, with the various loads for which individual moment expressions must be written, is shown

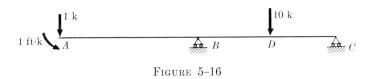

FIGURE 5-16

in Fig. 5-16. Note that in writing these moment expressions, it is essential that some arbitrary sign convention be adopted. In this case, any moment which will result in compression on the top side of the beam will be considered positive. However, the opposite convention could have been adopted without changing the results. As a matter of fact, the convention can be reversed from section to section so long as the signs for all moments within each section are governed by the particular sign convention adopted for that section.

In Table 5-2, M is the moment caused by the real 10 k load, m_{Av} is the moment due to the 1 k vertical load (either real or fictitious) at A, and m_{Am} is the moment resulting from the action of the 1 ft·k couple (either real or fictitious) at A.

TABLE 5-2.

Section	$x = 0$ at	x increasing	M	m_{Av}	m_{Am}
AB	A	A to B	0	$-x$	-1
BD	B	B to D	$+5x$	$+x - 10$	$+\dfrac{x}{10} - 1$
CD	C	C to D	$+5x$	$-x$	$-\dfrac{x}{10}$

By virtual work,

$$EI\,\Delta_A = \int m_{Av}M\,dx$$

$$= \int_0^5 (+x - 10)(+5x)\,dx + \int_0^5 (-x)(+5x)\,dx = -625,$$

$$EI\theta_A = \int m_{Am}M\,dx$$

$$= \int_0^5 \left(+\frac{x}{10} - 1\right)(+5x)\,dx + \int_0^5 \left(-\frac{x}{10}\right)(+5x)\,dx = -62.5,$$

$$EI\,\delta_{AA} = \int (m_{Av})^2\,dx$$

$$= \int_0^{10} (-x)^2\,dx + \int_0^5 (+x - 10)^2\,dx + \int_0^5 (-x)^2\,dx$$

$$= +666.7,$$

$$EI\alpha_{AA} = \int (m_{Am})^2\,dx$$

$$= \int_0^{10} (-1)^2\,dx + \int_0^5 \left(+\frac{x}{10} - 1\right)^2 dx + \int_0^5 \left(-\frac{x}{10}\right)^2 dx$$

$$= +13.34,$$

$$EI\,\delta'_{AA} = EI\alpha'_{AA} = \int (m_{Av})(m_{Am})\,dx$$

$$= \int_0^{10} (-x)(-1)\,dx + \int_0^5 (+x - 10)\left(+\frac{x}{10} - 1\right) dx$$

$$+ \int_0^5 (-x)\left(-\frac{x}{10}\right) dx = +83.34.$$

Substituting in the deflection condition equations yields

$$-625 + 666.7 R_A + 83.34 M_A = 0,$$

$$-62.5 + 83.34 R_A + 13.34 M_A = 0.$$

Simultaneous solution gives

$$R_A = +1.6 \text{ k} \quad \text{and} \quad M_A = -5.3 \text{ ft·k}.$$

The positive sign for R_A indicates that it has the same sense as the unit load assumed to act at A, that is, a downward sense. If the unit load at A had been assumed with an upward sense, then the sign of the answer for R_A would have been negative and the indicated true sense of R_A still would have been down. The negative answer for M_A indicates a true sense opposite to that assumed for the unit couple at A, that is, the true sense is clockwise.

EXAMPLE 5–4. Find the internal moment at B (see Fig. 5–17) by the general method. Use the conjugate beam to find the necessary deflection components. The moment of inertia is constant.

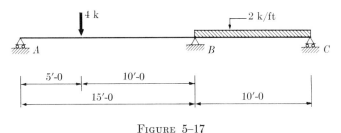

FIGURE 5–17

Since M_B has been specified as the redundant, the cut-back structure will consist of two simple beams, as shown in Fig. 5–18. These beams will

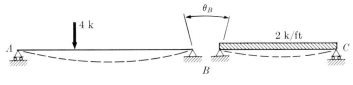

FIGURE 5–18

deflect as indicated by the dashed lines. The conjugate beam for this arrangement is shown in Fig. 5–19:

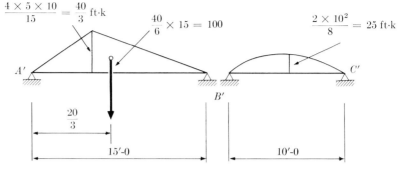

FIGURE 5-19

$$EI\theta_B = 100 \times \tfrac{20}{3} \times \tfrac{1}{15} + \tfrac{1}{2} \times \tfrac{2}{3} \times 25 \times 10 = 127.8.$$

If the real loads are removed and unit couples are simultaneously applied to the B ends of the two simple spans, they will deflect as shown in

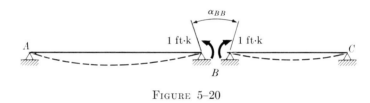

FIGURE 5-20

Fig. 5-20. The conjugate beam is shown in Fig. 5-21, and from this conjugate beam,

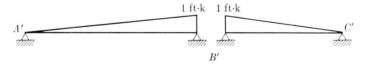

FIGURE 5-21

$$EI\alpha_{BB} = \tfrac{2}{3}[\tfrac{15}{2} \times 1 + \tfrac{10}{2} \times 1] = 8.33.$$

In the actual structure no relative rotational deflection actually exists between the two spans. Consequently, the deflection condition equation is

$$\theta_B + M_B \cdot \alpha_{BB} = 0.$$

Substitution in this equation, after multiplying all terms by EI, yields

$$127.8 + 8.33 M_B = 0,$$

from which

$$M_B = -15.3 \text{ ft·k}.$$

Problems

5-5. Find the reaction at C (see Fig. 5–22) using the general method. Determine the various deflection components by virtual work, and then check by the conjugate beam. [*Ans.:* 6.3 k up.]

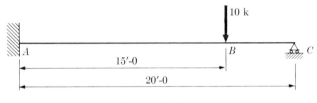

Figure 5–22

5-6. Determine the value of the reaction at B in Fig. 5–23. [*Ans.:* 3.5 k up.]

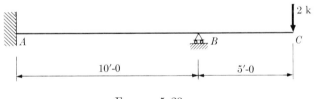

Figure 5–23

5-7. Find the reaction at C in Fig. 5–24. [*Ans.:* 3.5 k up.]

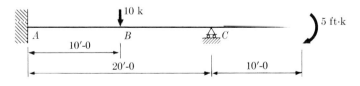

Figure 5–24

5-8. In the beam of Fig. 5–24 evaluate the external moment at A directly by the general method. [*Ans.:* 35 ft·k counterclockwise.]

5-9. In Fig. 5-25, evaluate the reaction at A by the general method. [*Ans.:* 4.4 k up.]

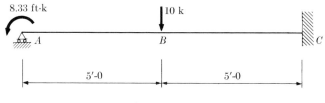

FIGURE 5-25

5-10. Find moments at A and B (see Fig. 5-26) by the general method. Use the conjugate beam method to find necessary deflection components. I is constant. [*Ans.:* $M_A = -21.6$ ft·k, $M_B = -16.9$ ft·k.]

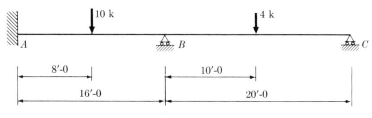

FIGURE 5-26

5-11. Given the fixed end beam AB of Fig. 5-27, with a constant moment of inertia. The moment M is applied as indicated. Using the general method, determine the expressions for the exterior end moments at A and B. [*Ans.:* $M_A = (Mb/L^2)(2a - b)$ clockwise, $M_B = (Ma/L^2)(2b - a)$ clockwise.]

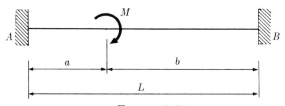

FIGURE 5-27

5-3 Analysis of articulated structures. Articulated indeterminate structures can be analyzed as easily by the general method as by any other procedure. (An alternate method is provided by Castigliano's second theorem and will be discussed in Chapter 6.) The application of the general method in the analysis of articulated structures will be demonstrated by several examples.

EXAMPLE 5–12. Find the stress T in the member AC of Fig. 5–28. All members are of steel, with cross-sectional areas of 1 in^2. The structure is cut back to a condition of determinateness and stability by cutting AC, arbitrarily selected as the redundant member. Any other member could have been taken as the redundant, but this would not give a direct solution for the stress in AC.

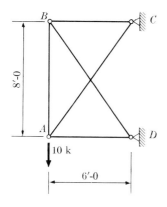
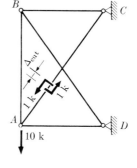

FIGURE 5–28 FIGURE 5–29

The procedure is first to find the sense and the magnitude of the relative movement of the ends of AC, at the cut, resulting from the strain in the other members as caused by the 10 k load. In other words, it is necessary to know the amount of the separation, or the slipping past each other, of the cut ends of the redundant member. This relative movement is evaluated by virtual work, and thus a pair of unit fictitious *tensile* forces are applied at the two cut ends. Figure 5–29 shows these unit fictitious forces in position.

By virtual work,

$$\Delta_{\text{cut}} = \sum \frac{uSL}{AE},$$

where S is the stress in any member caused by the real 10 k load and u is the stress in the same member caused by the pair of unit fictitious tensile forces. It will also be necessary to know the relative movement of the cut ends as caused by a pair of unit tensile real forces acting on the cut ends in the same manner as the fictitious forces shown in Fig. 5–29. This separation would be given by

$$\delta_{\text{cut}} = \sum \frac{u^2 L}{AE}.$$

In the actual structure the stress T in the redundant AC must be of sufficient magnitude to prevent any separation of the cut ends. The proper deflection condition equation is therefore

$$\Delta_{\text{cut}} + T \cdot \delta_{\text{cut}} = 0.$$

The two deflection components required can most easily be evaluated by an arrangement as shown in Table 5–3.

TABLE 5–3.

Member	Stress-S (k)	u (k)	L (ft)	uSL	u^2L
AB	$+10$	-0.8	8	-64	5.12
BC	$+7.5$	-0.6	6	-27	2.16
AD	0	-0.6	6	0	2.16
BD	-12.5	$+1.0$	10	-125	10.00
AC	0	$+1.0$	10	0	10.00
				-216	$+29.44$

(Since A and E are constant for the structure, they have been omitted from the table.) From this table,

$$AE\,\Delta_{\text{cut}} = -216 \quad \text{and} \quad AE\,\delta_{\text{cut}} = +29.44.$$

If the deflection condition equation is multiplied through by AE, the above values may be substituted and the result is

$$-216 + 29.44T = 0,$$

from which

$$T = +7.3 \text{ k}.$$

The positive sign indicates that the redundant stress T has the same sense as the 1 k forces assumed as acting on the ends adjacent to the cut, that is, a sense that will cause tension in the redundant.

Particular attention is called to the fact that unit fictitious compressive forces could have been assumed just as properly. The author prefers, however, to assume *tensile* fictitious forces for the reason that, in all problems in this text involving stresses in articulated structures, a positive sign has arbitrarily been used to designate a tensile axial stress in a member and a negative sign to indicate a compressive axial stress. When we solve for redundant stresses or reactions by the general method, a positive sign of an answer indicates that the particular redundant which the answer defines acts with the same sense as the unit fictitious action, which was previously assumed to act at the point of application of the redundant. Therefore, if unit fictitious *tensile* forces are always assumed as acting on the cut ends of a redundant member of an articulated structure, a positive answer for the stress in that redundant automatically signifies a tensile stress.

EXAMPLE 5–13. From Fig. 5–30 find the stresses T_a in member a, and T_d in member d, by the general method. Cross-sectional areas and the cantilever moment of inertia are shown in the illustration. E is constant at 30,000 k/in^2.

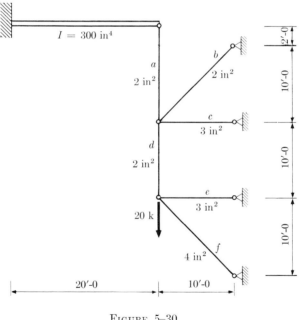

FIGURE 5–30

Assume that members a and d are cut. The structure is thus reduced to stable determinateness. If, while the 20 k load is acting on the structure, forces of magnitude T_a act on the two ends of a adjacent to the cut in a, and, at the same time, forces of magnitude T_d act on the two ends of d adjacent to the cut in d, then no relative movement of the two cut ends of a or of d can occur. The two deflection condition equations necessary for a solution are written to express zero relative movement of the ends adjacent to the two cuts. These are

$$\Delta_a + T_a \cdot \delta_{aa} + T_d \cdot \delta_{ad} = 0,$$

$$\Delta_d + T_a \cdot \delta_{da} + T_d \cdot \delta_{dd} = 0,$$

where Δ_a and Δ_d represent relative movements of the cut ends of members a and d, respectively, as caused by the real loads of 20 k and as shown in Fig. 5–31. If 1 k forces are assumed to act on the cut ends of member a so as to cause tension in that member, as in Fig. 5–32, then the relative movement of the cut ends of a is represented by δ_{aa} and the simultaneous relative movement of the cut ends of d is represented by δ_{da}. If, as in Fig. 5–33, 1 k forces are assumed to act on the cut ends of member d, the relative movement of the cut ends of this member is represented by δ_{dd} and the simultaneous relative movement of the cut ends of member a is represented by δ_{ad}.

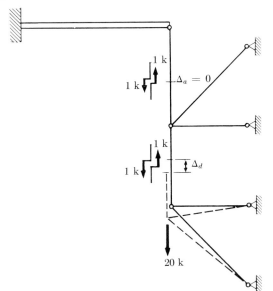

FIGURE 5–31

The cut-back structure, deflected by the 20 k load, is shown in Fig. 5–31. Unit fictitious forces, considered in the deflection computations to act as necessary, are shown in position adjacent to each of the cuts. Moments and axial stresses caused by the real load are represented in the general expressions for deflection by M and S, respectively. Either or both of these may be zero in the various parts of the structure. Moments and axial stresses as caused by the unit forces on the cut ends of member a are represented by m_a and u_a, respectively, and as caused by the unit forces on the cut ends of member d, by m_d and u_d. The expressions for the relative deflections of the cut ends, as caused by the real load of 20 k, are

$$E \Delta_a = \int m_a \frac{M\,dx}{I} + \sum u_a \frac{SL}{A} = 0 + 0,$$

$$E \Delta_d = \int m_d \frac{M\,dx}{I} + \sum u_d \frac{SL}{A} = 0 + \sum u_d \frac{SL}{A}.$$

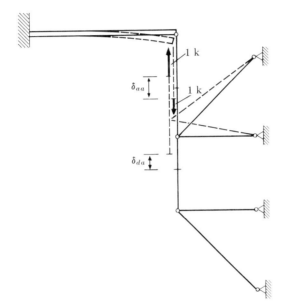

FIGURE 5-32

If unit real loads are applied to the cut ends of member a, the structure will deflect as shown in Fig. 5-32. The unit fictitious forces, shown in Fig. 5-31, are again considered to act in Fig. 5-32 as required for the computation of deflections. The relative deflections of the cut ends of members a and d are given by

$$E \delta_{aa} = \int m_a^2 \frac{dx}{I} + \sum u_a^2 \frac{L}{A},$$

$$E \delta_{da} = \int m_d m_a \frac{dx}{I} + \sum u_d u_a \frac{L}{A} = 0 + \sum u_d u_a \frac{L}{A}.$$

If unit real loads are applied to the cut ends of member d, the structure will deflect as shown in Fig. 5–33.

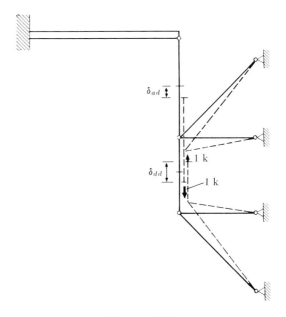

FIGURE 5–33

The unit fictitious forces of Fig. 5–31 are again considered to act as necessary. The relative deflections of the cut ends of members d and a are given by

$$E\,\delta_{dd} = \int m_d^2 \frac{dx}{I} + \sum u_d^2 \frac{L}{A} = 0 + \sum u_d^2 \frac{L}{A},$$

$$E\,\delta_{ad} = \int m_a m_d \frac{dx}{I} + \sum u_a u_d \frac{L}{A} = 0 + \sum u_a u_d \frac{L}{A}.$$

The terms in the above expressions for deflection, which do not involve a moment, are most easily evaluated by an arrangement of computations as shown in Tables 5–4 and 5–5.

ANALYSIS OF ARTICULATED STRUCTURES

TABLE 5-4.

Member	L (ft)	A (in^2)	L/A	S (k)	u_a (k)	u_d (k)
a	12	2	6	0	$+1$	0
b	14.14	2	7.07	0	-1.414	$+1.414$
c	10	3	3.33	0	$+1.0$	-1.0
d	10	2	5	0	0	$+1.0$
e	10	3	3.33	$+20$	0	-1.0
f	14.14	4	3.54	-28.28	0	$+1.414$

TABLE 5-5.

Member	$u_d \dfrac{SL}{A}$	$u_a^2 \dfrac{L}{A}$	$u_d^2 \dfrac{L}{A}$	$u_a u_d \dfrac{L}{A}$
a	—	$+6$	—	—
b	—	$+14.14$	$+14.14$	-14.14
c	—	$+3.33$	$+3.33$	-3.33
d	—	—	$+5.00$	—
e	-66.6	—	$+3.33$	—
f	-141.8	—	$+7.08$	—
Σ	-208.4	$+23.47$	$+32.88$	-17.47

$$\int m_a^2 \frac{dx}{I} = \int_0^{20} \frac{x^2 \, dx}{300} = \frac{8000 \times 144}{900} = 1280 \text{ ft·k}^2/\text{in}^2.$$

Note that the use of the number 144 in the numerator above is to adjust the units of the answer obtained by integration to be the same as the units of the summations in Table 5-5.

$$E \, \delta_{aa} = \int m_a^2 \frac{dx}{I} + \sum u_a^2 \frac{L}{A} = +1280 + 23 = +1303.$$

Substituting in the two deflection condition equations, after multiplying all terms by E, yields

$$0.0 + 1303 T_a - 17.47 T_d = 0,$$

$$-208.4 - 17.47 T_a + 32.88 T_d = 0.$$

Simultaneous solution of these two equations will give

$$T_a = +0.09 \text{ k},$$

$$T_d = +6.4 \text{ k}.$$

EXAMPLE 5-14. Outline the analysis for the reaction at B (see Fig. 5-34) and the stresses in the redundant bars a, b, and c, by the general method.

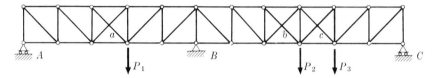

FIGURE 5-34

The structure is cut back and, under the action of the real loads, will deflect to give the various deflection components shown in Fig. 5-35. The dashed lines represent members in the deflected truss.

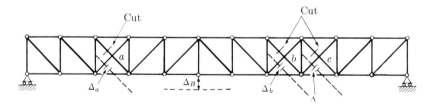

FIGURE 5-35

The expressions for the deflection components are

$$\Delta_B = \sum u_B S \frac{L}{AE}, \qquad \Delta_a = \sum u_a S \frac{L}{AE},$$

$$\Delta_b = \sum u_b S \frac{L}{AE}, \qquad \Delta_c = \sum u_c S \frac{L}{AE}. \tag{5-1}$$

If we assume that a pair of 1 k forces act on the ends of member a adjacent to the cut, the relative movements of cut ends in members a, b, and c, and the deflection at B resulting therefrom, will be

$$\delta_{Ba} = \sum u_B u_a \frac{L}{AE}, \qquad \delta_{aa} = \sum u_a^2 \frac{L}{AE},$$
$$\delta_{ba} = \sum u_b u_a \frac{L}{AE} = 0, \qquad \delta_{ca} = \sum u_c u_a \frac{L}{AE} = 0. \tag{5-2}$$

Transferring the pair of 1 k forces to the cut ends of member b yields the following deflections:

$$\delta_{Bb} = \sum u_B u_b \frac{L}{AE}, \qquad \delta_{bb} = \sum u_b^2 \frac{L}{AE},$$
$$\delta_{ab} = \sum u_a u_b \frac{L}{AE} = 0, \qquad \delta_{cb} = \sum u_c u_b \frac{L}{AE}. \tag{5-3}$$

If the pair of 1 k forces act on the cut ends of member c, the resulting deflections are

$$\delta_{Bc} = \sum u_B u_c \frac{L}{AE}, \qquad \delta_{cc} = \sum u_c^2 \frac{L}{AE},$$
$$\delta_{ac} = \sum u_a u_c \frac{L}{AE} = 0, \qquad \delta_{bc} = \sum u_b u_c \frac{L}{AE}. \tag{5-4}$$

Finally, if a 1 k force is considered to act down at B, the deflections are

$$\delta_{BB} = \sum u_B^2 \frac{L}{AE}, \qquad \delta_{aB} = \sum u_a u_B \frac{L}{AE},$$
$$\delta_{bB} = \sum u_b u_B \frac{L}{AE}, \qquad \delta_{cB} = \sum u_c u_B \frac{L}{AE}. \tag{5-5}$$

All the above deflection components having been evaluated, they are substituted in the following deflection condition equations:

$$\Delta_a + T_a \cdot \delta_{aa} + T_b \cdot \delta_{ab} + T_c \cdot \delta_{ac} + R_B \cdot \delta_{aB} = 0,$$
$$\Delta_b + T_a \cdot \delta_{ba} + T_b \cdot \delta_{bb} + T_c \cdot \delta_{bc} + R_B \cdot \delta_{bB} = 0,$$
$$\Delta_c + T_a \cdot \delta_{ca} + T_b \cdot \delta_{cb} + T_c \cdot \delta_{cc} + R_B \cdot \delta_{cB} = 0,$$
$$\Delta_B + T_a \cdot \delta_{Ba} + T_b \cdot \delta_{Bb} + T_c \cdot \delta_{Bc} + R_B \cdot \delta_{BB} = 0. \tag{5-6}$$

The simultaneous solution of these equations will give the desired values of the redundants.

The solution of simultaneous equations often causes some difficulty. Various methods are available, as, for example, iteration or determinants. Each of these, in certain cases, may be used to advantage, but in other cases may cause considerable difficulty. The following simple tabular method will usually, however, give the desired solution with the minimum chance of error.

To illustrate, consider the following equations:

(1) $\quad 2x + 3y - z - 15 = 0,$

(2) $\quad 3x - 2y + 4z - 2 = 0,$

(3) $\quad x - 5y - 2z + 20 = 0.$

These equations are entered and solved simultaneously in Table 5-6.

TABLE 5-6.

Equation	Operation	x	y	z	Constant	Check equation	Check operation
(1)		+2	+3	−1	−15	−11	
(2)		+3	−2	+4	−2	+3	
(3)		+1	−5	−2	+20	+14	
(4)	4(1) + (2)	+11	+10	—	−62	−41	−41
(5)	(2) + 2(3)	+5	−12	—	+38	+31	+31
(6)	1.2(4) + (5)	+18.2	—	—	−36.4	−18.2	−18.2

The value entered in the "Check equation" column is the algebraic summation of the coefficients and the constant term for each equation. The ways by which previously obtained equations are combined to obtain new equations with a reduced number of terms is indicated in the column entitled "Operation." Each operation is checked immediately by comparing the value in the "Check equation" column with the value in the "Check operation" column. If the values are equal, no error has been made in the execution of the corresponding operation. The entry in the "Check operation" column is obtained by performing the indicated operation on corresponding values in the "Check equation" column. For example, to check equation (4), the coefficients of x and y and the constant term are added algebraically and the number -41 is entered in the "Check equation" column. Then the indicated operation, specifically, $4(1) + (2)$, is performed on the "Check equation" values for equations (1) and (2). The result is $4(-11) + 3 = -41$. This value is entered in the "Check equation" column opposite equation (4), and the new equation is correct except for possible compensating errors.

Solving equation (6), the result is

$$x = +2.$$

The above result should be substituted in both equations (4) and (5) to determine if the same value of y is obtained from each. In this case, $y = +4$. Finally, the values of x and y should be substituted in equations (1), (2), and (3) to be certain that the same value of z will result from each of these. The value of z in this problem is $+1$.

The advantage of this method for solving simultaneous equations is that each operation is checked immediately after it is performed and, presumably, any errors are immediately detected. A computing machine should be used for best results, although a computing machine will not, of course, give greater accuracy than can be justified by the accuracy of the numbers used in writing the simultaneous equations. In the preceding illustrative solution, all coefficients and constant terms of the initial equations are assumed to be exact; that is, an approximate number was not rounded off to obtain coefficients or constant terms. In most cases in practice, of course, these will be approximate numbers. A brief review of the few principles involved in determining the accuracy of the results of computations involving approximate numbers might be helpful.

The reader is probably aware that any number is composed of digits, a digit being any one of the ten Arabic numerals. Each digit in a number, unless it is used solely to indicate the location of the decimal point, is a significant digit or figure. Consequently, the digits 1 through 9 are always significant. The digit 0 is significant if it appears in a number between any two of the digits 1 through 9, or if it appears to the right of the decimal point at the end of a number. Consider, for example, the following numbers: (a) 3206, (b) 0.0326, (c) 32.06, (d) 32.60, and (e) 32,600. In (a) there are four significant figures; in (b) the last three digits are significant; (c) has four significant figures; and in (d) there are four significant figures. In (e) the two zeros may or may not be significant. Actually, from the way the number is written, it is impossible to determine whether it is accurate to the nearest unit, or ten units, or one hundred units. Most readers probably would assume that the two zeros serve to indicate the position of the decimal point and are not significant. If the ambiguity is to be completely removed, the number should be expressed in powers of ten or with subdigits. Thus, 32.600×10^3 definitely means that all five digits are significant or 326_{00} definitely indicates that the last two digits are uncertain.

Approximate numbers are rounded off to reduce the number of significant figures in accordance with the following rule: If the digits to be dropped represent a value less than 5 in the first discarded place, the digit in the last retained place is not changed; if they represent a value greater than 5, the digit in the last retained place is increased by one; if the portion to be dropped is 5 followed by zeros, the digit in the last retained place should have the nearest even value. In accordance with this rule, and rounding to three significant figures,

47.352 will be written as 47.4,

47.350 will be written as 47.4,

47.450 will be written as 47.4,

47.349 will be written as 47.3.

In accordance with the last paragraph above, the approximate number 3206 may have an actual value from and including 3205.5 up to and including 3206.5. The approximate number 0.0326 may have a true value from and including 0.03255 up to and including 0.03265. The true value of 3205 might be anything from, but not including, 3204.5 up to, but not including, 3205.5.

When two or more numbers are multiplied together, the accuracy of the product, that is, the number of significant figures in the product, can be no greater than the least accurate of any one of the numbers. Thus, to write that

$$32.736 \times 1.421 = 46.517856$$

is incorrect because an accuracy is implied which does not exist. The answer should be written as 46.52. The same rule applies to division of approximate numbers, to roots, and to powers. To illustrate, several operations are indicated below with the incorrect and correct form of the answer:

Incorrect	Correct
$24.672 \div 13.79 = 1.7891225$	$= 1.789$
$176^3 = 5,451,776$	$= 5,450,000$ or 545×10^4
$\sqrt{47.28} = 6.87604$	$= 6.876$

When two or more approximate numbers are to be added or subtracted, the precision, that is, the number of significant figures to the right of the decimal point, must be considered. The sum of two or more approximate numbers, or the difference between two approximate numbers, is not correct to more places to the right of the decimal than the least precise of the numbers added or subtracted.

Thus, for the following,

$$\begin{array}{r} 17.623 \\ 298.1 \\ \underline{47.29} \\ 363.013 \end{array} :$$

Answer = 363.0,

or the extra digits to the right of the decimal may be dropped before adding, to give

$$\begin{array}{r} 17.6 \\ 298.1 \\ \underline{47.3} \\ 363.0 \end{array}$$

The accuracy of the final results obtained from simultaneous equations may be determined by applying the above rules to each step of the solution.

Problems

5–15. Find the reaction at E (see Fig. 5–36) and the stress in the redundant BC by the general method. All members have cross-sectional areas of 4 in^2 and E is 30,000 k/in^2. [*Ans.:* R_E = 2.8 k up, BC = 5.3 k compression.]

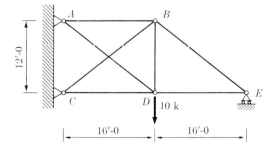

Figure 5–36

5-16. Find the stresses in the redundants BC and BE of Fig. 5-37 by the general method. The cross-sectional areas of the various members, in in^2, are shown circled on the illustration. [Ans.: $BC = 24.2$ k compression, $BE = 18.3$ k tension.]

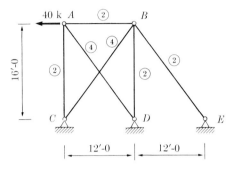

FIGURE 5-37

5-17. Find the stress in member CD of Fig. 5-38. The cantilever beam is assumed to be aluminum, with $E = 10,000$ k/in^2; the vertical CD is a wire rope, with $E = 20,000$ k/in^2; and members AC and BC are steel, with $E = 30,000$ k/in^2. Cross-sectional areas and the I of the beam are noted on the illustration. [Ans.: $CD = 0.54$ k tension.]

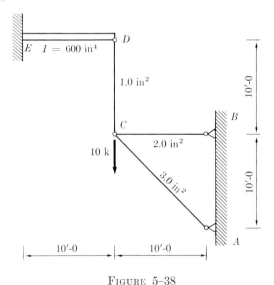

FIGURE 5-38

Note that additional articulated structures, which may be solved by the general method, are included in Chapter 6.

5–4 Analysis of continuous frames. The general method probably would never be selected for the analysis of a multi-story continuous frame. A solution can be more easily effected by moment distribution or by slope deflection. If, however, the analysis concerns a single-story continuous frame, either single or multiple bay, a combination of the conjugate structure and the general method will be found to be very effective. This is especially true when the final members of the frame have varying moments of inertia, since the effects of progressively varying member sizes, as the design proceeds, can be easily introduced by making adjustments to the first analysis. The method can also be used to advantage in the case of single-bay, single-story frames of unusual shape.

EXAMPLE 5–18. Find values for the horizontal and vertical reaction components at A (see Fig. 5–39) by the general method. Use virtual work to find the deflections.

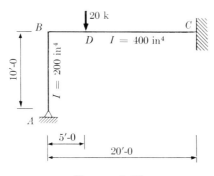

FIGURE 5–39

The structure is cut back as shown in Fig. 5–40. Unit loads, considered to be either real or fictitious as required, and assumed to act simultaneously with the real load or with each other as necessary, are shown at A of this figure. The various moments existing in the cut-back struc-

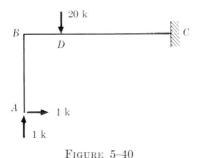

FIGURE 5–40

ture due to the several loads are shown in Table 5–7. A moment causing tension on the outside of the frame is considered to be positive.

TABLE 5–7.

Section	$x = 0$ at	x increasing	M (ft·k)	m_h (ft·k)	m_v (ft·k)
AB	A	A to B	—	$+x$	—
BD	B	B to D	—	$+10$	$-x$
DC	D	D to C	$+20x$	$+10$	$-5 - x$

The deflection condition equations are

$$\Delta_h + H\delta_{hh} + V\delta_{hv} = 0,$$
$$\Delta_v + H\delta_{vh} + V\delta_{vv} = 0. \tag{5-7a}$$

By virtual work,

$$E\,\Delta_h = \int m_h M \frac{dx}{I} = \int_0^{15} (+10)(+20x)\frac{dx}{400} = +56.25,$$

$$E\,\Delta_v = \int m_v M \frac{dx}{I} = \int_0^{15} (-5 - x)(+20x)\frac{dx}{400} = -84.38,$$

$$E\,\delta_{hh} = \int m_h^2 \frac{dx}{I} = \int_0^{10} x^2 \frac{dx}{200} + \int_0^5 (+10)^2 \frac{dx}{400} + \int_0^{15} (+10)^2 \frac{dx}{400}$$
$$= +6.667,$$

$$E\,\delta_{vv} = \int m_v^2 \frac{dx}{I} = \int_0^5 (-x)^2 \frac{dx}{400} + \int_0^{15} (-5 - x)^2 \frac{dx}{400} = +6.667,$$

$$E\,\delta_{vh} = E\,\delta_{hv} = \int m_h m_v \frac{dx}{I} = \int_0^5 (+10)(-x)\frac{dx}{400}$$
$$+ \int_0^{15} (+10)(-5 - x)\frac{dx}{400} = -5.00.$$

Multiplying each term of the condition equations by E and substituting the above values yields

$$+56.25 + 6.667H - 5.00V = 0,$$
$$-84.38 - 5.00H + 6.667V = 0. \tag{5-7b}$$

A simultaneous solution will result in

$$H = +2.4 \text{ k} \quad \text{and} \quad V = +14.5 \text{ k}.$$

The positive signs of the answers indicate that the true sense of H and of V is the same as the corresponding 1 k loads assumed at A.

EXAMPLE 5–19. Find the horizontal, vertical, and moment reaction components at A (see Fig. 5–41) by the general method. Use the conjugate structure for finding deflection components. Moment of inertia is constant. Draw the moment diagram, showing all ordinates on the tension side of members.

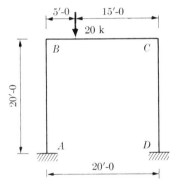

FIGURE 5–41

The structure is cut back by removing the support at A. The required deflection condition equations express the fact that, under the action of the 20 k load and the three reaction components at A, the net horizontal, vertical, and rotational deflections at A must be zero. These equations are

$$\Delta_h + H\,\delta_{hh} + V\,\delta_{hv} + M\,\delta'_{hm} = 0,$$
$$\Delta_v + H\,\delta_{vh} + V\,\delta_{vv} + M\,\delta'_{vm} = 0, \tag{5-8a}$$
$$\theta + H\alpha'_{mh} + V\alpha'_{mv} + M\alpha_{mm} = 0.$$

The conjugate structure for the evaluation of the deflection components of A, resulting from the action of the 20 k load, is shown in Fig. 5–42.

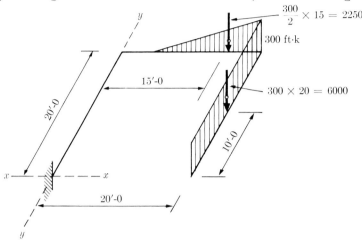

FIGURE 5–42

In the computations which follow, relating to Fig. 5–42 and the next three conjugate structures, the values of the various elastic loads have, for convenience, been divided by 1000.

$$EI \, \Delta_h = 2.25 \times 20 + 6 \times 10 = +105,$$

$$EI \, \Delta_v = 2.25 \times 15 + 6 \times 20 = +153.75,$$

$$EI \theta = 2.25 + 6 = +8.25.$$

If a 1 k horizontal force is assumed to act alone and to the right at A, the conjugate structure of Fig. 5–43 results.

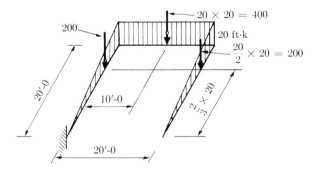

FIGURE 5–43

The deflection components of A are found to be:

$$EI\delta_{hh} = 2 \times 0.2 \times \tfrac{2}{3} \times 20 + 0.4 \times 20 = +13.33,$$

$$EI\delta_{vh} = 0.4 \times 10 + 0.2 \times 20 = +8.00,$$

$$EI\alpha'_{mh} = 2 \times 0.2 + 0.4 = +0.80.$$

With a 1 k force acting down at A, the conjugate structure will be as shown in Fig. 5–44.

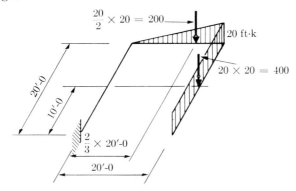

FIGURE 5–44

From this structure,

$$EI\delta_{hv} = 0.2 \times 20 + 0.4 \times 10 = +8.00,$$

$$EI\delta_{vv} = 0.2 \times \tfrac{2}{3} \times 20 + 0.4 \times 20 = +10.67,$$

$$EI\alpha'_{mv} = 0.2 + 0.4 = +0.60.$$

Finally, if a 1 ft·k counterclockwise couple acts alone at A, the resulting conjugate structure will be as shown in Fig. 5–45. From this structure, the

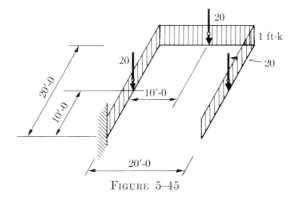

FIGURE 5–45

deflection components of point A are found to be

$$EI\delta'_{hm} = 2 \times 0.02 \times 10 + 0.02 \times 20 = +0.80,$$

$$EI\delta'_{vm} = 0.02 \times 20 + 0.02 \times 10 = +0.60,$$

$$EI\alpha_{mm} = 3 \times 0.02 = +0.06.$$

Note that Maxwell's reciprocal deflection theorem may be used to check several of the above values.

Substitution in the deflection condition equations, after all terms have been multiplied by EI, yields

$$+ 105 + 13.33H + 8.00V + 0.80M = 0,$$

$$+ 153.75 + 8.00H + 10.67V + 0.60M = 0, \qquad (5\text{-}8b)$$

$$+ 8.25 + 0.80H + 0.60V + 0.06M = 0.$$

Simultaneous solution will result in

$$H = +1.9 \text{ k}, \qquad V = -15.3 \text{ k}, \qquad M = -9.6 \text{ ft·k}.$$

The positive sign in the first answer above indicates that H actually acts with the same sense as the 1 k horizontal force assumed at A. The negative signs for V and M indicate that each is actually opposite in sense to the vertical 1 k force and the 1 ft·k couple assumed at A. The bending moment diagram is shown in Fig. 5-46.

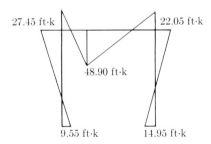

FIGURE 5-46

EXAMPLE 5-20. Determine the magnitudes and senses of the horizontal, vertical, and moment reaction components at A (see Fig. 5-47) by the general method. Use the conjugate structure to find deflections.

5–4]   ANALYSIS OF CONTINUOUS FRAMES   219

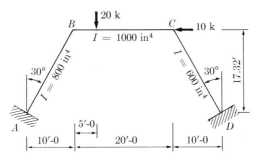

FIGURE 5–47

The cut-back structure is shown in Fig. 5–48, with the assumed senses of the unit forces and couple acting at A.

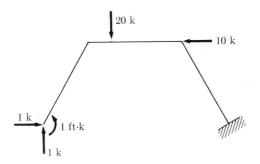

FIGURE 5–48

The required deflection condition equations are

$$\Delta_h + H\delta_{hh} + V\delta_{hv} + M\delta'_{hm} = 0,$$

$$\Delta_v + H\delta_{vh} + V\delta_{vv} + M\delta'_{vm} = 0, \qquad (5\text{--}9)$$

$$\theta + H\alpha'_{mh} + V\alpha'_{mv} + M\alpha_{mm} = 0.$$

The deflection components of point A as caused by the real loads are computed from Fig. 5–49:

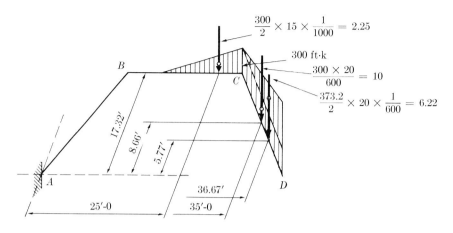

FIGURE 5-49

$$E \Delta_h = 2.25 \times 17.32 + 10 \times 8.66 + 6.22 \times 5.77 = +161.5,$$

$$E \Delta_v = 2.25 \times 25 + 10 \times 35 + 6.22 \times 36.67 = +634.25,$$

$$E\theta = 2.25 + 10 + 6.22 = +18.47.$$

For the 1 k horizontal force at A,

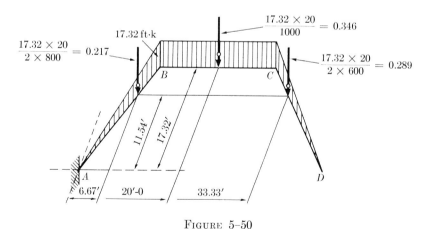

FIGURE 5-50

$$E\delta_{hh} = 0.217 \times 11.54 + 0.346 \times 17.32 + 0.289 \times 11.54 = +11.84,$$

$$E\delta_{vh} = 0.217 \times 6.67 + 0.346 \times 20.00 + 0.289 \times 33.33 = +18.00,$$

$$E\alpha'_{mh} = 0.217 + 0.346 + 0.289 = +0.852.$$

For the 1 k vertical force at A,

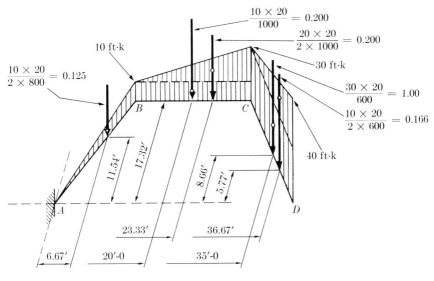

Figure 5–51

$$E\delta_{hv} = 0.125 \times 11.54 + (0.200 + 0.200)\,17.32 + 1.00 \times 8.66$$
$$+\, 0.166 \times 5.77 = +17.99,$$

$$E\delta_{vv} = 0.125 \times 6.67 + 0.200\,(20.00 + 23.33) + 1.00 \times 35.00$$
$$+\, 0.166 \times 36.67 = +50.59,$$

$$E\alpha'_{mv} = 0.125 + 0.200 + 0.200 + 0.166 + 1.000 = +1.691.$$

With a 1 ft·k counterclockwise couple acting at A on the cut-back structure (see Fig. 5–52),

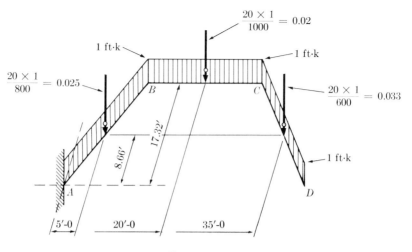

Figure 5-52

$E\delta'_{hm} = +0.025 \times 8.66 + 0.020 \times 17.32 + 0.033 \times 8.66 = +0.85,$

$E\delta'_{vm} = +0.025 \times 5.0 + 0.020 \times 20.0 + 0.033 \times 35.0 = +1.68,$

$E\alpha_{mm} = +0.025 + 0.020 + 0.033 = +0.078.$

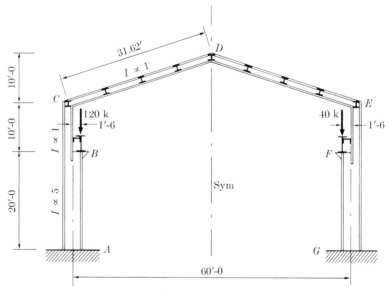

Figure 5-53

When all terms of the deflection condition equations have been multiplied by E, a substitution of the above values and a subsequent simultaneous solution will give

$$H = +13.5 \text{ k}, \qquad V = -16.8 \text{ k}, \qquad M = -20.5 \text{ ft·k}.$$

EXAMPLE 5–21. Find the reaction components at A in the stepped-column gable bent (see Fig. 5–53) as caused by the indicated eccentric crane loads. Use the conjugate structure with the general method. Note that, as indicated on the illustration, the moment of inertia of the heavy lower column section is five times the moment of inertia of the remainder of the bent.

The cut-back structure, with equivalent real moments applied and with the assumed unit forces and couple at A, is shown in Fig. 5–54. The condition equations again express the fact that no deflection will occur at A.

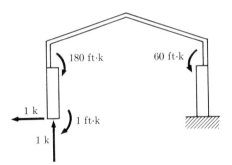

FIGURE 5–54

As in Examples 5–19 and 5–20, these equations are

$$\Delta_h + H\delta_{hh} + V\delta_{hv} + M\delta'_{hm} = 0,$$

$$\Delta_v + H\delta_{vh} + V\delta_{vv} + M\delta'_{vm} = 0, \qquad (5\text{–}10)$$

$$\theta + H\alpha'_{mh} + V\alpha'_{mv} + M\alpha_{mm} = 0.$$

The conjugate structure for evaluating the deflection components of A on the cut-back structure due to the two real couples is shown in Fig. 5–55:

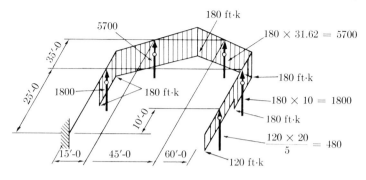

FIGURE 5-55

$E \Delta_h = 480 \times 10 + 2 \times 1800 \times 25 + 2 \times 5700 \times 35 = +493{,}800,$

$E \Delta_v = 5700 \,(15 + 45) + 1800 \times 60 + 480 \times 60 = +478{,}800,$

$E \theta = 480 + 2 \times 1800 + 2 \times 5700 = +15{,}480.$

For the 1 k horizontal force acting to the left at A (see Fig. 5-56),

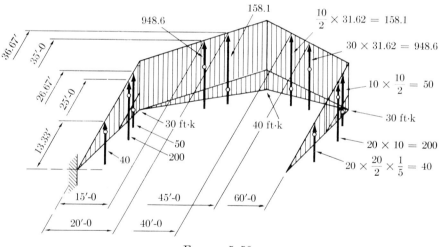

FIGURE 5-56

$E \,\delta_{hh} = 2(40 \times 13.33 + 200 \times 25.00 + 50 \times 26.67 + 948.6 \times 35.00 \\ + 158.1 \times 36.67) = +91{,}730,$

$E \,\delta_{vh} = (2 \times 158.1 + 2 \times 948.6)30 + 60(50 + 200 + 40) = +83{,}800,$

$E \alpha'_{mh} = 2(40 + 200 + 50 + 948.6 + 158.1) = +2793.$

For the 1 k vertical force acting up at A (see Fig. 5-57),

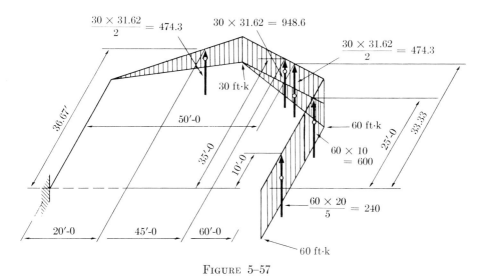

Figure 5-57

$$E\,\delta_{hv} = 474.3 \times 36.67 + 948.6 \times 35 + 474.3 \times 33.33 + 600 \times 25$$
$$+ 240 \times 10 = +83\,800$$

$$E\,\delta_{vv} = 474.3 \times 20 + 948.6 \times 45 + 474.3 \times 50 + 60(600 + 240)$$
$$= +126\,290,$$

$$E\alpha'_{mv} = 474.3 + 948.6 + 474.3 + 600 + 240 = +2737.$$

For the 1 ft·k counterclockwise couple acting at A (see Fig. 5–58),

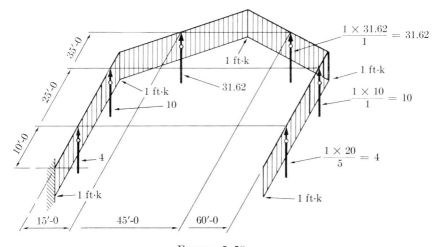

Figure 5-58

$$E \delta'_{hm} = 2(4 \times 10 + 10 \times 25 + 31.62 \times 35) = +2793,$$

$$E \delta'_{vm} = 2 \times 31.62 \times 30 + (10 + 4)60 = +2737,$$

$$E \alpha_{mm} = 2(4 + 10 + 31.62) = +91.24.$$

When all terms of the condition equations have been multiplied by E, the above values are substituted therein and a subsequent simultaneous solution will give the following answers:

$$H = -3.14 \text{ k}, \qquad V = -0.33 \text{ k}, \qquad M = -63.7 \text{ ft·k}.$$

EXAMPLE 5-22. It is required to design a two-span continuous rigid frame to support a uniform live load of 2 k/ft on the horizontal projection (see Fig. 5-59). The frame, as ultimately designed, will have a varying

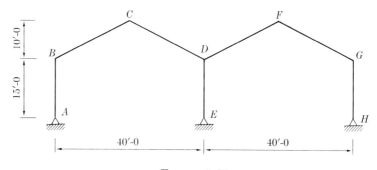

FIGURE 5-59

moment of inertia. The required variations of the moments of inertia are unknown and, consequently, a first analysis is to be made on the assumption of a uniform moment of inertia. The results of this first analysis will be used to proportion the first trial frame with nonprismatic members.

The frame is first cut back to a statically determinate structure as shown in Fig. 5-60. Deflections caused by the 2 k/ft load are as shown by the dotted outline and as labeled. There are 1 k forces applied individually and successively at A, first horizontally and then vertically, and then horizontally at E. These loads, and the resulting deflections, are shown in Fig. 5-61(a), (b), and (c).

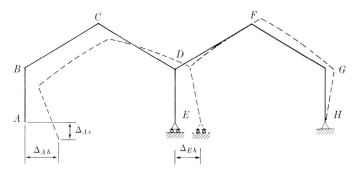

FIGURE 5–60

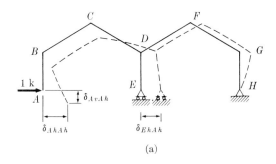

(a)

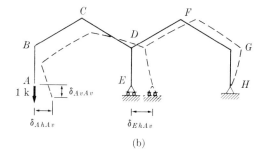

(b)

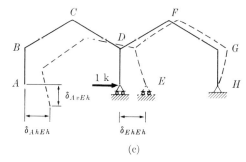

(c)

FIGURE 5–61

The three required deflection condition equations state that in the actual loaded structure the horizontal deflection at A, the vertical deflection at A, and the horizontal deflection at E are zero when the redundant reactions H_A, V_A, and H_E are acting. These redundant reactions necessarily are assumed to act in the same direction as the unit forces shown in Fig. 5-61. The three condition equations are

$$H_A\,\delta_{AhAh} + V_A\,\delta_{AhAv} + H_E\,\delta_{AhEh} + \Delta_{Ah} = 0,$$

$$H_A\,\delta_{AvAh} + V_A\,\delta_{AvAv} + H_E\,\delta_{AvEh} + \Delta_{Av} = 0, \qquad (5\text{--}11)$$

$$H_A\,\delta_{EhAh} + V_A\,\delta_{EhAv} + H_E\,\delta_{EhEh} + \Delta_{Eh} = 0.$$

Since the frame as finally designed will be composed of members with varying moments of inertia, it is advisable to divide the structure into segments (see Fig. 5-62). The segments in the columns are 5 ft long and,

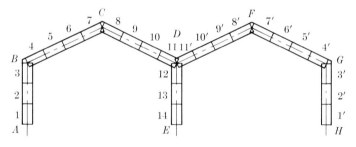

FIGURE 5-62

in the sloping girders, 5.59 ft long. Note that if the frame were to be fabricated with prismatic members, the use of segments would entail a great deal of unnecessary computation. It would be much easier, in this case, to consider the entire elastic load for each member as a single concentration acting through the centroid of the elastic load on that member.

The bending moments at the centers of the girder segments resulting from the 2 k/ft load acting on the determinate structure of Fig. 5-60 are shown in Table 5-8, as well as the relative values of the elastic weights of these segments. The actual value of the elastic weight for each segment would be $M\,\Delta S/EI$, where M is the moment at the center of each segment, ΔS is the segment length, I is the moment of inertia at the center of each segment, and E is the modulus of elasticity. Since, in the present case, E and I are constant, the elastic load can be used as the relative value $M\,\Delta S$. The vertical reaction at E in Fig. 5-60 is 160 k.

Table 5-8.

Segment	x_A	$M = \dfrac{2x_A^2}{2}$	$M = \dfrac{2x_A^2}{2} - 160(x_A - 40)$	ΔS	$M\,\Delta S$
4	2.5	6.25		5.59	35
5	7.5	56.25			314
6	12.5	156.25			873
7	17.5	306.25			1712
8	22.5	506.25			8230
9	27.5	756.25			4227
10	32.5	1056.25			5904
11	37.5	1406.25			7861
11'	42.5		1806.25 — 400 = 1406.25		7861
10'	47.5		2256.25 — 1200 = 1056.25		5904
9'	52.5		2756.25 — 2000 = 756.25		4227
8'	57.5		3306.25 — 2800 = 506.25		2830
7'	62.5		3906.25 — 3600 = 306.25		1712
6'	67.5		4556.25 — 4400 = 156.25		873
5'	72.5		5256.25 — 5200 = 56.25		314
4'	77.5		6006.25 — 6000 = 6.25		35

Table 5-9.

Segment	x_E	y_E	$M\,\Delta S$	$M\,\Delta S \cdot x_E$	$M\,\Delta S \cdot y_E$	$M\,\Delta S(y_E - 15)$
11'	2.5	16.25	7,861	19,653	127,741	9,826
10'	7.5	18.75	5,904	44,280	110,700	22,140
9'	12.5	21.25	4,227	52,837	89,824	26,418
8'	17.5	23.75	2,830	49,525	67,212	24,762
7'	22.5	23.75	1,712	38,520	40,660	14,980
6'	27.5	21.25	873	24,007	18,551	5,456
5'	32.5	18.75	314	10,205	5,887	1,177
4'	37.5	16.25	35	1,312	568	44
			23,756	240,339	461,143	104,803

$$R_H = \frac{240{,}339}{40} = 6008, \qquad \Delta_{Eh} = +461{,}143,$$

$$\Delta_{Dh} = 104{,}803 + 6008 \times 15 = +194{,}923,$$

$$\theta_D = 23{,}756 - 6008 = +17{,}748.$$

The conjugate structure for span *EH* of Fig. 5–60, shown in Fig. 5–63, is loaded with the elastic loads computed in Table 5–8 for segments 11' through 4'. The deflection components Δ_{Eh}, Δ_{Dh}, and θ_D are computed in Table 5–9.

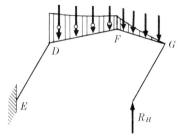

FIGURE 5–63

Vertical and horizontal deflection components of *A* (Δ_{Av} and Δ_{Ah}) in Fig. 5–60 are found from the conjugate structure of Fig. 5–64.

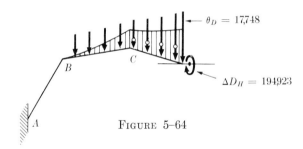

FIGURE 5–64

TABLE 5–10.

Segment	x_A	y_A	$M \Delta S$	$M \Delta S \cdot x_A$	$M \Delta S \cdot y_A$
4	2.5	16.25	35	87	568
5	7.5	18.75	314	2,355	5,887
6	12.5	21.25	873	10,912	18,551
7	17.5	23.75	1,712	29,960	40,660
8	22.5	23.75	2,830	63,675	67,212
9	27.5	21.25	4,227	116,242	89,824
10	32.5	18.75	5,904	191,880	110,700
11	37.5	16.25	7,861	294,787	127,741
				709,898	461,143

$$\Delta_{Av} = +709{,}898 + 17{,}748 \times 40 = +1{,}419{,}818,$$

$$\Delta_{Ah} = +461{,}143 + 17{,}748 \times 15 + 194{,}923 = +922{,}286.$$

Deflection components of A and E resulting from the 1 k horizontal load at A, as shown in Fig. 5–61(a), are next computed. The relative elastic loads for this loading, which are symmetrical about the vertical centerline, are computed for the left half of the frame in Table 5–11.

TABLE 5–11.

Segment	$M = y_A$	ΔS	$M \Delta S$
1	2.5	5.0	12.5
2	7.5	5.0	37.5
3	12.5	5.0	62.5
4	16.25	5.59	90.8
5	18.75	5.59	104.8
6	21.25	5.59	118.8
7	23.75	5.59	132.8
8	23.75	5.59	132.8
9	21.25	5.59	118.8
10	18.75	5.59	104.8
11	16.25	5.59	90.8

The conjugate structure for the right span is similar to that shown in Fig. 5–63, but is loaded with elastic loads as computed in Table 5–11. The necessary computations are shown in Table 5–12.

TABLE 5–12.

Segment	x_E	y_A	$M \Delta S$	$M \Delta S \cdot x_E$	$M \Delta S \cdot y_A$	$M \Delta S(y_A - 15)$
1'	40.0	2.5	12.5	500	31	−156
2'	40.0	7.5	37.5	1,500	281	−281
3'	40.0	12.5	62.5	2,500	781	−156
4'	37.5	16.25	90.8	3,405	1,475	+113
5'	32.5	18.75	104.8	3,406	1,965	+393
6'	27.5	21.25	118.8	3,267	2,524	+742
7'	22.5	23.75	132.8	2,988	3,154	+1,162
8'	17.5	23.75	132.8	2,324	3,154	+1,162
9'	12.5	21.25	118.8	1,485	2,524	+742
10'	7.5	18.75	104.8	786	1,965	+393
11'	2.5	16.25	90.8	227	1,475	+113
			1,006.9	22,388	19,329	4,227

$$R'_H = \frac{22{,}388}{40} = 559.7, \qquad \delta_{EhAh} = +19{,}329,$$

$$\delta_{DhAh} = 4227 + 559.7 \times 15 = +12{,}623,$$

$$\alpha_D = 1006.9 - 559.7 = +447.2.$$

The conjugate structure for the left span is similar to Fig. 5-64 and the elastic loads along the girders are similar to those used in Table 5-12, with the addition of the concentration α_D and the couple δ_{DhAh} at the free end of the conjugate structure. The computations to obtain δ_{AvAh} and δ_{AhAh} are given in Table 5-13.

TABLE 5-13.

Segment	x_A	y_A	$M\,\Delta S$	$M\,\Delta S \cdot x_A$	$M\,\Delta S \cdot y_A$
1	0	2.5	12.5	—	31
2	0	7.5	37.5	—	281
3	0	12.5	62.5	—	781
4	2.5	16.25	90.8	227	1,475
5	7.5	18.75	104.8	786	1,965
6	12.5	21.25	118.8	1,485	2,524
7	17.5	23.75	132.8	2,324	3,154
8	22.5	23.75	132.8	2,988	3,154
9	27.5	21.25	118.8	3,267	2,524
10	32.5	18.75	104.8	3,406	1,965
11	37.5	16.25	90.8	3,405	1,475
			1,006.9	17,888	19,329

$$\delta_{AvAh} = +17{,}888 + 447.2 \times 40 = +35{,}776,$$

$$\delta_{AhAh} = +19{,}329 + 447.2 \times 15 + 12{,}623 = +38{,}660.$$

A load of 1 k applied vertically at A, as in Fig. 5-61(b), will result in the deflections shown therein. The relative elastic loads are computed in Table 5-14 for the left half of the frame; those for the right half of the frame are similar.

TABLE 5-14.

Segment	$M = x_A$	ΔS	$M\,\Delta S$
4	2.5	5.59	14.0
5	7.5	5.59	41.9
6	12.5	5.59	69.9
7	17.5	5.59	97.8
8	22.5	5.59	125.8
9	27.5	5.59	153.7
10	32.5	5.59	181.7
11	37.5	5.59	209.6

The conjugate structure to determine δ_{EhAv} is similar to that of Fig. 5-63, and the necessary computations are given in Table 5-15.

TABLE 5-15.

Segment	x_E	y_A	$M\,\Delta S$	$M\,\Delta S \cdot x_E$	$M\,\Delta S \cdot y_A$	$M\,\Delta S(y_A - 15)$
4'	37.5	16.25	14.0	525	227	17
5'	32.5	18.75	41.9	1,362	786	157
6'	27.5	21.25	69.9	1,922	1,485	437
7'	22.5	23.75	97.8	2,201	2,323	856
8'	17.5	23.75	125.8	2,202	2,988	1,101
9'	12.5	21.25	153.7	1,921	3,266	961
10'	7.5	18.75	181.7	1,363	3,407	681
11'	2.5	16.25	209.6	524	3,406	262
			894.4	12,020	17,888	4,472

$$R'_H = \frac{12,020}{40} = 300.5, \qquad \delta_{EhAv} = +17,888,$$

$$\delta_{DhAv} = +4472 + 300.5 \times 15 = +8979,$$

$$\alpha_D = 894.4 - 300.5 = +593.9.$$

The deflection components δ_{AvAv} and δ_{AhAv} are computed (see Table 5-16) with a conjugate structure similar to Fig. 5-64. This is loaded with the elastic loads of Table 5-15 computed for segments 4 through 11, with the addition of a couple at the free end equal to δ_{DhAv}, as well as a concentration equal to α_D.

TABLE 5-16.

Segment	x_A	y_A	$M \Delta S$	$M \Delta S \cdot x_A$	$M \Delta S \cdot y_A$
4	2.5	16.25	14.0	35	227
5	7.5	18.75	41.9	314	786
6	12.5	21.25	69.9	874	1,485
7	17.5	23.75	97.8	1,712	2,323
8	22.5	23.75	125.8	2,831	2,988
9	27.5	21.25	153.7	4,227	3,266
10	32.5	18.75	181.7	5,905	3,407
11	37.5	16.25	209.6	7,860	3,406
				23,758	17,888

$$\delta_{AvAv} = +23{,}758 + 593.9 \times 40 = +47{,}514,$$

$$\delta_{AhAv} = +17{,}888 + 593.9 \times 15 + 8979 = +35{,}776.$$

A load of 1 k acting horizontally at E, as shown in Fig. 5–61(c), will result in the relative elastic loads computed in Table 5–17.

TABLE 5-17.

Segment	$M = y_A$	ΔS	$M \Delta S$
14	2.5	5.0	12.5
13	7.5	5.0	37.5
12	12.5	5.0	62.5
11′	16.25	5.59	90.8
10′	18.75	5.59	104.8
9′	21.25	5.59	118.8
8′	23.75	5.59	132.8
7′	23.75	5.59	132.8
6′	21.25	5.59	118.8
5′	18.75	5.59	104.8
4′	16.25	5.59	90.8
3′	12.5	5.0	62.5
2′	7.5	5.0	37.5
1′	2.5	5.0	12.5

The deflection δ_{EhEh} is computed (see Table 5–18) by means of a conjugate structure which is, again, similar to Fig. 5–63 but which is loaded with the relative elastic loads computed in Table 5–17.

TABLE 5–18.

Segment	x_E	y_A	$M\,\Delta S$	$M\,\Delta S \cdot x_E$	$M\,\Delta S \cdot y_A$	$M\,\Delta S(y_A - 15)$
14	—	2.5	12.5	—	31	—
13	—	7.5	37.5	—	281	—
12	—	12.5	62.5	—	781	—
11′	2.5	16.25	90.8	227	1,476	114
10′	7.5	18.75	104.8	786	1,965	393
9′	12.5	21.25	118.8	1,485	2,525	743
8′	17.5	23.75	132.8	2,324	3,154	1,162
7′	22.5	23.75	132.8	2,988	3,154	1,162
6′	27.5	21.25	118.8	3,267	2,525	743
5′	32.5	18.75	104.8	3,406	1,965	393
4′	37.5	16.25	90.8	3,405	1,476	114
3′	40.0	12.5	62.5	2,500	781	−156
2′	40.0	7.5	37.5	1,500	281	−281
1′	40.0	2.5	12.5	500	31	−156
Σ(11′ through 1′) =			1,006.9	22,388	20,426	4,231

$$R'_H = \frac{22{,}388}{40} = 559.7, \qquad \delta_{EhEh} = +20{,}426,$$

$$\delta_{DhEh} = +4231 + 559.7 \times 15 = +12{,}627,$$

$$\alpha_D = 1006.9 - 559.7 = +447.2.$$

The deflections δ_{AvEh} and δ_{AhEh} are found by use of a conjugate structure similar to Fig. 5–64. This is loaded at its free end with a couple equal to δ_{DhEh} and a concentration equal to α_D.

$$\delta_{AhEh} = 447.2 \times 15 + 12{,}627 = +19{,}335,$$

$$\delta_{AvEh} = 447.2 \times 40 = +17{,}888.$$

Attention is called to the fact that, by Maxwell's reciprocal theorem,

$$\delta_{AhEh} = \delta_{EhAh}, \qquad \delta_{AvEh} = \delta_{EhAv},$$

and

$$\delta_{AvAh} = \delta_{AhAv}.$$

Comparison of the computed relative values for these deflections will show that the above equalities exist. This is a valuable check on the correctness of the values for these deflections.

The computed relative values for all deflection components are now substituted in the deflection condition equations previously written:

(1) $\quad +38{,}660 H_A + 35{,}776 V_A + 19{,}335 H_E + 922{,}286 = 0,$

(2) $\quad +35{,}776 H_A + 47{,}514 V_A + 17{,}888 H_E + 1{,}419{,}818 = 0,$

(3) $\quad +19{,}329 H_A + 17{,}888 V_A + 20{,}426 H_E + 461{,}143 = 0.$

The above equations are solved simultaneously in Table 5–19.

TABLE 5–19.

Eq.	Operation	H_A	V_A	H_E	Constant	Equation sum	Check operation
(1)		+38,660	+35,776	+19,335	+922,286	+1,016,057	
(2)		+35,776	+47,514	+17,888	+1,419,818	+1,520,996	
(3)		+19,329	+17,888	+20,426	+461,143	518,786	
(4)	(1) − 1.0806(2)	—	−15,568	+5	−611,969	−627,532	−627,531
(5)	(1) − 2.0001(3)	—	—	−21,519	−46	−21,565	−21,567

From (5), for all practical purposes, $H_E = 0$. The symmetry of the structure indicates that this is the correct value for H_E.

From (4), $V_A = -39.3$ k. The negative sign indicates that the true sense of V_A is opposite to the sense of the 1 k vertical load assumed in Fig. 5–61(b).

From (3), $H_A = +12.5$ k. The positive sign indicates that the true sense of H_A is the same as that of the 1 k horizontal load in Fig. 5–61(a).

The above redundant reactions having been determined, we can now compute moments, thrusts, and shears throughout the structure. Since (from the statement of the example) the frame is to be fabricated with a varying moment of inertia, the above analysis is preliminary. The moments just computed, combined with the thrusts if necessary, are used to determine the required dimensions at critical sections. The outline of the frame is arranged to effect a smooth and graceful transition from one critical section to another. The resulting structure will be the first trial nonprismatic frame.

After the outline of the first trial nonprismatic frame has been established, the first analysis must be modified to determine the effects of the variation in moment of inertia. This can be done quite easily. It will be recalled that relative values of elastic loads were used in the first analysis to simplify computations. In other words, instead of using the absolute value $M \, \Delta S/EI$, the relative value $M \, \Delta S$ was used. This was possible because E and I were constant. Now, to include the effect of a varying moment of inertia, it is only necessary to use a new relative elastic weight, specifically, $M \, \Delta S/I$. This is accomplished by dividing all computations in the preceding tables, which involve $M \, \Delta S$, by the value for I at the center of each segment, and resolving the three deflection condition equations. Critical sections of the first trial nonprismatic frame are rechecked for adequacy on the basis of this second analysis, and revisions and additional modifications of the first analysis are made as necessary.

There is always a question as to whether or not the results of any analysis are correct. In this case it is apparent, from the symmetry of the frame and loading, that the horizontal reaction at E should be zero. The fact that the computed value of H_E is zero (or practically so) is an indication, but not conclusive proof, that the analysis is correct. A complete check on the first analysis is possible by selecting a different set of redundant reactions and making a second set of computations. An alternate check can be accomplished by applying the final moment diagrams as load intensities on the conjugate structure to determine whether or not the indicated deflections are consistent with the constraints. That is, if the cut-back frame of Fig. 5–60 is subjected to the same moments as the loaded frame of Fig. 5–59, the computed horizontal deflection component of E and the computed horizontal and vertical deflection components of A should be zero.

Problems

5–23. Find the vertical and horizontal reaction components at A (see Fig. 5–65). The moment of inertia is constant. Use virtual work to find the required deflection components. [*Ans.:* $V = 14.8$ k up, $H = 1.9$ k left.]

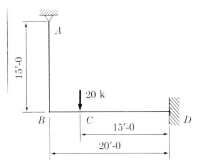

Figure 5–65

5-24. Find the vertical and horizontal reaction components at A (see Fig. 5-66). Evaluate the necessary deflection components by virtual work. [*Ans.:* $V = 5.0$ k up, $H = 3.4$ k right.]

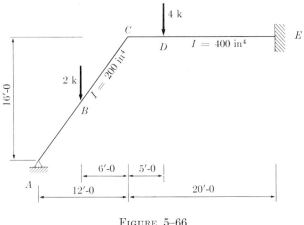

FIGURE 5-66

5-25. Evaluate the horizontal, vertical, and moment reaction components at A (see Fig. 5-67). Use virtual work to find the required deflection components and check by the conjugate structure. [*Ans.:* $H = 4.1$ k right, $V = 14.8$ k up, $M = 13.8$ ft·k clockwise.]

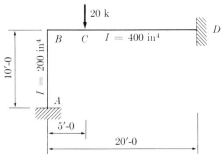

FIGURE 5-67

5-26. Find the reaction components at A (see Fig. 5-68). Use virtual work to evaluate the necessary deflection components and check by the conjugate structure. Moment of inertia is constant. [*Ans.:* $H = 3.7$ k right, $V = 16.3$ k up, $M = 32.5$ ft·k clockwise.]

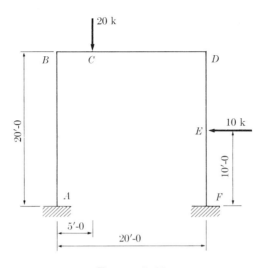

FIGURE 5-68

5-27. Using the conjugate structure and the general method, find (from Fig. 5-69) the values for the following two combinations of redundant reaction components:

(a) The horizontal and vertical reaction components at D.

(b) The horizontal and moment reaction components at A. [*Ans.:* (a) $H_D = 2.6$ k right, $V_D = 5.4$ k up; (b) $H_A = 5.4$ k right, $M_A = 22$ ft·k clockwise.]

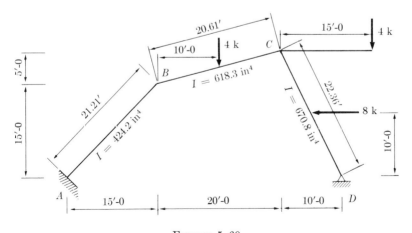

FIGURE 5-69

5–28. Find the reaction components at E (see Fig. 5–70). The moment of inertia is constant. [*Ans.*: $H = 3.9$ k left, $V = 3.0$ k up, $M = 38$ ft·k counterclockwise.]

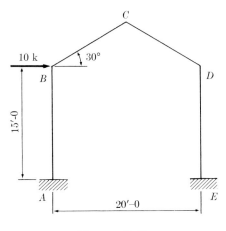

FIGURE 5–70

5–29. The stepped-column gable frame of Fig. 5–53 supports concentrations from the interior and peak purlins of 13.5 k at each framing point. The concentrations at C and E can be neglected in the analysis, since their effect on the frame is obvious. Note that the purlins are spaced at 7.5 ft on the horizontal projection. Find the reaction components at A resulting from the purlin concentrations. Attention is called to the fact that the deflection components of A in the cut-back structure, due to 1 k forces acting vertically and horizontally at A and a 1 ft·k couple at A, have already been evaluated in Example 5–21 and need not be recomputed. [*Ans.*: $V_A = 47.3$ k up, $H_A = 29.4$ k right, $M_A = 529$ ft·k clockwise.]

5–5 The elastic center. In 1866, Carl Culmann (1821–1881), a German, published his book on graphic statics. In this work he demonstrated that the analysis of hingeless arches can be considerably simplified if the redundant reaction components are considered to act at the centroid of the elastic areas. This point is now known as the *elastic center* or, sometimes, as the *neutral point*.

The method of the elastic center, which is merely a special application of the general method to single-span frames or arches, is seldom used in practice for anything except the analysis of a single-span hingeless arch. It is limited, in direct application, to single-span frames or arches with a maximum of three redundant reaction components. It may be applied indirectly, however, as demonstrated by McCullough and Thayer (1), in the analysis of an arch group on elastic piers.

It is convenient to develop the method with the analysis of the frame shown in Fig. 5-71.

If an analysis is to be made by the general method, it will be remembered that the first step may be to remove the support at A. Vertical, horizontal, and rotational deflection components of end A are evaluated—first, as caused by the real load, and then as caused by successive and individual applications of unit horizontal, vertical, and rotational actions at A. The following deflection condition equations, in which all deflection components are for A, will apply:

$$\Delta_h + H \cdot \delta_{hh} + V \cdot \delta_{hv} + M \cdot \delta'_{hm} = 0,$$

$$\Delta_v + H \cdot \delta_{vh} + V \cdot \delta_{vv} + M \cdot \delta'_{vm} = 0, \qquad (5\text{-}12)$$

$$\theta + H \cdot \alpha'_{mh} + V \cdot \alpha'_{mv} + M \cdot \alpha_{mm} = 0.$$

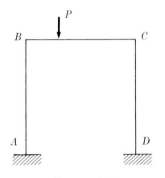

FIGURE 5-71

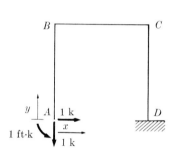

FIGURE 5-72

The unit actions mentioned above are considered to act at A as shown in Fig. 5-72. The following expressions, in which dA represents a differential elastic area, will result from an application of virtual work:

$$\delta'_{hm} = \alpha'_{mh} = \int m_\alpha m_h \frac{ds}{EI} = \int y\, dA,$$

$$\delta'_{vm} = \alpha'_{mv} = \int m_\alpha m_v \frac{ds}{EI} = \int x\, dA, \qquad (5\text{-}13)$$

$$\delta_{hv} = \delta_{vh} = \int m_v m_h \frac{ds}{EI} = \int xy\, dA.$$

The first two of the above expressions represent the first moment of the total elastic area of the frame about the x- and about the y-axes, respectively. The third expression is in the usual form for the product of inertia of an area. If, therefore, the reference axes are so chosen that the first moments and the product of inertia of the elastic area are zero, the deflection condition equations may be written for a point located at the centroid of the elastic areas and will be simplified to

$$\Delta_h + H_0\,\delta_{hh} = 0,$$
$$\Delta_v + V_0\,\delta_{vv} = 0, \qquad (5\text{-}14)$$
$$\theta + M_0\alpha_{mm} = 0.$$

Note that the redundant reactions are now designated as H_0, V_0, and M_0, since they act at the elastic center. They are considered to be applied at the end of a rigid bracket as shown in Fig. 5–73.

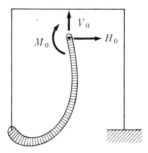

FIGURE 5–73

Obviously two conditions must be satisfied before Eqs. (5–14) will apply. The first condition is that the first moments of the elastic areas must be zero. This means that the reference axes must pass through the centroid of the elastic areas, and these reference axes are usually rectangular. In addition, it is necessary that the product of inertia shall be zero. If either or both of the centroidal rectangular reference axes are axes of symmetry, this requirement is satisfied. (The y centroidal axis is often an axis of symmetry in a hingeless arch.) If, however, no axis of symmetry exists, then rectangular centroidal reference axes cannot be used if the product of inertia is to be zero. Instead, one centroidal reference axis may be retained as either horizontal or vertical, but the other (either vertical or horizontal) must be rotated through a certain angle in order to reduce the product of inertia to zero. Nonrectangular reference axes, with respect to which the product of inertia is zero, are called conjugate axes of inertia.

Conjugate axes of inertia are rather annoying to work with, however, and the analyst may prefer to use rectangular centroidal reference axes even when the y centroidal axis is not an axis of symmetry. When this is done, Eqs. (5–12) reduce to

$$\Delta_h + H_0\,\delta_{hh} + V_0\,\delta_{hv} = 0,$$
$$\Delta_v + H_0\,\delta_{vh} + V_0\,\delta_{vv} = 0, \qquad (5\text{–}15)$$
$$\theta + M_0\alpha_{mm} = 0.$$

It is apparent that M_0 may be determined directly from the last of these equations; H_0 and V_0 must be found by a simultaneous solution of the other two equations.

The application of the method will be demonstrated with the following example.

EXAMPLE 5–30. Using the method of the elastic center, find the redundant reaction components at A of the frame of Fig. 5–74. The moment of inertia is constant. (Note that this is the same frame that is analyzed in Example 5–19.)

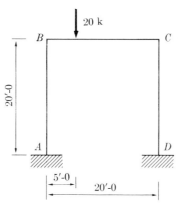

FIGURE 5–74

The elastic center will obviously be on a vertical line located midway between the two columns. The distance y_0 from a horizontal line through A and D up to this elastic center will be

$$y_0 = \frac{2 \times 20 \times 10 + 20 \times 20}{60} = 13.33 \text{ ft.}$$

The widths of the several elastic areas do not appear in the above computation, since they are constant and equal.

A rigid bracket is considered to be attached to the frame at A (see Fig. 5–75) and to extend up to the elastic center at O. The redundant reaction components act at the elastic center on the end of this bracket. The immediate object of the analysis is to determine the sense and magnitude of each of these components.

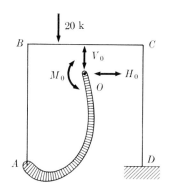

FIGURE 5–75

The three deflection components of the point O, resulting from the action of the real load, are computed by means of the conjugate structure (see Fig. 5–76):

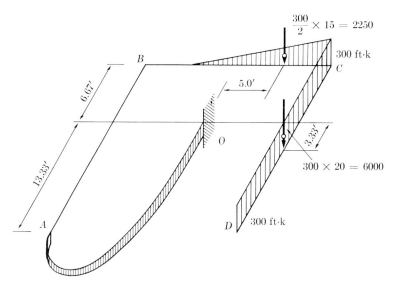

FIGURE 5–76

$$\Delta H = 2250 \times 6.67 - 6000 \times 3.33 = -5000,$$

$$\Delta V = 2250 \times 5 + 6000 \times 10 = +71{,}250,$$

$$\theta = 2250 + 6000 = +8250.$$

For a 1 k force acting to the right at O, the conjugate structure and the only resulting deflection component of O are as shown in Fig. 5–77:

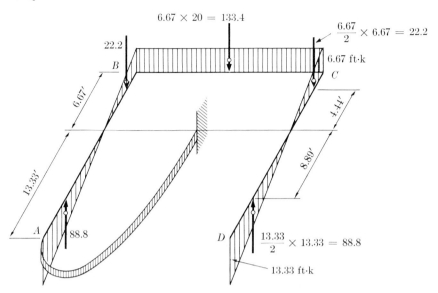

FIGURE 5-77

$\delta_{hh} = 133.4 \times 6.67 + 2 \times 22.2 \times 4.44 + 2 \times 88.8 \times 8.89 = +2667.$

The conjugate structure for a 1 k force acting down at O, and the only deflection component of O resulting from this force, are indicated in Fig. 5–78:

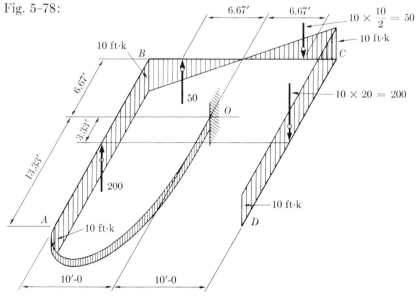

FIGURE 5-78

$\delta_{vv} = 2(200 \times 10 + 50 \times 6.67) = +4666.$

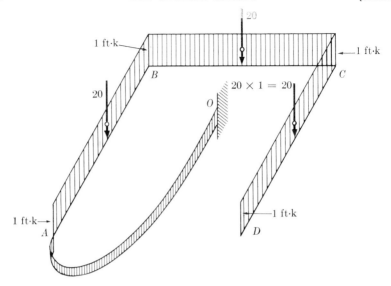

FIGURE 5-79

Finally, the conjugate structure for a 1 ft·k counterclockwise couple acting at O is as shown in Fig. 5-79. The only deflection component of O resulting from the action of the 1 ft·k couple will be

$$\alpha_{mm} = 20 + 20 + 20 = +60.$$

Substituting in Eqs. (5-14),

$$H_0 = -\frac{\Delta H}{\delta_{hh}} = -\frac{-5000}{+2667} = +1.87 \text{ k (to the right)},$$

$$V_0 = -\frac{\Delta V}{\delta_{vv}} = -\frac{+71{,}250}{+4666} = -15.27 \text{ k (up)},$$

$$M_0 = -\frac{\theta}{\alpha_{mm}} = -\frac{+8250}{+60} = -137.5 \text{ ft·k (clockwise)}.$$

The redundant reaction components acting at A are easily determined by statics as follows:

$$H_A = H_0 = 1.87 \text{ k to the right},$$

$$V_A = V_0 = 15.27 \text{ k up},$$

$$M_A = M_0 + 13.33 H_0 - 10 V_0 \quad \text{(clockwise is +)}$$

$$= +137.5 + 13.33 \times 1.87 - 10 \times 15.27$$

$$= +9.7 \text{ ft·k}.$$

Attention is called to the fact that, since the frame in question is symmetrical, it can also be solved by cutting the member BC at its midpoint and by assuming that a rigid bracket is attached to the part of the frame on each side of the cut. Both rigid brackets extend down to the elastic center as shown in Fig. 5–80.

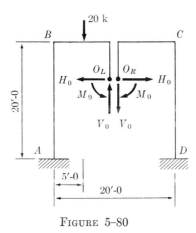

Figure 5–80

The first step in this analysis is to find the three components of the deflection of O_L. This is most easily accomplished with the conjugate structure as shown in Fig. 5–81:

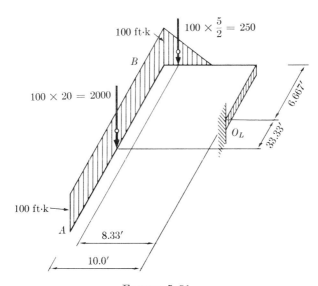

Figure 5–81

$$\Delta_{O_L h} = +2000 \times 3.333 - 250 \times 6.667 = +5000 \text{ to right,}$$

$$\Delta_{O_L v} = +2000 \times 10 + 250 \times 8.333 = +22{,}083 \text{ down,}$$

$$\theta_{O_L} = +2000 + 250 = +2250 \text{ clockwise.}$$

Note that a positive sign for the horizontal deflection component signifies a movement of O_L to the right relative to O_R, or a movement of O_R to the left relative to O_L. A positive sign for the vertical deflection component signifies a downward movement of O_L relative to O_R, or an upward movement of O_R relative to O_L. Finally, a positive sign for the rotation θ_{O_L} indicates a clockwise rotation of O_L relative to O_R, or a counterclockwise rotation of O_R relative to O_L.

It is now necessary to determine the components of the relative deflections of the two free ends of the rigid brackets, as caused by unit actions applied at these free ends. First, apply unit horizontal forces of 1 k to each bracket as shown in Fig. 5–82(a). From the conjugate structure [shown in Fig. 5–82(b)], taking moments about the x-axis, we find the relative horizontal displacement of the free ends of the two rigid brackets will be

$$\delta_{OhOh} = -2(66.67 \times 6.667 + 22.22 \times 4.444 + 88.89 \times 8.889)$$

$$= -2667.$$

Inspection of the conjugate structure will indicate that the relative vertical and the relative rotational deflection components of O_L and O_R will be zero. The actual rotational deflection components of O_L and O_R also will be zero. The actual vertical deflection components of O_L and O_R will have values, but these will be equal and therefore the relative value is zero.

From the above, the required condition equation for the relative horizontal deflection component of O_L and O_R, in the loaded structure of Fig. 5–80, may be written as

$$\Delta_{O_L h} + H_0 \, \delta_{OhOh} = 0,$$

from which

$$H_0 = -\frac{\Delta_{O_L h}}{\delta_{OhOh}} = -\frac{+5000}{-2667} = +1.87 \text{ k.}$$

The positive sign of the answer indicates that the correct senses of the forces H_0 are the same as the senses of the unit horizontal forces assumed in Fig. 5–82(a).

5-5] THE ELASTIC CENTER 249

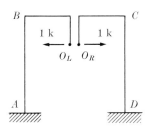

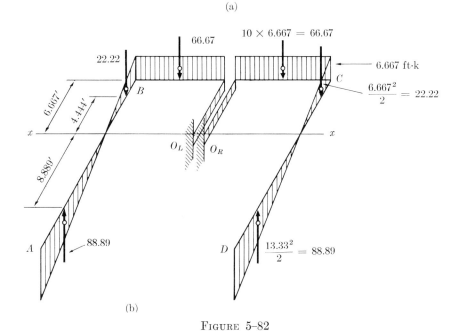

Figure 5-82

Unit vertical forces are now considered to act at O_L and O_R as shown in Fig. 5–83(a). The conjugate structure is shown in Fig. 5–83(b). If we take moments about the y-axis, the relative vertical deflection component of O_L and O_R, as caused by these unit vertical forces, will be

$$\delta_{O_vO_v} = 2(200 \times 10 + 50 \times 6.667) = +4667.$$

Relative horizontal and rotational deflection components of O_L and O_R will be zero.

The required condition equation for the relative vertical deflection component of O_L and O_R, for the loaded structure of Fig. 5–80, will be

$$\Delta_{O_Lv} + V_0\,\delta_{O_vO_v} = 0,$$

from which

$$V_0 = -\frac{\Delta_{O_L v}}{\delta_{O_v O_v}}$$

$$= -\frac{+22{,}083}{+4667} = -4.73 \text{ k.}$$

The negative sign of the answer indicates that the correct senses of the two redundants V_0 are opposite to the senses of the unit vertical forces as assumed in Fig. 5–83(a).

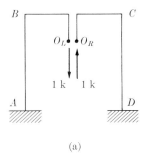

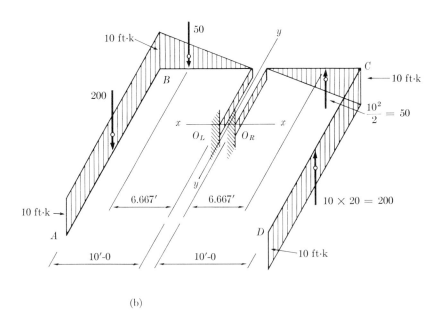

FIGURE 5–83

As the final step in the analysis, assume that unit couples are applied at O_L and O_R as indicated in Fig. 5–84(a). It is apparent from the conjugate structure of Fig. 5–84(b) that the relative rotational deflection component of O_L and O_R will be

$$\alpha_{O_m O_m} = +2(20 + 10) = +60.$$

Relative horizontal and vertical deflection components of O_L and O_R will be zero.

The required condition equation for the relative rotational deflection component of O_L and O_R, for the loaded structure of Fig. 5–80, will be

$$\theta_{O_L} + M_O \alpha_{O_m O_m} = 0,$$

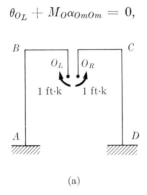

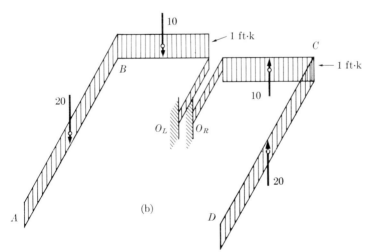

FIGURE 5–84

from which

$$M_0 = -\frac{\theta_{OL}}{\alpha_{OmOm}}$$

$$= -\frac{+2250}{+60} = -37.5 \text{ ft·k.}$$

The true senses of the redundant couples M_0 will be opposite to the senses of the unit couples as assumed in Fig. 5–84.

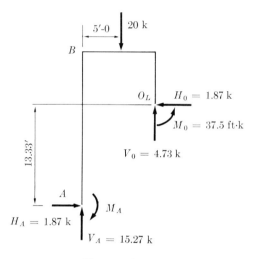

FIGURE 5–85

The values of the redundant reactions at A may now be computed by statics. H_A and V_A are determined by inspection to be as shown in Fig. 5–85. The value of M_A is computed from an equation which expresses the fact that $\Sigma M = 0$ for the free body ABO_L, with A as the center of moments. For the purpose of writing the equation, M_A is assumed to be clockwise and clockwise moments are assumed to be positive:

$$M_A + 20 \times 5 - 1.87 \times 13.33 - 4.73 \times 10 - 37.5 = 0,$$

from which

$$M_A = +9.7 \text{ ft·k.}$$

Particular attention is called to the fact that this last method of analysis will apply only to *symmetrical* hingeless frames and arches, and that it is particularly valuable in the case of symmetrical hingeless arches.

General References

1. ANDERSEN, P., *Statically Indeterminate Structures*. Chap. 4. New York: Ronald, 1953.

2. DELL, G. H., "Solution of Equations in Structural Analysis by Converging Increments," *Trans. Am. Soc. Civ. Engrs.*, **104**, 1543–54, 1939.

3. PARCEL, J. I. and MOORMAN, R. B. B., *Analysis of Statically Indeterminate Structures*. New York: Wiley, 1955.

4. SCHALBE, W. L., "*Simultaneous Equations in Mechanics Solved by Iteration,*" *Trans. Am. Soc. Civ. Engrs.*, **102**, 939–56, 1937.

5. WANG, C. K., *Statically Indeterminate Structures*. Chaps. 4 and 5. New York: McGraw-Hill, 1953.

CHAPTER 6

THE METHOD OF LEAST WORK

6-1 General. The method of least work is credited to Alberto Castigliano (1847–1884), an engineer of the Italian railways, who presented it in a thesis for a diploma in engineering at Turin in 1873. A paper published by him in 1876 presented his method for finding deflections as a first theorem, with the method of least work as a corollary and second theorem. Consequently, the method is sometimes designated as *Castigliano's second theorem*.

It is interesting to note that Menabrea, an Italian general, had stated the principle of least work in a paper in 1858, but had given no satisfactory proof. It may also be recalled, from Chapter 1, that Leonard Euler (1707–1783) used a form of the method of least work in developing his treatment of the buckling of columns. Considerable credit for this goes to Daniel Bernoulli (1700–1782), who suggested to Euler that the true elastic curve might be that which would cause the total internal elastic strain energy to be a minimum.

Castigliano's second theorem provides a powerful method for analyzing indeterminate structures, and is very effective in the analysis of articulated indeterminate structures. In the case of continuous beams or frames, however, most analysts prefer some other method, usually that of moment distribution. Castigliano's second theorem cannot be used to determine stresses caused by errors in fabrication, temperature changes, or the settlement of supports.

6-2 Development of Castigliano's second theorem. Consider the truss of Fig. 6-1. As the load P is applied to the truss, point 1 moves to $1'$ and point C moves to C', as shown in the illustration of bar 1–C in Fig. 6-2.

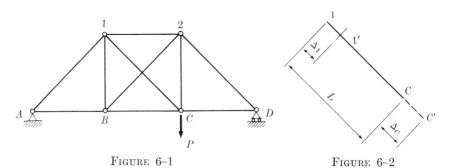

FIGURE 6-1   FIGURE 6-2

Rotation of the bar is neglected. If the cross-sectional area of the bar is A, the length is L, the modulus of elasticity is E, and if the stress in the bar as caused by the load P is S, then the elongation of the bar in the loaded truss will be SL/AE.

Assume that the redundant bar is removed from the truss and that its effect is replaced by two forces, T_1 and T_C, as indicated in Fig. 6-3.

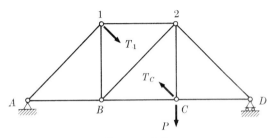

FIGURE 6-3

Although the forces T_1 and T_C are equal, they will at first be assumed to differ in magnitude. The truss is now stable and determinate. Consequently, the deflections of points 1 and C, resulting from the action of T_1, T_C, and P, can be evaluated by Castigliano's first theorem.

Let W_1 represent the total internal elastic strain energy in the structure, with T_1, T_C, and P acting but with the redundant bar omitted. By Castigliano's first theorem,

$$\Delta_1 = \frac{\partial W_1}{\partial T_1} \quad \text{and} \quad \Delta_C = \frac{\partial W_1}{\partial T_C}.$$

Note that Δ_C is intrinsically negative, since the sense of T_C is obviously opposite to the sense of the deflection component of C along the line of 1-C. The difference between Δ_1 and Δ_C may therefore be expressed algebraically as

$$\Delta_1 + \Delta_C = \frac{\partial W_1}{\partial T_1} + \frac{\partial W_1}{\partial T_C}.$$

It is apparent that, in the original loaded truss of Fig. 6-1, the elastic strain in the member 1-C must be equal in magnitude to $\Delta_1 + \Delta_C$, since otherwise the member will not extend between joints 1 and C. Consequently, the algebraic summation of Δ_1, Δ_C, and the elastic strain in the member 1-C must be zero. This is expressed as

$$\frac{\partial W_1}{\partial T_1} + \frac{\partial W_1}{\partial T_C} + \frac{SL}{AE} = 0. \qquad (6\text{-}1)$$

Actually, however, both T_1 and T_C are equal in magnitude to S. Consequently, $\partial W_1/\partial S$ will be equivalent to $\partial W_1/\partial T_1 + \partial W_1/\partial T_C$, for the reason that the single mathematical operation indicated by $\partial W_1/\partial S$, with forces S acting at joints 1 and C, will give the difference between (or algebraic sum of) the deflections Δ_1 and Δ_C. Equation (6–1) may therefore be written as

$$\frac{\partial W_1}{\partial S} + \frac{SL}{AE} = 0. \tag{6–2}$$

As previously mentioned, W_1 does not include the internal strain energy of the redundant bar. The internal strain energy in this redundant bar is given by $S^2L/2AE$, and thus

$$\frac{\partial}{\partial S}\left(\frac{S^2L}{2AE}\right) = \frac{SL}{AE}. \tag{6–3}$$

It is evident, therefore, that the left side of Eq. (6–2) is actually the first partial derivative of the total internal work of the structure, including the redundant bar. Consequently, the condition equation which must be satisfied by the stress S in the redundant bar may be written as

$$\frac{\partial W}{\partial S} = 0, \tag{6–4}$$

where W represents the total internal strain energy of the entire structure, including the redundant bar. This condition requires that the internal strain energy shall be a minimum. Therefore *Castigliano's second theorem* may, in effect, be stated as follows:

In any loaded indeterminate structure the values of the redundants must be such as to make the total elastic internal strain energy, resulting from the application of a given system of loads, a minimum.*

Equation (6–4) can be derived very easily and directly by use of the general method. Consider that the redundant 1–C is cut as indicated in Fig. 6–4. With the load P acting, the ends of the member adjacent to the cut will tend to separate by some amount Δ. If, however, tensile forces equal in magnitude to the true stress S act on the ends adjacent to the cut, the value of Δ must be zero.

---

* The preceding derivation and the statement of the theorem are not as presented by Castigliano. The reader should refer to Southwell (1) and Timoshenko (2) for a discussion of the original work.

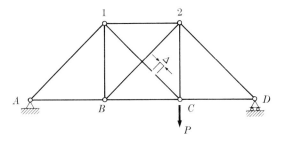

FIGURE 6-4

Expressing Δ by Castigliano's first theorem yields

$$\Delta = \frac{\partial W}{\partial S} = 0,$$

and this is the same as Eq. (6-4).

It is apparent from the foregoing that Castigliano's second theorem (the method of least work) and the general method accomplish the same thing. Aside from variations in detail in the computations, the difference is essentially one of concept. In the case of the method of least work, the condition equations are considered to be requirements which must be satisfied to ensure that the total internal elastic strain energy will be a minimum. If, however, the general method is being used, these same condition equations are considered to be requirements that the distortions of the points of application of the redundants shall be consistent with the constraints of the structure.

That the two methods are essentially the same can be shown very easily in connection with Fig. 6-5. By the general method the condition equation for R_B is

FIGURE 6-5

$$\Delta_B + R_B \delta_{BB} = 0.$$

By virtual work this equation becomes

$$\int M m_B \frac{dx}{EI} + R_B \int m_B^2 \frac{dx}{EI} = 0, \qquad (6\text{-}5)$$

where M is the moment caused by the real load and m_B is the moment resulting from a unit vertical force at B.

If a solution is desired by Castigliano's second theorem, the condition equation is

$$\frac{\partial W}{\partial R_B} = 0. \tag{6-6}$$

But

$$\frac{\partial W}{\partial R_B} = \int M' \frac{\partial M'}{\partial R_B} \frac{dx}{EI},$$

where $M' = M + R_B m_B$. Substitution in Eq. (6-6) results in

$$\int M m_B \frac{dx}{EI} + R_B \int m_B^2 \frac{dx}{EI} = 0,$$

which is the same as Eq. (6-5).

Thus it is seen that the condition equations necessary for a solution by the general method are, in effect, the same as those required to analyze the same structure by the method of least work.

6-3 Analysis of continuous beams and frames. As previously indicated, the method of least work is not usually selected for the analysis of continuous beams and frames. Its use normally involves considerably more labor than moment distribution, slope deflection, or the general method when used with the conjugate structure. Two examples will be presented, however, to demonstrate the method.

EXAMPLE 6-1. Determine the reaction at C in Fig. 6-6 by the method of least work.

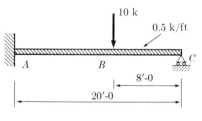

FIGURE 6-6

The expressions for bending moment in the two different sections of the beam, and the first derivatives thereof with respect to the redundant R_C, are shown in Table 6-1. For the purpose of the analysis, R_C is assumed to act upward. A moment causing compression on the top side of the beam is considered to be positive.

TABLE 6-1.

Section	$x = 0$ at	M (ft·k)	$\dfrac{\partial M}{\partial R_C}$
C–B	C	$R_C x - \dfrac{0.5x^2}{2}$	$+x$
B–A	B	$R_C(8+x) - \dfrac{0.5(8+x)^2}{2} - 10x$	$+8+x$

Substitution in the condition equation

$$\frac{\partial W}{\partial R_C} = \int M \frac{\partial M}{\partial R_C} \frac{dx}{EI} = 0$$

yields

$$\int_0^8 \left[R_C x - \frac{0.5x^2}{2} \right] x\, dx$$

$$+ \int_0^{12} \left[R_C(8+x) - \frac{0.5(8+x)^2}{2} - 10x \right](8+x)\, dx = 0.$$

Note that E and I have been dropped, since they are constant.

When we perform the indicated integration, the condition equation becomes

$$2667 R_C - 21{,}520 = 0,$$

from which

$$R_C = +8.1 \text{ k}.$$

The positive sign of the answer indicates that the true sense of R_C is the same as that assumed, that is, up.

EXAMPLE 6-2. Find the horizontal and vertical reaction components at F for the continuous frame of Fig. 6-7.

For the purpose of the analysis, the horizontal and vertical reaction components at F are assumed to act with the senses shown

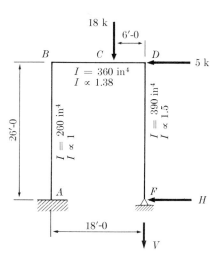

FIGURE 6-7

in the illustration. The expressions for the moments resulting from the action of these two reaction components and the loads, and the required partial derivatives, are shown in Table 6–2. Any moment tending to cause tension on the outside of the frame is considered to be positive.

TABLE 6–2.

Section	$x = 0$ at	M (ft·k)	$\dfrac{\partial M}{\partial H}$	$\dfrac{\partial M}{\partial V}$
F-D	F	$+Hx$	$+x$	0
D-C	D	$+26H + Vx$	$+26$	$+x$
C-B	C	$+26H + V(6+x) + 18x$	$+26$	$+6+x$
B-A	B	$+H(26-x) + 18V - 5x + 216$	$+26-x$	$+18$

Two condition equations are required for a solution, the first of which is

$$\frac{\partial W}{\partial H} = \int M \frac{\partial M}{\partial H} \cdot \frac{dx}{I} = 0.$$

Observe that the modulus of elasticity E has been omitted from the denominator above, since it is constant for the entire frame. Substituting from Table 6–2, and using relative values for I, the result is

$$\int_0^{26} (+Hx)(+x) \frac{dx}{1.5}$$

$$+ \int_0^6 (+26H + Vx)(+26) \frac{dx}{1.38}$$

$$+ \int_0^{12} [+26H + V(6+x) + 18x][+26] \frac{dx}{1.38}$$

$$+ \int_0^{26} [+H(26-x) + 18V - 5x + 216][+26-x] \, dx = 0.$$

After we perform the indicated integrations, the first condition equation, in its final form, is

$$+18{,}560H + 9140V + 82{,}760 = 0. \tag{6-7}$$

The second condition equation is

$$\frac{\partial W}{\partial V} = \int M \frac{\partial M}{\partial V} \cdot \frac{dx}{I} = 0.$$

When we substitute from Table 6–2, the above becomes

$$\int_0^6 (+26H + Vx)(+x) \frac{dx}{1.38}$$

$$+ \int_0^{12} [+26H + V(6+x) + 18x][+6+x] \frac{dx}{1.38}$$

$$+ \int_0^{26} [+H(26-x) + 18V - 5x + 216][+18] \, dx = 0.$$

Integration will give the second condition equation as

$$+9140H + 9830V + 83{,}800 = 0 \qquad (6\text{–}8)$$

and simultaneous solution of Eqs. (6–7) and (6–8) will give

$$V = -8.1 \text{ k} \quad \text{and} \quad H = -0.46 \text{ k}.$$

The solution outlined above, it should be noted, does not create a true impression of the considerable amount of work involved in evaluating the integrals.

Problems

6–3. Find the reaction at B in Fig. 6–8 by the method of least work. Moment of inertia and E are constant. [*Ans.:* 5.1 k up.]

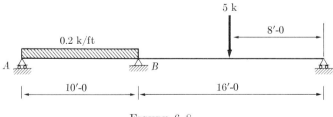

Figure 6–8

6-4. Find the reactions at B and C (see Fig. 6-9) by the method of least work. Moment of inertia and E are constant. [*Ans.*: $R_B = 7.7$ k up, $R_C = 3.9$ k up.]

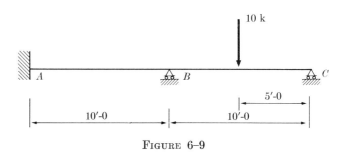

FIGURE 6-9

6-5. Determine the value of the reaction component at A in Fig. 6-10. The modulus of elasticity is constant. [*Ans.*: 4.3 k left.]

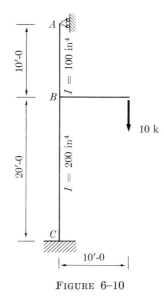

FIGURE 6-10

6-6. Find the horizontal reaction component at A (see Fig. 6-11) by the method of least work. E is constant throughout the frame. [*Ans.*: 0.56 k.]

6-7. Determine the values of the three reaction components at A in Fig. 6-12. E and I are constant. [*Ans.*: $H = 1.9$ k right, $V = 15.3$ k up, $M = 9.8$ ft·k clockwise.]

6-8. Find the magnitude of the horizontal reaction component at F in Fig. 6-13. E and I are constant. [*Ans.*: 4.3 k left.]

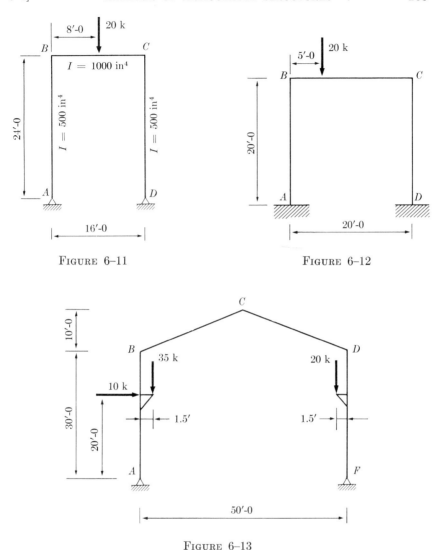

FIGURE 6-11

FIGURE 6-12

FIGURE 6-13

6-4 Analysis of articulated structures. Castigliano's second theorem provides an excellent method for the analysis of structures which are either partially or completely articulated. As previously pointed out, however, the general method is equally effective. Any choice between the two methods is, for most problems, usually a matter of personal preference and not because one method is in fact easier or better than the other. The following examples will illustrate the application of the method of least work to structures which are partially or entirely articulated.

EXAMPLE 6-9. Find the stress T in the redundant AD of Fig. 6-14. E. is constant.

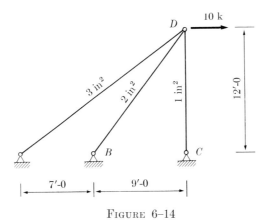

FIGURE 6-14

The required condition equation, by Castigliano's second theorem, is

$$\frac{\partial W}{\partial T} = \sum S \frac{\partial S}{\partial T} \cdot \frac{L}{A} = 0.$$

The modulus E is omitted, since it is constant for the entire structure. The information for writing this equation in the form necessary for a solution is given in Table 6-3. The redundant T is assumed to be tension.

TABLE 6-3.

Member	L (ft)	A (in²)	Stress S (k)	$\dfrac{\partial S}{\partial T}$	$S \cdot \dfrac{\partial S}{\partial T} \cdot \dfrac{L}{A}$
AD	20	3	$+T$	$+1$	$+6.67T$
BD	15	2	$+16.7 - 1.33T$	-1.33	$-167 + 13.33T$
CD	12	1	$-13.3 + 0.47T$	$+0.47$	$-75 + 2.65T$

A summation of the last column of the table will give the required condition equation, that is,

$$-242 + 22.65T = 0,$$

from which

$$T = +10.7 \text{ k (tension)}.$$

Note that either BD or CD could have been used alternately as the redundant.

EXAMPLE 6-10. Find the stress in the redundant FC of Fig. 6-15. E is constant. Cross-sectional areas are shown on the illustration.

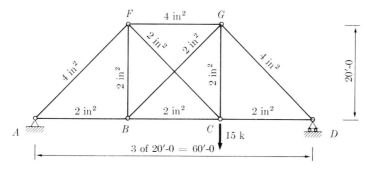

FIGURE 6-15

The condition equation is

$$\frac{\partial W}{\partial T} = \sum S \frac{\partial S}{\partial T} \cdot \frac{L}{A} = 0.$$

The stress T in the redundant is assumed to be tension. It should be noted that the only members for which $\partial S/\partial T$ will have a value are those within and enclosing the panel $FGCB$. In other words, these are the only members whose stresses are influenced by the magnitude of the stress T; consequently, only six members appear in Table 6-4.

TABLE 6-4.

Member	L (ft)	A (in²)	Stress S (k)	$\frac{\partial S}{\partial T}$	$\frac{SL}{A}\left(\frac{\partial S}{\partial T}\right)$
BC	20	2	$+10 - 0.707T$	-0.707	$-7.07(+10 - 0.707T)$
FG	20	4	$-5 - 0.707T$	-0.707	$-3.53(-5 - 0.707T)$
FB	20	2	$+5 - 0.707T$	-0.707	$-7.07(+5 - 0.707T)$
BG	28.28	2	$-7.07 + 1.00T$	$+1.00$	$+14.14(-7.07 + 1.000T)$
GC	20	2	$+15 - 0.707T$	-0.707	$-7.07(+15 - 0.707T)$
FC	28.28	2	$0 + 1.00T$	$+1.00$	$+14.14(+1.000T)$

The required condition equation is obtained by adding the values in the last column of this table. Thus,

$$-294.4 + 45.78T = 0,$$

from which

$$T = +6.4 \text{ k (tension)}.$$

EXAMPLE 6-11. The welded tower supporting the end of an aerial conveyor system consists of two vertical trusses, one of which is indicated in Fig. 6-16. The cross-sectional areas of the members in square inches are shown in parentheses on the illustration. The original load on each truss was 15 k, acting as shown. Installation of new equipment will increase the load to 50 k. It is proposed to reinforce each truss with a spun wire strand $\frac{1}{2}$ inch in diameter, with a cross-sectional steel area of 0.362 in^2 and $E = 20{,}000$ k/in^2. The wire strand is to be located as shown by the dashed line and has a working strength of 16.2 k.

The strands are to have small initial stresses in them when erected, so

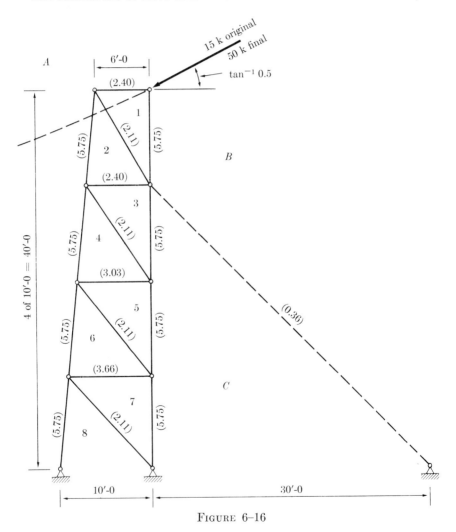

FIGURE 6-16

that in the analysis they may be considered straight. The modulus of elasticity E for the truss is 30,000 k/in². If the truss compression members have an allowed working stress of 12 k/in² and the tension members have a working stress of 20 k/in², is the proposed arrangement satisfactory from the standpoint of individual member strength and the strength of the strand? It is assumed that some additional welding may have to be placed at the joints to reinforce them.

The required condition equation, as in Examples 6–9 and 6–10, is

$$\frac{\partial W}{\partial T} = \sum S \frac{\partial S}{\partial T} \cdot \frac{L}{A} = 0,$$

where T is the unknown tensile stress in the wire strand and S is the stress in the members of the structure resulting from the action of the 50 k load and from the assumed tensile stress T in the strand. The values of S for the several members are most easily determined with two Maxwell-Cremona diagrams. For this reason the Bow system of notation has been used in Fig. 6–16.

All computations are tabulated in Table 6–5.

TABLE 6–5.

(1) Member	(2) L (ft)	(3) A (in²)	(4) Stress S (k)	(5) $\frac{dS}{dT}$	(6) $\frac{L}{A} \cdot \frac{dS}{dT} \frac{S}{\text{Rel } E}$	(7) Final S (k)	(8) Member capacity (k)
B8	10.05	5.75	−182.0 + 2.16T	+2.16	−688 + 8.16T	−152.0	−69.0
B6	10.05	5.75	−152.0 + 1.60T	+1.60	−432 + 4.55T	−129.8	−69.0
B4	10.05	5.75	−114.0 + 0.89T	+0.89	−180 + 1.41T	−101.6	−69.0
B2	10.05	5.75	−66.4	—	—	−66.4	−69.0
B1	6.00	2.40	−44.4	—	—	−44.4	−28.8
C1	10.0	5.75	−22.0	—	—	−22.0	−69.0
23	7.0	2.40	−38.0 + 0.71T	+0.71	−79 + 1.47T	−28.2	−28.8
45	8.0	3.03	−33.6 + 0.63T	+0.63	−56 + 1.05T	−24.8	−36.4
67	9.0	3.66	−30.0 + 0.56T	+0.56	−42 + 0.77T	22.2	13.0
D3	10.0	5.75	+44.0 − 0.71T	−0.71	−54 + 0.88T	+34.1	+115.0
D5	10.0	5.75	+91.4 − 1.60T	−1.60	−254 + 4.45T	+69.2	+115.0
D7	10.0	5.75	+129.2 − 2.30T	−2.30	−518 + 9.20T	+97.2	+115.0
12	11.7	2.11	+76.0	—	—	+76.0	+42.2
34	12.1	2.11	+58.0 − 1.09T	−1.09	−362 + 6.82T	+42.8	+42.2
56	12.8	2.11	+48.4 − 0.91T	−0.91	−268 + 5.02T	+35.8	+42.2
78	13.5	2.11	+40.0 − 0.76T	−0.76	−194 + 3.70T	+29.4	+42.2
CD	42.4	0.362	— + 1.00T	+1.00	+117.2T (+58.6T)	+14.0	+16.2

268 THE METHOD OF LEAST WORK [CHAP. 6

Attention is called to the fact that since E is not constant it must somehow appear in the computations. It is convenient, in most cases, to use a relative value for E, as has been done in column (6) of Table 6–5. A relative E of unity has been used for all members except the wire strand, for which the value is two-thirds. Observe that the last entry in column (6) is ($+58.6T$). Since it is one-half the value of the preceding entry for CD, this is the necessary adjustment for a relative E of two-thirds.

The summation of all values in column (6) is the required condition equation and is

$$-3127 + 223.0T = 0,$$

from which

$$T = +14.0 \text{ k}.$$

Note that the values for S in column (7) are obtained by substituting this result for T in the various expressions for S in column (4).

Comparison of the values in columns (7) and (8) will show that five members will be overstressed under the proposed plan. An alternate method for reinforcing the tower is suggested in Problem 6–21.

EXAMPLE 6–12. Find the stress in the redundants a and d (see Fig. 6–17) by least work. Assume that T_a and T_d are tension. E is constant.

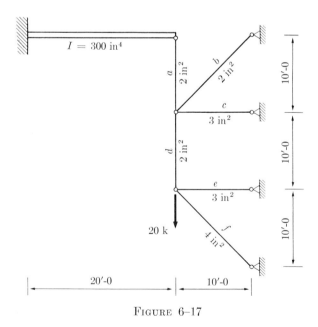

FIGURE 6–17

The two condition equations are

$$\frac{\partial W}{\partial T_a} = \sum S \cdot \frac{\partial S}{\partial T_a} \cdot \frac{L}{A} + \int M \frac{\partial M}{\partial T_a} \cdot \frac{dx}{I} = 0, \qquad (6\text{-}9)$$

$$\frac{\partial W}{\partial T_d} = \sum S \cdot \frac{\partial S}{\partial T_d} \cdot \frac{L}{A} + \int M \frac{\partial M}{\partial T_d} \cdot \frac{dx}{I} = 0. \qquad (6\text{-}10)$$

All computations necessary for the evaluation of the first term of the right side of each equation are shown in Table 6–6.

TABLE 6–6.

(1) Member	(2) L (ft)	(3) A (in^2)	(4) Stress S (k)			(5) $\dfrac{\partial S}{\partial T_a}$	(6) $\dfrac{\partial S}{\partial T_d}$
			Real	$+\ T_a$	$+\ T_d$		
a	10	2	0	$+1.000 T_a$	0	$+1.0$	0
b	14.14	2	0	$-1.414 T_a$	$+1.414 T_d$	-1.414	$+1.414$
c	10	2	0	$+1.000 T_a$	$-1.000 T_d$	$+1.0$	-1.0
d	10	2	0	0	$+1.000 T_d$	0	$+1.0$
e	10	3	$+20.00$	0	$-1.000 T_d$	0	-1.0
f	14.14	4	-28.28	0	$+1.414 T_d$	0	$+1.414$

Member	(7) $S\dfrac{\partial S}{\partial T_a} \cdot \dfrac{L}{A}$		(8) $S\dfrac{\partial S}{\partial T_d} \cdot \dfrac{L}{A}$	
a	$+5.0 T_a$		0	
b	$+14.14 T_a$	$-\ 14.14 T_d$	$-14.14 T_a$	$+\ 14.14 T_d$
c	$+3.33 T_a$	$-\ 3.33 T_d$	$-3.33 T_a$	$+\ 3.33 T_d$
d	0			$+\ 5.0 T_d$
e	0	-66.6		$+\ 3.33 T_d$
f	0	-141.8		$+\ 7.08 T_d$

The expression for M in the cantilever beam, with $x = 0$ at the free end, is

$$M = T_a \cdot x,$$

and therefore

$$\frac{\partial M}{\partial T_a} = +x \quad \text{and} \quad \frac{\partial M}{\partial T_d} = 0.$$

Therefore,

$$\int M \frac{\partial M}{\partial T_a} \cdot \frac{dx}{I} = \int_0^{20} T_a x^2 \frac{dx}{300} = \frac{T_a x^3}{900}\bigg]_0^{20} = \frac{8000 \times T_a \times 144}{900}$$

$$= 1280 T_a.$$

Note that the units of the result obtained directly from this integration would be $\text{ft}^3 \cdot \text{k}/\text{in}^4$. This is multiplied by 144 in order to reduce the units to $\text{ft} \cdot \text{k}/\text{in}^2$ to agree with the units of the several values in columns (7) and (8) of Table 6–6. This agreement in units is essential, since the result of this integration must be added to the summation of column (7) in order to give the first condition equation. This addition will result in

$$+1302.5 T_a - 17.47 T_d = 0. \tag{6-11}$$

Since $\partial M / \partial T_d = 0$, the second term of Eq. (6–10) reduces to zero. Consequently, the second condition equation is obtained by adding all terms in column (8), and is

$$-208.4 - 17.47 T_a + 32.88 T_d = 0. \tag{6-12}$$

A simultaneous solution will result in

$$T_a = +0.09 \text{ k} \quad \text{and} \quad T_d = +6.4 \text{ k}.$$

This problem was worked previously by the general method in Example 5–13. The two solutions should be compared.

EXAMPLE 6–13. Find the stresses in the backstay r and the redundants m and n (see Fig. 6–18) by least work. Cross-sectional areas of the various

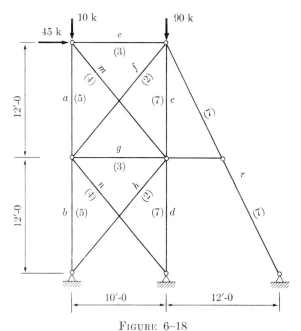

FIGURE 6–18

members in square inches are noted in parentheses on the illustration. E is constant. The condition equations are

$$\frac{\partial W}{\partial T_r} = \sum S \frac{\partial S}{\partial T_r} \cdot \frac{L}{A} = 0,$$

$$\frac{\partial W}{\partial T_m} = \sum S \frac{\partial S}{\partial T_m} \cdot \frac{L}{A} = 0, \qquad (6\text{–}13)$$

$$\frac{\partial W}{\partial T_n} = \sum S \frac{\partial S}{\partial T_n} \cdot \frac{L}{A} = 0.$$

Assuming that the stresses in the redundants are tension, the computations incidental to writing these equations in their final form are given in Tables 6–7 and 6–8.

TABLE 6–7.

Member	L (ft)	A (in^2)	Stress S (k)			
a	12	5	$-\ 10.0\ +$	0	$-\ 0.767 T_m\ +$	0
b	12	5	$+\ 43.9\ +\ 0.536 T_r\ +$	0		$-\ 0.767 T_n$
c	12	7	$-143.9\ -\ 1.429 T_r\ -$		$0.767 T_m\ +$	0
d	12	7	$-197.8\ -\ 1.965 T_r\ +$	0		$-\ 0.767 T_n$
e	10	3	$-\ 45.0\ +$	0	$-\ 0.639 T_m\ +$	0
f	15.62	2	$+\ 70.3\ +\ 0.698 T_r\ +\ 1.000 T_m\ +$			0
g	10	3	$-\ 45.0\ -\ 0.447 T_r\ -\ 0.639 T_m\ -\ 0.639 T_n$			
h	15.62	2	$+\ 70.3\ +\ 0.698 T_r\ +$	0		$+\ 1.000 T_n$
r	26.83	7	$0\ +\ 1.000 T_r\ +$	0	$+$	0
m	15.62	4	$0\ +$	0	$+\ 1.000 T_m\ +$	0
n	15.62	4	$0\ +$	0	$+$	$0\ +\ 1.000 T_n$

Member	$\dfrac{\partial S}{\partial T_r}$	$\dfrac{\partial S}{\partial T_m}$	$\dfrac{\partial S}{\partial T_n}$
a	0	-0.767	0
b	$+0.536$	0	-0.767
c	-1.429	-0.767	0
d	-1.965	0	-0.767
e	0	-0.639	0
f	$+0.698$	$+1.0$	0
g	-0.447	-0.639	-0.639
h	$+0.698$	0	$+1.0$
r	$+1.0$	0	0
m	0	$+1.0$	0
n	0	0	$+1.0$

TABLE 6–8.

	Member				
(1) $S\dfrac{\partial S}{\partial T_r}\cdot \dfrac{L}{A}$	a				
	b	$+23.6$	$+\ 0.288T_r$	$+\ \ \ 0$	$-\ 0.411T_n$
	c	$+205.5$	$+\ 2.040T_r$	$+\ 1.098T_m$	$+\ \ \ 0$
	d	$+388.0$	$+\ 3.860T_r$	$+\ \ \ 0$	$+\ 1.510T_n$
	e	0			
	f	$+49.0$	$+\ 0.486T_r$	$+\ 0.698T_m$	
	g	$+20.1$	$+\ 0.200T_r$	$+\ 0.286T_m$	$+\ 0.286T_n$
	h	$+49.0$	$+\ 0.486T_r$	$+\ \ \ 0$	$+\ 0.698T_n$
	r	0	$+\ 1.000T_r$	$+\ \ \ 0$	$+\ \ \ 0$
	m	0			
	n	0			
(2) $S\dfrac{\partial S}{\partial T_m}\cdot \dfrac{L}{A}$	a	$+7.7$	$+\ \ \ 0$	$+\ 0.589T_m$	$+\ \ \ 0$
	b	0			
	c	$+110.5$	$+\ 1.095T_r$	$+\ 0.589T_m$	$+\ \ \ 0$
	d	0			
	e	$+28.8$	$+\ \ \ 0$	$+\ 0.408T_m$	$+\ \ \ 0$
	f	$+70.3$	$+\ 0.698T_r$	$+\ 1.000T_m$	$+\ \ \ 0$
	g	$+28.8$	$+\ 0.286T_r$	$+\ 0.408T_m$	$+\ 0.408T_n$
	h	0			
	r	0			
	m	0	$+\ \ \ 0$	$+\ 1.000T_m$	$+\ \ \ 0$
	n				
(3) $S\dfrac{\partial S}{\partial T_n}\cdot \dfrac{L}{A}$	a	0			
	b	-33.7	$-\ 0.412T_r$	$+\ \ \ 0$	$+\ 0.590T_n$
	c	0			
	d	$+151.7$	$+\ 1.510T_r$	$+\ \ \ 0$	$+\ 0.589T_n$
	e	0			
	f	0			
	g	$+28.8$	$+\ 0.286T_r$	$+\ 0.408T_m$	$+\ 0.408T_n$
	h	$+70.3$	$+\ 0.698T_r$	$+\ \ \ 0$	$+\ 1.000T_n$
	r	0			
	m	0			
	n	0	$+\ \ \ 0$	$+\ \ \ 0$	$+\ 1.000T_n$

A summation of all entries in the three sections of Table 6–8, marked (1), (2), and (3), will result in the three required condition equations:

$$+8.360 T_r + 2.082 T_m + 2.083 T_n + 735.2 = 0,$$

$$+2.079 T_r + 3.994 T_m + 0.408 T_n + 246.1 = 0,$$

$$+2.082 T_r + 0.408 T_m + 3.587 T_n + 217.1 = 0.$$

A simultaneous solution will give

$$T_r = -80.6 \text{ k}, \qquad T_m = -18.4 \text{ k}, \qquad T_n = -11.4 \text{ k}.$$

Problems

6–14. Find the stress in member AD of Fig. 6–19 by least work. All members are of steel and cross-sectional areas are as indicated on the illustration. [*Ans.:* 6.0 k tension.]

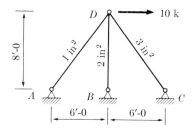

Figure 6–19

6–15. The timber beam shown in Fig. 6–20 is 20 in. deep and 10 in. wide, with $E = 1000 \text{ k/in}^2$. The steel rod has a cross-sectional area of 0.20 in^2 with $E = 30{,}000 \text{ k/in}^2$. Find the stress in the steel rod. [*Ans.:* 2.8 k tension.]

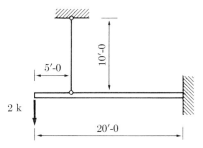

Figure 6–20

6–16. Find the stress in the redundant FC of Fig. 6–21 by least work. E is constant. Cross-sectional areas in square inches are shown in parentheses on the illustration. [*Ans.:* 29.6 k tension.]

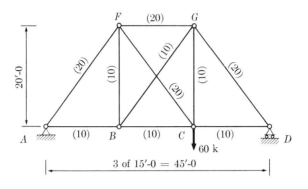

FIGURE 6–21

6–17. The timber beam shown in Fig. 6–22 is 10 in. wide and 24 in. deep, with $E = 1500$ k/in^2. The steel rod, with $E = 30{,}000$ k/in^2, has a cross-sectional area of 0.5 in^2. Find the stress in the rod. [*Ans.:* 9.0 k tension.]

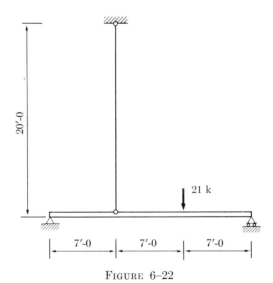

FIGURE 6–22

6-18. Determine the stress in the strut of the trussed beam shown in Fig. 6–23. The beam is a 20I70, with $I = 1214$ in^4 and $A = 20.42$ in^2. The strut has a cross-sectional area of 10 in^2 and the ties have cross-sectional areas of 8 in^2 each. E is constant. [*Ans.:* 82.5 k compression.]

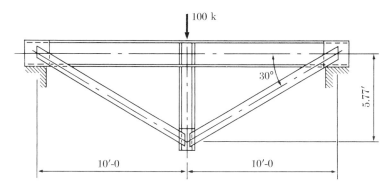

FIGURE 6–23

6-19. Find the stress T in the cable DC (see Fig. 6–24) by least work. The cross-sectional area of the cable is 1.5 in^2 with an E of 20,000 k/in^2. The timber beam AC is 12 in. $\times$ 12 in. in section, with $E = 1600$ k/in^2. Each member of the steel cantilever truss has a sectional area of 4 in^2, and E is 30,000 k/in^2. [*Ans.:* 5.1 k tension.]

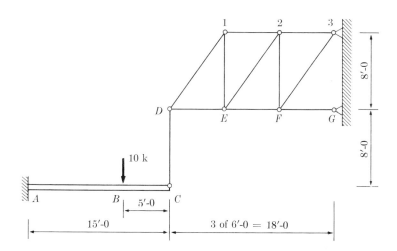

FIGURE 6–24

6–20. The towers for supporting the ends of the cable for a cable car crossing a gorge were built as shown in Fig. 6–25. It is now desired to install a new car and cables, resulting in increased loads. The towers are not strong enough as originally built. It is proposed to tie the top of each tower back to rock with two wire ropes, one in the plane of each side truss of the tower as indicated. Each rope is to have an area of 2 in^2 and an E of 10,000 k/in^2, and each rope acts with a set of loads and a vertical side truss as shown. Cross-sectional areas of the tower members in square inches are shown in parentheses on the illustration. The allowable working stress in the cable is 30,000 lb/in^2, 20,000 lb/in^2 in the tension members of the tower, and 14,000 lb/in^2 in the compression members. Is the proposed arrangement satisfactory? The rope will have a small initial stress so that in the analysis it may be regarded as straight. E for the tower is 30,000 k/in^2. [*Ans.:* The stress in the rope is found to be 58.5 k tension, which gives a stress intensity of 29.2 k/in^2. This is satisfactory. One member, CE, will be overstressed at 15.1 k/in^2 compression.]

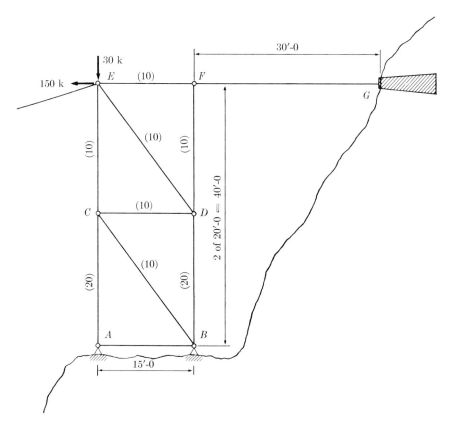

Figure 6–25

6-21. In Example 6-11 it was found that the suggested method for reinforcing the tower was not entirely satisfactory. It is now suggested that the diameter of the cable be increased to 1.25 in., with an area of 0.93 in^2 and a working strength of 40 k. E for the cable remains at 20,000 k/in^2. Furthermore, the cable will be stressed to 20 k tension when installed. Will this be a better solution than the proposal of Example 6-11? (See Fig. 6-16.) [*Ans.:* Final stress in cable = 27 k. Members $B8$, $B6$, $B4$, $B1$, and 12 are overstressed.]

6-22. Find the redundant reaction components R_D, M_D, and T_D (see Fig. 6-26) by least work. The welded pipe bracket ACD is located in a horizontal plane and the 4 k load is vertical. The ends at A and D are fixed. The pipe is 6-in. standard with a plane moment of inertia = 28.14 in^4, E = 30,000 k/in^2, and G = 15,000 k/in^2. [*Ans.:* R_D = -1.9 k (up), M_D = -1.8 ft·k (counterclockwise), T_D = $+11.8$ ft·k (clockwise).]

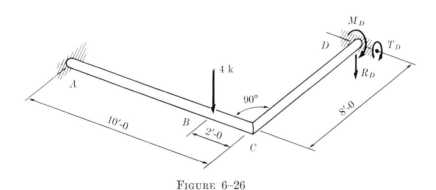

FIGURE 6-26

Specific References

1. SOUTHWELL, R. V., *An Introduction to the Theory of Elasticity*, p. 96. Oxford: Clarendon Press, 1936.
2. TIMOSHENKO, S. P., *History of Strength of Materials*, p. 289. New York: McGraw-Hill, 1953.

General References

3. ANDERSEN, P., *Statically Indeterminate Structures*. Chap. 4. New York: Ronald, 1953.
4. GRINTER, L. E., *Theory of Modern Steel Structures*. Vol. 2, Chap. 4. New York: Macmillan, 1949.
5. PARCEL, J. I. and MOORMAN, R. B. B., *Analysis of Statically Indeterminate Structures*. Chap. 3. New York: Wiley, 1955.

CHAPTER 7

THE COLUMN ANALOGY

7–1 General. The column analogy, introduced by Professor Hardy Cross about the time that his method for distributing moments was published (1930), provides a method of analysis which is applicable to indeterminate continuous frames with a redundancy up to and including the third degree. The method exists because the moments in a continuous frame, as caused by redundant reactions (up to a maximum of three in number), are analogous to the fiber stresses in an eccentrically loaded short column.

As applied to single-span rigid frames and arches, the column analogy is very similar to the method of the elastic center. There is very little difference, if any, in the amount of work or complexity of theory involved in the two methods, and a selection of one in place of the other is a matter of personal preference.

7–2 The column flexure formula. A proper understanding of the particular approach to the column analogy as presented herein depends on a knowledge of the *column flexure formula*. Consequently, this formula will be discussed before proceeding with the development of the analogy.

A column with an unsymmetrical cross section is shown in Fig. 7–1(a). (Any other unsymmetrical section could have been used.) The load P is eccentrically applied with respect to the two rectangular centroidal axes x–x and y–y. The fiber stresses in the column, under the action of P, are assumed to vary directly with the x- and y-coordinates of the various fibers. Therefore the equation for the fiber stress in the column is linear in x and y, or, in other words, it is the equation of a plane.

If the various values of fiber stress are plotted downward with reference to some base plane, such as 1-2-3-4, all the plotted values will fall on some other plane, such as 5-6-7-8, as indicated in Fig. 7–1(b). The traces of the vertical planes including the x- and y-axes are shown on the plane 5-6-7-8. If the slope of the trace of the vertical plane including the x-axis (the xf-plane) is represented by m and the slope of the trace of the vertical plane including the y-axis (the yf-plane) is represented by n, then it is apparent from the illustration that the equation for the stress on any particular fiber, such as z, will be

$$f_z = c + mx_z + ny_z.$$

The general expression for any fiber will be

$$f = c + mx + ny. \tag{7–1}$$

7–2]   THE COLUMN FLEXURE FORMULA   279

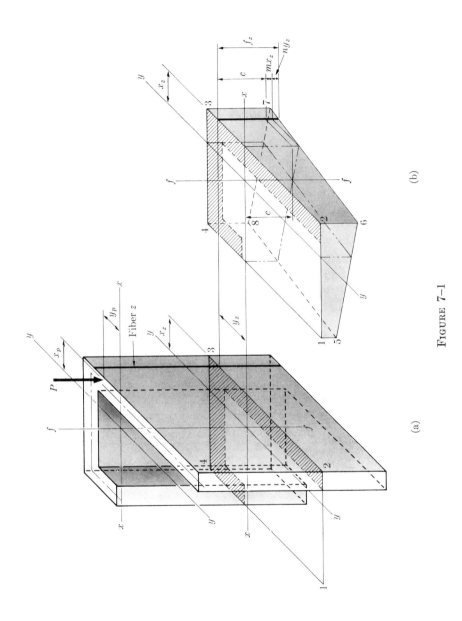

FIGURE 7–1

Now, if the part of the column above the plane 1-2-3-4 is considered as a free body, the following condition equations must be satisfied:

$$P = \int f \cdot dA, \qquad (7\text{–}2)$$

$$P \cdot y_p = \int f \cdot dA \cdot y = M_x, \qquad (7\text{–}3)$$

$$P \cdot x_p = \int f \cdot dA \cdot x = M_y. \qquad (7\text{–}4)$$

In the above equations, f is the internal fiber stress at the bottom end (plane 1-2-3-4) of the free body. Note that either side of Eq. (7–3) may be designated as M_x. The left side of this equation is M_x in terms of the external force P and the right side, the expression $\int f \cdot dA \cdot y$, is M_x in terms of the internal stresses in the column. Similarly, the left side of Eq. (7–4) is M_y in terms of the external force P and the right side, the expression $\int f \cdot dA \cdot x$, is M_y in terms of the internal stresses.

The internal stresses are of particular interest. Substituting the value of f from Eq. (7–1) in Eqs. (7–2), (7–3), and (7–4) yields

$$P = c\int dA + m\int x \cdot dA + n\int y \cdot dA, \qquad (7\text{–}5)$$

$$M_x = c\int y \cdot dA + m\int x \cdot y \cdot dA + n\int y^2 \cdot dA, \qquad (7\text{–}6)$$

$$M_y = c\int x \cdot dA + m\int x^2 \, dA + n\int xy \, dA. \qquad (7\text{–}7)$$

If reference axes through the centroid of the cross section are used, the above equations reduce to

$$P = c \cdot A, \qquad (7\text{–}8)$$

$$M_x = m \cdot I_{xy} + n \cdot I_x, \qquad (7\text{–}9)$$

$$M_y = m \cdot I_y + n \cdot I_{xy}. \qquad (7\text{–}10)$$

Solving these equations simultaneously gives the following values for c, m, and n:

$$c = \frac{P}{A}, \qquad m = \frac{M_y I_x - M_x I_{xy}}{I_x I_y - I_{xy}^2}, \qquad n = \frac{M_x I_y - M_y I_{xy}}{I_x I_y - I_{xy}^2}.$$

Substituting the above values in Eq. (7-1) yields

$$f = \frac{P}{A} + \left[\frac{M_y I_x - M_x I_{xy}}{I_x I_y - I_{xy}^2}\right] x + \left[\frac{M_x I_y - M_y I_{xy}}{I_x I_y - I_{xy}^2}\right] y, \qquad (7\text{-}11)$$

and this is the required *column flexure formula*. Attention is called to the fact that *this formula may be used only with centroidal rectangular reference axes*.

If one or both of the centroidal rectangular reference axes are axes of symmetry, then they are principal axes of inertia and the product of inertia I_{xy} of the cross section is zero. In this case, Eq. (7-11) reduces to

$$f = \frac{P}{A} + \frac{M_y}{I_y} x + \frac{M_x}{I_x} y. \qquad (7\text{-}12)$$

7-3 Development of the method. The column analogy will be derived with reference to the rigid frame of Fig. 7-2, although this should not be construed as limiting the method to frames of this kind. It may be applied to indeterminate single-span beams or to any type of single-span rigid frame or arch which may be reduced to a statically determinate and stable structure by removing up to three redundant reactions. The method may also be used to analyze closed rings, such as sewer sections.

Consider the frame of Fig. 7-2.

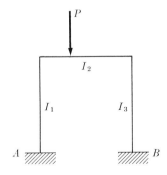

FIGURE 7-2

Arbitrarily regard the three reaction components at A as the redundants. If the support at A is removed, then the frame will deflect as indicated in Fig. 7–3, and point A will have three deflection components as shown.

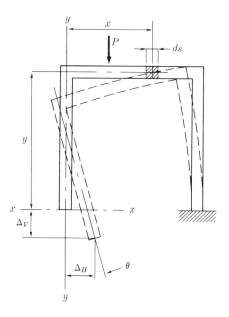

FIGURE 7–3

Let M_s represent the moment on any segment of this statically determinate, or "cut-back," structure as caused by the real load P. The flexural strain in the segment ds will be $M_s\,ds/EI$. The horizontal deflection of point A resulting from this flexural strain in the segment will be $(M_s\,ds/EI)y$, and the vertical deflection will be $(M_s\,ds/EI)x$. Therefore, if we sum the flexural strain effects for all segments, the three deflection components of A may be expressed as follows:

$$\theta = \int \frac{M_s\,ds}{EI}, \tag{7-13}$$

$$\Delta_H = \int \frac{M_s\,ds}{EI} y, \tag{7-14}$$

$$\Delta_V = \int \frac{M_s\,ds}{EI} x. \tag{7-15}$$

In the original structure, however, there is no deflection at A, all deflection being prevented by the three redundant reaction components. These three redundants, shown in Fig. 7-4, must cause three deflection compo-

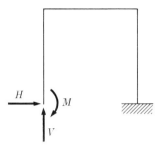

FIGURE 7-4

nents at A which are equal in magnitude and direction, but opposite in sense, to the deflection components at A caused by P in Fig. 7-3. These deflections are shown in Fig. 7-5.

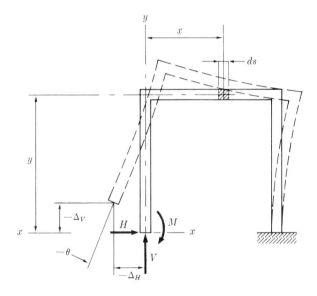

FIGURE 7-5

Let M_r represent the moment on any segment of the frame of Fig. 7-5 as caused by these redundant reaction components. The following expressions can then be written:

$$-\theta = \int \frac{M_r\,ds}{EI}, \tag{7-16}$$

$$-\Delta_H = \int \frac{M_r\,ds}{EI} y, \tag{7-17}$$

$$-\Delta_V = \int \frac{M_r\,ds}{EI} x. \tag{7-18}$$

It is important here to note that M_r is a linear function of x and y.

From Eqs. (7–13) and (7–16), (7–14) and (7–17), and from (7–15) and (7–18), it is apparent that arithmetically, for the entire structure,

$$\int \frac{M_s\,ds}{EI} = \int \frac{M_r\,ds}{EI}, \tag{7-19}$$

$$\int \frac{M_s\,ds}{EI} y = \int \frac{M_r\,ds}{EI} y, \tag{7-20}$$

$$\int \frac{M_s\,ds}{EI} x = \int \frac{M_r\,ds}{EI} x. \tag{7-21}$$

From the previous discussions of the conjugate beam and the conjugate structure the reader will recognize that the left side of Eq. (7–19) may be considered as an elastic load. It will also be apparent that the left side of Eq. (7–20) is the first moment of this elastic load about the x-axis, and that the left side of Eq. (7–21) is the first moment of this elastic load about the y-axis. Accordingly, it is possible to write

$$W = \int \frac{M_r\,ds}{EI}, \tag{7-22}$$

$$M_x = \int \frac{M_r\,ds}{EI} y, \tag{7-23}$$

$$M_y = \int \frac{M_r\,ds}{EI} x, \tag{7-24}$$

where W represents the elastic load, and M_x and M_y represent first moments of this elastic load about the x- and y-axes, respectively.

A comparison of Eqs. (7–22), (7–23), and (7–24) with Eqs. (7–2), (7–3), and (7–4) of Section 7–2 will show that the two groups of equations are analogous, and hence the column analogy. It is apparent that M_r is in every respect analogous to the fiber stress in a short column eccentrically loaded. Therefore, M_r may be considered to be the fiber stress in a short column, hereafter called the *analogous column*, the cross section of which has a center line similar in shape and length to the center line of the real structure. The width at any section is proportional to $1/EI$ of the corresponding section of the real structure. The load on any differential area of the analogous column will be $M_s\,ds/EI$; that is, M_s may be considered to be the intensity of the elastic load on the analogous column. Attention is again called to the fact that M_s at any section of the analogous column is the moment at the corresponding section of the cut-back structure with all loads acting. The length (or height) of the analogous column is not important and is considered to be some small unknown value.

As a result of the foregoing discussion, the analogous column for the frame in question would be as shown in Fig. 7–6.

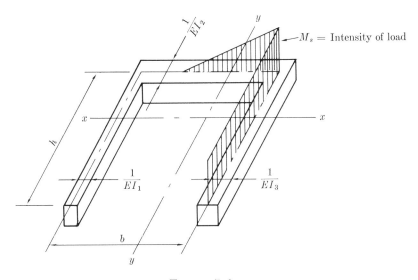

FIGURE 7–6

The reference axes which should be used for various types of structures will now be considered. If both supports are fixed, then rectangular reference axes through the centroid of the elastic areas should be used (see

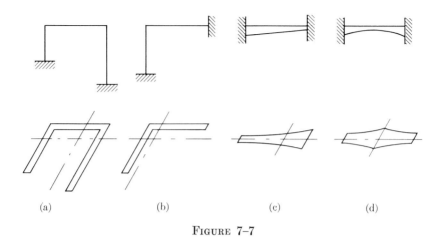

FIGURE 7-7

Fig. 7-7). If one support is hinged, then rectangular reference axes should be used through the point on the analogous column corresponding to the hinge on the real structure (see Fig. 7-8). This is because the area of the analogous column is infinite at this point, since I is zero, and consequently this is the centroid of the elastic areas. If both supports are hinged, then

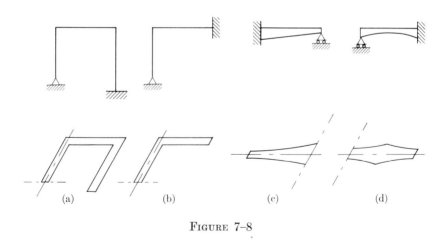

FIGURE 7-8

infinite elastic areas exist at the points on the analogous column corresponding to the hinges in the real structure (see Fig. 7-9). The axis through these two points of infinite elastic area will be a principal axis of inertia.

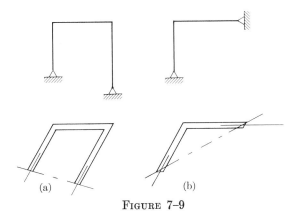

FIGURE 7-9

In Section 7-2 the column flexure formula was derived as Eq. (7-11) and may now be written as

$$f = M_r = \frac{W}{A} + \left[\frac{M_y I_x - M_x I_{xy}}{I_x I_y - I_{xy}^2}\right] x + \left[\frac{M_x I_y - M_y I_{xy}}{I_x I_y - I_{xy}^2}\right] y. \qquad (7\text{-}25)$$

When the rectangular centroidal reference axes are principal axes of inertia, I_{xy} is zero and the formula reduces to

$$M_r = \frac{W}{A} + \frac{M_y}{I_y} x + \frac{M_x}{I_x} y. \qquad (7\text{-}26)$$

An understanding of the physical significance of the various terms in Eq. (7-25) is important. From this equation the change in M_r with respect to a change in x, which must be the value of a redundant reaction V_0 acting along the y-axis, will be

$$\frac{\partial M_r}{\partial x} = \frac{M_y I_x - M_x I_{xy}}{I_x I_y - I_{xy}^2} = V_0. \qquad (7\text{-}27)$$

Similarly, the change in M_r with respect to a change in y, which must be the value of a redundant reaction H_0 acting along the x-axis, will be

$$\frac{\partial M_r}{\partial y} = \frac{M_x I_y - M_y I_{xy}}{I_x I_y - I_{xy}^2} = H_0. \qquad (7\text{-}28)$$

The first term of Eq. (7-25) is the value of a redundant moment reaction component, M_0, acting at the origin of the reference axes. It is apparent, therefore, that all three of the redundant reaction components which are represented in Eq. (7-25) actually act at the origin of the reference axes. Thus it is seen that they are *not* the actual redundant reaction components acting directly on the structure which is being analyzed. Instead, they must be considered as acting on the end of a rigid bracket which extends from one end to the origin of the reference axes. The other end of the structure is considered to be fixed. This arrangement, for an arch, is shown in Fig. 7-10. The redundant reaction components represented in Eq. (7-25) are designated in this figure as H_0, V_0, and M_0.

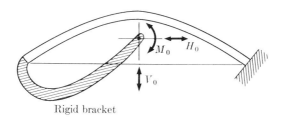

FIGURE 7-10

If desired, the rigid bracket could have been considered to be connected to the right end of the arch, with the left end fixed. If this arrangement had been assumed, the correct senses of H_0, V_0, and M_0 would have been opposite to those as determined with the bracket connected to the left end.

It is important to note that the indeterminate structure to be analyzed may be reduced to determinateness by using any combination of actual reaction components as the redundants. A different set of values for H_0, V_0, and M_0 will be found for each different set of redundant reaction components. In any case, H_0, V_0, and M_0 will constitute a system of actions which will be statically equivalent to the system of reaction components selected as the redundants. Consequently, when H_0, V_0, and M_0 have been determined for any given structure, the values of the actual redundant reaction components may be computed.

A brief discussion of the application of Eqs. (7-25) and (7-26) to the structures shown in Figs. 7-7, 7-8, and 7-9 will be helpful. Equation (7-25) applies in the case of Fig. 7-7(a) and (b). These structures are indeterminate to the third degree and all three terms of the equation will have values. The beams of Fig. 7-7(c) and (d), although theoretically indeterminate to the third degree, are considered to be indeterminate to the second degree, since flexure is caused by vertical loads only, or by vertical components of loads. Equation (7–26) applies, since the reference axes are principal axes of inertia. In addition, the last term of this equation disappears, because M_x is zero (since the elastic load is symmetrically applied with respect to the x-axis).

The structures of Fig. 7–8(a) and (b) are indeterminate to the second degree; Eq. (7–25) applies but the first term, W/A, is zero since the elastic area is infinite. The propped cantilever beams of Fig. 7–8(c) and (d) are indeterminate to the first degree; Eq. (7–26) applies, with the first and last terms reducing to zero.

The frames of Fig. 7–9(a) and (b) are indeterminate to the first degree; Eq. (7–26) applies, since the reference axes are principal axes of inertia. The first term reduces to zero, since the elastic area is infinite, and the second term is zero, since I_y is infinite. For this reason the location of the y-axis is immaterial.

The exact method of applying the column analogy can be easily inferred from the preceding discussions, but a recapitulation may be helpful. The step-by-step procedure is outlined below:

(1) The structure is cut back, as would be done for a solution by the general method. Any one reaction component, or any combination of two or three reaction components (depending on the degree of redundancy of the structure), may be selected as the redundants which are removed to cut back the structure to stability and determinateness. These reaction components do not all have to act at one and the same support. As previously indicated, however, the column analogy is inapplicable to structures with higher than the third degree of redundancy.

(2) The dimensions of the cross section of the analogous column are computed. The center line of this cross section is identical to the center line of the structure but is considered to be positioned in a horizontal plane. The width at any section is proportional to $1/EI$ of the corresponding section of the structure. The units to be used in expressing the length of the center line, E and I, will be discussed in the next section.

(3) The analogous column is loaded with the M_s diagram, that is, the bending moment diagram of the loaded cut-back structure. This M_s diagram is drawn on the center line of the column cross section and perpendicular to the plane of that cross section. Any ordinate of this diagram represents the *intensity of load* on the analogous column at that point. The units to be used in expressing these ordinates will be discussed in the next section.

(4) Rectangular reference axes are drawn in the plane of the column cross section, with their origin at the centroid of the elastic area of that

cross section. Any pin or hinge support on the original structure is represented on the column cross section as an infinite area concentrated at the corresponding point.

(5) The fiber stresses in the analogous column (which are at once the bending moments M_r in the structure as caused by the redundant reactions) are computed either by Eq. (7–25) or by Eq. (7–26).

(6) The bending moments M_r are combined with the bending moments M_s at corresponding sections to obtain the moments in the original indeterminate structure. The exact method of combining these moments can best be explained by an illustrative problem (see Example 7–1).

7–4 Units in the column analogy. Confusion often exists regarding the units which should be used to express the bending moment M_s, as well as E and I of the original structure, when setting up the analogous column. For example, if the moments M_s in the cut-back structure are expressed in ft·k, the modulus of elasticity E in k/in^2, and the moment of inertia I of the real structure in in^4, the question arises as to what units will apply to the moments in the real structure which are represented by the fiber stresses in the analogous column. This can be easily answered by a dimensional analysis of the previously derived expressions for these fiber stresses.

The simpler of these expressions is Eq. (7–26), which will be used for the discussion:

$$f = M_r = \frac{W}{A} + \frac{M_y}{I_y} x + \frac{M_x}{I_x} y.$$

In the above equation:

$f = $ the fiber stress in the analogous column.

$M_r = $ the moment at the section in the original structure corresponding to the fiber in the analogous column, defined by x and y, as caused by the redundant reactions.

$W = $ the algebraic summation of the elastic loads on the analogous column.

M_y (or M_x) $= $ the first moment of the total elastic load with respect to the y- (or x-) axis.

$A = $ the total elastic area of the analogous column.

I_y (or I_x) = the second moment of the elastic area of the analogous column with respect to the y- (or x-) axis.

x and y = coordinates of the fiber on which the stress f exists.

In Eq. (7–26) the following substitutions can be made:

$$M_y = We_y, \qquad M_x = We_x,$$

$$I_y = Ar_y^2, \qquad I_x = Ar_x^2,$$

where

e_y (or e_x) = the perpendicular distance from the y- (or x-) axis to the resultant W of the elastic loads;

r_y (or r_x) = the radius of gyration of the elastic area A with respect to the y- (or x-) axis.

Furthermore, *dimensionally* it is possible to write

$$W = M_s A,$$

where M_s, as previously defined, is the bending moment in the cut-back structure as caused by the system of real loads for which the structure is being analyzed.

When we have made all substitutions, Eq. (7–26) becomes

$$f = M_r = \frac{M_s A}{A} + \frac{M_s A e_y}{Ar_y^2} x + \frac{M_s A e_x}{Ar_x^2} y. \qquad (7\text{–}29)$$

The above expression is of no value except to check dimensions. It is immediately apparent that the units of f, and hence of M_r, are the same as those used for M_s, since all other dimensions on the right side of the equation cancel. Obviously it is immaterial what units are used to express E and I of the real structure or to express the length of the segments of the analogous column, since these will only affect the units for A, which, from the preceding, will cancel. The units for e, r, x, and y must, of course, be consistent within themselves so that they too will cancel.

7–5 Analysis of continuous frames. Only one problem will be presented to illustrate the application of the column analogy in the analysis of continuous frames, but this should be sufficient to demonstrate the method.

Moreover, since moment distribution, or slope-deflection, or the general method with the conjugate structure will usually require less work to effect a solution, greater emphasis is not justified. Note, however, that the analogous column is effective as a method for analyzing single-span arches.

Before we proceed with the illustrative problem, it would be well to discuss one particular computation which occurs quite frequently if the column analogy is used to any great extent. This is the evaluation of the moments of inertia and product of inertia of a long and relatively narrow rectangle with respect to rectangular axes which are not parallel to the major and minor centroidal axes. For example, in Fig. 7–11, the values of $I_{x'}$, $I_{y'}$, and $I_{x'y'}$ are required.

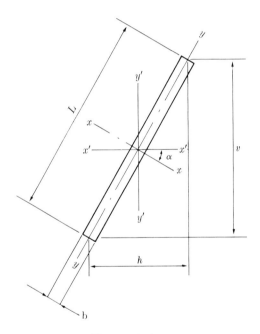

FIGURE 7–11

From mechanics,

$$I_{x'} = I_x \cos^2 \alpha - 2I_{xy} \sin \alpha \cos \alpha + I_y \sin^2 \alpha,$$

$$I_{y'} = I_x \sin^2 \alpha + 2I_{xy} \sin \alpha \cos \alpha + I_y \cos^2 \alpha,$$

$$I_{x'y'} = (I_x - I_y) \sin \alpha \cos \alpha + I_{xy} (\cos^2 \alpha - \sin^2 \alpha).$$

For the rectangle in question,

$$I_{xy} = 0,$$

$$I_x = bL^3/12,$$

and I_y is small and is neglected.

Substitution of these values in the above equations will result in

$$I_{x'} = \frac{bL^3}{12}\cos^2\alpha = \frac{Av^2}{12},$$

$$I_{y'} = \frac{bL^3}{12}\sin^2\alpha = \frac{Ah^2}{12},$$

$$I_{x'y'} = \frac{bL^3}{12}\sin\alpha\cos\alpha = \frac{Ahv}{12}.$$

EXAMPLE 7–1. Find the moments at points A, B, C, D, and E of Fig. 7–12 by the column analogy. Cut back the original structure by removing the support at A.

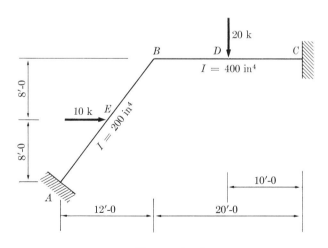

FIGURE 7–12

The centroid of the elastic areas must first be located; through this the rectangular reference axes are drawn. The necessary computations are shown in Table 7–1. The coordinates x and y are referred to point A.

TABLE 7-1.

Member	Relative $1/EI$	L (ft)	Relative elastic area A	x (ft)	y (ft)	Ax	Ay	x' (ft)	y' (ft)
AB	2	20	40	+6	+8	+240	+320	−5.33	−2.67
BC	1	20	20	+22	+16	+440	+320	+10.67	+5.33
			60			+680	+640		

$$x_0 = \frac{+680}{60} = +11.33 \text{ ft}, \qquad y_0 = \frac{+640}{60} = +10.67 \text{ ft}.$$

The reference axes, together with the sign conventions assumed for x' and y', are shown on the sketch of the analogous column in Fig. 7-13.

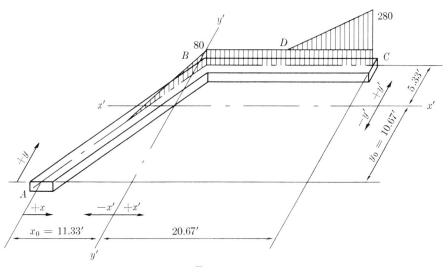

FIGURE 7-13

Having located the x' and y' reference axes as shown in this figure, we are now required to compute $I_{x'}$, $I_{y'}$, and $I_{x'y'}$. Certain preliminary work is necessary for AB. Segment AB of the analogous column is shown in Fig. 7-14.

ANALYSIS OF CONTINUOUS FRAMES

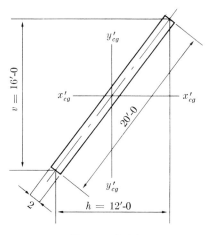

FIGURE 7-14

From the expressions previously derived,

$$I_{x'_{cg}} = \frac{Av^2}{12} = \frac{40 \times 16^2}{12} = 854,$$

$$I_{y'_{cg}} = \frac{Ah^2}{12} = \frac{40 \times 12^2}{12} = 480,$$

$$I_{x'y'_{cg}} = \frac{Ahv}{12} = \frac{40 \times 12 \times 16}{12} = 640.$$

The remainder of the computations for $I_{x'}$, $I_{y'}$, and $I_{x'y'}$ for the entire column cross section are shown in Table 7-2.

TABLE 7-2.

Member	$I_{x'_{cg}}$	Ay'^2	$I_{y'_{cg}}$	Ax'^2	$I_{x'y'_{cg}}$	$Ax'y'$
AB	854	285	480	1140	+640	+570
BC	Negligible	568	667	2275	0	+1140
Totals	854	853	1147	3415	+640	+1710

$$I_{x'} = 854 + 853 = 1707,$$

$$I_{y'} = 1147 + 3415 = 4562,$$

$$I_{x'y'} = +640 + 1710 = +2350.$$

The values for W, M_y, and M_x are computed in Table 7-3. Note that in writing M_s, a moment resulting in tension on the outside of the frame has been designated as positive.

TABLE 7-3.

Member	Average M_s	Loaded relative elastic area	Relative elastic load W	x'	y'	$M_y = Wx'$	$M_x = Wy'$
AB	+40	20	+800	−1.33	+2.66	−1,060	+2,130
BC	+80	20	+1,600	+10.67	+5.33	+17,070	+8,530
DC	+100	10	+1,000	+17.33	+5.33	+17,330	+5,330
			+3,400			+33,340	+15,990

The above values are used to compute W/A and the coefficients of x and y in Eq. (7-25) as follows:

$$\frac{W}{A} = +\frac{3400}{60} = +56.67 \text{ ft·k},$$

$$\frac{M_y I_x - M_x I_{xy}}{I_x I_y - I_{xy}^2} = \frac{(+33,340)(+1707) - (+15,990)(+2350)}{(+1707)(+4562) - (+2350)^2} = +8.54 \text{ k},$$

$$\frac{M_x I_y - M_y I_{xy}}{I_x I_y - I_{xy}^2} = \frac{(+15,990)(+4562) - (+33,340)(+2350)}{(+1707)(+4562) - (+2350)^2} = -2.39 \text{ k}.$$

Moments are required at several points around the frame. The computations for M_r at these several points, using Eq. (7-25), as well as the combination of M_r with M_s to obtain the final moment M, are now advantageously arranged as shown in Table 7-4.

TABLE 7-4.

Point	M_s (ft·k)	$\frac{W}{A}$	x' (ft)	$+8.54x'$	y' (ft)	$-2.39y'$	M_r (ft·k)	$M = M_s - M_r$ (ft·k)
A	—	+56.67	−11.33	−96.76	−10.67	+25.50	−14.59	+14.59
B	+80	+56.67	+0.67	+5.72	+5.33	−12.74	+49.65	+30.35
C	+280	+56.67	+20.67	+176.52	+5.33	−12.74	+220.45	+59.55
D	+80	+56.67	+10.67	+91.12	+5.33	−12.74	+135.05	−55.05
E	—	+56.67	−5.33	−45.52	−2.67	+6.38	+17.53	−17.53

An explanation is required relative to the heading for the last column of Table 7–4, $M = M_s - M_r$. The general expression would be

$$M = M_s \pm M_r.$$

The decision as to whether the positive or negative sign should be used for M_r must be made for each individual problem.

In the preceding problem the tabular values of M_s are correct for their respective points in the frame with the redundant reactions removed. When these reactions are replaced, the largest value of M_s (in this case, $+280$ at C) must be reduced. M_r is the internal moment resulting from the action of these redundant reaction components. Therefore the sign of M_r must be such that when added algebraically to the maximum value for M_s, the resulting value for M will be smaller than M_s. It is obvious in the foregoing problem that the correct sign for M_r in the expression for M will be negative. Particular attention is called to the fact that this adjustment will make the results of the analogous column consistent with the sign convention originally assumed for the M_s in the cut-back structure; that is, a moment causing tension on the outside of the frame is positive. This requirement that M shall be less than M_s applies only to the point of maximum M_s in the frame; M at other points may be larger than M_s.

The redundant reaction components which are represented by W/A and by the coefficients of x and y in Eqs. (7–25) and (7–26) *always* act at the origin of the reference axes; that is, they act at the centroid of the elastic areas. The redundant reaction components represented by the coefficients of x and y *always* act along the y- and x-axes, respectively. As previously suggested, these reaction components may be considered to act on the end of a rigid bracket which extends from one end of the structure to the origin of the reference axes. The other end of the structure is considered to be fixed when the moments caused throughout the structure by these reaction components are evaluated. This is true regardless of the combination of actual reaction or stress components selected as the redundants. Internal stress components may also be selected as the redundants, as, for example, the stress components at the crown of a fixed-end arch.

The particular redundant reaction components which are represented in Eqs. (7–25) and (7–26), however, are always the static equivalent of the actual redundant reaction or stress components selected for the purpose of the analysis.

In Example 7–1 the redundant moment component acting at the origin of the reference axes is given by W/A and is $+56.67$ ft·k. The redundant vertical component, the coefficient of x, is $+8.54$ k; the redundant horizontal component, the coefficient of y, is -2.39 k. These values are correct in magnitude but may be incorrect in sign.

It was previously determined that, in order to make the sign of M_r consistent with the sign convention assumed for M_s, all stresses in the analogous column should have their signs reversed, and this means that the signs of the above reaction components must be reversed. Consequently, the sign of the moment reaction component becomes negative, which signifies that this moment must have a sense such that it will cause compression on the outside of the frame; that is, it must be clockwise. The vertical reaction component becomes -8.54 k and must have a sense such that when it is multiplied by a positive x-coordinate, a negative moment (causing compression on the outside fibers) will result. The sense must therefore be up. The horizontal reaction component will be $+2.39$ k and must have a sense such that it will cause positive moment on all sections of the structure with a positive y-coordinate. The sense is therefore toward the right.

Having evaluated the redundant reaction components acting at the centroid of the elastic areas, we can determine the sense and magnitude of each of the redundant reactions acting directly on the cut-back structure. This is possible because *the two sets of reaction components are statically equivalent.* Several sketches will serve to illustrate this point.

In Fig. 7–15 the reactions acting at C on the cut-back structure are indicated. In Fig. 7–16 are shown the redundant reaction components which, by the analogous column, have been determined as acting at the centroid of the elastic areas. Observe that reactions are induced at C and also that the rigid bracket could have been assumed to be connected to end C of the structure with end A fixed. In this case, the redundants acting on the end of the bracket would be of the same magnitude as determined above but would be reversed in sense.

By statics, the redundant reactions at A may be computed and are shown in Fig. 7–17. The induced reaction components at C remain as in Fig. 7–16. If the loads and reaction components of Fig. 7–15 are added to the reaction components of Fig. 7–17, the loaded structure, with the final value for all reaction components, will result as shown in Fig. 7–18.

As previously indicated, if the real structure is externally redundant to the second degree, that is, if one support is fixed and the other is pinned, then the centroid of the elastic areas will coincide with the pin. If the real structure is redundant to the first degree, which will be the case if both supports are pinned, then the centroid of the elastic areas will be located on the line joining the two pins and midway between them. The single redundant reaction will act along this line.

7-5] ANALYSIS OF CONTINUOUS FRAMES 299

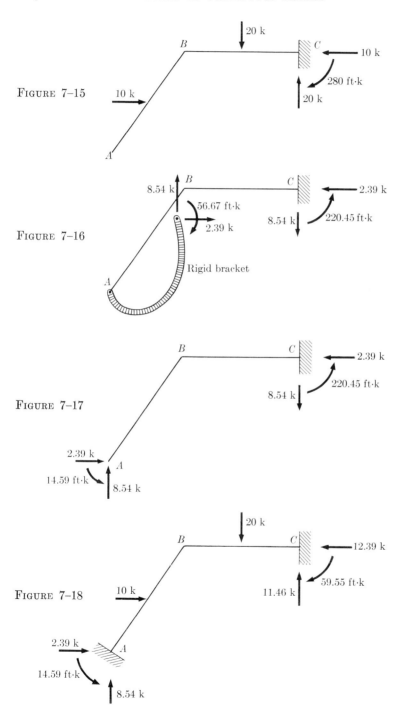

Figure 7-15

Figure 7-16

Figure 7-17

Figure 7-18

Problems

7-2. Determine the moments at B and C (see Fig. 7-19) by the column analogy. E and I are constant. [*Ans.:* $B = 43$ ft·k tension on outside, $C = 79$ ft·k tension on bottom.]

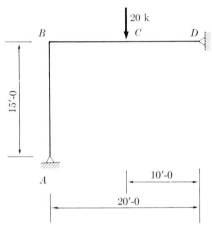

FIGURE 7-19

7-3. For Fig. 7-20 find the moments at B, C, and D by the column analogy. E is constant. [*Ans.:* $B = 0.8$ ft·k tension on outside, $C = 4.1$ ft·k tension on bottom, $D = 1.1$ ft·k tension on outside.]

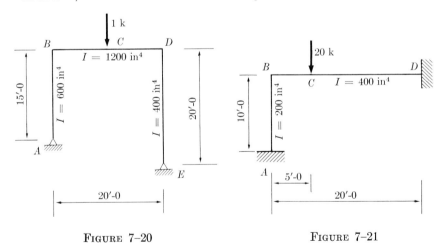

FIGURE 7-20   FIGURE 7-21

7-4. Using the column analogy, find the reaction components acting at A in Fig. 7-21. [*Ans.:* $H = 4.1$ k right, $V = 14.8$ k up, $M = 13.8$ ft·k clockwise.]

7-5. For Fig. 7-22 find the moments at critical points by the analogous column. E and I are constant. Consider that a moment causing tension on the outside of the frame is positive. [*Ans.*: $A = -32$ ft·k, $B = +42$ ft·k, $C = -39$ ft·k, $D = +16$ ft·k, $E = -22$ ft·k, $F = +41$ ft·k.]

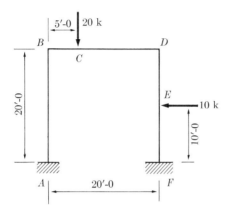

FIGURE 7-22

GENERAL REFERENCES

1. CROSS, H. and MORGAN, N. D., *Continuous Frames of Reinforced Concrete*. Chap. 3. New York: Wiley, 1932.
2. GRINTER, L. E., *Theory of Modern Steel Structures*. Vol. 2. New York: Macmillan, 1949.
3. PARCEL, J. I. and MOORMAN, R. B. B., *Analysis of Statically Indeterminate Structures*. Chap. 4. New York: Wiley, 1955.
4. SUTHERLAND, H. and BOWMAN, H. L., *Structural Theory*. Chaps. 8 and 10. New York: Wiley, 1950.
5. WANG, C. K., *Statically Indeterminate Structures*. Chap. 9. New York: McGraw-Hill, 1953.
6. WILLIAMS, C. D., *Analysis of Statically Indeterminate Structures*. Chap. 10. Scranton: International, 1943.

CHAPTER 8

INTRODUCTION TO MOMENT DISTRIBUTION

8-1 General. The method of moment distribution was introduced by Professor Hardy Cross, who began teaching it to his students at the University of Illinois in 1924. It was given to the profession in the 1932 Transactions of the American Society of Civil Engineers in a paper (1) entitled "Analysis of Continuous Frames by Distributing Fixed-End Moments."

Moment distribution attracted immediate attention. During the several years immediately following its introduction, the journals of various professional societies in the United States published numerous papers on the method, and many of these represented significant contributions. Thus the number of those who assisted in bringing the *Cross method of moment distribution* to its present state of refinement is very great. This in itself is a notable tribute to the splendid teacher and engineer through whose perception the method was conceived.

There are some who believe that moment distribution constitutes the greatest single contribution ever made to indeterminate structural theory. This, of course, is a matter of opinion but the fact remains that the method has provided a means whereby many types of continuous frames which were formerly designed by empirical rules or approximate methods can now be analyzed with accuracy and comparative ease. The method has only slight application, however, to the analysis of articulated structures.

8-2 Sign convention for moments. An understanding of the sign convention to be used throughout the discussion of moment distribution is necessary. Any moment considered at the end of a member will always be the moment the member exerts on the joint or support. *If the member tends to rotate the support clockwise, the moment will be considered to be positive. If the member tends to rotate the support counterclockwise, then the moment will be considered to be negative.*

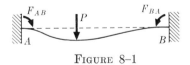

FIGURE 8-1

In Fig. 8-1 the fixed-end moment at A, F_{AB}, is positive but that at B, F_{BA}, is negative. In the frame shown in Fig. 8-2 the fixed-end moment F_{AB} is positive, F_{BA} is negative, F_{BC} is positive, and F_{CB} is negative.

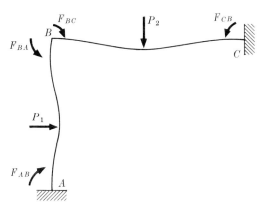

FIGURE 8-2

8-3 Absolute stiffness and distribution factor. It is necessary to have a measure of the capacity of a member to resist the rotation of one end when a moment is applied at that end. Consider Fig. 8–3. If the support at A rotates through some angle, the distorted beam will be as shown in Fig. 8–4. So far as the internal stresses in the beam are concerned, the left support can be represented as a simple support and the effect of the rotating wall can be replaced by an external couple M. The result is shown in Fig. 8–5, and from this figure the measure of the capacity of a

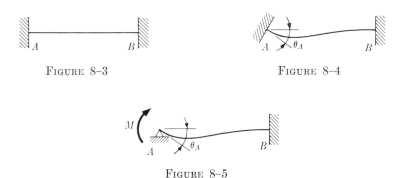

FIGURE 8-3   FIGURE 8-4

FIGURE 8-5

member to resist rotation of one end can be readily understood. If the value of θ is 1 radian, then the value of M required to produce this unit rotation is arbitrarily called the *absolute stiffness*. Therefore the following definition will apply: *Absolute stiffness is the value of the moment, applied at a simply supported end of a member, necessary to produce a rotation of 1 radian of this simply supported end, no translation of either end being permitted and the far end being either simply supported, restrained, or fixed.*

The application of absolute stiffness to moment distribution may be understood to some degree by reference to Fig. 8–6. Under the action of the applied moment M, the frame will deflect as in Fig. 8–7.

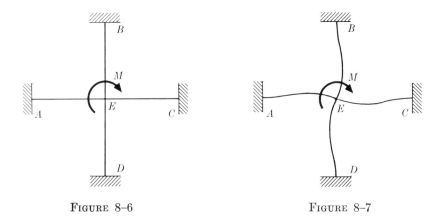

FIGURE 8–6   FIGURE 8–7

It is necessary to know what internal moments are developed in the ends of the various members at E. Suppose that the values of the absolute stiffness for the various members (in inch kilopounds) are as indicated in Fig. 8–8. It should be clear that the summation of the absolute stiffnesses

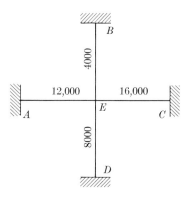

FIGURE 8–8

of all the members framing at E will be a measure of the resistance of joint E to rotation. It should be equally clear that the action of any applied moment M will be resisted by each member in proportion to that member's capacity to resist rotation of its end at E. Each member, then, will supply a proportion of the total resisting moment necessary to satisfy the condition that $\Sigma M = 0$ at the joint. Thus M_{EA} (the internal moment at end E of the member EA) would be $(12,000/40,000) \times M$, M_{EC} would be $(16,000/40,000) \times M$, and M_{EB} and M_{ED} can be expressed in the same way. The value of $(12,000/40,000)$ is called the *distribution factor* for EA; $(16,000/40,000)$ is the distribution factor for EC. The following definition then results: *The distribution factor for any member at a joint is equal to the stiffness of the member divided by the sum of the stiffnesses of all members at the joint.*

8–4 Relative stiffness of a member. It is apparent that the foregoing values for absolute stiffness are rather cumbersome numbers to handle; obviously they would be much easier to use in computations if reduced to lowest terms by a common divisor. When such a division is made, the resulting numbers will all be related in some fixed ratio to corresponding values of absolute stiffness and are therefore called *relative stiffnesses.*

It will presently be demonstrated that the absolute stiffness of a prismatic member with the far end fixed is $4EI/L$. Thus when a frame is composed entirely of prismatic members, it is most convenient and customary to use values of I/L for the relative stiffness of each member. If, however, one or more members in the frame is nonprismatic, and since all relative stiffnesses must be related to the absolute stiffnesses in the same ratio, care must be used in selecting the relative stiffnesses for the prismatic members. It is probable that the value of I/L will not be correct.

For the purpose of the discussion, assume that all members of the frame under consideration are prismatic. In this case, the relative stiffnesses may be obtained by dividing the absolute stiffnesses by $4E$. If E is taken as 1000 k/in^2, the relative stiffnesses will be as shown in Fig. 8–9. The values of the distribution factors are grouped around the joint.

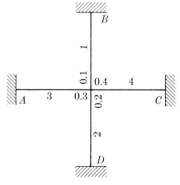

FIGURE 8–9

8–5 Carry-over factor. Consideration of the distorted frame of Fig. 8–7 will make it apparent that when a resisting moment develops in the end of a member at E, then a moment is also induced at the far end of that member. This induced moment at the fixed or restrained end of any member, when the opposite end is rotated, always has a certain definite relationship to the resisting moment developed at the rotated end. *The carry-over factor is that factor by which the developed moment at the rotated end of a member may be multiplied (the other end being fixed or restrained) to give the induced moment at the fixed or restrained end.*

It will now be demonstrated that *this carry-over factor for prismatic members is always $+\frac{1}{2}$ if the far end is fixed.* Figure 8–10 shows a propped cantilever beam with a moment M applied at the simply supported end.

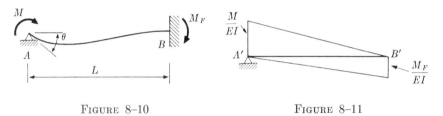

FIGURE 8–10   FIGURE 8–11

This moment causes the rotation of this end through the angle θ, and the moment M_F is induced at the fixed end. The conjugate beam for this situation is shown in Fig. 8–11. For this conjugate beam, ΣM must equal zero about A', from which

$$\frac{M}{EI} \cdot \frac{L}{2} \cdot \frac{L}{3} - \frac{M_F}{EI} \cdot \frac{L}{2} \cdot \frac{2}{3}L = 0,$$

which results in

$$M_F = +\frac{M}{2}.$$

The significance of the positive sign in the above equation is that the assumption made as to the sign of M_F when loading the conjugate beam (that is, that M_F would cause tension on the top fibers of the beam) is correct. Note that the sign convention used thus far in this derivation is necessarily that which was arbitrarily adopted for the conjugate beam; that is, a moment causing compression on the top fibers of the real beam is considered to be a positive moment and is drawn on the top of the conjugate beam. The above expression for M_F, then, aside from having the significance mentioned, also means that if the moment applied at the free end causes compression on the top side of the real beam, then the induced moment at the fixed end will cause tension in the top side.

It is necessary to interpret these findings in terms of the sign convention for moment distribution. If the moments exerted by the beam are indicated, the sketch of Fig. 8–12 will result. It is now obvious that the

FIGURE 8–12

moment exerted by the beam at the fixed end will act with the same sense as the moment exerted by the beam at the simply supported end. Therefore, according to the moment distribution sign convention which has been adopted, the carry-over factor for a prismatic beam with the far end fixed will be $+\frac{1}{2}$.

8–6 Evaluation of absolute stiffness of prismatic members. The value of the angle θ in terms of the applied moment M, in Fig. 8–10, is equal to the shear in the conjugate beam at A', in Fig. 8–11. Therefore,

$$\theta_A = \left(M \cdot \frac{L}{2} - M_F \cdot \frac{L}{2}\right)\frac{1}{EI}$$

$$= \left(M \cdot \frac{L}{2} - \frac{M}{2} \cdot \frac{L}{2}\right)\frac{1}{EI} = \frac{ML}{4EI},$$

from which

$$M = \frac{4EI\theta_A}{L}.$$

Then, with $\theta_A = 1$, the value for the absolute stiffness becomes

$$K_{AB} = \frac{4EI}{L}.$$

Another relationship that results in a considerable simplification when the end of a span rests on a simple support (as at the simply supported ends of a continuous beam) will now be developed.

Suppose that a moment M is applied at the left end of a simple beam and results in a rotation θ_A of the left end, as shown in Fig. 8–13.

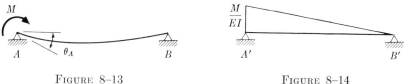

FIGURE 8–13   FIGURE 8–14

The conjugate beam is represented by Fig. 8–14. The value of θ_A is again given by the shear at A' of the conjugate beam:

$$\theta_A = \frac{M}{EI} \cdot \frac{L}{2} \cdot \frac{2L}{3} \cdot \frac{1}{L} = \frac{ML}{3EI},$$

from which

$$M = \frac{3EI\theta_A}{L},$$

and the absolute stiffness for this beam is

$$K_{AB} = \frac{3EI}{L}.$$

In other words, the absolute stiffness (or the relative stiffness) of a simply supported prismatic member is three-fourths of the absolute stiffness (or the relative stiffness) the same member would have had if the far end had been fixed instead of simply supported.

8–7 Fixed-end moments induced by displaced supports of prismatic members. It is necessary to derive the expression for the moments induced in the ends of a prismatic member, fixed at one or both ends, when one end B is displaced a distance Δ relative to the other end A. The effects of such a displacement on a member fixed at both ends are shown in Fig. 8–15.

FIGURE 8–15

The reactions indicated are those exerted by the supports on the member. It is apparent that the condition $\Sigma M = 0$ for the conjugate beam (represented by Fig. 8–16) can only be satisfied by applying an external moment Δ to the end B'.

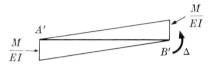

FIGURE 8–16

The moment must be applied at this end, since it has been stated that B deflects relative to A. Consequently,

$$\Delta = \frac{M}{EI} \cdot \frac{L}{2} \cdot \frac{L}{3},$$

from which

$$M = \frac{6EI\Delta}{L^2}.$$

If one end of the member is simply supported, the moment at the fixed end will be given by $M = 3EI\Delta/L^2$.

8–8 Why moment distribution works. A simple example will serve to illustrate the *how* and *why* of the moment distribution method. Consider the continuous beam of Fig. 8–17. Here the relative values of K are

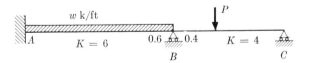

FIGURE 8–17

shown, as are the distribution factors for joint B. The spans and loads are such that the values of the internal fixed-end moments are as follows: $F_{AB} = +20$, $F_{BA} = -20$, $F_{BC} = +40$, and $F_{CB} = -30$. These are the fixed-end moments (in ft·k) for the beam ends indicated by the subscripts and as caused by the loads shown; and these fixed-end moments can exist only if joint rotation is prevented at all joints. This can only be true at B and C if some temporary external moment is applied at each of these joints to satisfy the condition that $\Sigma M = 0$. These temporary external moments, designated as M_{B1} and M_{C1}, are shown in Fig. 8–18.

```
                    M_B1 = 20        M_C1 = 30
   A         0.6  ⩔  0.4
                     B                  C
   |+20        -20 | +40           -30 |
```

FIGURE 8–18

It is apparent, however, that M_{B1} and M_{C1} cannot be permitted to remain, for this would alter the conditions of the problem. Accordingly, an external moment M_{C2} is introduced at C to cancel the effect of M_{C1}. (See Fig. 8–19.) The action of M_{C2} will develop an internal moment

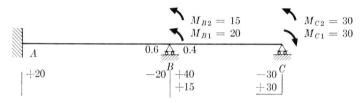

FIGURE 8–19

acting clockwise at C which will resist the rotation of joint C as caused by M_{C2} and which will maintain the condition for the joint that $\Sigma M = 0$. This developed internal moment will thus be $+30$.

It has been previously established that the above moment will induce at the other end of the span an internal moment of the same sign and with one-half its value. In other words, as indicated by application of the carry-over factor of $+\frac{1}{2}$, there will be an induced internal moment at B of $+15$. However, this moment induced at B can exist only if an additional temporary external moment M_{B2} is applied to satisfy $\Sigma M = 0$; obviously, M_{B2} will have a value of 15 and act counterclockwise. The external moments M_{C2} and M_{B2} are shown as additional vectors in Fig. 8–19. The internal moments are added, with their correct signs, in the tabulation under the span, and a horizontal line is drawn under the $+30$ entry at C to indicate that the joint is balanced, that is, that all temporary external moments are removed at the joint.

Joint B, however, is not in balance. In order to remove all temporary external moments at this joint, we must apply an additional temporary external moment M_{B3} having a value 35 and acting in a clockwise direction. (See Fig. 8–20.) As before, this external moment will develop internal moments in BA and BC which will act in a counterclockwise direction. The

8-8] WHY MOMENT DISTRIBUTION WORKS

[Figure 8-20 diagram with beam ABC, fixed at A, supports at B and C]

```
                    M_B3 = 35
                    M_B2 = 15         M_C2 = 30
                    M_B1 = 20         M_C1 = 30
   A          0.6       0.4
                     B              C
  +20        -20  +40           -30
                  +15           +30
             -21  -14
```

FIGURE 8-20

magnitudes of these moments will depend on the relative stiffnesses of the members, or, in other words, on the distribution factors at the joint. The developed internal moment in BA will be $0.6 \times 35 = 21$, and in BC it will be $0.4 \times 35 = 14$; since both of these act in a counterclockwise direction, they will be negative. These developed moments are entered in the tabulation under the beam of Fig. 8-20, and a horizontal line is again drawn to indicate that all temporary external moments have been removed at the joint, or, in other words, that the joint is in balance.

Now, however, these developed internal moments will result in induced moments at the opposite ends of the two spans, each one equal to $+\frac{1}{2}$ of the respective developed moment. The induced moments can exist only if rotation of the joints of their induction is prevented. Since A is a fixed support, the required internal moment can be induced at this joint without difficulty, and its value would be $+\frac{1}{2}(-21)$ or -10.5 ft·k. At joint C, however, rotation must be prevented by the application of a new temporary external moment equal in magnitude and opposite in sense to the induced internal moment, whose value would be $+\frac{1}{2}(-14)$ or -7, indicating an action on the joint in a counterclockwise direction. Consequently, the new temporary external moment M_{C3} (shown in Fig. 8-21, together with the complete tabulation of internal moments up to this point) would have a value of 7 ft·k acting clockwise.

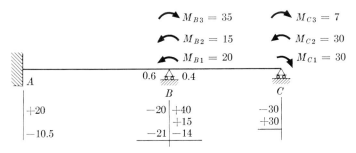

FIGURE 8-21

The complete cycle as previously described is now repeated as many times as required to yield results of the desired accuracy, the corrections in successive cycles becoming smaller and smaller until they no longer have practical significance. Figure 8–22 shows the complete distribution.

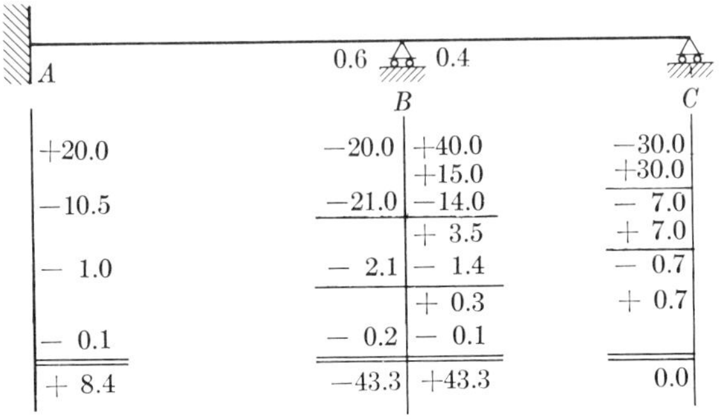

	A		0.6 B 0.4		C
	+20.0	−20.0	+40.0	−30.0	
			+15.0	+30.0	
	−10.5	−21.0	−14.0	− 7.0	
			+ 3.5	+ 7.0	
	− 1.0	− 2.1	− 1.4	− 0.7	
			+ 0.3	+ 0.7	
	− 0.1	− 0.2	− 0.1		
	+ 8.4	−43.3	+43.3	0.0	

FIGURE 8–22

It should be pointed out that consideration of the temporary external moments is entirely unnecessary in the actual working of a problem, but it is absolutely necessary to consider them if the reader is to get a real understanding of why the method works. However, such an understanding, although desirable, is not necessary to enable one to use the method. As a matter of fact, many who use moment distribution do not actually understand the *why* of it; they merely know how to perform the balancing operation. This whole operation can be largely automatic and can require very little thought. In order to balance any joint, it is necessary only to determine the magnitude and sign of the unbalanced internal moment that must be distributed to the various members intersecting at that joint to make ΣM for the internal moments equal zero. This unbalanced internal moment, with sign reversed, is distributed to the members intersecting at the joint in accordance with the distribution factors. The carry-over operation then performed will usually upset the previous balance of one or more joints, and these must be rebalanced. The carry-over from these joints will upset other joints, and so on, the corrections becoming smaller and smaller until they are no longer significant.

The balancing operation can be carried to any required degree of accuracy. The order in which the various joints are balanced is unimportant insofar as the final results are concerned. In most cases, however, the author prefers to balance next that joint with the largest unbalanced moment. He also prefers to perform the carry-over operation immediately after balancing each joint. The final answer for the internal moment at any member end is, of course, the algebraic sum of all internal moments tabulated for that member end.

8-9 Fixed-end moments for various loads. Prismatic members.

Occasionally the designer is confronted with unusual loads on beams for which the expressions for fixed-end moments are not readily available. Accordingly, the following sketches of beam loads and moment diagrams are presented. The loading shown in Fig. 8–23(a) is not unusual, of course, but is included here because of its importance. Note that the expressions given for the fixed-end moments for the single load P can be used to find fixed-end moments for any arrangement of uniform load by substituting $w\,dx = P$, $a = x$, and $b = L - x$, and integrating between the proper limits.

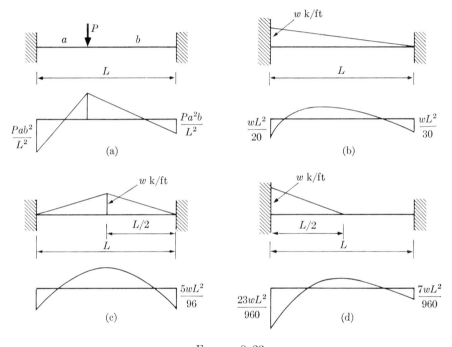

FIGURE 8–23

The formula for the fixed-end moments for the case of Fig. 8–24, where any number of equal loads are spaced equally across the span, is at times very useful.

The expression for this fixed-end moment is

$$F = \frac{(N^2 - 1)PL}{12N}.$$

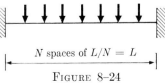

FIGURE 8–24

8-10 Continuous beams with prismatic members. The application of moment distribution to the analysis of continuous beams can be most effectively explained by illustrative examples, several of which follow.

EXAMPLE 8–1. Solve for moments at A and B (see Fig. 8–25) by moment distribution, using (a) the ordinary method, and (b) the simplified method.

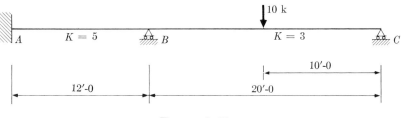

FIGURE 8–25

NOTE. In the application of moment distribution, it has been almost universal procedure to show the actual balancing computations directly on a sketch of the structure, as in the illustrations in Section 8–8. This arrangement is satisfactory when continuous beams are being analyzed. However, when the structure is composed of horizontal and vertical members, with perhaps a few of these inclined, the method of placing the balancing computations in columns perpendicular to the members results in columns of figures extending in all directions, often interfering with one another. It is therefore recommended that the balancing computations be placed in a table, entirely separate from a sketch of the structure. For the majority of cases this tabular arrangement will be advantageous.

For this example the fixed-end moments are computed as follows:

$$F_{BC} = \frac{PL}{8} = \frac{10 \times 20}{8} = +25.0 \text{ ft·k},$$

$$F_{CB} = -25.0 \text{ ft·k}.$$

Since the relative stiffness is given in each span, the distribution factors are

$$DF_{AB} = \frac{K_{AB}}{\Sigma K} = \frac{5}{\infty + 5} = 0,$$

$$DF_{BA} = \frac{K_{BA}}{\Sigma K} = \frac{5}{8} = 0.625,$$

$$DF_{BC} = \frac{K_{BC}}{\Sigma K} = \frac{3}{8} = 0.375,$$

$$DF_{CB} = \frac{K_{CB}}{\Sigma K} = \frac{3}{3} = 1.$$

The balancing computations are shown in Table 8–1.

TABLE 8–1.

Joint	A	B		C
Member	AB	BA	BC	CB
K	5	5	3	3
Distribution factor $= K/\Sigma K$	0	0.625	0.375	1
			+25.0	−25.0
			+12.5	+25.0
	−11.7	−23.4	−14.1	−7.0
			+3.5	+7.0
	−1.1	−2.2	−1.3	−0.6
			+0.3	+0.6
	−0.1	−0.2	−0.1	
Total	−12.9	−25.8	+25.8	0

The above solution is that referred to in part (a) of the example statement as the *ordinary method*, so named to designate the manner of handling the balancing at the simple support at C. It is known, of course, that the final moment must be zero at this support because it is simple. Consequently, the first step is to balance the fixed-end moment at C to zero. The carry-over is then made immediately to B. When B is balanced, however, a carry-over must be made back to C simply because the relative stiffness of BC is based on end C of this span being fixed. It is apparent, however, that the moment carried back to C (in this case, -7.0) cannot exist at this joint. Accordingly, it is immediately balanced out, and a carry-over is again made to B, this carry-over being considerably smaller than the first. Now B is again balanced, and the process continues until the numbers involved become too small to have any practical value.

In part (b) of Example 8–1 a solution by use of the *simplified method* is specified. It was shown in Section 8–6 that if the support at C is simple and a moment is applied at B, then the resistance of the span BC to this moment is reduced to three-fourths of the value it would have had with C fixed. Consequently, if the relative stiffness of span BC is reduced to three-fourths of the value given, it will not be necessary to carry over to C.

TABLE 8-2.

Joint	A	B		C
Member	AB	BA	BC	CB
K	5	5	$\frac{3}{4} \times 3 = 2.25$	3.00
Distribution factor $= K/\Sigma K$	0	0.69	0.31	1
			+25.0	−25.0
			+12.5	+25.0
	−12.9	−25.8	−11.7	
Total	−12.9	−25.8	+25.8	0

From the standpoint of work involved, the advantage of the simplified method is obvious. It should always be used when the external (terminal) end of a member rests on a simple support, but it does not apply when a structure is continuous at a simple support, except as illustrated in Example 8–2. Attention is called to the fact that when the opposite end of the member is simply supported, the reduction factor for stiffness is always $\frac{3}{4}$ for a prismatic member but a variable quantity for nonprismatic members. (The expression for this reduction factor for nonprismatic members will be derived in Section 10–5.)

One valuable feature of the tabular arrangement in Tables 8–1 and 8–2 is that of dropping down one line for each balancing operation and making the carry-over on the same line. This practice clearly indicates the order of balancing the joints, which in turn makes it possible to check back in the event of error. Moreover, the placing of the carry-over on the same line with the balancing moments definitely decreases the chance of omitting a carry-over.

The correctness of the answers may in a sense be checked by verifying that $\Sigma M = 0$ at each joint. However, *even though the final answers satisfy this equation at every joint, this is in no way a check on the initial fixed-end moments.* These fixed-end moments, therefore, should be checked with great care before beginning the balancing operation. Moreover, it occasionally happens that compensating errors are made in the balancing, and these errors will not be apparent when checking $\Sigma M = 0$ at each joint. A better check on the correctness of the balancing operation will be discussed in Section 8–12.

EXAMPLE 8–2. Using the simplified method of moment distribution, find the moments in the continuous beam of Fig. 8–26. The values of I, as indicated by the various values of K, are different for the various spans. Determine the values of reactions, draw the shear and bending moment diagrams, and sketch the deflected structure.

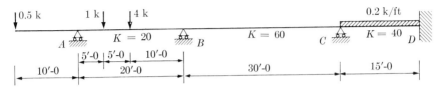

FIGURE 8–26

Fixed-end moments:

$$F_{AO} = -0.5 \times 10 = -5.0 \text{ ft·k}.$$

For the 1 k load:

$$F_{AB} = \frac{Pab^2}{L^2} = \frac{1 \times 5 \times 15^2}{20^2} = +2.8 \text{ ft·k},$$

$$F_{BA} = \frac{Pa^2b}{L^2} = \frac{1 \times 5^2 \times 15}{20^2} = -0.9 \text{ ft·k}.$$

For the 4 k load:

$$F_{AB} = \frac{PL}{8} = \frac{4 \times 20}{8} = +10.0 \text{ ft·k},$$

$$F_{BA} = -10.0 \text{ ft·k}.$$

For the uniform load:

$$F_{CD} = \frac{wL^2}{12} = \frac{0.2 \times 15^2}{12} = +3.8 \text{ ft·k},$$

$$F_{DC} = -3.8 \text{ ft·k}.$$

The balancing operation is shown in Table 8–3.

TABLE 8-3.

Joint	A		B		C		D
Member	AO	AB	BA	BC	CB	CD	DC
K	0	20	$\frac{3}{4} \times 20 = 15$	60	60	40	40
Distribution factor $= K/\Sigma K$	0	1	0.2	0.8	0.6	0.4	0
	−5.0	+12.8 −7.8	−10.9 −3.9			+3.8	−3.8
			+2.9	+11.9	+5.9		
				−2.9	−5.8	−3.9	−1.9
			+0.6	+2.3	+1.1		
				−0.3	−0.7	−0.4	−0.2
			+0.1	+0.2			
Total	−5.0	+5.0	−11.2	+11.2	+0.5	−0.5	−5.9

The only new point in this example is the method of handling the overhanging end. It is obvious that the final internal moment at A must be 5.0 ft·k and, accordingly, the first step is to balance out 7.8 ft·k of the fixed-end moment at AB, leaving the required 5.0 ft·k for the internal moment at AB. Since the relative stiffness of BA has been reduced to three-fourths of its original value, to permit considering the support at A as simple in the balancing, no carry-over from B to A is required.

The easiest way to determine the reactions is to consider each span as a free body. End shears are first determined as caused by the loads alone on each span and, following this, the end shears caused by the end moments are computed. These two shears are added algebraically to obtain the net end shear for each span. An algebraic summation of the end shears at any support will give the total reaction. (The procedure is illustrated in Fig. 8–27.)

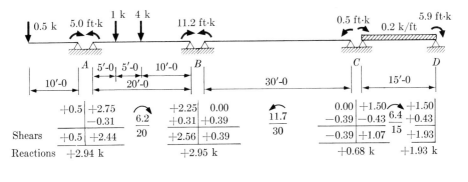

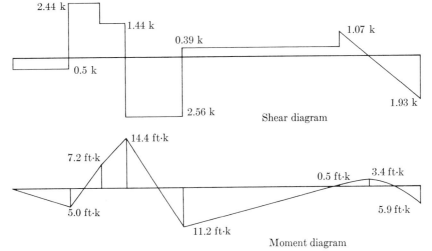

FIGURE 8-27

EXAMPLE 8-3. For the beam of Fig. 8-28 find the moments at A, B, and C by moment distribution. The support at C settles by 0.1 in. Use $E = 30{,}000 \text{ k/in}^2$.

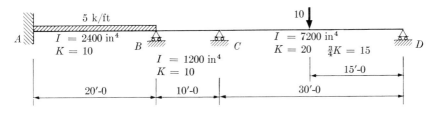

FIGURE 8-28

Fixed-end moments:

(1) Uniform load:

$$F_{AB} = \frac{wL^2}{12} = \frac{5 \times 20^2}{12} = +167 \text{ ft·k}, \qquad F_{BA} = -167 \text{ ft·k},$$

(2) Concentrated load:

$$F_{CD} = \frac{PL}{8} = \frac{10 \times 30}{8} = +37.5 \text{ ft·k}, \qquad F_{DC} = -37.5 \text{ ft·k}.$$

Moments caused by deflection:

$$F_{BC} = \frac{6EI\Delta}{L^2} = \frac{6 \times 30{,}000 \times 1200 \times 0.1}{(120)^2} = +1500 \text{ in·k} = +125 \text{ ft·k},$$

$$F_{CB} = +125 \text{ ft·k},$$

$$F_{CD} = \frac{6EI\Delta}{L^2} = \frac{6 \times 30{,}000 \times 7200 \times 0.1}{(360)^2} = 1000 \text{ in·k} = -83 \text{ ft·k},$$

$$F_{DC} = -83 \text{ ft·k}.$$

TABLE 8–4.

Joint	A	B		C		D
Member	AB	BA	BC	CB	CD	DC
K	10	10	10	10	$\frac{3}{4} \times 20 = 15$	20
Distribution factor = $K/\Sigma K$	0	0.5	0.5	0.4	0.6	1
	+167	−167			+38	−38
			+125	+125	−83	−83
					+60	+121
			−28	−56	−84	
	+17	+35	+35	+17		
			−3	−7	−10	
	+1	+2	+1			
Total	+185	−130	+130	+79	−79	0

Problems

8-4. Determine by moment distribution the moment at B in Fig. 8-29. [$Ans.$: 96.7 ft·k.]

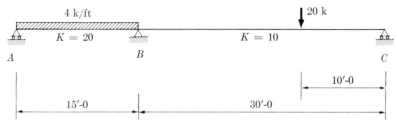

FIGURE 8-29

8-5. Find moments at A and B (see Fig. 8-30) by moment distribution. [$Ans.$: $M_A = 11.9$ ft·k, $M_B = 34.7$ ft·k.]

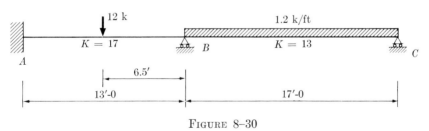

FIGURE 8-30

8-6. Find the moments at A, B, and C in Fig. 8-31. Determine the reactions, draw the shear and moment diagrams, and sketch the deflected structure. [$Ans.$: $M_A = 5.6$ ft·k, $M_B = 11.4$ ft·k, $M_C = 50.1$ ft·k, $R_A = 1.7$ k down, $R_B = 7.1$ k up, $R_C = 38.8$ k up, $R_D = 19.8$ k up.]

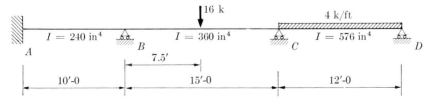

FIGURE 8-31

8-7. Find the moments at A, B, C, and D in Fig. 8-32. Determine the reactions, draw the shear and bending moment diagrams, and sketch the deflected structure. [Ans.: $M_A = 27.4$ ft·k, $M_B = 17.1$ ft·k, $M_C = 11.9$ ft·k, $M_D = 18.0$ ft·k, $R_A = 12.9$ k, $R_B = 14.9$ k, $R_C = 10.6$ k, $R_D = 8.6$ k.]

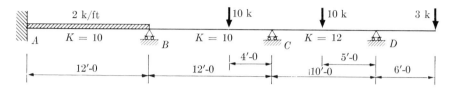

FIGURE 8-32

8-8. Find the moments at A, B, and C (see Fig. 8-33) caused by a vertical settlement of 0.2 in. at B and by a rotation of 0.1° clockwise with a vertical settlement of 0.1 in. at A. $E = 30{,}000$ k/in². [Ans.: $M_A = 53$ ft·k, $M_B = 94$ ft·k, $M_C = 114$ ft·k.]

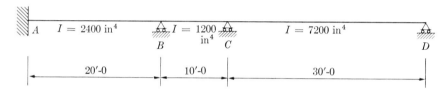

FIGURE 8-33

8-11 Symmetry and antisymmetry. When a symmetrical continuous beam has a center span, the applied loads may cause a condition of *symmetry* or *antisymmetry* in the bending moment diagram for this span. If either of these conditions exist, an adjustment can be made to the stiffness of the center span that will permit the analyst to work with only half the structure while balancing moments.

Consider the symmetrical beam of Fig. 8-34 with symmetrical loads

FIGURE 8-34

applied. The moment diagram for the span CD will be symmetrical. The loaded conjugate beam for this span will be as shown in Fig. 8-35, and the slope at the end of this span must be

$$\theta = \frac{ML}{2EI},$$

from which

$$M = \frac{2EI\theta}{L}.$$

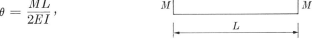

FIGURE 8-35

Obviously, then, the effect of the condition of symmetry is to reduce the effective stiffness of the center span to one-half the value normally used to compute distribution factors at C and D. Consequently, when symmetry of the kind illustrated by Fig. 8-34 exists, it is permissible to use one-half the usual stiffness of the center span when computing the distribution factors for the joints at its ends, and then to balance moments for half the structure.

A condition of antisymmetry in the center span will result from an antisymmetrical loading, as shown in Fig. 8-36.

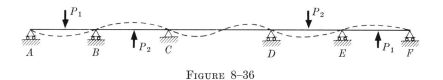

FIGURE 8-36

The loaded conjugate beam for the span CD is shown in Fig. 8-37, and from this beam the slope at the end of the span must be

$$\theta = \frac{ML}{6EI},$$

from which

$$M = \frac{6EI\theta}{L}.$$

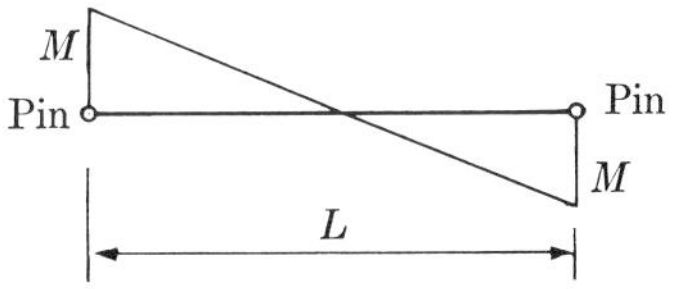

FIGURE 8-37

It is apparent, therefore, that antisymmetry in the bending moment diagram for the center span increases the effective stiffness of that span to one and one-half the stiffness normally used to compute the distribution factors for the joints at its ends. Consequently, in the case of antisymmetry the stiffness of the center span is increased to one and one-half its usual value and moments are balanced for half the beam.

Certain unsymmetrical loading systems on symmetrical structures can be replaced by a symmetrical and an antisymmetrical system. Consider the loaded beam of Fig. 8–38(a). The two loads are symmetrically placed

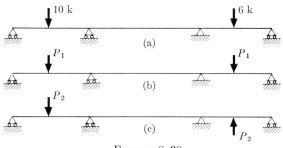

FIGURE 8–38

but are unequal in magnitude. These loads are to be replaced by the symmetrical system of 8–38(b) and the antisymmetrical system of 8–38(c). Consequently,

$$P_1 + P_2 = 10 \text{ k} \quad \text{and} \quad P_1 - P_2 = 6 \text{ k},$$

from which

$$P_1 = 8 \text{ k} \quad \text{and} \quad P_2 = 2 \text{ k}.$$

The moments resulting from these two separate loading systems may be found as indicated above and the results added. It is doubtful, however, that such a solution would be as easy as the usual moment distribution applied to the initial loading of Fig. 8–38(a).

8–12 Check on results of moment distribution. Prismatic members. As previously indicated, the internal moments resulting from a balancing operation may be verified to some extent by adding these moments algebraically at the ends of all members intersecting at each joint to ensure that the algebraic sum is zero; that is, by making certain that $\Sigma M = 0$ at each joint. This, however, is only a partial check, since it proves merely that probably no error has been made in the balancing operation and in the addition of moments. Not only is it possible to make compensating errors which will not be detected by this method, but also this check will not indicate an error made in computing the original fixed-end moments. Probably the best complete check, after having determined the internal moments by moment distribution, is to solve for all reaction components, and then to make certain that $\Sigma H = 0$, $\Sigma V = 0$, and $\Sigma M = 0$ for the entire structure. However, even this method is not infallible.

An alternate and interesting check on the correctness of the balancing operation is given below. It should be used immediately after ascertaining that the final internal moments in the members at each joint satisfy $\Sigma M = 0$. When this check is used, it is quite unlikely that compensating errors in the balancing will go undetected.

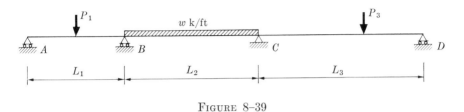

FIGURE 8-39

A loaded continuous beam is shown in Fig. 8-39. Under the action of the loads the beam will deflect as shown in Fig. 8-40, and the final internal

FIGURE 8-40

moment at a given end of any span will be given by the superposition of three effects: (1) the internal fixed-end moment at the given end resulting from loads acting directly on the span, (2) the internal moment at the given end resulting directly from any rotation of the given end of the span, and (3) the internal moment at the given end resulting from any rotation of the far end of the span.

The following expressions for the final internal end moments, M_{AB} and M_{BA} for span AB, can be written:

$$M_{AB} = F_{AB} - K_{AB}\theta_A - C_{BA}K_{BA}\theta_B = F_{AB} + \Delta F_{AB}, \quad (8\text{-}1)$$

$$M_{BA} = F_{BA} - K_{BA}\theta_B - C_{AB}K_{AB}\theta_A = F_{BA} + \Delta F_{BA}. \quad (8\text{-}2)$$

In the above equations:

F_{AB} and F_{BA} = initial fixed-end moments caused by any loads on span AB.

ΔF_{AB} and ΔF_{BA} = increments of M_{AB} and M_{BA} due to θ_A and θ_B.

K_{AB} and K_{BA} = absolute stiffnesses at the two ends of span AB, equal to $4EI/L$ if AB is prismatic.

C_{AB} and C_{BA} = carry-over factors for the two ends of span AB. Each is equal to $+\frac{1}{2}$ if AB is prismatic.

θ_A and θ_B = the rotation in radians of ends A and B of span AB. A clockwise rotation is positive.

From Eq. (8–1),

$$\Delta F_{AB} = -K_{AB}\theta_A - C_{BA}K_{BA}\theta_B. \tag{8-3}$$

From Eq. (8–2),

$$\Delta F_{BA} = -K_{BA}\theta_B - C_{AB}K_{AB}\theta_A. \tag{8-4}$$

From Eq. (8–3),

$$\theta_A = \frac{-C_{BA}K_{BA}\theta_B - \Delta F_{AB}}{K_{AB}}.$$

If the value of θ_A, obtained from Eq. (8–4), is substituted in this last equation, a solution for θ_B will result in

$$\theta_B = \frac{\Delta F_{BA} - C_{AB}\Delta F_{AB}}{C_{AB}C_{BA}K_{BA} - K_{BA}}. \tag{8-5}$$

Use of the same procedure for span BC yields

$$\theta_B = \frac{\Delta F_{BC} - C_{CB}\Delta F_{CB}}{C_{CB}C_{BC}K_{BC} - K_{BC}}. \tag{8-6}$$

Obviously the end rotation of span BA must be equal to the end rotation of span BC. Therefore the right sides of Eqs. (8–5) and (8–6) may be set equal to each other, with the result that

$$\frac{\Delta F_{BA} - C_{AB}\Delta F_{AB}}{C_{AB}C_{BA}K_{BA} - K_{BA}} = \frac{\Delta F_{BC} - C_{CB}\Delta F_{CB}}{C_{CB}C_{BC}K_{BC} - K_{BC}}. \tag{8-7}$$

If more than two members meet at a joint (which will be the case in many continuous frames) similar expressions may be written for each member. The value of θ obtained from all members framing at the joint must be equal in magnitude and sign.

Each side of Eq. (8-7) is the rotation in radians of joint B. The expressions apply for nonprismatic and prismatic members; however, if all members in a frame are prismatic, then $K_{AB} = K_{BA} = 4EI/L = 4E$ (Rel K_{AB}), and $C = +\frac{1}{2}$. Substitution of these values in Eq. (8-7) yields

$$\frac{\Delta F_{AB} - 2\,\Delta F_{BA}}{6E \text{ (Rel } K_{AB})} = \frac{\Delta F_{CB} - 2\,\Delta F_{BC}}{6E \text{ (Rel } K_{BC})}. \tag{8-8}$$

Each side of Eq. (8-8) will still give the rotation of joint B in radians provided that all quantities have been expressed in the proper units. For the purpose of checking a balancing operation in moment distribution, however, the $6E$ may be omitted from the denominator and any units may be used to express relative K. This procedure simplifies the check considerably.

From the above discussion, the following rule will hold: *For all prismatic members intersecting at a joint of a rigid frame, the change in moment at the far end of each member minus twice the change in moment at the joint end of the member, all divided by the stiffness of the member, must be a constant value.*

The above check will apply only when joints are perfectly rigid. It can be used in regard to any single balancing operation, but it cannot be used, by one application, as a check on the final moments in a frame when these moments are the combined results of two or more balancing operations, as is the case when structures lurch sideways under load. In such a case the check can be used to verify the results of each balancing operation separately, and thus is still very effective. The reader is again cautioned however, that satisfaction of this check in a given case is in no way an indication of the correctness of the initial fixed-end moments.

The check will now be applied in connection with the example which follows. Observe that two additional horizontal lines are drawn at the bottom of Table 8-5, and also that the value which appears in the table as relative θ is determined in each case by applying the rule stated above for checking prismatic members.

EXAMPLE 8-9. Analyze the frame of Fig. 8-41 by moment distribution. Find all reaction components, draw the moment diagram, and sketch the deflected structure.

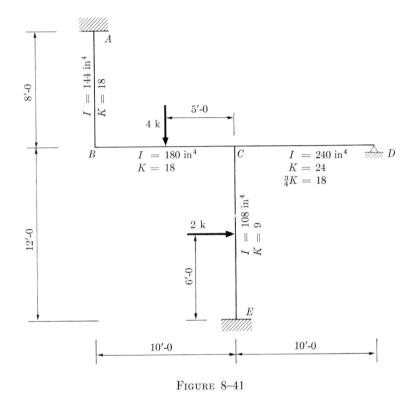

FIGURE 8-41

From Fig. 8-41,

$$F_{BC} = \frac{PL}{8} = \frac{4 \times 10}{8} = +5.0 \text{ ft·k}, \qquad F_{CB} = -5.0 \text{ ft·k},$$

$$F_{CE} = \frac{2 \times 12}{8} = -3.0 \text{ ft·k}, \qquad F_{EC} = +3.0 \text{ ft·k}.$$

In the balancing operation (see Table 8-5) three significant figures have been used. This will usually be necessary when a close agreement in the values of relative θ is desired. The use of only two significant figures, on the other hand, will result in discrepancies of perhaps 5 to 7% in the values of relative θ for a given joint.

From the values of relative θ the direction and magnitude of the rotation of each joint may be determined, and such information will occasionally be very useful.

TABLE 8-5.

Joint	A	B		C			D	E
Member	AB	BA	BC	CB	CE	CD	DC	EC
K	18	18	18	18	9	$\tfrac{3}{4}K = 18$	24	9
Distribution factor $= K/\Sigma K$	0	0.5	0.5	0.4	0.2	0.4	1	0
			+5.00 +1.60	−5.00 +3.20	−3.00 +1.60	+3.20		+3.00 +0.8
	−1.65	−3.30	−3.30	−1.65				
			+0.33	+0.66	+0.33	+0.66		
		−0.16	−0.16	−0.08	+0.02	+0.03		+0.16
				+0.03				
Moment	−1.73	−3.46	+3.47	−2.84	−1.05	+3.89		+3.96
ΔF	−1.73	−3.46	−1.53	+2.16	+1.95	+3.89	0	+0.96
Relative θ	0	+0.288	+0.290	−0.326	−0.326	−0.324	+0.162	0

For example, if E is 30,000 k/in^2,

$$\theta_B = \frac{\Delta M_{CB} - 2\Delta M_{BC}}{6EK_{BC}} = +\frac{0.290}{6E}$$

$$= +\frac{0.290 \times 144}{6 \times 30{,}000} = +2.32 \times 10^{-4} \text{ rad.}$$

In the above equation multiplication by 144 is necessary to adjust units, and the positive sign of the answer signifies clockwise rotation.

The determination of the reaction components for the example in question is managed by considering various parts of Fig. 8–41 as free bodies (see Figs. 8–42 through 8–47).

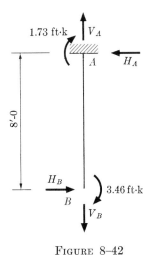

FIGURE 8–42

From Fig. 8–42,

$$\Sigma M_B = 0 \text{ (counterclockwise +),}$$

$$8H_A - 1.73 - 3.46 = 0,$$

$$H_A = +0.65 \text{ k,}$$

and the sense is correct as assumed.

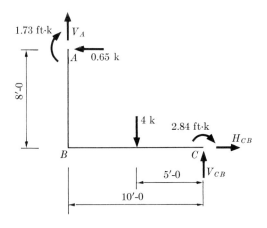

FIGURE 8-43

From Fig. 8-43,

$$\Sigma M_C = 0 \text{ (clockwise +)},$$

$$10V_A + 1.73 - 0.65 \times 8 - 4 \times 5 + 2.84 = 0,$$

$$V_A = +2.06 \text{ k},$$

and the sense is correct as assumed.

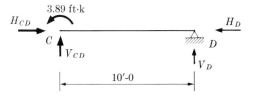

FIGURE 8-44

From Fig. 8-44,

$$\Sigma M_C = 0 \text{ (clockwise +)},$$

$$-10V_D - 3.89 = 0,$$

$$V_D = -0.39 \text{ k},$$

and the sense is incorrect as assumed (V_D is actually down).

From Fig. 8–45,

$\Sigma M_C = 0$ (clockwise +),

$+1.05 - 2 \times 6 - 3.96 + 12H_E = 0,$

$H_E = +1.24$ k,

and the sense is correct as assumed.

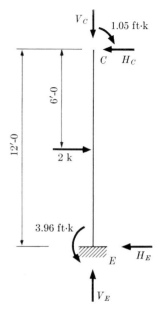

FIGURE 8–45

From Fig. 8–46,

$\Sigma H = 0$ (to the left +),

$+0.76 - 0.65 - H_D = 0,$

$H_D = +0.11$ k,

and the sense is to the left, as assumed.

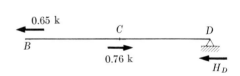

FIGURE 8–46

From Fig. 8–47,

$\Sigma V = 0$ (down is +),

$1.94 + 0.39 - V_E = 0,$

$V_E = +2.33$ k,

and the sense is up, as assumed.

FIGURE 8–47

The moments at the two concentrated loads must be computed before the complete moment diagram can be drawn. Thus M under the 4 k load on BC is given by

$$M_4 = 2.06 \times 5 - 3.47 = +6.83 \text{ ft·k (compression on top)},$$

and M at the 2 k load on CE will be

$$M_2 = 1.24 \times 6 - 3.96 = +3.48 \text{ ft·k (compression on left)}.$$

The moment diagram is shown in Fig. 8–48(a). The ordinates are plotted on the tension side of the members because, when so plotted, the diagram will approximate the deflection curve to some extent. The deflected structure is shown in Fig. 8–48(b).

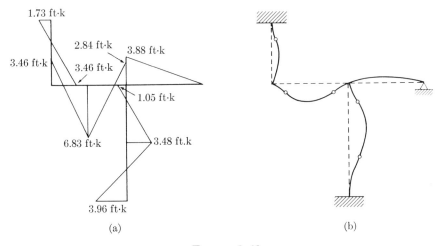

FIGURE 8–48

The points of contraflexure are indicated by the moment diagram. The direction of rotation of some joints may be predicted by the sign of the original fixed-end moments at the joints, with a positive sign indicating clockwise rotation. Thus in the above case, the clockwise rotation of B and the counterclockwise rotation of C would have been predicted. Occasionally, however, these initial fixed-end moments may give the wrong indication as to joint rotation. Definite information as to the direction of joint rotations may be obtained, of course, by computing relative θ as illustrated in the foregoing example.

Problems

8-10. Find moments at B and C (see Fig. 8–49) by moment distribution. Draw the moment diagram and sketch the deflected structure. [*Ans.:* $M_B = 24.1$ ft·k, $M_C = 34.9$ ft·k.]

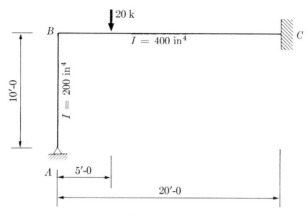

FIGURE 8–49

8-11. Find moments at all joints in Fig. 8–50. Draw the moment diagram and sketch the deflected structure. [*Ans.* (in ft·k): $M_A = -5.7$, $M_B = 11.5$, $M_{CB} = -18.6$, $M_{CE} = +27.1$, $M_{CD} = -8.5$, $M_{DC} = +58.2$, $M_{EC} = -0.2$.]

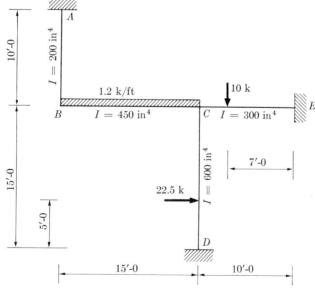

FIGURE 8–50

8-12. Find all moments at joints in Fig. 8-51. Draw the moment diagram and sketch the deflected structure. [Ans. (in ft·k): $M_{AB} = -6.8$, $M_B = 14.1$, $M_{CB} = -36.0$, $M_{CD} = +13.8$, $M_{CE} = +22.2$.]

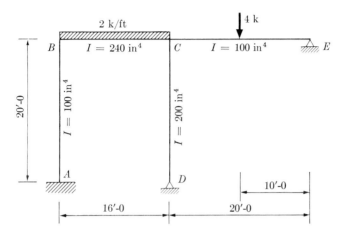

FIGURE 8-51

8-13. The support at A in Fig. 8-52 rotates 0.01 rad clockwise. Support D settles vertically by 1 in. Find all joint moments by moment distribution. $E = 30,000$ k/in^2. [Ans. (in ft·k): $M_A = -47.0$, $M_{BA} = -31.5$, $M_{BC} = -53.6$, $M_{BD} = +85.1$, $M_D = 50.0$, $M_{CB} = +1.0$.]

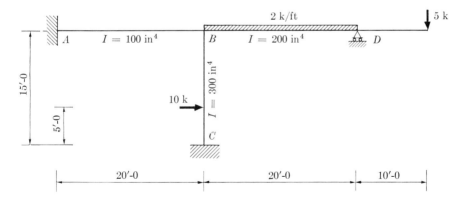

FIGURE 8-52

8-14. Find all joint moments in Fig. 8-53 by moment distribution. The support at A settles vertically 0.1 in. The support at F rotates 0.001 rad clockwise. $E = 30{,}000$ k/in^2. [Ans. (in ft·k): $M_{AB} = +18.8$, $M_{BA} = -20.2$, $M_{BE} = -21.6$, $M_{BC} = +76.6$, $M_{BF} = -25.9$, $M_{BG} = -8.9$, $M_{FB} = -31.7$, $M_{CB} = -40.0$, $M_{BG} = -4.4$.]

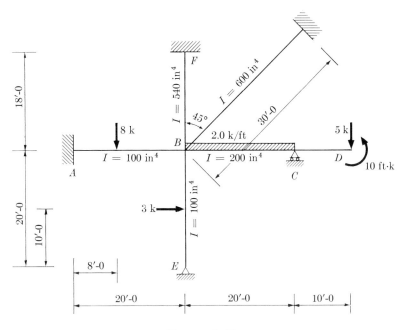

FIGURE 8-53

8-15. Find all joint moments in Fig. 8-54 by moment distribution. [Ans. (in ft·k): $M_{AB} = -40.0$, $M_{BA} = -22.7$, $M_{DB} = -78.8$, $M_{DC} = +6.5$, $M_{DE} = +54.9$, $M_{DF} = +17.4$, $M_{FD} = +8.6$, $M_{ED} = -68.6$.]

8-16. Find all joint moments in Fig. 8-55 by moment distribution. Determine all reaction components, draw the bending moment diagram, and sketch the deflected structure. [Ans. (in ft·k): $M_{BA} = -11.8$, $M_{BE} = -0.4$, $M_{BC} = +12.2$, $M_{EB} = -6.5$, $M_{EF} = +12.5$, $M_{EG} = -6.0$, $M_{FE} = -13.7$, $M_{CB} = +6.0$.]

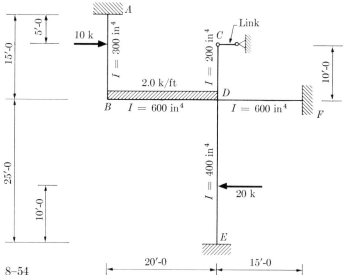

FIGURE 8–54

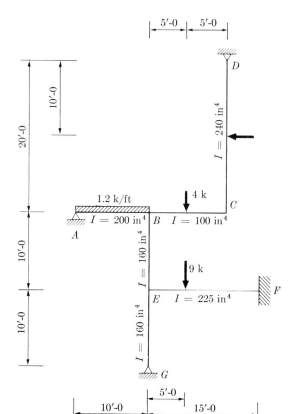

FIGURE 8–55

8–13 Frames with one degree of freedom of joint translation. Inspection of the problems of the last section will show that the supports of all the frames considered there are arranged so that translation of joints is impossible. In certain cases a predetermined translation was cited as being due to settlement of foundations, but the effects could be introduced immediately as additional fixed-end moments because the amount of the translation was definitely known. In many frames encountered in practice, however, the magnitude and sense of the translations of joints is unknown at the outset, and thus the evaluation of these translations and their effects becomes an important part of the analysis.

Two methods are available for including the effects of joint translations in an analysis. The first of these, known as the method of *successive shear corrections*, was originally presented by C. T. Morris in his discussion of Professor Hardy Cross' paper (1), and is also described in some detail by L. E. Grinter (2). The author believes, however, that the alternate method, requiring the superposition of the effects of separate joint displacements, is, in most cases, easier to apply. This method results in a direct solution, which is not possible by successive shear corrections. Accordingly, all examples involving joint translation given here will be solved by the method of separate joint displacements. It is true, of course, that some analysts prefer the method of successive shear corrections, and in the case of a structure with a large number of degrees of freedom of joint translation this method may give the easier solution. This will depend, to a great extent, on the experience of the analyst.

It is convenient to classify frames in accordance with the number of degrees of freedom of joint translation. For a given frame, one degree of freedom exists for each possible translation of a joint or system of interconnected joints that can occur as the result of flexing of the members, and independently of the translations of any other joints or systems of interconnected joints.

The simplest type of frame, of course, is that with one degree of freedom, several examples of which are shown in Fig. 8–56. In each of these examples, if the translation of one joint is known, the translations of all other joints are either known or can be determined. In Fig. 8–56(a) and (c) the horizontal displacements of all column tops will be equal. A displacement

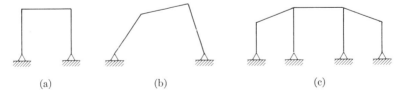

Figure 8–56

diagram provides an easy method for finding the relationship between the translations of the two joints in Fig. 8–56(b). Displacement diagrams will be discussed in Examples 8–28 and 8–29.

Three examples of frames with two degrees of freedom are shown in Fig. 8–57. The arrangement of the structure of 8–57(a) is such that

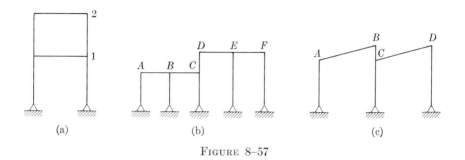

FIGURE 8–57

the translations of levels 1 and 2 are not required to be in any fixed ratio as different loadings are applied. In 8–57(b) the translations of ABC will not have any fixed relationship to the translations of DEF for different loading systems. In 8–57(c) the translation of girder AB can be different from the translation of CD.

Consider the simple rectangular frame of Fig. 8–58. An analysis of

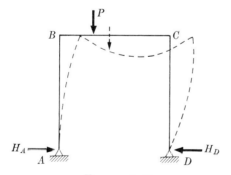

FIGURE 8–58

this frame by moment distribution (with no side lurch allowed) will result in M_{BA} being larger than M_{CD}. This means that H_A must be larger than H_D, which is impossible since ΣH must equal zero for the loaded frame. Actually, as the load P is applied, the frame will lurch to the right, and as it sways in this direction the moment M_{BA} is decreased and M_{CD} is increased until they become equal; H_A will then be equal to H_D, and the frame will come to rest in a position approximating the dashed lines of Fig. 8–59. Obviously the magnitude of the lurch cannot be predetermined, and in many cases the sense cannot be predicted. A description of the method of correcting for this lurch, as well as that for determining its magnitude and sense, follows.

It has been proved that when one end of a prismatic fixed-end member is displaced a distance Δ with respect to the other end in a direction perpendicular to the axis of the member, then the moments developed in the ends of the member are given by $M = 6EI\,\Delta/L^2$. This condition is shown in Fig. 8–59.

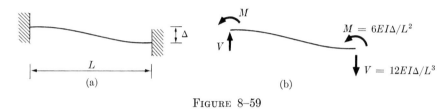

FIGURE 8–59

Solving in the usual way for the vertical shear yields $V = 12EI\,\Delta/L^3$. In a prismatic member fixed at both ends, it is evident that the end moments that are induced by a deflection Δ of one end relative to the other in a direction perpendicular to the member axis are proportional to I/L^2. Furthermore, the shear necessary to cause this deflection, or resulting from it, is proportional to I/L^3.

In the case of a member with a pin at one end (as in the beam of Fig. 8–60, which has the same E, I, and L as the beam of Fig. 8–59),

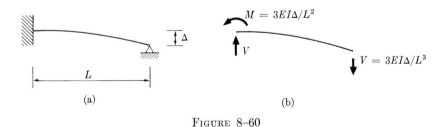

FIGURE 8–60

$V = 3EI\,\Delta/L^3$, and therefore $M = 3EI\,\Delta/L^2$. Thus in a member fixed at one end and pinned at the other, the moment induced by a deflection Δ of one end relative to the other in a direction perpendicular to the member axis is also proportional to I/L^2. The shear V necessary to cause this deflection Δ is proportional to I/L^3.

For moment distribution the relative stiffness of a member in a frame composed solely of prismatic members is usually taken as I/L, designated, for convenience, by K. Two additional relative stiffnesses are now introduced. The first, designated by K_s, will be called the *relative shearing-stiffness* of a member. *The relative shearing-stiffness is a number which is proportional to, or a measure of, the internal shear induced in a member as it resists the displacement of one end relative to the other in a direction perpendicular to the axis of the member.*

The second new relative stiffness, designated by K_m, is closely related to the first, and will be called the *relative shearing-moment stiffness*. *The relative shearing-moment stiffness is a number which is proportional to, or a measure of, the end moments (or moment) induced in the ends (or end) of a member as it resists the displacement of one end relative to the other in a direction perpendicular to the axis of the member.*

The absolute shearing-stiffness is defined as the shear in a member that results when one end of the member is displaced a unit distance with respect to the other end, the displacement being in a direction perpendicular to the member axis, and the rotation of any rigid joint at the ends of the member being prevented.

The absolute shearing-moment stiffness is defined as the moment induced at the ends (or end, if the member is pinned at one end) of a member when one end of the member is displaced a unit distance with respect to the other end, the displacement being in a direction perpendicular to the member axis, and the rotation of any rigid joint at the ends of the member being prevented.

When a frame is arranged as shown in Fig. 8–61, with the column base

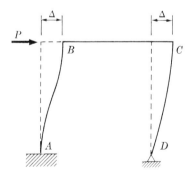

FIGURE 8–61

at A fixed and that at D pinned, and the top of the frame forced to the right a distance Δ under the action of P without rotation of the top joints, structural sense will dictate that the two columns will have different shears induced in them. Obviously, the relative shearing-stiffness and the relative shearing-moment stiffness can be made proportional to the absolute shearing-stiffness and the absolute shearing-moment stiffness in any ratio desired. Arbitrarily, then, the following will be adopted:

$$\text{Rel } K_m \text{ for } AB \text{ is } \frac{I}{L^2},$$

$$\text{Rel } K_s \text{ for } AB \text{ is } \frac{I}{L^3}.$$

Inspection of the formulas given in connection with Figs. 8–59 and 8–60 will indicate that, to be consistent,

$$\text{Rel } K_m \text{ for } CD \text{ must be } \frac{I}{2L^2},$$

$$\text{Rel } K_s \text{ for } CD \text{ must be } \frac{I}{4L^3}.$$

The above values of K_s indicate that if no rotation of rigid joints is permitted, then a pin at one end of a prismatic member will reduce the shear induced therein to one-quarter of what it would have been if the pinned end had been fixed. The moment at the fixed end will be one-half the end moments which would have been induced if both ends had been fixed.

EXAMPLE 8–17. Find all moments by moment distribution for the frame of Fig. 8–62. Draw the bending moment diagram and the deflected structure.

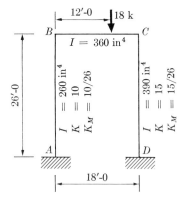

FIGURE 8–62

8-13] FRAMES WITH ONE DEGREE OF FREEDOM

The first step is to perform the usual moment distribution. The reader should fully understand that this balancing operation adjusts the internal moments at the ends of the members by a series of corrections as the joints are considered to rotate, until $\Sigma M = 0$ at each joint. The reader should also realize that *during this balancing operation no translation of any joint is permitted.*

The fixed-end moments are

$$F_{BC} = \frac{18 \times 12 \times 6^2}{18^2} = +24 \text{ ft·k},$$

$$F_{CB} = \frac{18 \times 6 \times 12^2}{18^2} = -48 \text{ ft·k}.$$

TABLE 8-6.

A	B		C		D
AB	BA	BC	CB	CD	DC
10	10	20	20	15	15
0	0.333	0.667	0.571	0.429	0
		+24.0	−48.0		
		+13.7	+27.4	+20.6	+10.3
−6.3	−12.6	−25.1	−12.5		
		+3.6	+7.1	+5.4	+2.7
−0.6	−1.2	−2.4	−1.2		
		+0.3	+0.7	+0.5	+0.2
	−0.1	−0.2			
−6.9	−13.9	+13.9	−26.5	+26.5	+13.2

The final moments listed in Table 8-6 are correct only if there is no translation of any joint. It is therefore necessary to determine whether or not, with the above moments existing, there is any tendency for side lurch of the top of the frame.

If the frame of Fig. 8-62 is divided into three free bodies, the result will be as shown in Fig. 8-63. Inspection of this sketch indicates that if the moments of the first balance exist in the frame, there is a net force of $1.53 - 0.80 = 0.73$ k tending to sway the frame to the left. In order to

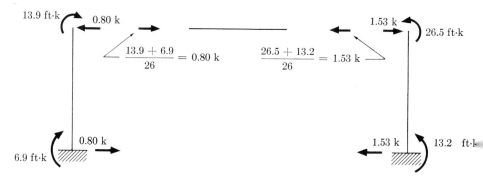

FIGURE 8-63

prevent side-sway, and thus allow these moments to exist (temporarily, for the purpose of the analysis), it is necessary that an imaginary horizontal force be considered to act to the right at B or C. This force is designated as the *artificial joint restraint* (abbreviated as AJR) and is shown in Fig. 8-64. This illustration now shows the complete load system which

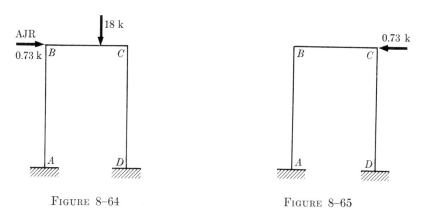

FIGURE 8-64   FIGURE 8-65

would have to act on the structure if the final moments of the first balance are to be correct. The AJR, however, cannot be permitted to remain, and thus its effect must be cancelled. This may be accomplished by finding the moments in the frame resulting from a force equal but opposite to the AJR and applied at the top, as shown in Fig. 8-65.

Although it is not possible to make a direct solution for the moments resulting from this force, they may be determined indirectly. Assume that some unknown force P acts on the frame, as shown in Fig. 8-66, and

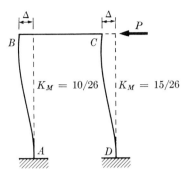

FIGURE 8-66

causes it to deflect laterally to the left, without joint rotation, through some distance Δ. Now, regardless of the value of P and the value of the resulting Δ, the fixed-end moments induced in the ends of the columns must be proportional to the respective values of K_M. These fixed-end moments could, for example, have the values of -10 and -15 ft·k, or -20 and -30, or -30 and -45, or any other combination so long as the above ratio is maintained. The proper procedure is to choose values for these fixed-end moments of approximately the same order of magnitude as the original fixed-end moments due to the real loads. This will result in the same accuracy for the results of the balance for the side-sway correction that was realized in the first balance for the real loads. Accordingly, it will be assumed that P, and the resulting Δ, are of such magnitudes as to result in the fixed-end moments shown in Fig. 8-67. Obviously, $\Sigma M = 0$ is not satisfied for joints B and C in this deflected frame. Therefore these joints must rotate until equilibrium is reached. The effect of this rotation is determined in the distribution in Table 8-7.

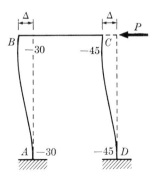

FIGURE 8-67

TABLE 8-7.

A	B		C		D
AB	BA	BC	CB	CD	DC
10	10	20	20	15	15
0	0.333	0.667	0.571	0.429	0
−30.0	−30.0			−45.0	−45.0
		+12.9	+25.8	+19.2	+9.6
+2.8	+5.7	+11.4	+5.7		
		−1.6	−3.3	−2.4	−1.2
+0.2	+0.5	+1.1	+0.5		
			−0.3	−0.2	−0.1
−27.0	−23.8	+23.8	+28.4	−28.4	−36.7

During the rotation of joints B and C, as represented by the above distribution, *the value of Δ has remained constant, with P varying in magnitude as required to maintain Δ.*

It is now possible to determine the final value of P simply by adding the shears in the columns. The shear in any member, without external loads applied along its length, is obtained by adding the end moments algebraically and dividing by the length of the member. The final value of P is the force necessary to maintain the deflection of the frame after the joints have rotated. In other words, it is the force which will be consistent with the displacement and internal moments of the structure as determined by the second balancing operation. Hence this final value of P will be called the *consistent joint force* (abbreviated as CJF).

The consistent joint force is given by

$$\text{CJF} = \frac{+27.0 + 23.8 + 28.4 + 36.7}{26} = 4.45 \text{ k},$$

and inspection clearly indicates that the CJF must act to the left.

Obviously, then, the results of the last balance above are moments which will exist in the frame when a force of 4.45 k acts to the left at the top level. It is necessary, however, to determine the moments resulting from a force of 0.73 k acting to the left at the top level, and some as yet unknown factor "z" times 4.45 will be used to represent this force acting to the left.

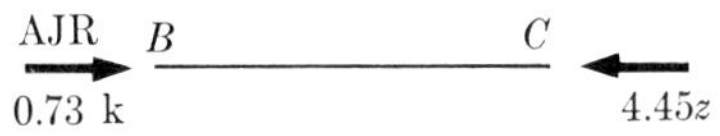

FIGURE 8-68

```
         AJR   B                        C
         ───▶                          ◀───
        0.73 k                         4.45z
```

FIGURE 8-68

The free body for the member BC is represented by Fig. 8-68. $\Sigma H = 0$ must be satisfied for this figure, and if forces to the left are considered as positive, the result is $4.45z - 0.73 = 0$, and

$$z = +0.164.$$

If this factor $z = +0.164$ is applied to the moments obtained from the second balance, the result will be the moments caused by a force of 0.73 k acting to the left at the top level. If these moments are then added to the moments obtained from the first balance, the result will be the final moments for the frame, the effect of the AJR having been cancelled. This combination of moments is shown in Table 8-8.

TABLE 8-8.

Joint	A	B		C		D
Member	AB	BA	BC	CB	CD	DC
Moments from first balance	−6.9	−13.9	+13.9	−26.5	+26.5	+13.2
$z \times$ moments from second balance	−4.4	−3.9	+3.9	+4.7	−4.7	−6.0
Final moments	−11.3	−17.8	+17.8	−21.8	+21.8	+7.2

If the final moments are correct, the shears in the two columns of the frame should be equal and opposite to satisfy $\Sigma H = 0$ for the entire frame.

This check is expressed as

$$\frac{+11.3 + 17.8}{26} + \frac{-21.8 - 7.2}{26} = 0,$$

and

$$+1.12 - 1.11 = 0 \text{ (nearly)}.$$

The signs of all moments taken from Table 8–8 have been reversed to give the correct signs for the end moments external to the columns. It will be remembered that the moments considered in moment distribution are always internal for each member. However, the above check actually considers each column as a free body and so external moments must be used.

In Fig. 8–69 the moment under the 18 k load is obtained by treating BC as a free body:

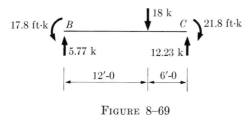

FIGURE 8–69

$$M_{18} = 5.77 \times 12 - 17.8 = +51.5 \text{ ft·k}.$$

The direction of side-lurch may be determined from the obvious fact that the frame will always lurch in a direction opposite to the AJR. If required, the magnitude of this side lurch may be found. The procedure which follows will apply.

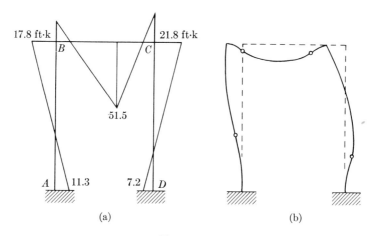

FIGURE 8–70

In Fig. 8–67 a force P of sufficient magnitude to result in the indicated column moments and the lurch Δ was applied to the frame. During the second balance this value of Δ was held constant as the joints B and C

rotated, and the value of P was considered to vary as necessary. The final value of P was found to be 4.45 k. Since Δ was held constant, however, its magnitude may be determined from the equation $M = 6EI\,\Delta/L^2$, where M is the fixed-end moment for either column in Fig. 8–67, I is the moment of inertia of that column, and L is the length. This Δ will be the lurch for 4.45 k acting at the top level. For any other force acting horizontally, Δ would vary proportionally and thus the final lurch of the frame would be the factor z multiplied by the Δ determined above.

EXAMPLE 8–18. Find by moment distribution the moments in the frame of Fig. 8–71.

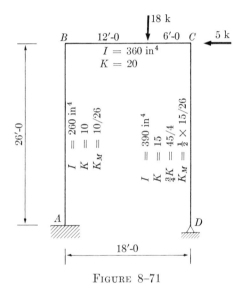

FIGURE 8–71

The first balance will give the results shown in Table 8–9.

TABLE 8–9.

AB	BA	BC	CB	CD	DC
−7.2	−14.6	+14.6	−22.5	+22.5	0

A check of the member BC as a free body for $\Sigma H = 0$ (see Fig. 8–72), will indicate that an AJR is necessary as follows:

$$\text{AJR} + 0.84 - 0.87 - 5.0 = 0,$$

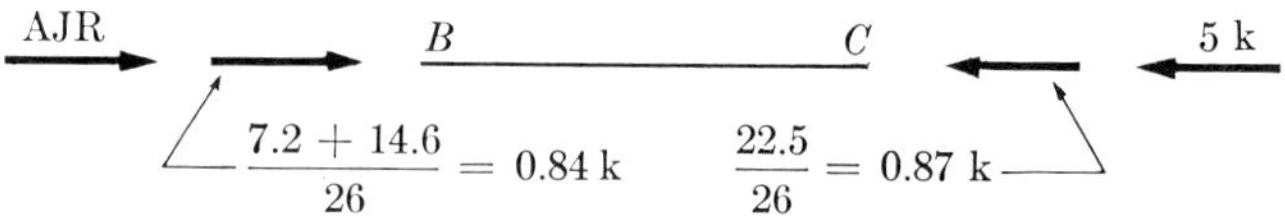

350   INTRODUCTION TO MOMENT DISTRIBUTION   [CHAP. 8

$$\text{AJR} \rightarrow \xrightarrow{} \underset{\dfrac{7.2 + 14.6}{26} = 0.84 \text{ k}}{B} \underset{\dfrac{22.5}{26} = 0.87 \text{ k}}{C} \xleftarrow{} 5 \text{ k}$$

FIGURE 8–72

from which

$$\text{AJR} = +5.03 \text{ in the direction assumed.}$$

The values of K_M for the two columns are shown in Fig. 8–71, with K_M for column CD being $K/2L$ because of the pin at the bottom. The horizontal displacement Δ of the top of the frame, as shown in Fig. 8–73, is

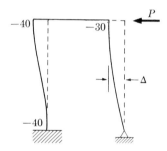

FIGURE 8–73

assumed to cause the fixed-end moments shown there. These moments are proportional to the values of K_M and of approximately the same order of magnitude as the original fixed-end moments due to the real loads. The results of balancing out these moments are given in Table 8–10.

TABLE 8–10.

AB	BA	BC	CB	CD	DC
−34.4	−28.4	+28.4	+23.6	−23.6	0

$$\text{CJF} = \frac{+34.4 + 28.4 + 23.6}{26} = 3.32 \text{ k,}$$

and

$$5.03 - z \times 3.32 = 0,$$

from which

$$z = +1.52.$$

The final results are given in Table 8–11.

TABLE 8–11.

	AB	BA	BC	CB	CD	DC
First balance	−7.2	−14.6	+14.6	−22.5	+22.5	0
$z \times$ second balance	−52.1	−43.0	+43.0	+35.8	−35.8	0
Final moments	−59.3	−57.6	+57.6	+13.3	−13.3	0

If these final moments are correct, the sum of the column shears will be 5.0 k:

$$\text{Sum of column shears} = \frac{59.3 + 57.6 + 13.3}{26} = 5.01 \text{ k}.$$

The 5 k horizontal load acting at C enters into the problem only in connection with the determination of the AJR. If this load had been applied to the column CD between the ends, it would have resulted in initial fixed-end moments in CD and these would be computed in the usual way. In addition, such a load would have entered into the determination of the AJR, since the horizontal reaction of CD against the right end of BC would have been computed by treating CD as a free body.

Problems

8–19. Find all joint moments in Fig. 8–74 by moment distribution. Draw the moment diagram and sketch the deflected structure. [*Ans.:* $M_{AB} = -4.7$ ft·k, $M_B = 21.0$ ft·k, $M_C = 15.2$ ft·k, $M_{DC} = +10.5$ ft·k.]

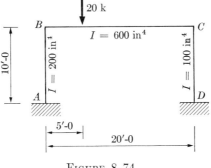

FIGURE 8–74

8-20. Find the moments at B and C in Fig. 8-75 by moment distribution. Draw the moment diagram and sketch the deflected structure. [Ans.: M_B = 21.4 ft·k, M_C = 21.4 ft·k.]

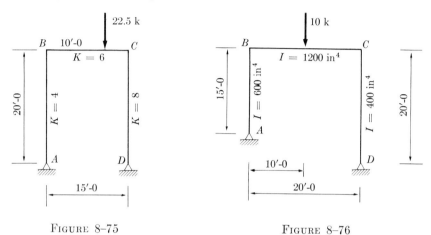

FIGURE 8-75

FIGURE 8-76

8-21. Determine the moments at B and C in Fig. 8-76. Draw the moment diagram and sketch the deflected structure. [Ans.: M_B = 8.0 ft·k, M_C = 10.7 ft·k.]

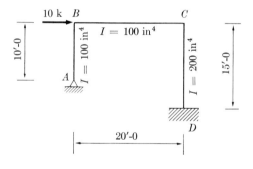

FIGURE 8-77

8-22. Find all moments in Fig. 8-77 by moment distribution. [Ans.: M_B = 31.4 ft·k, M_C = 33.8 ft·k, M_{DC} = +69.8 ft·k.]

8-23. Analyze the frame of Fig. 8-78 by moment distribution. Draw the moment diagram and sketch the deflected structure. [*Ans.:* $M_B = 25.4$ ft·k, $M_C = 14.8$ ft·k, $M_{DC} = -50.9$ ft·k.]

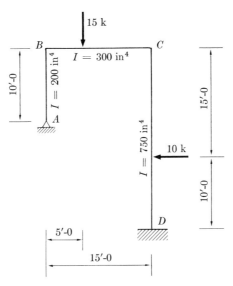

FIGURE 8-78

8-24. Evaluate joint moments in Fig. 8-79 by moment distribution. Draw the moment diagram and sketch the deflected structure. [*Ans.:* $M_B = 14.1$ ft·k, $M_{CB} = -21.9$ ft·k, $M_{CD} = +11.9$ ft·k, $M_{DC} = +9.1$ ft·k.]

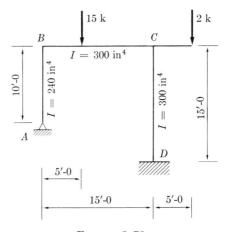

FIGURE 8-79

8-25. Find all moments in Fig. 8-80 by moment distribution. Draw the moment diagram and sketch the deflected structure. [Ans. (in ft·k): $M_{AB} = +53.8$, $M_{BA} = +14.2$, $M_{BC} = -21.3$, $M_{BD} = +7.1$, $M_{DB} = -70.6$, $M_{DE} = -9.4$.]

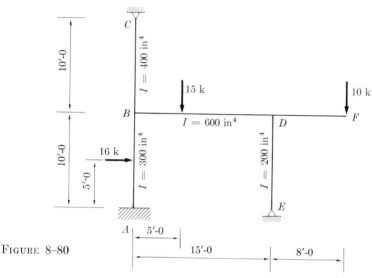

FIGURE 8-80

8-26. Find all moments in Fig. 8-81. Draw the moment diagram and sketch the deflected structure. [Ans. (in ft·k): $M_{AB} = +10.6$, $M_{BA} = +8.7$, $M_{BD} = +9.1$, $M_{BC} = -17.8$, $M_D = 0.6$, $M_{ED} = -6.0$.]

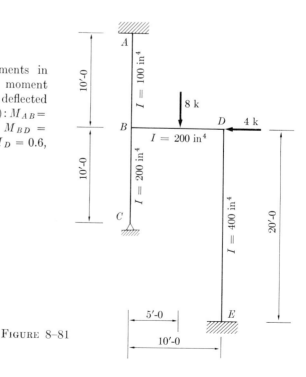

FIGURE 8-81

8-27. Find all moments in Fig. 8-82 by moment distribution. Determine the sense and magnitude of the rotation of joints B and C, as well as that of the lurch of level BC. [Ans.: $M_{AB} = +33.5$ ft·k, $M_B = 2.2$ ft·k, $M_{CB} = -106.8$ ft·k, $M_{CD} = +66.8$ ft·k, $\theta_B = +4.3 \times 10^{-3}$ rad, $\theta_C = +0.18 \times 10^{-3}$ rad, lurch = 0.67 in. to right.]

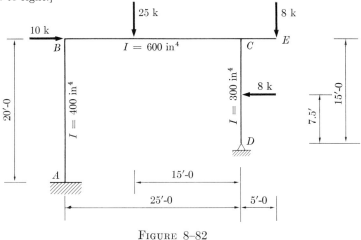

FIGURE 8-82

The examples which follow illustrate the method of analyzing frames with sloping legs.

EXAMPLE 8-28. Find all moments in Fig. 8-83 by moment distribution. $E = 30,000$ k/in².

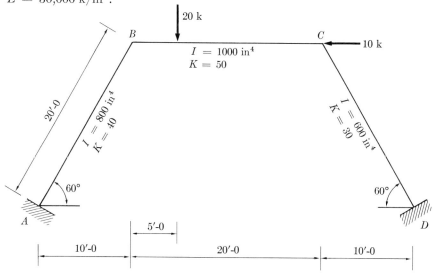

FIGURE 8-83

The fixed-end moments for the 20 k load are computed and the frame is balanced in the usual way, side-lurch being prevented by an AJR to be later evaluated. The results of the first balance are given in Table 8–12.

TABLE 8–12.

AB	BA	BC	CB	CD	DC
−15.1	−30.2	+30.2	−14.1	+14.1	+6.9

The evaluation of the AJR is somewhat more complicated than in preceding examples because the column thrusts have horizontal components. As a first step, treat BC as a free body (see Fig. 8–84) in order to determine

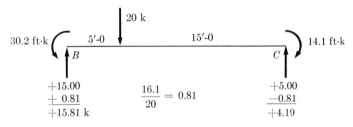

FIGURE 8–84

end shears. Next, consider each column as a free body, using the shears determined above.

From Fig. 8–85,

$\Sigma M_A = 0$ (clockwise is +),

$-17.32 \, H_{BA} + 15.1 + 30.2$

$+ 15.81 \times 10 = 0$,

from which

$H_{BA} = +11.73$ k.

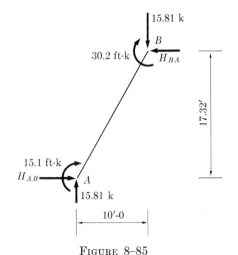

FIGURE 8–85

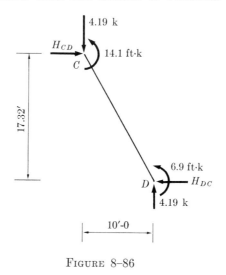

FIGURE 8-86

From Fig. 8-86,

$$\Sigma M_D = 0 \text{ (clockwise is +)},$$

$$17.32\, H_{CD} - 4.19 \times 10 - 14.1 - 6.9 = 0,$$

from which

$$H_{CD} = +3.62 \text{ k}.$$

The member BC is again drawn as a free body in Fig. 8-87, with only horizontal forces being shown, and the result is:

FIGURE 8-87

$$\Sigma H = 0 \quad \text{(to the right is positive)},$$

$$\text{AJR} + 11.73 - 3.62 - 10 = 0,$$

and

$$\text{AJR} = +1.89 \text{ in the direction assumed.}$$

It is now necessary to find the value of a CJF. The procedure, however, is different from that previously used. In this case, any horizontal displacement of BC will also result in vertical displacement of B and C. Consequently, there will be fixed-end moments in all members. Arbitrarily assume that BC is displaced by an unknown force P so that the horizontal components of the movement of B and C are 0.1 in. to the left. The point C will move about point D, with CD as a radius, until the horizontal component of its motion is 0.1 in. Point B will move in a similar manner about A, with AB as a radius. It is thus possible to determine the actual movements of points B and C, as well as the vertical components of their displacements, as shown in the displacement diagram of Fig. 8–88. The fixed-end moments resulting from these joint translations (with joint rotation being prevented) are computed as follows:

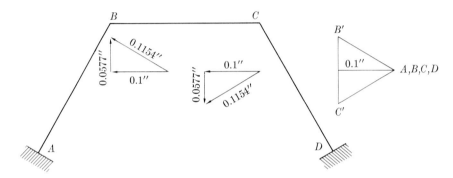

FIGURE 8–88

$$F_{AB} = F_{BA} = \frac{6EI\Delta}{L^2} = \frac{6 \times 30{,}000 \times 800 \times 0.1154}{20^2 \times 12^3} = -24 \text{ ft·k},$$

$$F_{BC} = F_{CB} = \frac{6 \times 30{,}000 \times 1000 \times (2 \times 0.0577)}{20^2 \times 12^3} = +30 \text{ ft·k},$$

$$F_{CD} = F_{DC} = \frac{6 \times 30{,}000 \times 600 \times 0.1154}{20^2 \times 12^3} = -18 \text{ ft·k}.$$

(All the above moments are proportional to $I\Delta/L^2$, which can be considered to be a *modified* K_M. It would have been easier to compute values for the modified K_M for all members of the frame and to assume fixed-end moments, resulting from an arbitrary but unknown lurch, in proportion to them. Such a procedure will be followed in the next problem.)

The moments in question are balanced, with the results given in Table 8–13.

TABLE 8–13.

AB	BA	BC	CB	CD	DC
−24.5	−25.1	+25.1	+22.3	−22.3	−20.1

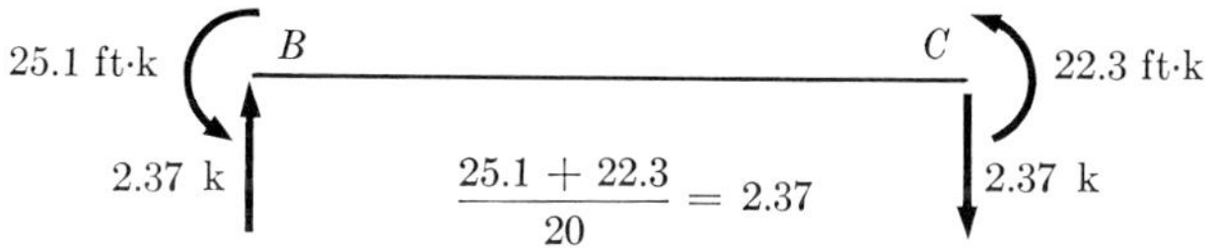

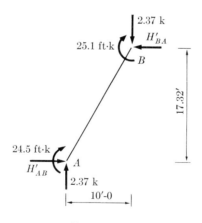

FIGURE 8–89

The value of the CJF is found in a manner similar to that employed for the AJR (see Figs. 8–89, 8–90, 8–91).

FIGURE 8–90

From Fig. 8–90,

$$\Sigma M_A = 0 \text{ (clockwise is +)},$$

$$-17.32 \, H'_{BA} + 2.37 \times 10 + 25.1 + 24.5 = 0,$$

from which

$$H'_{BA} = +4.23 \text{ k}.$$

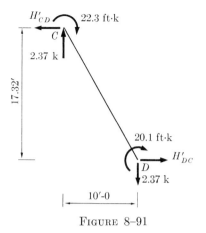

FIGURE 8-91

From Fig. 8-91,

$$\Sigma M_D = 0 \text{ (clockwise is +)},$$

$$-17.32\, H'_{CD} + 2.37 \times 10 + 22.3 + 20.1 = 0,$$

from which

$$H'_{CD} = +3.82 \text{ k}.$$

The CJF is obviously

$$4.23 + 3.82 = 8.05 \text{ k acting to the left}.$$

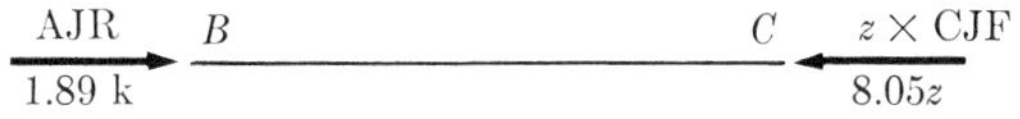

FIGURE 8-92

In Fig. 8-92, ΣH must be zero for BC, and thus

$$1.89 - 8.05z = 0,$$

from which

$$z = +0.235.$$

The final results are given in Table 8-14.

TABLE 8–14.

	AB	BA	BC	CB	CD	DC
First balance	−15.1	−30.2	+30.2	−14.1	+14.1	+6.9
$z \times$ second balance	−5.8	−5.9	+5.9	+5.3	−5.3	−4.7
Final moments	−20.9	−36.1	+36.1	−8.8	+8.8	+2.2

As a check on these moments, the frame of Fig. 8–83 may be considered to be split into three free bodies, a procedure previously followed in computing the AJR and the CJF. The vertical and horizontal reaction components may then be determined. All reaction components are shown in Fig. 8–93.

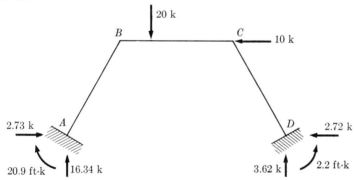

FIGURE 8–93

From this check, $\Sigma V = 0$ and $\Sigma H = 0$ are nearly satisfied, and the final moments may be considered to be correct.

EXAMPLE 8–29. Evaluate joint moments in Fig. 8–94 by moment distribution.

First, the fixed-end moments are computed as follows:

$$F_{BC} = +\frac{4 \times 20}{8} = +10 \text{ ft·k}, \qquad F_{CB} = -10 \text{ ft·k},$$

$$F_{CD} = +\frac{8 \times 20}{8} = +20 \text{ ft·k}, \qquad F_{DC} = -20 \text{ ft·k}.$$

These are balanced in the usual manner, with the resulting moments (in ft·k) shown in Table 8–15.

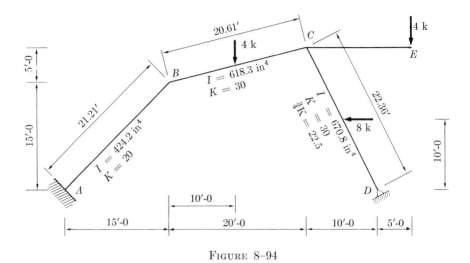

FIGURE 8-94

TABLE 8-15.

AB	BA	BC	CB	CD	CE
+2.7	+5.5	−5.5	−54.0	−6.0	+60.0

These moments will be correct only if no lurch of the frame is permitted. The value of the AJR which prevents this lurch will be computed by a different and shorter method than that used in the preceding example. Here it will be assumed that the AJR acts horizontally to the right at joint C. (It could have been assumed to act at joint B.) The value of the AJR is easily determined by taking moments at the intersection of the two sloping legs, as shown in Fig. 8-95:

$$\Sigma M_{\text{c.m.}} = 0 \text{ (clockwise is +)},$$

$$+0.39 \times 42.42 - 2.7 - 3.85 \times 33.54 - 4 \times 5 + 4 \times 20$$

$$+ 8 \times 20 - 10 \text{ AJR} = 0,$$

from which

$$\text{AJR} = +10.47 \text{ k to the right, as assumed.}$$

8-13] FRAMES WITH ONE DEGREE OF FREEDOM 363

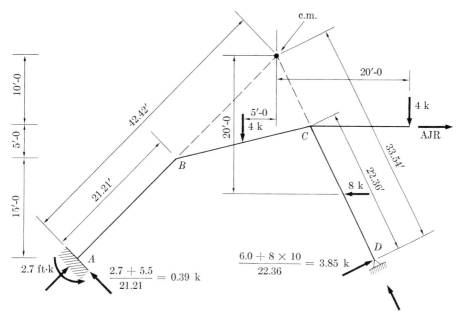

FIGURE 8-95

The effect of this AJR must be cancelled by a force of 10.47 k acting toward the left at C. The moments resulting from the action of this force must be evaluated indirectly. Assume that some force of unknown magnitude acts toward the left at C. This force will cause the frame to lurch to the left, with the amount of the lurch being arbitrarily set at some convenient value. In this case, let us assume that the horizontal projection of the lurch of B will be 1 in. to the left (see Fig. 8-96). The actual displacements of joints B and C, which must be consistent with the assumed

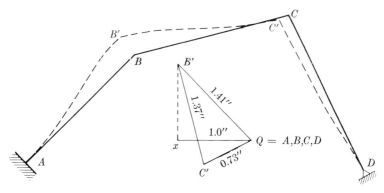

FIGURE 8-96

horizontal component of the displacement of joint B, are most easily determined by drawing a displacement diagram, as shown in Fig. 8-96. Here all members are considered to be reduced to zero length. Therefore joints $A, B, C,$ and D coincide at Q. Joint B must rotate about A, with AB as a radius. Therefore, B' is located at the intersection of a vertical line drawn 1 in. to the left of $Q = A$ with another line drawn through $Q = A$ but perpendicular to the member AB. QB' is the displacement of joint B.

During the lurch of the frame joint C must rotate about D, with CD as a radius, and at the same time must rotate about the translating joint B, with BC as a radius. The final position of joint C, represented by C', must therefore be at the intersection of a line drawn through $Q = D$ perpendicular to the original direction of the member CD, with another line drawn through B' perpendicular to the original direction of the member BC. (The line QX, representing the assumed 1-in. horizontal component of the displacement of joint B, may be drawn to any desired scale, but the use of an enlarged scale gives greater accuracy.) The magnitudes of the displacements are shown in Fig. 8-96, and from these displacements the various values for the modified K_M are computed. For prismatic members fixed at both ends the modified K_M is $I\Delta/L^2$; if the prismatic member is pinned at one end the value is $I\Delta/2L^2$. These are computed as follows:

$$\text{Mod } K_M \text{ for } AB = 424 \times \frac{1.41}{21.21^2} = -1.33,$$

$$\text{Mod } K_M \text{ for } BC = 618 \times \frac{1.37}{20.61^2} = +1.99,$$

$$\text{Mod } K_M \text{ for } CD = 671 \times \frac{0.73}{2} \times 22.36^2 = -0.49.$$

Any lurch of the frame, either to the right or to the left, will result in fixed-end moments which will be proportional to these values of the modified K_M. If the lurch is to the left (desirable in the present case), the signs will agree with those in the above equations. It will be assumed, therefore, that the lurch of the frame is to the left and that it is of sufficient magnitude to cause fixed-end moments (in ft·k) as shown in Table 8-16.

TABLE 8-16.

AB	BA	BC	CB	CD	CE
−13.3	−13.3	+19.9	+19.9	−4.9	0

The magnitude of the lurch which is consistent with the above moments is not known but could, if desired, be computed. These moments are balanced out in the usual way, and the moments (in ft·k) resulting from this second balance are given in Table 8–17.

TABLE 8–17.

AB	BA	BC	CB	CD	CE
−13.7	−14.3	+14.3	+11.0	−11.0	0

The force (CJF) necessary to hold the frame in the deflected position is evaluated by the same method used to find the AJR:

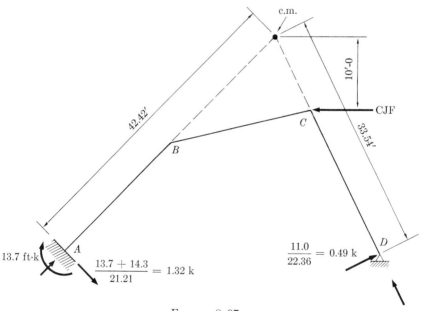

FIGURE 8 07

$$\Sigma M_{c.m.} = 0 \text{ (clockwise is +)},$$

$$+10 \times \text{CJF} + 13.7 - 1.32 \times 42.42 - 0.49 \times 33.54 = 0,$$

from which

$$\text{CJF} = +5.87 \text{ to the left, as assumed.}$$

To obtain the moments resulting from the action of a force of 10.47 k to the left at C, the moments resulting from the second balance must be multiplied by the factor

$$z = \frac{+10.47}{+5.87} = +1.78.$$

If this is done, and the results added to the moments obtained from the first balance, the final moments will be as shown in Table 8–18.

TABLE 8–18.

AB	BA	BC	CB	CD	CE
-21.7	-19.9	$+19.9$	-34.4	-25.6	$+60.0$

PROBLEMS

8–30. Determine all joint moments in Fig. 8–98. Draw the bending moment diagram and sketch the deflected structure. [Ans. (in ft·k): $M_B = 14.5$, $M_{CB} = -24.7$, $M_{CD} = -25.3$, $M_{DC} = -31.0$.]

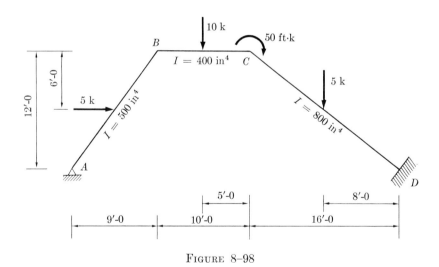

FIGURE 8–98

8–31. Find all joint moments in Fig. 8–99 by moment distribution. Draw the bending moment diagram and sketch the deflected structure. [Ans. (in ft·k): $M_{AB} = -65$, $M_{BA} = +18$, $M_{BC} = +123$, $M_C = 49$, $M_{DC} = -67$.]

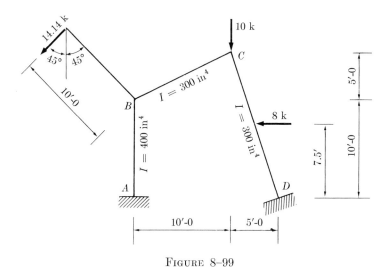

Figure 8-99

Specific References

1. Cross, H., "Analysis of Continuous Frames by Distributing Fixed-End Moments," *Trans. Am. Soc. Civ. Engrs.*, **96**, 1–10, 1932.
2. Grinter, L. E., *Theory of Modern Steel Structures*. New York: Macmillan, 1949.

General References

3. Andersen, P., *Statically Indeterminate Structures*. Chap. 6. New York: Ronald, 1953.
4. Cross, H. and Morgan, N. D., *Continuous Frames of Reinforced Concrete*. Chap. 4. New York: Wiley, 1932.
5. Maugh, L. C., *Statically Indeterminate Structures*. Chap. 3. New York: Wiley, 1946.
6. Parcel, J. I. and Moorman, R. B. B., *Analysis of Statically Indeterminate Structures*. Chap. 6. New York: Wiley, 1955.
7. Sutherland, H. and Bowman, H. L., *Structural Theory*. Chap. 8. New York: Wiley, 1950.
8. Wang, C. K., *Statically Indeterminate Structures*. Chap. 8. New York: McGraw-Hill, 1953.
9. Wilbur, J. B. and Norris, C. H., *Elementary Structural Analysis*. Chap. 14. New York: McGraw-Hill, 1948.
10. Williams, C. D., *Analysis of Statically Indeterminate Structures*. Chap. 6. Scranton: International, 1943.

CHAPTER 9

ADDITIONAL APPLICATIONS OF MOMENT DISTRIBUTION

9-1 General. This chapter includes discussions of the application of moment distribution to the analysis of continuous frames with two or more degrees of freedom, to the evaluation of secondary stresses in articulated structures, and to the analysis of frames with restrained joints.

With one exception, all the structures to be considered in this chapter are composed entirely of *prismatic* members. This exception, a rigid gabled frame with stepped columns to support a crane runway girder, can be solved with an interesting supplemental application of moment distribution. The analysis of frames with nonprismatic members will be the subject of the next chapter.

9-2 Frames with two degrees of freedom of joint translation. The structures shown in Fig. 9-1 are typical examples of frames with two degrees of freedom. The structure in 9-1(a) [similar to the frame of Fig. 8-57(a)] is occasionally found in light building frames and supports for industrial equipment, but is, to some extent, academic. However,

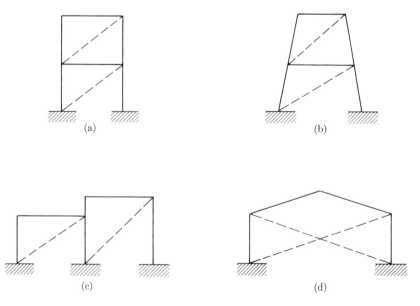

FIGURE 9-1

because its analysis is somewhat simpler than those of the other three frames, it will be used as a first demonstration. The frame in Fig. 9–1(b) is commonly used for the bents of highway and railway trestles, while the frame in (c) is quite common in industrial buildings. The gabled frame in (d) is the type ordinarily spoken of as a "rigid frame" and has wide usage in auditoriums, hangars, and industrial buildings.

In each frame of Fig. 9–1 diagonal dashed lines are shown. If all joints were pinned instead of rigid, these dashed members would have to be inserted in order to maintain the stability of the structure. When inserted, each such member cancels one degree of freedom, and this provides a method for determining the number of degrees of freedom of more complicated frames. All joints are considered to be pinned and diagonals are inserted until the structure is stable. The number of diagonals will usually (but not always) equal the number of degrees of freedom. Sometimes a single diagonal will cancel two degrees of freedom instead of the usual one, and so this method must be used with discretion.

EXAMPLE 9–1. Find all joint moments in the frame of Fig. 9–2 by moment distribution. Draw the bending moment diagram.

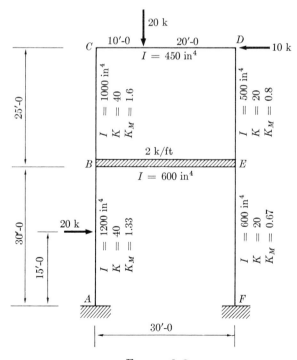

FIGURE 9–2

The fixed-end moments are computed in the usual way and (expressed in ft·k) are shown in Fig. 9–3. These moments are balanced out, with the

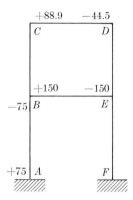

FIGURE 9–3

results shown in Fig. 9–4. The moments obtained from this first balance will be correct only if no side-sway is permitted of either CD or BE, and therefore side-sway is assumed to be prevented by AJR's acting at each level.

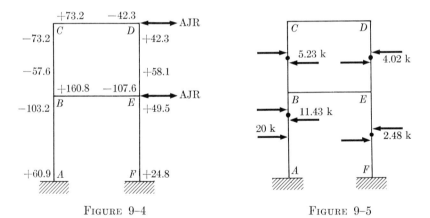

FIGURE 9–4   FIGURE 9–5

The values of these AJR's must be determined. In multi-story frames they are best obtained by first finding the shears in the columns and then placing these shears on a sketch, as shown in Fig. 9–5.

In this figure, dots indicate division points for separating the frame into free bodies, as will be presently demonstrated. The location of each of these dots is in no way related to the moments at the ends of the corresponding columns. That is, the dots do not represent points of contraflexure but are used to represent more clearly the actions of the column shears. It is important to note that the column shears, as shown in Fig. 9–5, are external to the column segments on which they act, and thus may be used directly on each free body.

An explanation of the method for determining the senses of these column shears is in order at this point. Consider, for example, the free body for CB, as shown in Fig. 9–6. Inspection indicates that the horizontal reactions external to this column section must act in the directions shown in Fig. 9–6, and the magnitude will be

$$\frac{73.2 + 57.6}{25} = 5.23 \text{ k}.$$

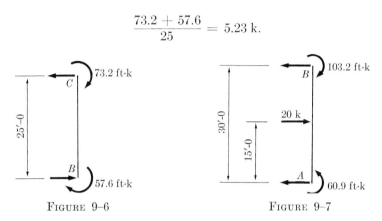

FIGURE 9–6   FIGURE 9–7

From the senses of these external horizontal reactions it is apparent that if any section or point is considered in the column, as, for example, the dot, then that part of the column below the dot will push to the right on the part above and, in turn, that part of the column above the dot will push to the left on the part below. Therefore the senses of shears external to the column segments adjacent to the dot will be as shown for CB in Fig. 9–5. A convenient name for this designation of column shears is *shear-couple*. In column CB, then, the shear-couple is said to be clockwise and in DE it is counterclockwise, as shown in Fig. 9–5.

Observe carefully the shear-couple shown in AB, the column which supports the interpanel load of 20 k. As would be expected, the value of this shear-couple is determined by treating AB as a free body (see Fig. 9–7). Here only the direction of the shear-couple in the part of the column above the 20 k load is of interest, since this will be a part of the free body to be formed about the level of BE. Taking moments about A, we find that the upper horizontal reaction on AB is

$$\frac{20 \times 15 + 103.2 - 60.9}{30} = 11.43 \text{ k}.$$

The sense of the shear-couple in the upper part of AB will then be as shown in Fig. 9–5.

A rule to follow for the easy determination of the sense of the shear-couple in a member, with no interpanel loads acting, is: *If the sum of the internal end moments of the member (these moments, and their signs, are the direct results of the moment distribution) is negative, then the shear-couple is positive, or clockwise. If the sum of the moments is positive, then the shear-couple is negative, or counterclockwise.* This rule does not apply when there are interpanel loads on the member, for in that case the sense of the shear-couple should be determined as demonstrated for AB.

Having determined the shear-couples of Fig. 9–5, we now consider the

FIGURE 9–8

free body in Fig. 9–8. Here, $\Sigma H = 0$ must be satisfied, and thus, considering forces to the right as positive, the following expression results:

$$AJR + 5.23 - 10 - 4.02 = 0,$$

from which

$$AJR = +8.79 \text{ k in the direction assumed.}$$

The free body about the level BE would be as shown in Fig. 9–9. From this figure:

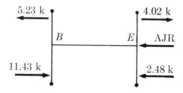

FIGURE 9–9

$$\Sigma H = 0 \text{ (to the left is positive)},$$

$$AJR + 5.23 + 2.48 - 11.43 - 4.02 = 0,$$

from which

$$AJR = +7.74 \text{ k in the direction assumed.}$$

The AJR's determined from Figs. 9–8 and 9–9 are shown in Fig. 9–10.

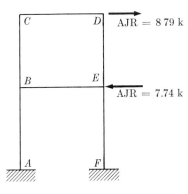

FIGURE 9-10

Forces equal and opposite to these AJR's are applied at the same levels and the moments caused by these forces are added to the moments obtained from the first balance to obtain the final moments in the frame. The following procedure will apply.

A horizontal force of unknown magnitude is assumed to act at the top level, while, at the same time, another horizontal force of unknown magnitude acts at level BE. Under the action of these forces level CD lurches to the left (the frame could have been assumed to lurch to the right), while level BE is held in position. (The sense of these forces will be determined later.) The distorted frame is shown in Fig. 9-11, with the resulting fixed-end moments. These moments are proportional to the K_M values of the distorted columns.

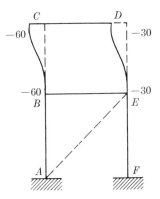

FIGURE 9-11

Moments are balanced in the usual manner. No additional translation of joints is permitted, but the two unknown forces change in magnitude as the joints are permitted to rotate during the balancing operation.

In their final value these forces are consistent with the moments resulting from the balancing operation. The resulting moments, the shear couples, and the CJF's are shown in Fig. 9–12.

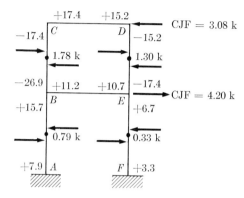

Figure 9–12

By another combination of horizontal forces acting at levels BE and CD the frame is next assumed to be forced into the distorted position shown in Fig. 9–13. The fixed-end moments, proportional to the K_M values of

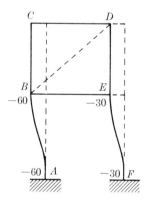

Figure 9–13

the distorted columns, are shown in this figure. These moments are balanced out, and the resulting moments, shear-couples, and CJF's are as shown in Fig. 9–14. The CJF's of Figs. 9–12 and 9–14, and the moments with which they are consistent, cannot be combined directly to find the moments resulting from two forces equal and opposite to the AJR's of Fig. 9–10. It is possible, however, to find some factor X by which all

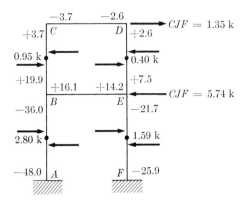

FIGURE 9–14

values shown in Fig. 9–12 may be multiplied, and another factor Y by which all values of Fig. 9–14 may be multiplied, such that an algebraic summation of the products will result in a set of moments consistent with forces acting equal and opposite to the AJR's of Fig. 9–10.

The two condition equations necessary to evaluate X and Y are obtained by simply expressing the fact that the superposition of X times the CJF's of Fig. 9–12 and Y times the CJF's of Fig. 9–14 and the AJR's of Fig. 9–10 must result in a net horizontal force of zero acting at each of the two levels. If we consider that forces acting to the right are positive, these equations are:

For level CD,
$$-3.08X + 1.35Y + 8.79 = 0.$$

For level BE,
$$+4.20X - 5.74Y - 7.74 = 0.$$

A simultaneous solution will result in
$$X = +3.34 \quad \text{and} \quad Y = +1.10.$$

If the moments of Fig. 9–12 are multiplied by X and those of Fig. 9–14 by Y, and if the products are added algebraically to the moments of Fig. 9–4, the final moments will result. (See Fig. 9–15.) In the solution above, if either or both X and Y had been negative, the multiplication would have been performed following the usual rules of algebra regarding the signs of products.

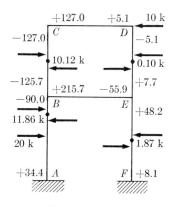
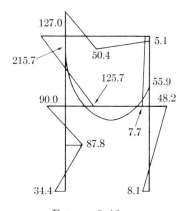

FIGURE 9-15   FIGURE 9-16

As a check on the moments, the final shear-couples are shown in Fig. 9-15. Here we see that the condition that $\Sigma H = 0$ is very nearly satisfied for the free body about CD and also for the free body about BE. The moment diagram (with moments expressed in ft·k) is shown in Fig. 9-16.

From the foregoing discussion it is apparent that in order to effect a solution of a frame with one or more degrees of freedom, it is necessary to have one set of consistent joint forces and moments for each degree of freedom. Obviously, to obtain each set, a distortion must be impressed which is consistent with the constraints of the structure. In this regard it is suggested that imaginary structural constraints be introduced in the form of dotted diagonals, each diagonal suppressing one degree of freedom. These imaginary diagonals are then removed (only one being omitted at any given time) in order to determine the various distortions to be impressed on the frame. The moments resulting from each distortion are selected in proportion to the K_M or modified K_M values of the members which are flexed as a result of each distortion. The analysis is then completed as shown above.

Frames which have a width of more than one bay are analyzed by this same method. In the case of additional stories, each extra story height will add an extra degree of freedom to the frame. Consequently, for each new story an additional distortion must be impressed, an additional set of CJF's must be evaluated, one more simultaneous equation must be written, and another constant similar to X and Y must be evaluated. Obviously the work involved in a complete analysis becomes quite extensive for even a small frame. Fortunately a "short cut" is possible in the application of moment distribution to building frames. This is known as the *two-cycle method of moment distribution*.*

---

* For an explanation of this method see *Continuity in Concrete Building Frames*, a booklet available from the Portland Cement Association.

EXAMPLE 9-2. A railway viaduct bent is proportioned as shown in Fig. 9-17. Lateral loads due to the nosing of the locomotive and wind on train and trestle result in a lateral force of 35 k at the top, as indicated. Find all joint moments.

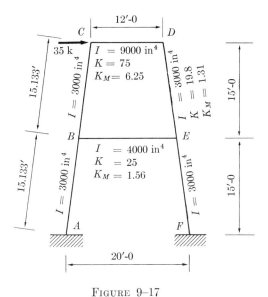

FIGURE 9-17

The AJR necessary to prevent lurch is as shown in Fig. 9-18. The effect of removing this AJR is determined by the following procedure.

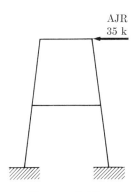

FIGURE 9-18

Since the frame has two degrees of freedom, two different distortions must be impressed and consistent joint forces and moments must be evaluated for each distortion. First, we assume that the frame is distorted as in Fig. 9-19(a), and then as shown in 9-19(b). To determine the several

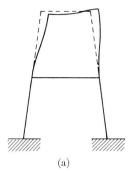
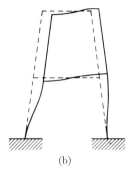

FIGURE 9-19

Δ's to be used in computing the values of modified K_M, it will be assumed in each case that the horizontal component of the displacement of each joint is 1 in. The displacement diagram is shown in Fig. 9-20.

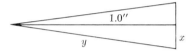

FIGURE 9-20

By similar triangles,

$$\frac{x}{1} = \frac{2}{15},$$

from which

$$x = 0.133 \text{ in.,}$$

and

$$\frac{y}{1} = \frac{15.133}{15},$$

from which

$$y = 1.009 \text{ in.}$$

The values for modified K_M are

$$AB = BC = DE = EF = K_M\Delta = 1.31 \times 1.009 = 1.320,$$

$$CD = 6.25 \times 0.266 = 1.670,$$

$$BE = 1.56 \times 0.266 = 0.415.$$

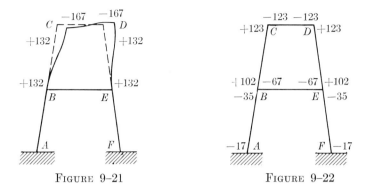

FIGURE 9-21   FIGURE 9-22

The top level of the frame is caused to lurch to the right, as shown in Fig. 9-21, resulting in the fixed-end moments (in ft·k) indicated. These moments are then balanced out, with the results given in Fig. 9-22.

The CJF at level CD is assumed to act to the right on the free body, as shown in Fig. 9-23. This assumption is verified, and the CJF is evaluated, by taking moments about the intersection of the two sloping legs:

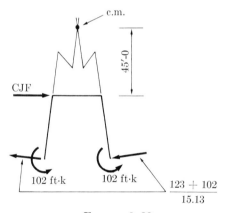

FIGURE 9-23

$$\Sigma M_{c.m.} = 0 \text{ (clockwise is +)},$$

$$2\left(\frac{123 + 102}{15.13}\right) \times 4 \times 15.13 - 204 - 45 \times \text{CJF} = 0,$$

from which

$$\text{CJF} = +35.47 \text{ k to the right, as assumed.}$$

The CJF at level BE is determined by means of the free body in Fig. 9-24:

380   APPLICATIONS OF MOMENT DISTRIBUTION   [CHAP. 9

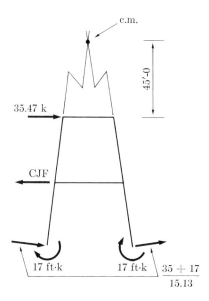

FIGURE 9–24

$\Sigma M_{\text{c.m.}} = 0$ (clockwise is $+$),

$$-2\left(\frac{35 + 17}{15.13}\right) \times 5 \times 15.13 + 34 - 35.47 \times 45 + 60\text{CJF} = 0,$$

from which

$$\text{CJF} = +34.70 \text{ k to the left, as assumed.}$$

Figure 9–25 shows the final moments resulting from the first distortion, with the consistent joint forces as just determined.

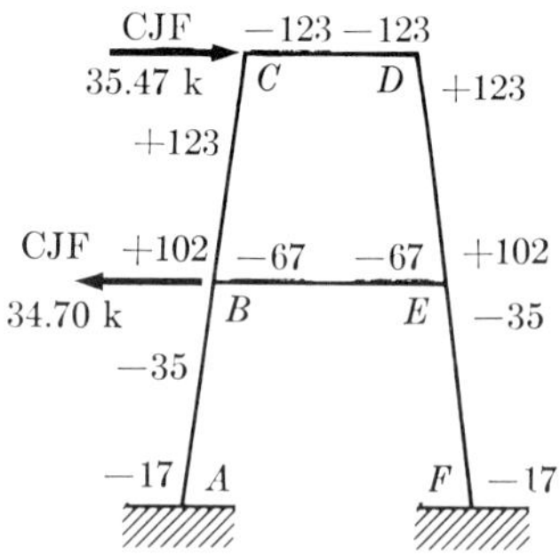

FIGURE 9–25

The second impressed distortion is shown in Fig. 9–26. The fixed-end moments are proportional to the modified K_M of each member. These moments are balanced out and the consistent joint forces are evaluated by the same method used for the first impressed distortion, with the results as shown in Fig. 9–27. It is now necessary to find some factor X by

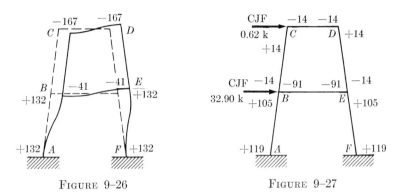

FIGURE 9–26   FIGURE 9–27

which all values of Fig. 9–25 may be multiplied, and to find another factor Y by which all values of Fig. 9–27 may be multiplied, such that an algebraic summation of the products will result in a set of moments which will be consistent with a force acting equal and opposite to the AJR of Fig. 9–18.

The two condition equations necessary to evaluate X and Y are again written, as in Example 9–1, to express the fact that the superposition of X times the CJF's of Fig. 9–25 and Y times the CJF's of Fig. 9–27 and the AJR of Fig. 9–18 must result in a net horizontal force of zero at each of the two levels. If we consider that forces acting to the right are positive, these equations are:

For level CD,
$$+35.47X + 0.62Y - 35.0 = 0.$$

For level BE,
$$-34.70X + 32.90Y = 0.$$

A simultaneous solution will result in
$$X = +0.97 \quad \text{and} \quad Y = +1.02.$$

If the values of Fig. 9–25 are multiplied by X and the values of Fig. 9–27 are multiplied by Y, and if the two sets of products are added algebraically, the results will be consistent with the original force of 35 k acting to the right at level CD. The resulting final moments (see Fig. 9–28) may be verified by taking moments about the intersection of the sloping legs for either of the free bodies previously used in the analysis.

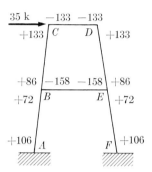

FIGURE 9–28

EXAMPLE 9–3. The frame of Fig. 9–29 is to be used in a small industrial building, with the necessary roof slope to be provided in the roof construction. The bays are 20 ft and the total roof load is 50 lb/ft². Wind, blowing from the left, is 20 lb/ft² on vertical surfaces. Find all joint moments resulting from the combination of roof and wind loads. (In practice, wind moments are computed separately.)

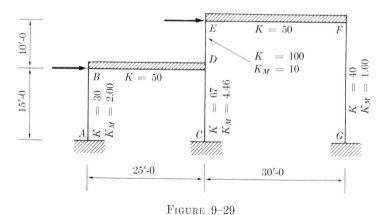

FIGURE 9–29

Wind concentration at

$$E = 5 \times 20 \times 0.02 = 2.0 \text{ k},$$

$$B = (7.5 + 5) 20 \times 0.02 = 5.0 \text{ k}.$$

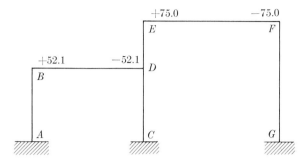

FIGURE 9–30

The fixed-end moments, computed and then indicated in Fig. 9–30, are balanced out, with the results given in Fig. 9–31. In the same figure are

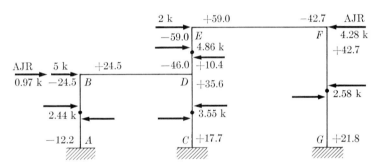

FIGURE 9–31

indicated the shear-couples for these moments and the AJR's. The first impressed distortion and the resulting fixed-end moments are shown in Fig. 9–32. These are balanced out, with the resulting moments, shear-

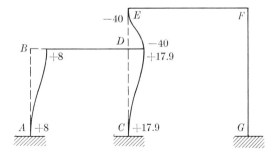

FIGURE 9–32

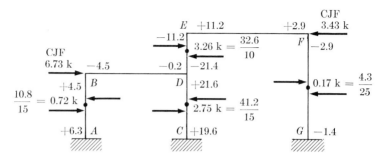

FIGURE 9-33

couples, and CJF's as indicated in Fig. 9-33. The second impressed distortion and the resulting fixed-end moments are as shown in Fig. 9-34.

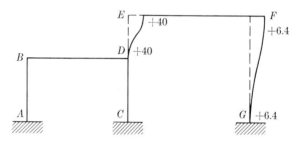

FIGURE 9-34

These are balanced out, with the resulting moments, shear-couples, and CJF's shown in Fig. 9-35.

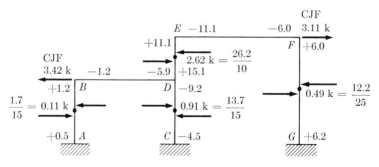

FIGURE 9-35

As in Examples 9–1 and 9–2, it is possible to find some factor X by which all values of Fig. 9–33 may be multiplied, and to find another factor Y by which all values of Fig. 9–35 may be multiplied, such that an algebraic summation of the products will result in a set of moments consistent with forces acting equal and opposite to the AJR's of Fig. 9–31. As before, the equations necessary to evaluate X and Y express the fact that the superposition of X times the CJF's of Fig. 9–33 and Y times the CJF's of Fig. 9–35 and the AJR's of Fig. 9–31 must result in a net horizontal force of zero at levels BD and EF. If we consider that forces acting to the right are positive, these equations are:

For level BD,

$$6.73X - 3.42Y + 0.97 = 0.$$

For level EF,

$$-3.43X + 3.11Y - 4.28 = 0.$$

A simultaneous solution will result in

$$X = +1.27 \quad \text{and} \quad Y = +2.77.$$

If the values of Fig. 9–33 are multiplied by X and the values of Fig. 9–35 are multiplied by Y, and if the products are added algebraically to the values of Fig. 9–31, the results will be the final moments in the frame. These final moments, together with the shear couples, are shown in Fig. 9–36. Note that because all computations were made with a slide rule, slight discrepancies exist between these shear-couples and the applied horizontal loads.

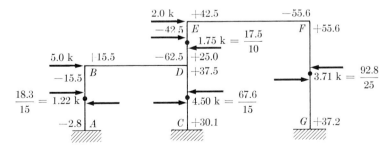

FIGURE 9–36

Problems

9-4. Find all moments in Fig. 9-37 by moment distribution. [*Ans.* (in ft·k): $M_{AB} = -1.9$, $M_{BA} = -15.6$, $M_{BE} = +47.3$, $M_{BC} = -31.7$, $M_{CB} = -37.1$, $M_{DC} = -54.1$, $M_{ED} = +14.9$, $M_{EB} = -23.3$, $M_{EF} = +8.4$, $M_{FE} = +8.5$.]

9-5. The frame sketched in Fig. 9-38 is proposed for a small industrial building. The necessary roof slope will be provided in the roof construction. The bays are 25 ft and the total roof load is 60 lb/ft². Wind, blowing from left to right, is assumed at 20 lb/ft² on a vertical surface, and is considered to cause a uniform load on the column AB. Find all joint moments by moment distribution. [*Ans.* (in ft·k): $M_{BA} = -32$, $M_{CD} = +101$, $M_{DC} = -18$, $M_{DF} = -40$, $M_{DE} = +58$, $M_{FG} = +108$.]

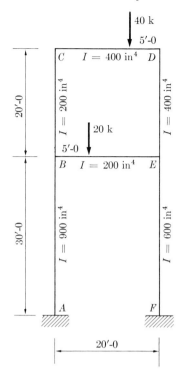

FIGURE 9-37

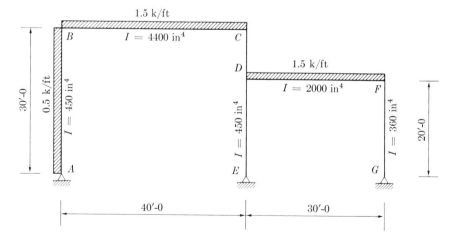

FIGURE 9-38

EXAMPLE 9–6. The rigid frame bents for a gymnasium are to be spaced at 25-ft centers (see Fig. 9–39). Dead and snow load, including an allowance for steel purlins and the frame itself, are assumed to be 65 lb/ft² of horizontal roof projection. Purlins are to be spaced at 7.14 ft center-to-center on the horizontal projection. The moment of inertia is constant. Find all joint moments by moment distribution.

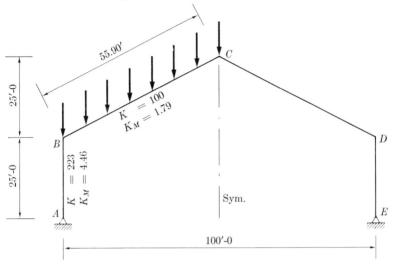

FIGURE 9–39

Each concentration = 25 × 7.14 × 0.065 = 11.6 k except at column tops, where the concentrations are 5.8 k. By the equation given in Section 8–9, the fixed-end moments in the sloping girders are:

$$F_{BC} = \frac{(N^2 - 1)PL}{12N} = \frac{(7^2 - 1)11.6 \times 50}{12 \times 7} = 331 \text{ ft·k.}$$

The initial fixed-end moments in the frame, therefore, are as shown in Fig. 9–40.

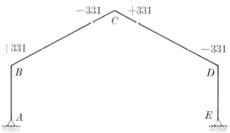

FIGURE 9–40

These are balanced out to give the moments shown in Fig. 9–41.

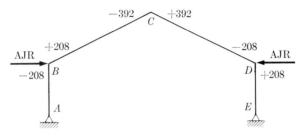

FIGURE 9–41

Joint translation is prevented by AJR's at B and D, and these AJR's must be determined. The AJR acting at B is most easily evaluated by use of the

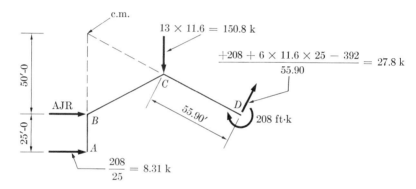

FIGURE 9–42

free body shown in Fig. 9–42. If $\Sigma M = 0$ is written with respect to the indicated center of moments (c.m.) and if clockwise moments are considered to be positive, the following equation results:

$$+150.8 \times 50 + 208 - 27.8 \times 111.80 - 8.31 \times 75 - 50\text{AJR} = 0,$$

from which

$$\text{AJR} = +80.46 \text{ k in the direction assumed.}$$

By a similar procedure, if we use free body $BCDE$, the AJR at D will be found to be 80.46 k to the left. (Actually it is unnecessary to compute this last value. Since we have computed the value of the AJR at B, that at D may be inferred from symmetry.)

The effects of the AJR's acting at B and D in Fig. 9-41 must be cancelled. This is accomplished by finding the moments caused by forces acting at B and D equal in magnitude but opposite in sense to the AJR's, and then by adding these moments to those in Fig. 9-41 to obtain the final moments in the frame. In order to accomplish this, one distortion must be impressed on the frame for each degree of freedom, and values of the modified K_M should be determined for each member. For the purpose of computing these values, it is assumed that joint B is displaced 1

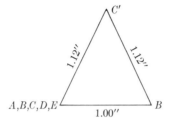

FIGURE 9-43

in. to the right, with joint D held in position. (See Fig. 9-43 for the displacement diagram.) The modified values of K_M are:

For AB,   Mod $K_M = \Delta \cdot K_M = 1 \times 4.46 = 4.46$.

For BC and CD,   Mod $K_M = \Delta \cdot K_M = 1.12 \times 1.79 = 2.00$.

The frame, with the first impressed distortion and the resulting fixed-end moments, is shown in Fig. 9-44. These moments are balanced out and the

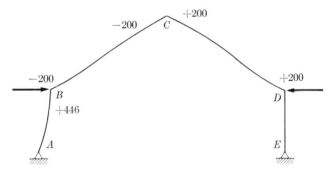

FIGURE 9-44

CJF's are evaluated by the use of two free bodies, $ABCD$ and $BCDE$ (a method previously explained in connection with Fig. 9-42), with the results

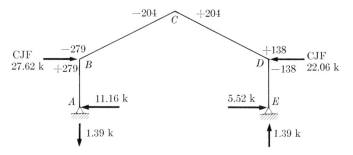

FIGURE 9–45

as shown in Fig. 9–45. Since the frame is symmetrical, this last figure may be drawn opposite hand (mirror image) to obtain the results of forcing D to the left while B is held fixed in position. Figure 9–46 is obtained in this way.

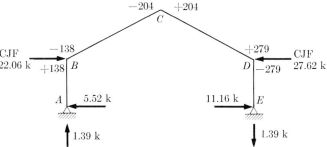

FIGURE 9–46

Two factors, X and Y, may be found such that when all values of Fig. 9–45 are multiplied by X and then added to the products obtained by multiplying corresponding values of Fig. 9–46 by Y, the resulting sums will be the effects of forces acting at B and D opposite to the AJR's. The two condition equations necessary for evaluating the factors X and Y are written from Figs. 9–41, 9–45, and 9–46. If we consider that forces acting to the right are positive, these equations are:

For joint B,
$$+80.46 + 27.62X + 22.06Y = 0.$$
For joint D,
$$-80.46 - 22.06X - 27.62Y = 0.$$

Obviously, X and Y must be equal, and a solution will result in $X = Y = -1.62$. The final moments are obtained by adding algebraically the moments of Fig. 9–41, the moments of Fig. 9–45 multiplied by factor X, and the moments of Fig. 9–46 multiplied by the factor Y. The results are shown in Fig. 9–47.

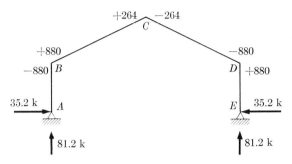

FIGURE 9–47

The method of analysis just demonstrated—that of *separate joint displacements*—is perfectly general and is applicable regardless of the type of loading applied to the frame. For the above vertical loads, however, the alternate solution which follows is easier. In addition, it is the appropriate method to use in order to determine the effect of spread of the column footings, temperature changes, or shortening of the girders due to axial stresses.

In this alternate solution the initial fixed-end moments caused by the applied loads, the first balance, and the resulting joint moments and AJR's are found exactly as explained in connection with Figs. 9–41 and 9–42. However, in this solution, in order to evaluate the effect of removing the AJR's, we first assume that a distortion is impressed on the frame by simultaneously forcing both joints B and D outward by equal amounts. The magnitude of the assumed enforced outward movement of joints B and D is arbitrary. For the purpose of drawing a displacement diagram, a movement of $\frac{1}{2}$ in. will be assumed. (See Fig. 9–48.) The modified values of K_M are:

For AB and DE,

Mod $K_M = \Delta \cdot K_M = 0.5 \times 4.46 = 2.23$.

For BC and CD,

Mod $K_M = \Delta \cdot K_M = 1.12 \times 1.79 = 2.00$.

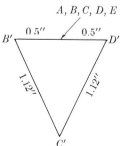

FIGURE 9–48

A distortion of the type described above is assumed to be impressed on the frame and to cause the fixed-end moments shown in Fig. 9–49. The

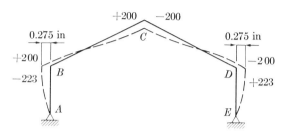

FIGURE 9–49

displacements of joints B and D, consistent with the fixed-end column moments of 223 ft·k, can be computed if the column sizes are known. For example, if the frame is fabricated of 36WF160 sections, with a moment of inertia of 9739 in^4, the consistent displacements of B and D will be

$$\Delta = \frac{ML^2}{3EI} = \frac{12 \times 223 \times 25^2 \times 144}{3 \times 30{,}000 \times 9739} = 0.275 \text{ in.}$$

The fixed-end moments of Fig. 9–49 are balanced, with the resulting moments and CJF's shown in Fig. 9–50. It is necessary to find the joint

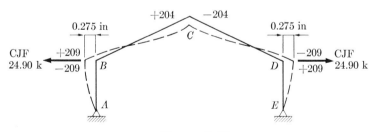

FIGURE 9–50

moments resulting from forces of 80.46 k acting outward at joints B and D on the frame of Fig. 9–41. These are readily obtained by multiplying the moments of Fig. 9–50 by a factor which is given by

$$z = \frac{80.46}{24.90} = 3.23.$$

If the moments of Fig. 9–50 are multiplied by 3.23 and the products are added to the moments of Fig. 9–41, the results will be as shown in Fig. 9–51. These are in good agreement with the moments shown in

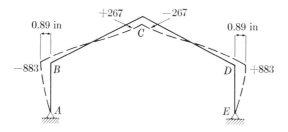

FIGURE 9–51

Fig. 9–47. (Any discrepancy is due to slide rule computations.) The final displacement of B or D will be z times the displacements shown in Fig. 9–50:

$$3.23 \times 0.275 = 0.89 \text{ in.}$$

EXAMPLE 9–7. Find the moments in the rigid frame bent of the preceding example (9–6) as caused by a wind with a velocity of 100 mi/hr. (See Fig. 9–52.)

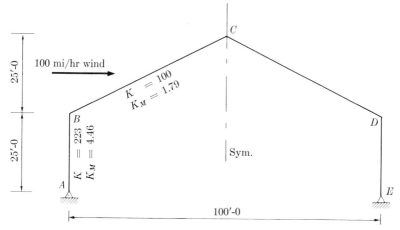

FIGURE 9–52

Probably the best recommendations relative to wind pressures for design for use in this country are those included in *Transactions of the American Society of Civil Engineers* on wind bracing in steel buildings, Vol. 105, p. 1713, 1940. The following recommendations are taken directly from these *Transactions*.

The report indicates that "... it is not necessary to divide the wind force into pressure and suction effects when dealing with the bracing of tall buildings. However, this may need to be done for structures with rounded roofs, such as armories, hangars, and drill sheds, and for mills or other buildings with large, open interiors and walls in which large openings may occur. For any building, possible high local suction effects, due to bad aerodynamic properties of the building profile, should be investigated in relation to secondary members and the attachments of roofing or siding."

External forces on windward and leeward walls. For surfaces which are perpendicular to the wind direction the report states that "... where q is the velocity pressure of the wind, an external pressure of $0.8q$ may be exerted on the windward wall of buildings of average ratio of height to width and an external suction of $0.5q$ on the leeward wall. For an air density of 0.07651 lb/ft^3, corresponding to 15°C, at 760 mm of mercury, and where the velocity V is expressed in miles per hour, the velocity pressure in pounds per square foot is given by $q = 0.002558V^2$." From the preceding, "it is seen that the total combined force on the outside of the windward and leeward vertical faces of an average building may be $1.3q$, or approximately $p = 0.0033V^2$." It may be determined from this last equation that a total wind force of 20 lb/ft^2 will result from a wind velocity of 77.8 mi/hr. The velocity pressure for this wind velocity is 15.5 lb/ft^2.

External forces on plane surfaces inclined to the wind. "For surfaces not more than 300 ft above the ground and for both symmetrical and unsymmetrical gable roofs, where α is the slope of the roof to the horizontal in degrees, the recommended wind forces are as follows:

(1) Windward slope—
  (a) For α not greater than 20°, a suction of 12 lb/ft^2
  (b) For α between 20° and 30°, a force of $p = 1.20\alpha - 36$, the negative value indicating suction.
  (c) For α between 30° and 60°, a force of $p = 0.30\alpha - 9$, the positive value indicating pressure.
  (d) For α greater than 60°, a pressure of 9 lb/ft^2.
(2) Leeward slope—
  For all values of α in excess of zero, a suction of 9 lb/ft^2."

External forces on rounded roof surfaces. "Considering the rounded roof as represented roughly by a circular arc passing through the two eaves and the ridge, or, where the roof slope starts from ground level, through the springings and the ridge, and letting the ratio of rise of the arc to its span be r, the recommended wind forces for surfaces not more than 300 ft above ground level are as follows:

(1) Windward quarter of roof arc—
  (a) For roofs resting on elevated vertical supports: when r is less than 0.20, a suction of 12 lb/ft^2; when r is greater than 0.20, a pressure of $p = 30r - 6$. Alternatively, when r is between 0.20 and 0.35, a force of $p = 80r - 28$, the negative sign indicating suction.
  (b) For roofs starting from ground level: for all values of r, a pressure of $p = 19r$.

(2) Central half of roof arc—
   (a) For roofs resting on elevated vertical supports, a suction of $p = 15r + 11$.
   (b) For roofs starting from ground level, a suction of 11 lb/ft^2.
(3) Leeward quarter of roof arc—
   For all values of r greater than zero, a suction of 9 lb/ft^2."

External forces on a flat roof. "For a flat roof a normal suction of not less than 12 b/ft^2 should be considered as applied to the entire roof surface."

External forces on walls parallel to the wind. "On walls parallel to the wind an external suction of 9 lb/ft^2 should be considered."

Internal wind forces. The wind force to be assumed in design is to be taken as the applicable external wind force (as outlined above) only in the event that the building is airtight. Such a condition will rarely arise, and therefore internal suction or pressure should also be considered.

"Normally, air leakage due to the usual small openings around windows, doors, skylights, and eaves will give rise to an internal pressure or suction of from $0.25q$ to $0.35q$, depending on whether the openings are chiefly in the windward or in the leeward surfaces.

"Where openings are of substantial size, the internal wind force may be of considerable magnitude. It may be pressure or suction, depending on whether the openings are in windward surfaces, in leeward surfaces, or in surfaces parallel to the wind." Very large internal pressures have been (and may be) built up because of the breaking of windows on the windward side of buildings by wind-carried objects. On the basis of various studies and investigations it appears that reasonable allowances for internal wind forces are as follows:

(1) "For buildings that, although nominally airtight, with closed doors and windows and unbroken glass, are nevertheless more or less 'air leaky' by reason of numerous distributed small openings, an internal pressure or suction of 4.5 lb/ft^2 normal to walls and roofs.
(2) For buildings with 30% or more of the wall surfaces open, an internal pressure of 12 lb/ft^2, or an internal suction of 9 lb/ft^2.
(3) For buildings that may have percentages n of openings varying from 0 to 30% of the wall space, an internal pressure of

$$p = 4.5 + 0.25n,$$

or an internal suction of

$$p = 4.5 + 0.15n."$$

Wind force on a series of roofs. "Where a series of roofs exists in one building, one roof lying behind another and being nominally masked by it, the structure as a whole is to be designed for the full wind load on the first roof and for 80% wind on the other roofs. Any one roof, however, is to be designed for the full wind load."

Returning to Example 9–7, the velocity pressure is computed as follows:

$$q = 0.002558 V^2 = 0.002558 \times 100^2 = 25.58 \text{ lb/ft}^2.$$

The pressure against the windward wall is

$$0.8 \times 25.58 = 20.4 \text{ (say 20) lb/ft}^2.$$

The suction on the leeward wall is

$$0.5 \times 25.58 = 12.79 \text{ (say 13) lb/ft}^2.$$

The action on the windward slope, where $\alpha = 26°40'$, is

$$1.20\alpha - 36 = 1.20 \times 26.7 - 36 = -4 \text{ lb/ft}^2 \text{ suction.}$$

The action on the leeward slope is -9 lb/ft^2 suction. Because of special features of construction, it is assumed that no internal pressure or suction need be considered. However it is usually advisable that, depending upon the location of glazed openings subject to breakage, either internal pressure or suction be assumed to exist. This internal effect would be added algebraically to the external forces as computed above.

For the problem at hand, the wind forces on the various surfaces are as shown in Fig. 9–53. Fixed-end moments as caused by the above loads are computed and balanced in the usual way, while the values of the AJR's at B and D are evaluated in a manner similar to that used for Fig. 9–42. The results are shown in Fig. 9–54, where it can be seen that ΣH and ΣV equal (or very nearly equal) zero.

If the moments of Fig. 9–54 are combined with moments resulting from the action of two horizontal forces applied at B and D, of the same magnitude but of a sense opposite to the AJR's at these joints, the required wind moments at the joints of the frame will be obtained. The moments resulting from two forces at B and D opposite to the AJR's may be very easily found by using the information given in Figs. 9–45 and 9–46. We can determine the values of two factors X and Y, such that when the values of Fig. 9–45 are multiplied by X and the values of Fig. 9–46 are multiplied by Y, the summation of the corresponding products will be the moments resulting from the simultaneous action of 15.16 k to the right at B and 0.42 k to the right at D. The required condition equations for evaluating X and Y (from Figs. 9–45, 9–46, and 9–54) are

$$+27.62X + 22.06Y - 15.16 = 0,$$

$$-22.06X - 27.62Y - 0.42 = 0,$$

and a simultaneous solution will give $X = +1.54$ and $Y = -1.25$. If these factors are applied as indicated above and the results combined with the values shown in Fig. 9–54, the final wind moments will result, as shown in Fig. 9–55. (Minor inconsistencies apparent in these results, chiefly in checking ΣH and $\Sigma V = 0$, are due to slide rule computations.)

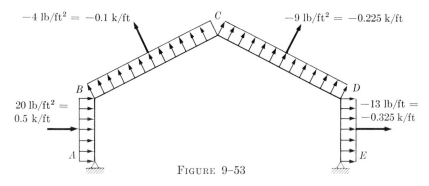

FIGURE 9–53

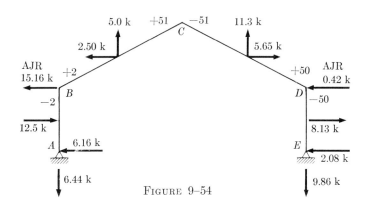

FIGURE 9–54

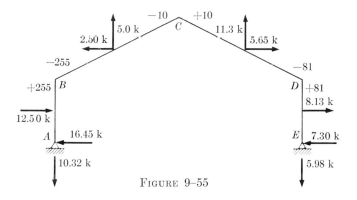

FIGURE 9–55

EXAMPLE 9–8. Find the effects of the elastic stretch of the tie, girder shortening, and temperature change for the loaded rigid frame bent of Example 9–6. Consider a temperature drop of 70°F and a temperature rise of 40°F. The frame is fabricated with 36WF160 sections throughout. (See Fig. 9–56.)

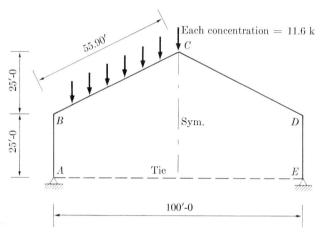

FIGURE 9–56

All the actions referred to in the example statement will cause a change in the horizontal distance between the knees of the bent relative to the distance between column bases. This change will induce moments in the frame, and these moments are to be evaluated.

In the design and construction of any rigid frame bent, it is important that any outward movement of the column bases be prevented. Unless the character of the foundation definitely ensures that such movement will not occur, it is advisable to use a structural tie between the column footings of each bent. This tie is usually a steel rod, suitably protected against corrosion and designed to provide all the required horizontal reaction components at the column bases. When the bent is loaded, the tie is assumed to stretch elastically and the column bases are assumed to move apart by the amount of the tie stretch.

Girder shortening results from the axial compression in the sloping girders of the bent and causes the knees of the frame to move toward each other. A drop in temperature will cause a similar movement of the knees, while a rise in temperature will cause the knees to spread.

In beginning the analysis, it is convenient to first determine the moments induced in the frame by an arbitrary outward movement of each column base. A lateral movement of 0.275 in. of one end of the 36WF160 column relative to the other end has been found (Fig. 9–49) to cause a fixed-end moment of 223 ft·k at the top end of the column. It will therefore be assumed that each column base moves outward by 0.275 in. Figure 9–57 shows these displacements and the fixed-end moments.

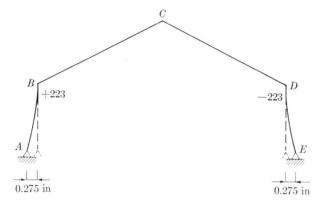

FIGURE 9–57

If these moments are balanced and the AJR's at B and D evaluated, the results are as shown in Fig. 9–58. We must now find the moments resulting

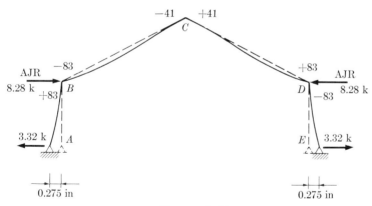

FIGURE 9–58

from horizontal forces acting at B and D, forces which are equal in magnitude and opposite in sense to the AJR's in Fig. 9–58. These moments are easily found by referring to Fig. 9–50, which shows moments resulting from horizontal forces of 24.90 k acting outward at B and D. By direct proportion, the moments and displacements resulting from horizontal forces of 8.28 k acting outward at B and D can be computed if we multiply the movements and displacements of Fig. 9–50 by a factor given by

$$z = \frac{8.28}{24.90} = 0.332.$$

When this is done, the results are as shown in Fig. 9–59.

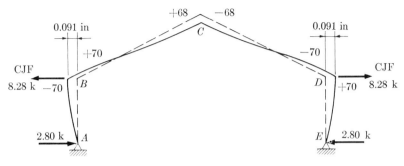

FIGURE 9–59

If the values of Fig. 9–59 are combined with those of Fig. 9–58, the effects of an outward displacement of 0.275 in. of each column base will result (see Fig. 9–60). The effects of elastic stretch of the tie, girder shortening, and temperature change are easily evaluated from Fig. 9–60. First, the influence of each of these actions in tending to cause the distance between column bases to become longer, or shorter, than the distance between the knees must be found. These influences are then combined to give the maximum differential lengthening or shortening. Finally, the resulting moments are determined, by direct proportion, from Fig. 9–60.

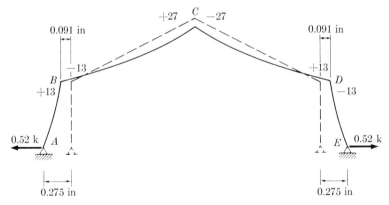

FIGURE 9–60

The effect of the elastic stretch of the tie will obviously be to tend to make the distance between column bases greater than the distance between column knees. Such a tendency will be considered to be positive. The tie will be designed for a working stress of 10 k/in² in order to provide for the possibility that the protective coating may not completely prevent corrosion. The elastic elongation of the tie will therefore be

$$\frac{10 \times 100 \times 12}{30{,}000} = +0.40 \text{ in.}$$

The example statement indicates that a temperature drop of 70°F and a temperature rise of 40°F should be considered. Since the tie is buried under the floor slab, it will be unaffected by changes in temperature. Consequently, a temperature drop will cause a positive length differential. Therefore the temperature length differentials are as follows:

70°F drop: $0.0000065 \times 70 \times 1200 = +0.55$ in.,

40°F rise: $\dfrac{40}{70} \times 0.545 = -0.31$ in.

Girder shortening, resulting from the axial compression of the girders, causes a positive length differential. The method of combining components will be evident from Fig. 9–61. Since the frame is loaded vertically, the

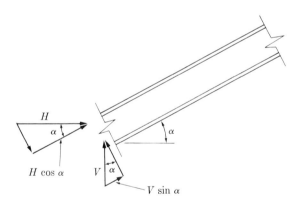

FIGURE 9–61

values of the horizontal thrust H are constant throughout the length of the girder. The vertical shear V, however, varies in magnitude from the vertical reaction component at the knee to a small value at the center. It is sufficiently accurate to use an average value of $V_A/2$. Consequently, the horizontal projection of the axial shortening of the girders may be expressed by

$$\left(H \cos \alpha + \frac{V_A}{2} \sin \alpha \right) \frac{2L \cos \alpha}{AE},$$

where L is the length in feet, from knee to ridge, of one girder and where A is the cross-sectional area of the girder in square inches. The length differential for girder shortening, for the actual values from Fig. 9–47, is

$$+ \left(35.2 \times 0.895 + \frac{81.2}{2} \times 0.447\right) \frac{2 \times 55.90 \times 12 \times 0.895}{47.09 \times 30{,}000} = +0.02 \text{ in.}$$

The combined positive length differentials are

$$+0.40 + 0.55 + 0.02 = +0.97 \text{ in.}$$

In order to determine the net negative length differential, the percentage of the dead load on the frame must be known. Such information is necessary because it is assumed that temperature rise can occur only in the summer when snow loading cannot exist. Consequently, the length differential due to temperature rise must be combined with length differentials caused by tie elongation and girder shortening when dead load alone is on the frame. Assume that 40% of the total design load is dead load. The net negative length differential will therefore be

$$+0.4\,(0.40 + 0.02) - 0.31 = -0.14 \text{ in.}$$

Actually the negative length differential is of little practical value, since it can occur only when dead load alone is on the structure and hence cannot be used to determine a critical design moment. The combined positive length differential, however, is necessary to determine the maximum positive moment at the center. Figure 9–60 shows the effects of an initial positive length differential of 0.55 in. The desired moments may be obtained by multiplying the values in this figure by a factor determined as

$$z = \frac{+0.97}{+0.55} = +1.76.$$

The resultant values are shown in Fig. 9–62. Comparison of these moments with those of Fig. 9–47 indicates that the secondary effects just considered will increase the positive moment at the peak of the frame but will decrease the moments at the knees. The wind moments of Fig. 9–55 will decrease the moments at both the peak and the knees. The maximum moments that can ever exist in the frame, for the loads shown, will therefore be taken as 880 ft·k at the knees (neglecting the relieving moment of 23 ft·k, as shown in Fig. 9–62, in the event that such relief does not materialize) and as $264 + 47 = 311$ ft·k at the peak. These tentative design moments are shown in Fig. 9–63.

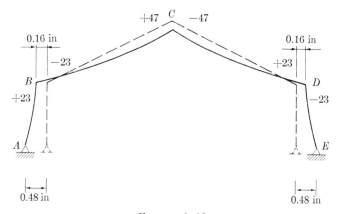

FIGURE 9-62

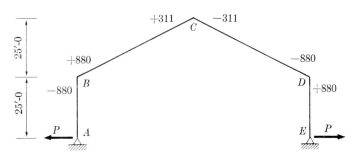

FIGURE 9-63

It should be pointed out here that the maximum moment at the knee can be considerably reduced by prestressing the frame. This procedure will increase the moment at the peak to a value which will be equal to the reduced moment at the knee, and this is desirable if the frame is to be fabricated with a constant moment of inertia. The frame would be prestressed by applying a horizontal force P, acting outward at each column base, as shown in Fig. 9-63. The value of P must be such as to make the maximum moment at the knee equal the maximum moment at the peak.

This condition is expressed as follows:

$$880 - 25P = 311 + 50P,$$

from which

$$P = 7.57 \text{ k}.$$

With this force acting on the frame, the moments at the knee and at the peak become
$$880 - 25 \times 7.57 = 690 \text{ ft·k},$$

and thus the maximum moment has been considerably reduced.

The required movement of each column base to accomplish this prestress may be readily computed by referring to Fig. 9-60:

$$\frac{7.57}{0.52} \times 0.275 = 4.0 \text{ in.}$$

This prestress would be very easily managed in the field by fabricating the bent so that each column base, with the bent erected but with no load acting, would be 4 in. inside the vertical line from the knee. The application of the dead load of the sloping girders would introduce nearly all the desired total spread of 8 in. between the column bases. Then the column bases could be easily pulled the remaining distance into position, the columns would be vertical, and the desired prestress would be introduced.

EXAMPLE 9-9. Find the moments in the frame of Fig. 9-64 as caused by the crane loads shown.

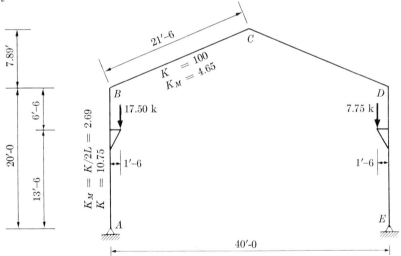

FIGURE 9-64

The first step in the solution of the problem is to find the fixed-end moments F_{BA} and F_{DE}. One method of computing these moments is to make use of the formulas developed in Problem 5-11. These formulas are shown in Fig. 9-65, with the signs reversed to give the action of the member

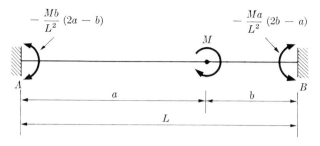

FIGURE 9-65

on the joint. The moment M applied to the column through the bracket will be considered to be positive if it is clockwise. A positive result obtained from either of the above formulas for the fixed-end moments will then indicate a tendency for the end of the member to rotate the support in a clockwise direction. If values for the column AB are substituted in the above formulas, assuming temporarily that the base A is fixed, the following fixed-end moments result:

$$F_{BA} = -\frac{Ma}{L^2}(2b - a) = \frac{-26.25 \times 13.5}{20^2}(13.0 - 13.5) = +0.44 \text{ ft·k},$$

$$F_{AB} = -\frac{Mb}{L^2}(2a - b) = \frac{-26.25 \times 6.5}{20^2}(27 - 6.5) = -8.75 \text{ ft·k}.$$

The fixed-end moment F'_{BA} at end B, with end A pinned, is easily found by balancing out the moment F_{AB} to zero and by carrying over one-half of the balancing moment. The result is that

$$F'_{BA} = +0.44 + 4.37 = +4.81 \text{ ft·k}.$$

The value of the fixed-end moment F'_{DE} is found by direct proportion and reversing the sign; thus:

$$F'_{DE} = \frac{7.75}{17.5} \times 4.81 = -2.13 \text{ ft·k}.$$

An interesting and useful alternate method for determining the fixed-end moments for a member loaded with a moment or couple, at any point along its length, is provided by moment distribution. Assume, for convenience, that the member AB is horizontal and that a support, or an AJR, is temporarily provided at X, the point of application of the couple.

The result is the two-span continuous beam shown in Fig. 9–66. The

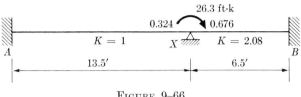

FIGURE 9–66

applied moment or couple is distributed in accordance with the indicated distribution factors, the carry-overs are made, and the results are as shown in Fig. 9–67.

FIGURE 9–67

Now consider that each span is a simple beam with the proper moments (as indicated by Fig. 9–67) acting on the ends as shown in Fig. 9–68. The external end reactions for each span are then found as indicated in this

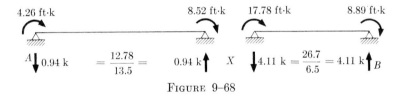

FIGURE 9–68

figure. The net external reaction, or AJR, at X in Fig. 9–68 is therefore $4.11 - 0.94 = 3.17$ k acting down. Since this reaction cannot be permitted to remain, we cancel its effect by loading the member AB with an

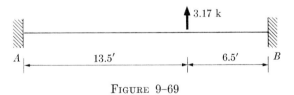

FIGURE 9–69

upward force of 3.17 k, as shown in Fig. 9–69. The fixed-end moments caused by this load are computed in the usual way, with

$$F_{AB} = \frac{Pab^2}{L^2} = \frac{3.17 \times 13.5 \times 6.5^2}{20^2} = -4.51 \text{ ft·k},$$

$$F_{BA} = \frac{Pa^2b}{L^2} = \frac{3.17 \times 13.5^2 \times 6.5}{20^2} = +9.37 \text{ ft·k}.$$

If the above end moments are added to the end moments of Fig. 9–67, the final fixed-end moments for the original member will be $F_{AB} = -8.77$ ft·k and $F_{BA} = +0.48$ ft·k. These values agree closely with those previously obtained by the formulas. The initial fixed-end moments

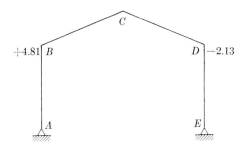

FIGURE 9–70

for the frame, before balancing, are therefore as shown in Fig. 9–70. These moments are balanced and the AJR's determined as described in Example 9–6, with the results shown on Fig. 9–71.

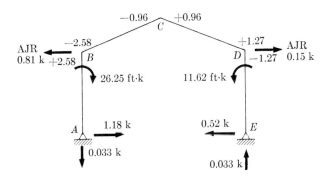

FIGURE 9–71

As in previous problems, it is now necessary to find the moments caused by forces acting equal and opposite to the AJR's. Since the procedure is exactly the same as described in Example 9–6, it is suggested that the reader make the detailed computations and find the CJF's and joint moments resulting from an impressed displacement of joint B of 1 in. to the right, joint D being held in position. (The results are shown in Fig. 9-72.)

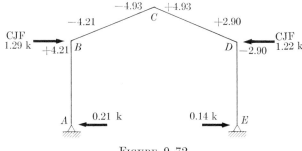

FIGURE 9-72

If Fig. 9-72 is drawn opposite hand, corresponding values will be obtained for a 1-in. impressed displacement of joint D toward the left, joint B being held in position. From these two figures and from Fig. 9-71, two factors X and Y are determined and applied as described in Example 9-6. The final results are shown in Fig. 9-73.

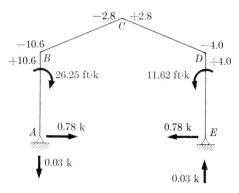

FIGURE 9-73

EXAMPLE 9-10. In Fig. 9-74 find all moments, as caused by the crane loads shown, by moment distribution. Column bases are fixed.

Stepped-columns are commonly used in industrial buildings with heavy crane installations. If these stepped-columns also support a gabled rigid frame, the analysis of the frame becomes slightly more involved than would otherwise be the case. This, of course, is because the columns are nonprismatic. Fixed-end moments, stiffness and carry-over factors must be computed for these columns, and this can be accomplished by moment distribution. (Two other methods will be presented in Chapter 10.)

9–2] FRAMES WITH TWO DEGREES OF FREEDOM 409

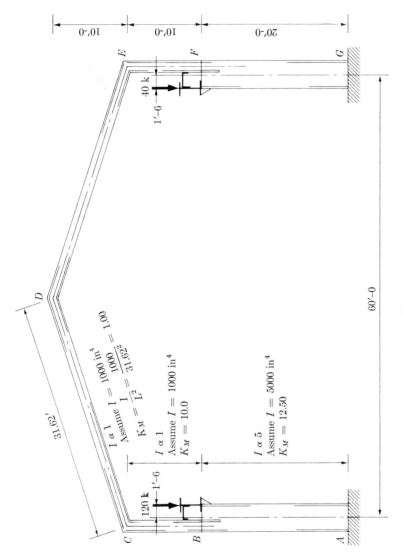

FIGURE 9-74

As a first step toward the solution of the problem, the fixed-end moments will be computed. For this purpose consider that the column ABC is placed on its side, with the ends fixed and with an AJR acting at B. The moment of $120 \times 1.5 = 180$ ft·k is applied to the member at B, and this moment is distributed to the members BA and BC in proportion to their distribution factors. One-half of each distributed amount is carried over to the opposite fixed end of each member. The resulting moments and the evaluation of the AJR are indicated in Fig. 9–75. The effect of this AJR

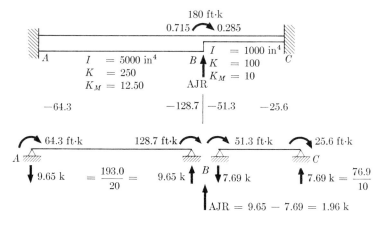

FIGURE 9–75

must be cancelled. To accomplish this, some unknown force is applied at B on the otherwise unloaded member, and point B is thereby displaced downward (without rotation) a distance sufficient to induce moments proportional to the values of K_M. This condition is shown, the moments are balanced, and the CJF evaluated in Fig. 9–76. The final

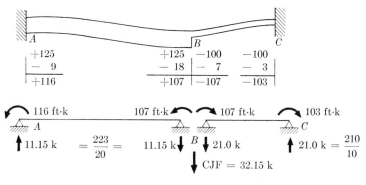

FIGURE 9–76

moments shown in this figure are thus found to be consistent with a CJF of 32.15 k acting down. It is now necessary to find the moments caused by a force of 1.96 k acting down, and these may be obtained by multiplying the final end moments of Fig. 9–76 by a factor

$$z = \frac{1.96}{32.15} = 0.061.$$

The final values of the fixed-end moments are obtained as follows:

	AB	CB
Moments of Fig. 9–75	−64.3	−25.6
(Moments of Fig. 9–76)z	+7.1	−6.3
Fixed-end moments	−57.2	−31.9

By inspection we see that the fixed-end moments F_{GF} and F_{EF} are opposite in sign to, and one-third the magnitude of, the corresponding values of F_{AB} and F_{CB}. Therefore,

$$F_{GF} = +\frac{57.2}{3} = +19.1 \text{ ft·k},$$

$$F_{EF} = +\frac{31.9}{3} = +10.6 \text{ ft·k}.$$

The carry-over factor C_{CA} is next computed. The member is first assumed to be arranged and loaded as shown in Fig. 9–77, and moments

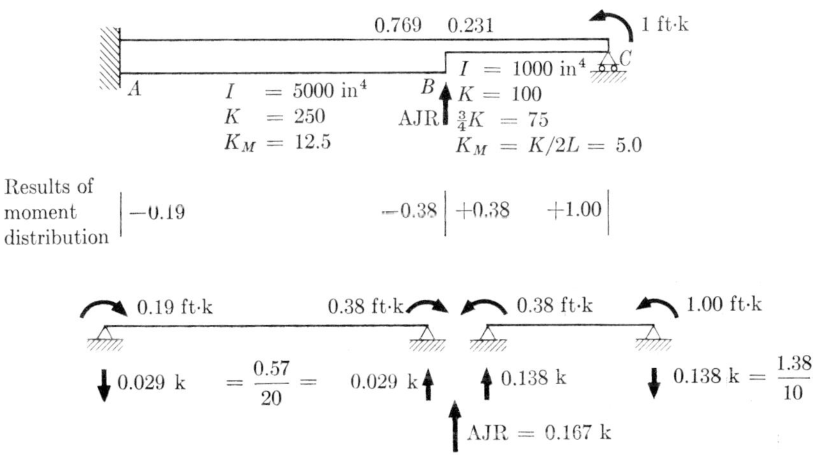

FIGURE 9–77

are balanced and the AJR evaluated as shown therein. Again it is necessary to find the moments resulting from the action of a force at B equal and opposite to the AJR. Assume that B is depressed sufficiently to induce end moments proportional to the values of K_M indicated in Fig. 9–77.

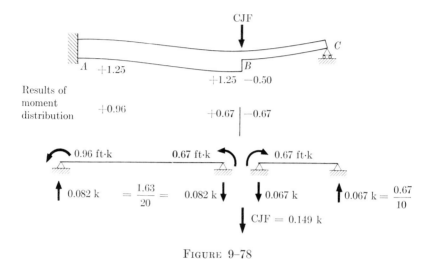

FIGURE 9–78

This condition, together with the results of the moment distribution and the evaluation of the CJF, is shown in Fig. 9–78. In order to find the moments resulting from a force of 0.167 k acting down at B, we simply multiply the moments of Fig. 9–78 by a factor

$$z = \frac{0.167}{0.149} = 1.12.$$

When the resulting moments are added to the moments of Fig. 9–77, the final moment at A will be 0.89 ft·k. The end moment at C will, of course, be 1.0 ft·k. The required carry-over factor is therefore

$$C_{CA} = \frac{0.89}{1.00} = +0.89.$$

The stiffness may be determined by first computing the rotation at end C resulting from the action of the 1 ft·k couple. When the rotation of end C caused by this unit couple is known, it is a simple matter to find the moment necessary to cause a rotation of 1 rad. This moment, by definition, will be the *absolute stiffness*.

The conjugate beam is shown in Fig. 9–79. It is clear that relative values of I have been used in computing the magnitudes of the elastic

FRAMES WITH TWO DEGREES OF FREEDOM

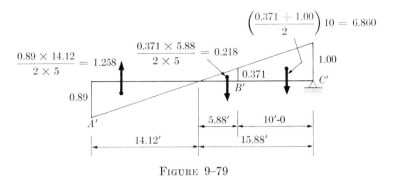

FIGURE 9-79

loads. Since the relative I for the segment BC of the column is taken as unity, then all relative moments of inertia are referred to the moment of inertia of BC. This will be designated as I_{ref}. The shear $R_{C'}$ in the conjugate beam at C' will be

$$R_{C'} = 6.860 + 0.218 - 1.258 = 5.820.$$

The slope at C will be given by

$$\theta_C = \frac{R_{C'}}{EI_{\text{ref}}}.$$

The absolute stiffness K_C at C will be:

$$\text{Abs } K_C = \frac{1}{\theta_C} = \frac{EI_{\text{ref}}}{R_{C'}}.$$

Since it is assumed that Fig. 9-79 represents an original design, the sizes of the rolled sections required in the frame are not known, and consequently, I_{ref} is not known. However, if E and I_{ref} are arbitrarily assigned values of unity, it is possible to obtain a relative value for K_{CA} which will be entirely satisfactory for the analysis. This value for relative K_{CA} is:

$$\text{Rel } K_{CA} = \frac{1}{5.82} = 0.172.$$

Care must be taken that the value used for the stiffness of the girder is related to the same datum as that of the relative stiffness of the stepped-column. The absolute stiffness for the girder CD is given by:

$$\text{Abs } K_{CD} = \frac{4EI_{\text{ref}}}{L}.$$

Consequently, since E and I_{ref} have been assigned values of unity, the correct relative stiffness will be given by:

$$\text{Rel } K_{CD} = \frac{4}{L} = \frac{4}{31.62} = 0.126.$$

The foregoing information, all of which is necessary for the first balancing operation, is assembled in Fig. 9–80. The moment distribution is performed and the AJR's computed as previously demonstrated, with the results shown in Fig. 9–81.

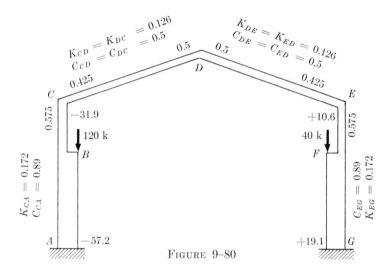

FIGURE 9–80

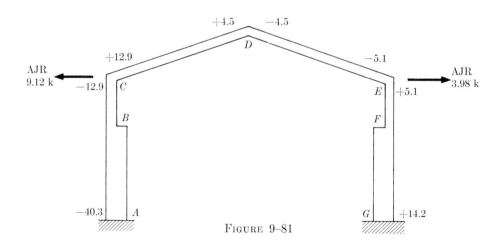

FIGURE 9–81

As in the preceding examples illustrating gable frame analysis, we must again find the moments induced in the frame by the action of forces equal and opposite to the AJR's. As a beginning, assume that joint C is moved to the right one unit while joints B, E, and F are held in position. (The relative displacement diagram is shown in Fig. 9–82.) The resulting fixed-end moments must be consistent with these relative displacements. It will be remembered that a set of consistent fixed-end moments can most easily be determined by first choosing a set of moments which are proportional to the values of K_M (indicated, in this example, in Fig. 9–74), and then by multiplying these moments for each member by the corresponding relative displacement between the two ends of that member. Thus, if we arbitrarily assume that the lateral displacement of C is sufficient to cause F_{BC} and F_{CB} to be 100 ft·k, then the magnitude of F_{CD}, F_{DC}, F_{DE}, and F_{ED} will be given by $1.58 \times 10 = 15.8$ ft·k. The distorted frame, these fixed-end moments, and the AJR's required to hold the frame in this distorted position are shown in Fig. 9–82.

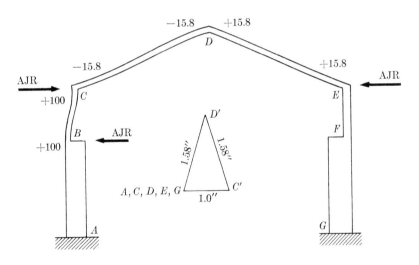

FIGURE 9–82

It is necessary to remove the AJR at B, and to do this it is convenient to consider the member ABC as placed on its side, as shown in Fig. 9–83. The fixed-end moments and distribution factors are indicated on the sketch. The unbalanced moment at B is distributed, carry-overs are made, and the AJR at B is evaluated. From Fig. 9–83 it is apparent that the effect of the AJR at B must be cancelled by a force of 21.09 k acting down at B on the fixed-end member ABC. The moments resulting from the action of

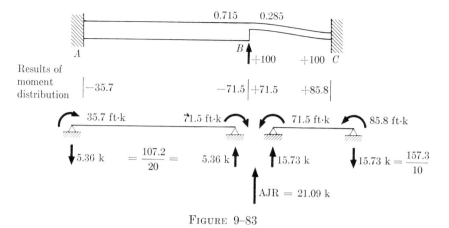

FIGURE 9-83

this force may be easily determined by multiplying the final moments of Fig. 9-76 by the factor

$$z = \frac{21.09}{32.15} = 0.657.$$

The final fixed-end moments F_{AC} and F_{CA} (with the AJR at B removed) are therefore obtained by combining the final end moments of Fig. 9-83 with the factor z times the end moments of Fig. 9-76:

$$F_{AC} = -35.7 + 0.657(+116) = +40.6 \text{ ft·k},$$
$$F_{CA} = +85.8 + 0.657(-103) = +18.2 \text{ ft·k}.$$

The AJR at B (indicated in Fig. 9-82) has now been removed. The final set of fixed-end moments, necessary before the balancing operation, are shown in Fig. 9-84, as are the AJR's. Moments are balanced and the

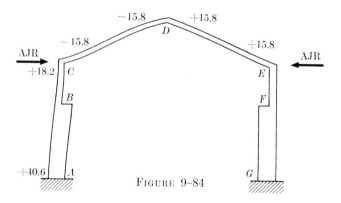

FIGURE 9-84

CJF's (the final values of the AJR's) are computed as before, with the results shown in Fig. 9–85. If Fig. 9–85 is drawn opposite hand, it will give the results of a similar enforced displacement of joint E to the left.

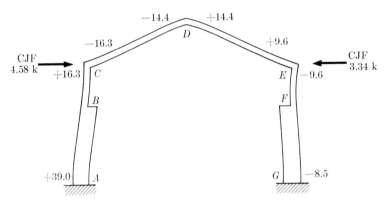

FIGURE 9–85

The information of this last figure, combined with that of Figs. 9–85 and 9–81, will permit the evaluation of factors X and Y and the completion of the analysis as previously demonstrated. The final results are shown in Fig. 9–86.

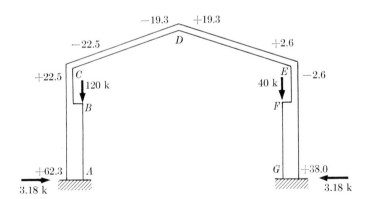

FIGURE 9–86

Problems

9-11. For the gable frame of Fig. 9-87 find all moments by moment distribution. Loads are 2 k/ft on the horizontal projection of the roof girders. Since it is assumed that this is to be a preliminary analysis, the moment of inertia will be taken as constant. [*Ans.*: $M_{BA} = -288$ ft·k, $M_{CB} = +239$ ft·k, $M_{DE} = +288$ ft·k.]

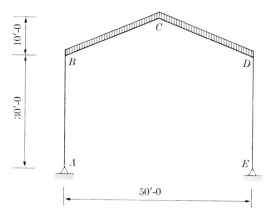

FIGURE 9-87

9-12. Using an assumed wind velocity of 80 mi/hr, find the moments caused by a wind blowing from left to right on the frame of Problem 9-11. Bay lengths are 25 ft. Use recommendations of the American Society of Civil Engineers for finding wind loads on the various surfaces of the bent, and neglect internal pressure. Note that it is difficult to obtain close accuracy in this type of problem because the moments resulting from the removal of the AJR's are much larger than the initial fixed-end moments. The coefficients X and Y should be obtained by use of a computing machine. [*Ans.* (approximate): $M_B = 156$ ft·k, $M_C = 26$ ft·k, $M_D = 90$ ft·k.]

9-13. The gable frame of Fig. 9-88 is subjected to crane loads, vertical and horizontal, as shown. The moment of inertia is constant. Find all moments by moment distribution. [*Ans.*: $M_{BA} = +124$ ft·k, $M_{CB} = -31$ ft·k, $M_{DC} = -99$ ft·k.]

9-14. In Fig. 9-89 find all moments in the unsymmetrical gable frame as caused by the vertical loads shown. The moment of inertia is constant. [*Ans.*: $M_{BA} = -69.7$ ft·k, $M_{CB} = +3.2$ ft·k, $M_{DE} = +69.7$ ft·k.]

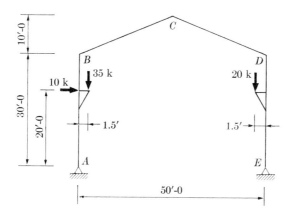

FIGURE 9–88

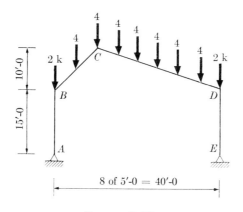

FIGURE 9–89

9–15. Find the moments which will result in the frame of Problem 9–14 from the action of 2 k at C and 5 k at B, both acting from left to right. [Ans.: M_{BA} = +56.6 ft·k, M_{CB} = +12.1 ft·k, M_{DE} = +48.5 ft·k.]

9–16. From Fig. 9–90 find the carry-over factor C_{CA} and the absolute stiffness K_{CA} in ft·k. $E = 30{,}000$ k/in². [Ans.: C_{CA} = +0.93, K_{CA} = 10,500 ft·k.]

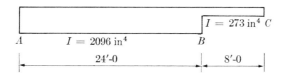

FIGURE 9–90

9–17. In Fig. 9–91 find the fixed-end moments caused by the 8 k load. In addition, find the carry-over factor C_{AC} and the absolute stiffness K_{AC} for the member. The beam is 1 ft wide. $E = 2000$ k/in^2. [*Ans.*: $F_{AC} = +13.6$ ft·k, $F_{CA} = -47.6$ ft·k, $C_{AC} = +1.20$, $K_{AC} = 43{,}400$ ft·k.]

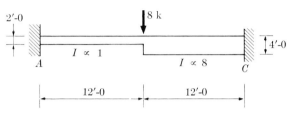

FIGURE 9–91

9–3 Frames with several degrees of freedom. The method of applying moment distribution to continuous frames with two degrees of freedom, demonstrated in the last section, can easily be extended to structures with several degrees of freedom.

EXAMPLE 9–18. Outline the analysis of the frame shown in Fig. 9–92 by moment distribution.

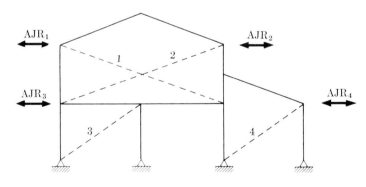

FIGURE 9–92

No loads are shown on the given frame because this method of analysis will apply to any loading system. Fixed-end moments must first be computed, and these are balanced in the usual manner. The moments resulting from this first distribution are valid only if all joint translation is prevented. As shown in Fig. 9–92, four diagonals would be necessary to prevent all joint translation, and thus four AJR's are required. These AJR's must be evaluated, which can be accomplished by a combination of the methods illustrated in previous examples. AJR_1 and AJR_2, for example, can be determined by computations similar to those demonstrated in connection with Fig. 9-42. AJR_3 and AJR_4, however, must be computed by using shear-couples, as illustrated in Example 9-1 or Example 9-3.

The effects of this initial set of AJR's must be cancelled, and the method previously demonstrated will apply. The joints of application of the initial AJR's are caused to translate, each joint in its turn. Consider, for example, the joint of application of AJR_3. This joint is caused to translate to the right some arbitrary amount, usually unknown in magnitude but possible of evaluation if desired. The frame will be distorted as shown in Fig. 9-93.

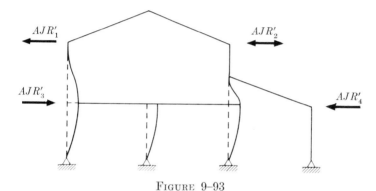

FIGURE 9-93

No rotation of joints is permitted during this translation. Fixed-end moments proportional to the K_M values (modified, if necessary) are induced in the members affected by the translation. The joints of the distorted frame are held in position by a new set of AJR's, the initial magnitudes of which are unimportant. The moments induced by the translation are now distributed; that is, the joints are permitted to rotate (without additional translation) to a position of moment balance at each joint. As the joints rotate, the AJR's change in magnitude and, in their final values, are consistent with the final moments. Hence these AJR's become the CJF's for the distorted and balanced frame of Fig. 9-93.

This process is repeated for each joint of application of an initial AJR, and four sets of balanced moments and CJF's result. These, together with the balanced moments and AJR's of Fig. 9-92, will permit the evaluation of four constants, W, X, Y and Z, by exactly the same procedure as was demonstrated for frames with two degrees of freedom. Applied in the same manner as before, these constants will permit completion of the analysis.

EXAMPLE 9–19. Outline the analysis of the continuous frame of Fig. 9–94 by moment distribution.

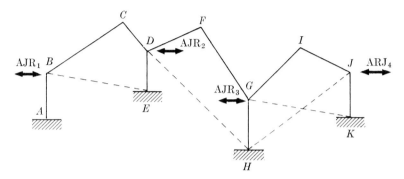

FIGURE 9–94

All fixed-end moments are computed in the usual way, and these moments are balanced, with joint translation being prevented by the AJR's shown in Fig. 9–94. The principal objective of this example is to explain the procedure for evaluating these AJR's.

In this sequence, the values of AJR_1 and V_A are computed from Fig. 9–95(a) and (b), respectively, by taking moments about the center indicated in each sketch. AJR_2 is next computed from Fig. 9–96, with the center of moments as indicated therein. Note that H_{DC} and V_{DC} may be

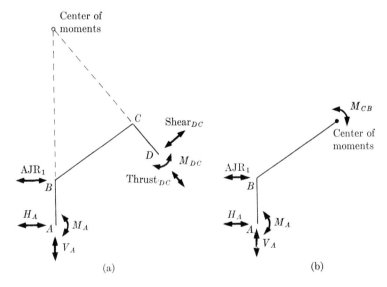

FIGURE 9–95

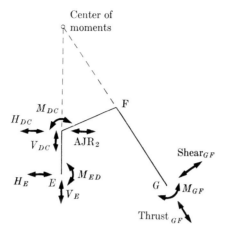

FIGURE 9-96

readily determined by statics after computing AJR_1 and V_A in the manner previously explained. The values of AJR_3 and AJR_4 are computed similarly.

As in all preceding problems the effects of the AJR's must be cancelled. Once again each joint of application of an AJR is, in its turn, caused to translate without rotation, while the other joints of application are fixed in position. As before, after each joint is caused to translate, the resulting fixed-end moments are balanced and the CJF's are computed. Thus four sets of moments and CJF's are obtained. These, together with the original balanced moments, will permit completion of the analysis.

EXAMPLE 9-20. Outline the analysis of the Vierendeel truss bent of Fig. 9-97 by moment distribution.

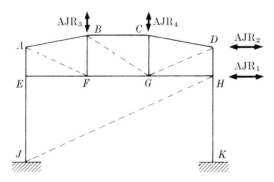

FIGURE 9-97

Fixed-end moments resulting from any interpanel loads are first computed, and these are balanced. AJR's (as indicated in the figure) are required to prevent joint translation. The magnitudes of these AJR's must be determined. The first step is to compute the end shears of all horizontal and vertical members, which can be done by statics, using the member end moments previously determined. The member AB is next treated as a free body, as shown in Fig. 9–98. H_{AB} may now be evaluated from the

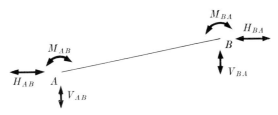

FIGURE 9–98

horizontal shear in AE (previously computed) and any horizontal load applied at A. H_{BA} is then computed from $\Sigma H = 0$ applied to the free body AB. The two moments M_{AB} and M_{BA} are known from the first balance. Finally, V_{BA} is determined by taking moments about end A.

It is now possible to determine the value of AJR_3 from the equation $\Sigma V = 0$ for the member BF treated as a free body. The vertical forces applied to this free body by the chord members AB, BC, FE, and FG can be computed by the method previously explained. AJR_4 can be evaluated similarly. All horizontal forces (the shears in the vertical members framing to the top chords and any external applied loads) acting on the top chord are now known. The value of AJR_2 can therefore be determined from the fact that ΣH must equal zero for the top chord. Similarly, the bottom chord may be treated as a free body in order to compute AJR_1.

It is now necessary to cancel the effects of these AJR's, and this is accomplished exactly as previously described. Each joint of application is, in its turn, caused to translate without rotation, while all other joints of application are fixed in position. Observe that a vertical movement of joint B will induce a horizontal displacement of joint A. The magnitude of this horizontal displacement may be easily determined with a displacement diagram. The resulting fixed-end moments are balanced and the CJF's evaluated, and thus four sets of CJF's are obtained. These, together with the AJR's, will permit the determination of four constants, W, X, Y, and Z, which are applied as demonstrated in previous problems, to complete the analysis.

Problems

9-21. Find all moments in Fig. 9-99 by moment distribution. Roof load is 2 k/ft on the horizontal projection. [*Ans.:* $M_{BA} = -187$ ft·k, $M_{CB} = +75$ ft·k, $M_{DC} = -213$ ft·k.]

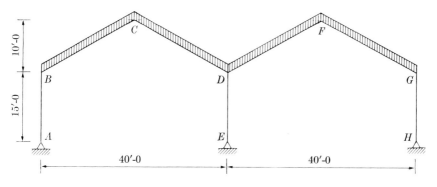

Figure 9-99

9-22. Analyze the frame of Fig. 9-100 by moment distribution. [*Ans.* (in ft·k): $M_{BA} = -18$, $M_{DC} = -5$, $M_{DB} = -23$, $M_{DE} = +28$, $M_{GF} = -1$, $M_{GJ} = +3$, $M_{GH} = -3$, $M_{ED} = -48$, $M_{FE} = -40$.]

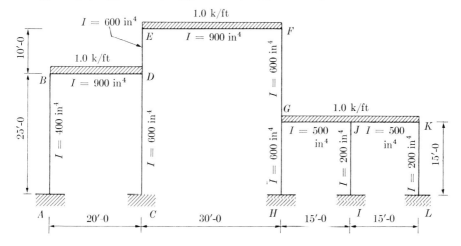

Figure 9-100

9-23. Assuming horizontal loads only of 2 k to the right at E and 8 k to the right at B on the frame of Problem 9-22, find all moments by moment distribution. [*Ans.* (in ft·k): $M_{BA} = +32$, $M_{DC} = +49$, $M_{DB} = -31$, $M_{DE} = -18$, $M_{GF} = +32$, $M_{GJ} = -27$, $M_{GH} = -5$.]

9–4 Secondary stresses by moment distribution. In analyzing articulated structures, it is customary to assume that the members are connected with frictionless pins at the joints. In other words, if the structure is loaded at panel points, the members are assumed to be subjected to axial loads only. The stresses resulting from these axial loads are called *primary stresses*. Actually, of course, in most articulated structures the members are quite rigidly connected at the joints, either by riveting or welding. Consequently, as the structure deflects and the joints tend to rotate through different angles, the various members are caused to flex and bending stresses result. These are called *secondary stresses*.

In the majority of articulated structures the members are not stiff enough to provide any significant flexural resistance to end rotation. In these cases the secondary stresses are small and are neglected. When loads are heavy and members are stiff, however, the secondary stresses in some members may amount to as much as 75 or 80% (or more) of the primary stresses. This is most likely to occur in long-span railway truss bridges with secondary systems, or in heavy trusses in tall building frames used to support the columns of the floors over banquet halls or foyers.

The first satisfactory solution of the secondary stress problem was presented by Heinrich Manderla in a series of papers published in 1879–80. In papers published in 1892–3, Otto Mohr presented his solution for the problem—the *slope-deflection method*—which will be discussed in Chapter 11. A solution can also be obtained by a combination of the Williot or Williot-Mohr diagram and moment distribution.

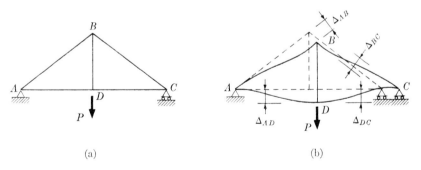

Figure 9–101

The truss of Fig. 9–101 is loaded as shown in sketch (a). Assuming that joints are not permitted to rotate, the loaded truss will deflect as shown in sketch (b). If the values of Δ, as indicated in (b), are known, then the fixed-end moments for the various members can be computed by $6EI\Delta/L^2$. The values of Δ are easily determined by the Williot or Williot-Mohr diagram. Some of the joints in the loaded truss will, of course, rotate.

9-4] SECONDARY STRESSES BY MOMENT DISTRIBUTION

The corrections to the fixed-end moments made necessary by this rotation of joints are applied by balancing moments in the usual way, and the resulting moments are used to compute the secondary stresses in the extreme fibers at the member-ends by means of the ordinary flexure formula. The application of the method is illustrated by the example which follows.

EXAMPLE 9-24. Steel trusses supporting the columns over a hotel foyer are proportioned as shown in Fig. 9-102. All WF sections are oriented with their flanges in vertical planes. Story heights above and below the trusses are 12 ft. Total strains (in inches) are indicated on the various members. Using the Williot diagram and moment distribution, determine secondary end moments and stresses in all truss members. $E = 30{,}000$ k/in^2. For the analysis, columns above and below the truss are assumed to be fixed at the far ends.

The Williot diagram, for determining the values of Δ, is shown in Fig. 9-102. Table 9-1 is used to compute the fixed-end moments and relative stiffnesses of the various members.

TABLE 9-1.

Member	I (in^4)	Δ (in.)	L (in.)	Fixed-end moment $\dfrac{6EI\Delta}{L^2}$ (in·k)	Fixed-end moment (ft·k)	$K = \dfrac{I}{L}$
1–2	1073	+0.765	192	+4000	+333	5.59
2–3	1073	+0.465	192	+2440	+204	5.59
A–B	568	+0.765	192	+2120	+177	2.95
B–C	568	+0.480	192	+1330	+111	2.95
1–A	1029	+0.310	144	+2760	+230	7.13
1–B	838	+0.735	240	+1930	+161	3.49
2–B	528	+0.230	144	+1050	+88	3.67
2–C	107	+0.385	240	+129	+11	0.45
3–C	350	0	144	—	—	2.43
1–Q	350	−0.190	144	−576	−48	2.43
2–R	350	−0.110	144	−334	−28	2.43
3–S	350	0	144	—	—	2.43
A–U	1029	−0.120	144	−1070	−89	7.13

The balancing operation, section moduli, and secondary stresses are shown in Table 9-2.

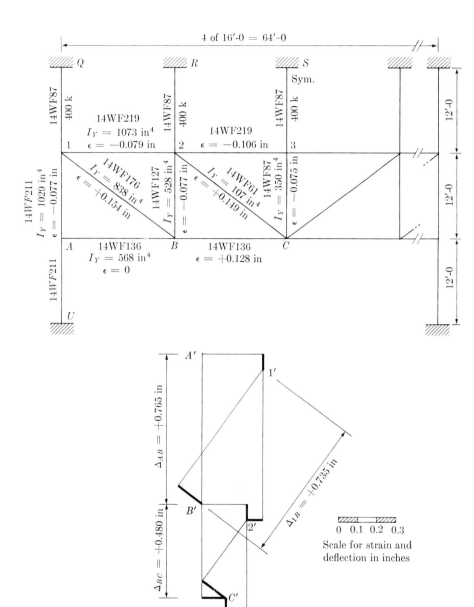

Figure 9-102

TABLE 9–2.

Joint	1				A			2		
Member	1–Q	1–2	1–B	1–A	A–1	A–B	A–U	2–1	2–R	2–3
Stiffness	2.43	5.59	3.49	7.13	7.13	2.95	7.13	5.59	2.43	5.59
Distribution factor	0.130	0.300	0.187	0.383	0.414	0.172	0.414	0.315	0.137	0.315
Fixed-end moment	−48	+333	+161	+230	+230	+177	−89	+333	−28	+204
.	−88	−203	−127	−258	−129			−101		
.			−63			−53				
.		−69						−139	−61	−139
.				−28	−56	−24	−56			
.	+21	+48	+30	+61	+30			+24		
.			+6			+4				
.		−5						−10	−4	−9
.				−7	−14	−6	−14			
.	+1	+2	+1	+2	+1			+1		
.										
M (ft·k)	−114	+106	+8	0	+62	+98	−159	+108	−93	+56
S_y (in^3)	48.2	135.6	107.1	130.2	130.2	77.0	130.2	135.6	48.2	135.6
Secondary stress (k/in^2)	28.4	9.4	0.9	0	5.7	15.3	14.6	9.6	23.2	5.0

Joint	2		B				3	C	
Member	2–C	2–B	B–A	B–1	B–2	B–C	3–2	C–2	C–B
Stiffness	0.45	3.67	2.95	3.49	3.67	2.95	5.59	0.45	2.95
Distribution factor	0.026	0.207	0.226	0.267	0.281	0.226	Fixed	Fixed	Fixed
Fixed-end moment	+11	+88	+177	+161	+88	+111	+204	+11	+111
.				−63					
.		−66	−107	−127	−133	−107			−53
.	−11	−91			−45		−69	−5	
.			−12						
.				+15					
.		+6	+9	+12	+12	+9			+4
.	−1	−6			−3		−4		
.			−3						
.									
.			+1	+2	+2	+1			
M (ft·k)	−1	−69	+65	0	−79	+14	+131	+6	+62
S_y (in^3)	21.5	71.8	77.0	107.1	71.8	77.0	135.6	21.5	77.0
Secondary stress (k/in^2)	0	11.5	10.1	—	13.2	2.2	11.6	3.3	9.7

Problems

9–25. Find the maximum secondary stress in members AB and AD of the welded truss of Fig. 9–103. Observe that the wide flange sections are placed with the flanges in vertical planes. $E = 30{,}000 \text{ k/in}^2$.

Section	A (in^2)	I_y (in^4)	S_y (in^3)
10WF39	11.48	45	11.2
10WF66	19.41	129	25.5
10WF72	21.18	142	27.9

[*Ans.:* $AB = 1.9 \text{ k/in}^2$, $AD = 8.8 \text{ k/in}^2$.]

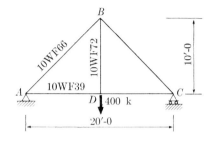

Figure 9–103

9–26. Find the maximum values of the secondary stresses in members AD, AB, 23, and 45 of the welded bent of Fig. 9–104. The wide flange sections are oriented with their flanges in vertical planes. Sizes of members and stresses in kilopounds are indicated, with a positive sign denoting tension. $E = 30{,}000 \text{ k/in}^2$.

Section	A (in^2)	I_y (in^4)	S_y (in^3)	Section	A (in^2)	I_y (in^4)	S_y (in^3)
10WF54	15.88	103.9	20.7	10WF37	10.88	42.2	10.6
10WF45	13.24	53.2	13.3	10WF23	6.77	11.3	3.9
10WF41	12.06	47.7	11.9	10WF21	6.19	9.7	3.4

[*Ans.:* $AD = 7.0 \text{ k/in}^2$, $BA = 3.3 \text{ k/in}^2$, $32 = 4.8 \text{ k/in}^2$, $54 = 5.6 \text{ k/in}^2$.]

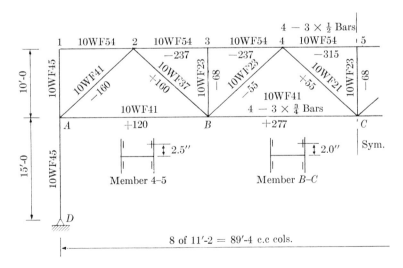

Figure 9–104

9-5 Frames with restrained joints.

All preceding discussions of continuous frame analyses have been based on two assumptions that are not entirely consistent with conditions often encountered in practice. Although the resulting errors are not usually considered to be significant, it is advisable for us to recognize that they do exist.

First, the supports of all structures considered thus far have been assumed to be either frictionless (pinned) or completely fixed. In practice, supports of these types are rare. Second, the connections of adjacent members have been assumed to give complete continuity. This can be obtained at the splices of adjacent spans of continuous girders by the use of groove welds. However, in the case of beam or girder connections to columns (such as exist in most frames), it will often be impractical to try to obtain this complete continuity.

The usual "moment resistant" beam to column connection may give 70, or 80, or 90% of complete continuity, but not 100%. This may or may not be economically advantageous. For example, if a beam supports a uniform load or concentrated loads at the third points, bending moments will be a minimum if the end connections and the frame are so arranged that the end restraints will be the equivalent of 75% of complete fixity. If, however, a concentrated load is applied at the center of the span, 100% of complete fixity is required at the beam ends for minimum moments in the beam.

These facts become evident when the moment values shown in Fig. 9–105 are compared. The left-hand column in the figure shows the fixity in

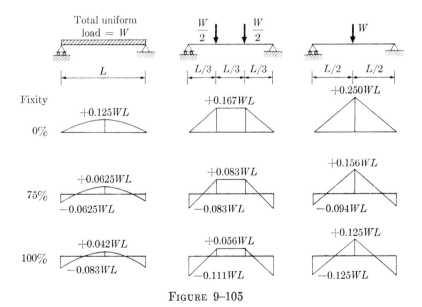

FIGURE 9–105

percent. For example, a beam with 75% fixity (fixation factor $f = 0.75$) will have end moments which are 0.75 times the end moments which would exist if the beam were completely fixed at the ends. The ends of this beam will rotate 0.25 times the end rotation of the same beam if it were simply supported.

The practical value of a method for analyzing frames with restrained joints is questionable. Probably it is practically impossible, at the time of this writing, to fabricate a beam to column connection to have a pre-designated fixity factor. Obviously, unless fixation factors can be specified by the engineer with definite assurance that end connections can be fabricated to have the designated values, the analysis is of little practical importance. The following method is therefore presented with the assumption that at some future time it may be of value.

Consider the beam of Fig. 9–106. The applied moment M_A is of sufficient magnitude to cause a rotation of 1 rad of the end of the beam at A.

FIGURE 9–106

All the values shown in the sketch have been derived previously. Observe that the value of f, the fixation factor at B, is unity.

If the propped cantilever of Fig. 9–106 is reduced to a simple beam, as shown in Fig. 9–107, the values indicated in this last figure apply. Here the fixation factor at B is zero. Now assume that an external moment M'_B is applied at B, in Fig. 9–107, of sufficient magnitude to reduce the rotation

FIGURE 9–107

at B to $\theta_A/8$, the value of θ_A being held at unity by any change necessary in the magnitude of M_{A1}. The deflected beam will then correspond to a fixation factor of 0.75 at B, since the rotation at B is now one-quarter of the rotation for a simple support at B. The final deflected beam and total indicated end moments are as shown in Fig. 9–108. The general case will

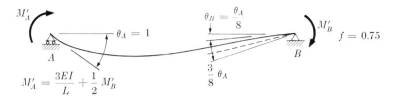

FIGURE 9-108

be as shown in Fig. 9-109. It has already been demonstrated that the moment necessary to give a rotation θ to the simply supported end of a

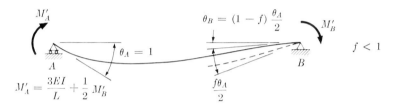

FIGURE 9-109

beam, with the opposite end being held against rotation, is $M = 4EI\theta/L$. This formula may be used to evaluate M'_B because M_A is varied to a value of M'_A (in order to maintain $\theta_A = 1$) as M'_B is applied. Consequently,

$$M'_B = \frac{4EI}{L} \cdot \frac{f\theta_A}{2}$$

$$= \frac{2EIf}{L}, \quad \text{since } \theta_A = 1,$$

and

$$M'_A = \frac{3EI}{L} + \frac{1}{2} \cdot \frac{2EIf}{L} = (3+f)\frac{EI}{L}.$$

The carry-over factor from A to B is

$$C_{AB} = \frac{M'_B}{M'_A} = \frac{2EIf/L}{(3+f)(EI/L)} = \frac{2f}{3+f}.$$

The value of M'_A determined above is the absolute stiffness. It is obvious that this stiffness is $(3+f)/4$ times the stiffness of a prismatic member fixed at the far end.

EXAMPLE 9–27. For the continuous beam of Fig. 9–110, find moments by moment distribution. Fixation factors are indicated on the sketch. The individual spans are prismatic, but I is not constant throughout the structure.

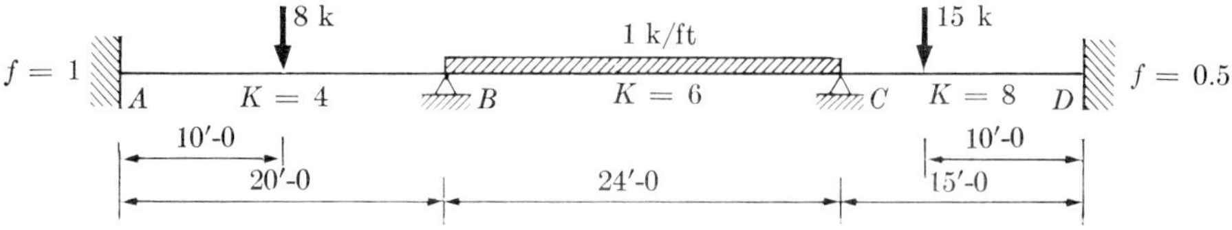

FIGURE 9–110

Fixed-end moments are:

$$F_{AB} = \frac{8 \times 20}{8} = +20 \text{ ft·k}, \qquad F_{BA} = -20 \text{ ft·k},$$

$$F_{BC} = \frac{1 \times 24^2}{12} = +48 \text{ ft·k}, \qquad F_{CB} = -48 \text{ ft·k},$$

$$F_{CD} = \frac{15 \times 5 \times 10^2}{15^2} = +33.3 \text{ ft·k},$$

$$F_{DC} = \frac{15 \times 5^2 \times 10}{15^2} = -16.7 \text{ ft·k}.$$

Relative stiffness for CD is:

$$K_{CD} = \frac{(3+f)k}{4} = \frac{3+0.5}{4} \times 8 = 7.$$

Carry-over factor for CD is:

$$C_{CD} = \frac{2f}{3+f} = \frac{2 \times 0.5}{3+0.5} = \frac{2}{7} = 0.286.$$

The example under discussion indicates the method of solution for the simple case of a structure with one terminal reaction having a fixation factor between zero and unity. It is to be noted that three modifications of the method of moment distribution as illustrated in previous examples are necessary here:

9–5]   FRAMES WITH RESTRAINED JOINTS   435

1. The relative stiffness of CD is modified to correct for the fixation factor of 0.5 at D This, in turn, affects the distribution factors at C.

2. The carry-over factor for CD is altered as required by the fixation factor of 0.5 at D.

3. The first balancing moment (+8.3) is introduced at D. This moment represents the correction to the original fixed-end moment at D (computed in the usual way for a fixed-end beam) for the effect of the end rotation of the beam at D. This end rotation amounts to $(1 - f_D)$ times the end rotation which would have occurred if the support at D had been simple. Since the end rotation is proportional to the change in end moment accompanying or causing it, then the correction to the original fixed-end moment is $(1 - f_D)$ times this original fixed-end moment (in this case, +8.3). No subsequent balancing of D is necessary, since the carry-over moments from C to D are determined by a carry-over factor corrected for the effect of the fixation factor of 0.5 at D.

TABLE 9–3.

Joint	A	B		C		D
Member	AB	BA	BC	CB	CD	DC
K	4	4	6	6	7	8
Distribution factor $= K/\Sigma K$	0	0.4	0.6	0.462	0.538	
Carry-over factor	—	0.5	0.5	0.5	0.286	
	+20.0	−20.0	+48.0	−48.0	+33.3 +4.1	−16.7 +8.3
	−5.6	−11.2	−16.8	−8.4		
			+4.4	+8.8	+10.2	+2.9
	−0.9	−1.8	−2.6	−1.3		
			+0.3	+0.6	+0.7	+0.2
Total	+13.5	−33.0	+33.0	−48.3	+48.3	−5.3

(Note that in the table a second horizontal line has been added above the balancing computations for the carry-over factors.)

EXAMPLE 9–28. Find all moments at joints in the frame of Fig. 9–111. Values of fixation factors are as shown.

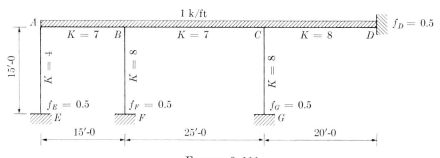

FIGURE 9-111

The procedure here will be to compute the modified values of the relative stiffnesses K_{AE}, K_{BF}, K_{CG}, and K_{CD}, and then to compute the various distribution factors on the basis of these modified values and the given values of relative stiffness for AB and BC. The modified values for the carry-over factors C_{AE}, C_{BF}, C_{CG}, and C_{CD} will then be determined. The moment distribution is managed in a manner similar to that indicated for the previous example, with the first step that of correcting the original fixed-end moment at D for the rotation of this support.

Modified relative stiffnesses are:

$$\text{Mod } K_{AE} = \frac{(3 + f_E)}{4} K_{AE} = \frac{(3 + 0.5)}{4} 4 = 3.5,$$

$$\text{Mod } K_{BF} = \frac{(3 + 0.5)}{4} 8 = 7,$$

$$\text{Mod } K_{CG} = \text{Mod } K_{CD} = 7.$$

Carry-over factors are:

$$C_{AE} = \frac{2f_E}{3 + f_E} = \frac{2 \times 0.5}{3 + 0.5} = \frac{2}{7} = 0.286,$$

$$C_{BF} = C_{CG} = C_{CD} = 0.286.$$

Fixed-end moments are:

$$F_{AB} = \frac{wL^2}{12} = \frac{1 \times 15^2}{12} = +18.7 \text{ ft·k}, \qquad F_{BA} = -18.7 \text{ ft·k},$$

$$F_{BC} = \frac{1 \times 25^2}{12} = +52.0 \text{ ft·k}, \qquad F_{CB} = -52.0 \text{ ft·k},$$

$$F_{CD} = \frac{1 \times 20^2}{12} = +33.3 \text{ ft·k}, \qquad F_{DC} = -33.3 \text{ ft·k}.$$

TABLE 9-4.

Joint	E	A		F	B			G	C			D
Member	EA	AE	AB	FB	BA	BF	BC	GC	CB	CG	CD	DC
K		3.5	7		7	7	7		7	7	7	
Distribution factor $= K/\Sigma K$		0.333	0.667		0.333	0.333	0.333		0.333	0.333	0.333	
Carry-over factor		0.286	0.500		0.500	0.286	0.500		0.500	0.286	0.286	0.500
			+18.7		−18.7		+52.0		−52.0		+33.3	−33.3
											+8.3	+16.6
			−5.5	−3.2	−11.1	−11.1	−11.1		−5.5			
	−1.3	−4.4	−8.8		−4.4		+2.6	+1.5	+5.3	+5.3	+5.3	+1.5
			+0.3	+0.2	+0.6	+0.6	+0.6		+0.3			
		−0.1	−0.2						−0.1	−0.1	−0.1	
Total	−1.3	−4.5	+4.5	−3.0	−33.6	−10.5	+44.1	+1.5	−52.0	+5.2	+46.8	−15.2

In the preceding examples fixation factors between zero and unity existed only at terminals of the frame. A more general condition would be the case where the beams are connected to the columns with joints which are not perfectly rigid, the fixity factors ranging from low values up to perhaps 0.95. When this condition exists, it is necessary to alter the method of analysis as follows:

1. The method for finding the various initial end moments must be modified to account for the fact that the ends of members are restrained instead of fixed.

2. The distribution factors must be computed on the basis of relative stiffnesses modified in accordance with the fixity factors at both ends of each member.

In connection with item 1 above, we will first develop the method for determining the initial end moments. Consider the span AB in Fig. 9–112

FIGURE 9–112

supporting certain loads (not shown) which result in the *external fixed-end moments* F_{AB} and F_{BA}. The beam ends are considered to be fixed when determining these moments. End B, however, is not fixed and actually it will rotate in a direction so as to decrease the value of F_{BA}. The value of this decrease will be $(1 - f_B)F_{BA}$, and this may be shown as an external moment at B (see Fig. 9–113). The application of this moment at B will induce a moment at A which will have a value of $2f_A/(3 + f_A)$ times the applied moment at B, in accordance with the carry-over factor previously derived. This corrective external moment is also shown in Fig. 9–113.

$\left[\dfrac{2f_A}{3 + f_A}\right](1 - f_B)F_{BA}$ $\qquad\qquad\qquad\qquad\qquad$ $(1 - f_B)F_{BA}$

FIGURE 9–113

Similarly, end A will rotate so as to decrease the end moment at A, and the corrective external moments which result are shown in Fig. 9–114.

$(1 - f_A)F_{AB}$ $\qquad\qquad\qquad\qquad\qquad$ $\left[\dfrac{2f_B}{3 + f_B}\right](1 - f_A)F_{AB}$

FIGURE 9–114

If we add the moments of Figs. 9–112, 9–113, and 9–114 at each end of the beam, the final expressions for initial end moments will be as shown in Fig. 9–115.

$$f_A F_{AB} + \left[\frac{2f_A}{3+f_A}\right](1-f_B)F_{BA} \qquad f_B F_{BA} + \left[\frac{2f_B}{3+f_B}\right](1-f_A)F_{AB}$$

FIGURE 9–115

As indicated in item 2, the relative stiffness I/L for each member must be modified in order to correct for the fixation factors at the two ends. It has been demonstrated that the relative stiffness for a prismatic member, to be used in determining the distribution factors for a joint at one end of the member, may be corrected for the fixation factor at the far end by multiplying the value of I/L by $(3+f_F)/4$, where f_F is the fixation factor at the far end. If, now, the fixation factor at the near end is designated by f_N, then the completely modified relative stiffness will be

$$f_N \frac{(3+f_F)}{4} \cdot \frac{I}{L}.$$

The analysis of this type of frame will be demonstrated in the example which follows. For the purpose of comparing maximum moments in the various members, the frame of Example 9–28 will be modified by assuming that the beams are joined to the columns with welded connections which have fixation factors of 0.75.

EXAMPLE 9–29. Using moment distribution, find all moments at joints in the frame of Fig. 9–116. Values of fixation factors are as shown. All members are prismatic.

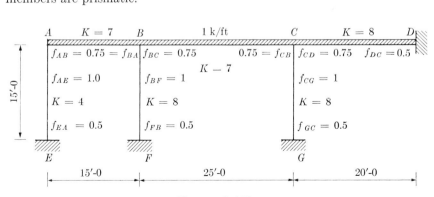

FIGURE 9–116

The values of fixed-end moments are as determined in Example 9–28. The initial end moments (abbreviated as IEM) are:

$$\text{IEM}_{AB} = f_A F_{AB} + \left[\frac{2f_A}{3 + f_A}\right](1 - f_B)F_{BA}$$

$$= 0.75 \times 18.7 + \left[\frac{2 \times 0.75}{3 + 0.75}\right](1 - 0.75)18.7$$

$$= 0.75 \times 18.7 + 0.1 \times 18.7 = +15.9 \text{ ft·k},$$

$$\text{IEM}_{BA} = -15.9 \text{ ft·k},$$

$$\text{IEM}_{BC} = 0.75 \times 52.0 + 0.1 \times 52.0 = +44.2 \text{ ft·k},$$

$$\text{IEM}_{CB} = -44.2 \text{ ft·k},$$

$$\text{IEM}_{CD} = f_C F_{CD} + \left[\frac{2f_C}{3 + f_C}\right](1 - f_D)F_{DC}$$

$$= 0.75 \times 33.3 + \left[\frac{2 \times 0.75}{3 + 0.75}\right](1 - 0.50)33.3 = +31.7 \text{ ft·k},$$

$$\text{IEM}_{DC} = f_D F_{DC} = \left[\frac{2f_D}{3 + f_D}\right](1 - f_C)F_{CD}$$

$$= 0.5 \times 33.3 + \left[\frac{2 \times 0.50}{3 + 0.50}\right](1 - 0.75)33.3 = +19.1 \text{ ft·k}.$$

At points A, B, and C of Fig. 9–116, the theoretical joints are located at the intersection of the axes of the various members; that is, these joints are located on the columns themselves. This accounts for the fact that at each joint the fixation factors of the beams are designated as 0.75, but the fixation factors for the columns are shown as 1.0. The computations for the various modified relative stiffnesses are as follows:

$$\text{Mod } K_{AE} = f_{AE}\left[\frac{3+f_{EA}}{4}\right]K_{AE} = 1\left[\frac{3+0.5}{4}\right]4 = 3.5,$$

$$\text{Mod } K_{AB} = f_{AB}\left[\frac{3+f_{BA}}{4}\right]K_{AB} = 0.75\left[\frac{3+0.75}{4}\right]7 = 4.92,$$

$$\text{Mod } K_{BA} = 4.92,$$

$$\text{Mod } K_{BF} = 1\left[\frac{3+0.5}{4}\right]8 = 7.0,$$

$$\text{Mod } K_{BC} = 0.703 \times 7 = 4.92,$$

$$\text{Mod } K_{CB} = 4.92,$$

$$\text{Mod } K_{CG} = 1\left[\frac{3+0.5}{4}\right]8 = 7.0,$$

$$\text{Mod } K_{CD} = f_{CD}\left[\frac{3+f_{DC}}{4}\right]K_{CD} = 0.75\left[\frac{3+0.5}{4}\right]8 = 5.25.$$

Modified carry-over factors are:

$$\text{Mod } C_{AE} = \frac{2f_{EA}}{3+f_{EA}} = \frac{2 \times 0.5}{3+0.5} = 0.286,$$

$$\text{Mod } C_{BF} = \text{Mod } C_{CG} = \text{Mod } C_{CD} = 0.286,$$

$$\text{Mod } C_{AB} = \frac{2f_{BA}}{3+f_{BA}} = \frac{2 \times 0.75}{3+0.75} = 0.4,$$

$$\text{Mod } C_{BA} = \text{Mod } C_{BC} = \text{Mod } C_{CB} = 0.4.$$

The final moments are shown in Table 9–5.

442   APPLICATIONS OF MOMENT DISTRIBUTION   [CHAP. 9

TABLE 9-5.

Joint	E	A		F	B			G	C			D
Member	EA	AE	AB	FB	BA	BF	BC	GC	CB	CG	CD	DC
Modified K		3.5	4.92		4.92	7.0	4.92		4.92	7.0	5.25	
Distribution factor $= K/\Sigma K$		0.416	0.584		0.292	0.416	0.292		0.286	0.407	0.307	
Modified carry-over factor		0.286	0.400		0.400	0.286	0.400		0.400	0.286	0.286	
			+15.9 −3.3		−15.9 −8.3	−11.7	+44.2 −8.3	+1.8	−44.2 −3.3	+6.4	+31.7	+19.1
	−1.5	−5.3	−7.3	−3.3	−2.9 +0.3	+0.5	+1.8 +0.3		+4.5		+4.9	+1.4
	−1.5	−5.3	+5.3	−3.3	−26.8	−11.2	+38.0	+1.8	−43.0	+6.4	+36.6	+20.5

Fixation factors less than unity will result in a considerable reduction in the maximum moments in the various beams of a frame for certain types of loads. This is clearly indicated in Table 9-6, where the maximum positive moments are shown for the beams of Examples 9-28 and 9-29. The results of Example 9-28 are tabulated in the top horizontal column for the case of beam to column connections with fixation factors of unity. The second horizontal column shows corresponding values for Example 9-29 with beam to column connections having fixation factors of 0.75, and the third column shows the percentage reduction in the maximum moments in each span.

Table 9-6 shows that the maximum negative moment in each span, for the fixation factors of 0.75, is considerably in excess of the maximum positive moment. This indicates that still greater economy could be realized by a further reduction in fixation factors. The optimum values for these factors may be determined by several trials.

TABLE 9-6.

AB	Maximum positive moment AB	BA	BC	Maximum positive moment BC	CB	CD	Maximum positive moment CD	DC
−4.5	+11.0	−33.6	−44.1	+29.9	−52.0	−46.8	+20.2	−15.2
−5.3	+13.2	−26.8	−38.0	+37.6	−43.0	−36.6	+21.8	−20.5
		−20.2%			−17.3%	−21.8%		

9-6 Comments on the method of moment distribution. There are some who believe that moment distribution constitutes the greatest single contribution ever made to indeterminate structural theory. While this would be difficult to establish, the fact remains that most of the structure types discussed in this and in the preceding chapter are, in the opinion of many, most easily analyzed by this method. It is probable that moment distribution, when it is applicable in the analysis of indeterminate structures, is used far more than any other available method.

It should be pointed out here that although all the structures considered in this chapter have been composed of members which were straight, or could be considered to be straight, moment distribution can be applied to frames containing members with curved axes. The difficulty is, however, that the evaluation of fixed-end moments, stiffness, and carry-over factors for curved members will in many cases require so much work that no over-all advantage is realized by using moment distribution. If, however, tables of these properties are available, moment distribution will be found to have very real advantages for this type of problem.

It should also be pointed out that for certain types of structures an analysis by moment distribution will be much more laborious than an alternate solution by the general method using the conjugate structure. Two examples are shown in Fig. 9–117(a) and (b). The frame of sketch (a)

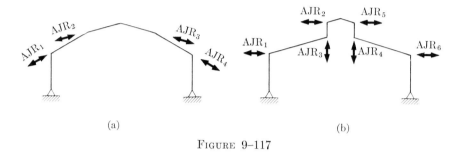

FIGURE 9–117

will require four AJR's, as indicated, while that of (b) will require six. Obviously considerable work would be involved in cancelling the effects of these AJR's in order to obtain the final moments for the frame. A solution by the general method using the conjugate structure would be much easier.

A point of particular importance here is that in the analysis of many structures by moment distribution, the moment corrections which result from cancelling the effects of the AJR's will be much larger than the initial fixed-end moments. (A note of caution in this regard appears in Problem 9–12.) Unless great care is used in evaluating the moment corrections in cases of this kind, the final results will be unreliable. Hence a computing machine should be used.

CHAPTER 10

ANALYSIS OF FRAMES WITH NONPRISMATIC MEMBERS BY MOMENT DISTRIBUTION

10-1 General. Basically, the application of moment distribution in the analysis of a continuous frame composed wholly or partly of nonprismatic members is exactly the same as if the frame were composed entirely of prismatic members. The detailed computations prerequisite to the actual distribution of moments are quite different, however.

Fixed-end moments, stiffnesses, and carry-over factors for nonprismatic members are evaluated by methods and formulas quite unlike those which are applicable to prismatic members. Also, the change in the stiffness at one end of a nonprismatic member, when the far end is reduced from a fixed condition to a pinned support, is not the same as in the case of a prismatic member. Finally, when one end of a nonprismatic member is displaced laterally relative to the other end, the induced fixed-end moments must be determined differently than for a prismatic member.

Methods and formulas necessary for computing these data will be developed in the sections which follow.

10-2 Stiffness and carry-over factors by the column analogy. In the opinion of the writer, one of the most useful applications of the column analogy is in the evaluation of the stiffnesses, carry-over factors, and fixed-end moments for nonprismatic members. It is this particular use of the column analogy which will first be explained.

Assume that the stiffness K_A and the carry-over factor C_A are to be computed for the beam of Fig. 10–1 by use of the column analogy. If

FIGURE 10–1

end A is caused to rotate through 1 rad, the developed moment at A will by definition be the absolute stiffness K_A, and the induced moment at B will be $C_A K_A$. The deformed beam will be as shown in Fig. 10–2. (The rotation at A is not drawn to scale.)

445

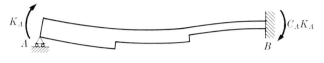

FIGURE 10–2

The analogous column should be loaded with the flexural strains of this distorted beam if the stiffness and carry-over factors are to be evaluated. The difficulty is that K_A and C_A are unknown, and thus the intensity of load on the column cannot be determined. Fortunately, however, an equivalent load can be substituted.

Since in Fig. 10–2 end B is fixed, the algebraic summation of all flexural strains must be 1 rad. In addition, because there is no vertical movement of end A, then by the moment-area method the first moment of all flexural strains about A must be zero. Consequently, the equivalent load on the analogous column will be a concentration of 1 rad placed at end A. Therefore, regardless of the shape of the beam, whenever stiffness and carry-over factors are required, a concentration of 1 rad is placed on the analogous column at the end corresponding to the rotated end of the beam. In this case it is not necessary to cut the real structure back to determinateness, since the equivalent loading on the analogous column can be completely determined, as explained above, without this cutback.

If the analogous column is proportioned as shown in Fig. 10–3, that is,

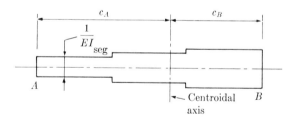

FIGURE 10–3

if the width of every segment of the column is taken as $1/EI_{\text{seg}}$, where I_{seg} is the moment of inertia of the corresponding segment of the beam, and if a unit load of 1 rad is applied at A, then

$$\text{Abs } K_{AB} = f_A = \frac{1}{A} + \frac{1 \times e \times c_A}{I}. \qquad (10\text{--}1)$$

In the above expression:

f_A = the fiber stress in the column at A.
A = the area of the analogous column.

I = the moment of inertia of the elastic area of the column with respect to the centroidal axis of that area.

e = the eccentricity of the unit load.

Equation (10–1) may be written as

$$f_A = \frac{1}{\Sigma(1/EI_{seg} \times L_{seg})}$$

$$+ \frac{1 \times e \times c_A}{\Sigma \left[\frac{1/EI_{seg} \times L_{seg}^3}{12} + (1/EI_{seg} \times L_{seg}) x_{seg}'^2 \right]} \quad (10\text{--}2)$$

$$= \frac{1}{\Sigma \left[\left(\dfrac{1}{EB\, d_{seg}^3/12} \right) L_{seg} \right]}$$

$$+ \frac{1 \times e \times c_A}{\Sigma \left[\dfrac{\left(\dfrac{1}{EB\, d_{seg}^3/12} \right) L_{seg}^3}{12} + \left(\left(\dfrac{1}{EB\, d_{seg}^3/12} \right) L_{seg} \right) x_{seg}'^2 \right]} \quad (10\text{--}3)$$

In the above equation:

I_{seg} = the moment of inertia of the cross section of the real beam about the centroidal axis of the beam cross section at the center of the real beam segment.

L_{seg} = the length of the segment.

B = the width of the beam at the center of the segment.

d_{seg} = the depth of the beam at the center of the segment.

x' = the distance from the centroidal axis of the analogous column to the center of each segment.

Substitution in Eq. (10–3) will give the value of the absolute stiffness.

The use of the width of the analogous column as $1/EI_{seg}$, however, will result in very troublesome decimals, and thus the width is modified by the multiplier $EB\, d_{ref}^3/12$, which is the equivalent of dividing Eq. (10–3) by $EB\, d_{ref}^3/12$. The result is:

448   ANALYSIS OF FRAMES WITH NONPRISMATIC MEMBERS   [CHAP. 10

$$f'_A = \frac{f_A}{EB\, d_{\text{ref}}^3/12} \tag{10-4}$$

$$= \frac{1}{\sum \left[\dfrac{d_{\text{ref}}^3}{d_{\text{seg}}^3} \cdot L_{\text{seg}}\right]} + \frac{1 \times e \times c_A}{\sum \left[\left(\dfrac{d_{\text{ref}}^3}{d_{\text{seg}}^3} \cdot L_{\text{seg}}^3\right)\dfrac{1}{12} + \left(\dfrac{d_{\text{ref}}^3}{d_{\text{seg}}^3} \cdot L_{\text{seg}}\right)x'^2_{\text{seg}}\right]}. \tag{10-5}$$

In the above equation, d_{ref} signifies a particular depth (not necessarily an actual depth of the beam) used in the multiplier $EB\, d_{\text{ref}}^3/12$ to obtain the modified width of the analogous column. The value of this multiplier is unimportant, since it is cancelled in the final step when the absolute stiffness is computed. Consequently, any depth may be used as d_{ref}. The modified or relative width of any segment of the analogous column will be $d_{\text{ref}}^3/d_{\text{seg}}^3$.

It is apparent, then, that f'_A is merely a number which is proportional to the absolute stiffness. Obviously,

$$\text{Abs } K_A = f_A = f'_A \times \frac{EB\, d_{\text{ref}}^3}{12}. \tag{10-6}$$

The value of the absolute stiffness for a member is usually a large number and cumbersome to handle in computations. Accordingly, it is more convenient to use a relative stiffness which is proportional to the absolute stiffness in the ratio of $12/E$, and the expression for this stiffness is

$$\text{Rel } K_A = f'_A \times B\, d_{\text{ref}}^3. \tag{10-7}$$

The effects of varying the units in which the different terms of Eq. (10-3) are expressed may be easily determined by substituting dimensions in the first term of that equation. For example, if the unit of length is taken as feet, with E expressed in k/ft^2, a substitution of these units will result in:

$$\text{Units of } f_A = \frac{k}{\text{ft}^2} \times \text{ft}^4 \times \frac{1}{\text{ft}} = \text{ft·k}.$$

If inches and k/in^2 are used, the result is:

$$\text{Units of } f_A = \frac{k}{\text{in}^2} \times \text{in}^4 \times \frac{1}{\text{in.}} = \text{in·k}.$$

If, however, E is expressed in k/in^2, I in in^4, and all other dimensions in feet, then the result is:

$$\text{Units of } f_A = \frac{\text{k}}{\text{in}^2} \times \text{in}^4 \times \frac{1}{\text{ft}} = \frac{\text{in}^2 \cdot \text{k}}{\text{ft}}.$$

In this last case, the answer obtained from the analogous column must be divided by 12 in order to have the units in in·k, and by 144 to obtain ft·k.

The carry-over factor is easily computed by the analogous column. Equation (10–1) was expressed as

$$\text{Abs } K_A = f_A = \frac{1}{A} + \frac{1 \times e \times c_A}{I}.$$

By substituting c_B for c_A, this becomes

$$f_B = \frac{1}{A} + \frac{1 \times e \times c_B}{I}, \qquad (10\text{--}8)$$

and the value of the carry-over factor C_A will obviously be

$$C_A = \frac{f_B}{f_A}.$$

In the above f_A and f_B are absolute values. Actually, of course, they may be relative values and the computed C_A will still be correct.

The actual application of the column analogy in the computation of fixed-end moment, stiffness, and carry-over factors can best be explained by an illustrative problem.

EXAMPLE 10–1. Compute the fixed-end moments, the absolute and relative stiffnesses, and the carry-over factors for the beam of Fig. 10–4, which is of reinforced concrete and which has a uniform width of 1 ft. E is 288,000 k/ft^2.

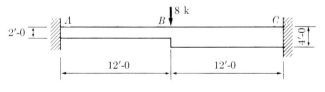

FIGURE 10–4

The beam will be cut back by removing the moments at each end. The M_s diagram is shown in Fig. 10–5 as the load intensity on the cross

450   ANALYSIS OF FRAMES WITH NONPRISMATIC MEMBERS   [CHAP. 10

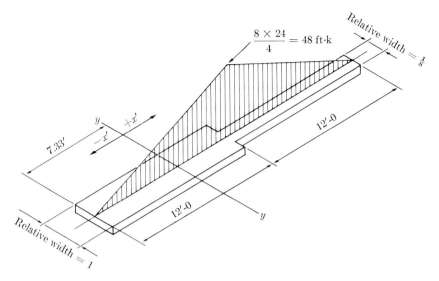

FIGURE 10–5

section of the analogous column. Moments causing compression on the top side of the beam will be considered to be positive. The 2-ft depth will be used as the reference depth.

The computations for locating the centroidal axis and for computing the moment of inertia of the elastic area about this axis are shown in Table 10–1.

TABLE 10–1.

Segment	L (ft)	Relative width $\dfrac{d_{\text{ref}}^3}{d_{\text{seg}}^3}$	A	x (ft)	Ax	x' (ft)	I_{cg}	Ax'^2
AB	12	1	12	6	72	-1.33	144	21.2
BC	12	0.125	1.5	18	27	$+10.67$	18	170.4
Σ			13.5		99		162	191.6

$$x_0 = \frac{99}{13.5} = 7.33 \text{ ft}, \qquad I_y = 162 + 191.6 = 353.6.$$

Table 10–2 is used to compute the total elastic load and the moment of this load about the y-axis.

TABLE 10-2.

Segment	A	Average M_s (ft·k)	W	x'_W	$M_{y'}$
AB	12	24	288	+0.67	+193
BC	1.5	24	36	+8.67	+312
Σ	13.5		324		+505

Using the information from the above table,

$$M_r = \frac{W}{A} + \frac{M_y}{I_y}x = \frac{324}{13.5} + \frac{(+505)}{353.6}x = +24.0 + 1.43x \text{ ft·k},$$

$$M_A = +24.0 + 1.43(-7.33) = +24.0 - 10.5 = +13.5 \text{ ft·k},$$

$$M_C = +24.0 + 1.43(+16.67) = +24.0 + 23.8 = +47.8 \text{ ft·k}.$$

Obviously, the stress on all fibers of the column will have a positive sign. The sign of the column fiber stress M_r at B must be negative so that when it is combined with $M_s(+48 \text{ ft·k})$ at B, the resulting value for M will be less than M_s. It is apparent, then, that the signs must be reversed, and the correct values for moments are $M_A = -13.5 \text{ ft·k}$ and $M_C = -47.8 \text{ ft·k}$. These signs are consistent with the original assumption that a moment causing compression on the top side of the beam would be considered positive.

With the correct signs, the expression for M_r is now

$$M_r = -24.0 - 1.43x.$$

Observe that the redundant reaction components represented in this expression will act as shown in Fig. 10–6. Either end of the beam may be considered fixed, with the rigid bracket attached to the opposite end. If, however, end A had been considered as fixed, with the bracket connected at C, the sense of the 24.0 ft·k couple would have been clockwise and the 1.43 k force would have acted up.

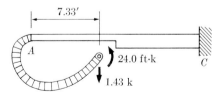

FIGURE 10-6

452   ANALYSIS OF FRAMES WITH NONPRISMATIC MEMBERS   [CHAP. 10

An interesting point here is that if the original beam had been cut back by removing all support at the left end, the M_s diagram would be as shown in Fig. 10–7(a) and the redundant reactions represented in the equation for M_r would have the values and could be considered to act either as shown in Fig. 10–7(b) or as in Fig. 10–7(c). If, however, the original beam had been cut back by removing all support at C, the M_s diagram would be as shown in Fig. 10–8(a), and the redundant reactions acting at the centroid of the elastic areas could be considered to act either as shown in Fig. 10–8(b) or as shown in Fig. 10–8(c).

In order to determine K_A and C_A, a concentration of 1 rad is considered to act down at A on the analogous column. Relative values for the fiber stresses in the column at A and C are computed as follows:

$$f'_A = \frac{1}{13.5} + \frac{(-7.33)(-7.33)}{353.6} = +0.226,$$

$$f'_c = \frac{1}{13.5} + \frac{(-7.33)(+16.67)}{353.6} = -0.272,$$

$$C_A = \frac{-0.272}{+0.226} = -1.20.$$

The negative sign indicates that the sign of the stress in the top fiber of the beam at C caused by the induced moment at C will be the opposite of the sign of the stress in the top fiber at A caused by the developed moment at A.

$$\text{Rel } K_A = f'_A \times B\, d^3_{\text{ref}} = 0.226 \times 1 \times 2^3 = 1.81 \text{ ft}^3,$$

$$\text{Abs } K_A = f'_A \times \frac{EB\, d^3_{\text{ref}}}{12} = \text{Rel } K_{AC} \times \frac{E}{12}$$

$$= 1.81 \times \frac{288{,}000}{12} = 43{,}400 \text{ ft·k}.$$

To compute K_C and C_C, the concentration of 1 rad is moved to C on the column. The necessary computations are:

$$f'_c = \frac{1}{13.5} + \frac{(+16.67)(+16.67)}{353.6} = +0.858,$$

$$f'_A = \frac{1}{13.5} + \frac{(+16.67)(-7.34)}{353.6} = -0.272,$$

$$C_C = \frac{-0.272}{+0.858} = -0.316,$$

10–2]   ANALYSIS BY THE COLUMN ANALOGY   453

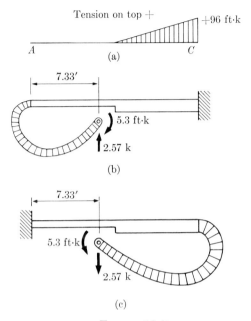

Figure 10–7

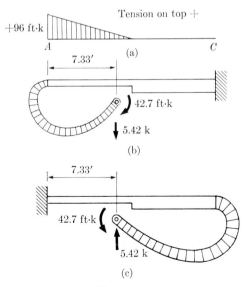

Figure 10–8

and

$$\text{Rel } K_C = f'_C \times B\, d_{\text{ref}}^3 = 0.858 \times 1 \times 2^3 = 6.86 \text{ ft}^3,$$

$$\text{Abs } K_C = f'_C \times \frac{EB\, d_{\text{ref}}^3}{12} = \text{Rel } K_C \times \frac{E}{12}$$

$$= 6.86 \times \frac{288{,}000}{12} = 165{,}000 \text{ ft·k}.$$

A valuable check on computations of this kind is provided by the fact that (as demonstrated in Example 3–2)

$$C_A K_A = C_C K_C.$$

In other words, the moment induced at C by a rotation of 1 rad at A (C being fixed) is equal to the moment induced at A by a rotation of 1 rad at C (A being fixed). That this is true will also be apparent from a comparison of the computations for f'_C for a load of 1 rad at A with the computations for f'_A for a load of 1 rad at C. In the above problem,

$$1.20 \times 1.81 = 0.316 \times 6.86,$$

from which

$$2.17 = 2.17,$$

and the computed values are satisfactory.

10–3 Curves for stiffness and carry-over factors. When it is necessary to compute stiffness and carry-over factors for nonprismatic members, it is obvious that considerable labor is involved preliminary to an analysis by moment distribution. Fortunately for the structural analyst, certain information has been assembled and made available by the Portland Cement Association, information which, in many cases, will eliminate the need for long and detailed computations of the kind just presented. This information will now be considered.

In this discussion the dimension of the column cross section corresponding to the span of the flexural member will be designated as the length of the analogous column. In Eq. (10–5), L_{seg} and c will vary directly with the length of the column, since it is assumed that the number of segments will be held constant as, and if, the length of the analogous column is arbitrarily varied. Moreover, if the widths at proportional points along the length of the analogous column are held constant as the length is arbitrarily varied, then x'_{seg} and e will also vary directly with the length of the column. Under these conditions, Eq. (10–5) indicates that f'_A will be inversely proportional to the length of the analogous column. Consequently, if f'_{A1} represents the stress in the analogous column with a unit length, then

$$f'_{A1} = Lf'_A \quad \text{or} \quad f'_A = f'_{A1}/L.$$

It should be apparent, therefore, that if values of f'_{A1} are computed on a basis of a length of the analogous column of unity, for the various proportional changes in shape of any desired standard type of nonprismatic member, the length L of any member may be eliminated as a variable in computing standard information.

One additional simplification is possible. We already know that any depth may be taken as d_{ref}. In order to make any standard information valid, it is necessary, of course, to use a definite depth as the reference depth. This is arbitrarily taken as the minimum depth of the member (designated as $d_{\min}$). The original expression for f'_A (Eq. 10-4),

$$f'_A = \frac{f_A}{EB\,d^3_{\text{ref}}/12},$$

now becomes

$$f'_A = \frac{f'_{A1}}{L} = \frac{f_A}{EB\,d^3_{\min}/12}.$$

Therefore,

$$\text{Abs } K_A = f_A = \frac{f'_{A1}}{L} \times \frac{EB\,d^3_{\min}}{12}.$$

Inspection of this last expression shows that the value of the absolute stiffness is directly proportional to $d^3_{\min}$, and to B. Thus it is apparent that if a modified value of f'_{A1} were to be determined for a value of $d^3_{\min} = 1$ and $B = 1$ (that is, assuming a $d_{\min}$ of the beam as 1 in. or 1 ft and width of 1 in. or 1 ft), this value of f'_{A1} could be easily modified to suit any beam having a minimum depth other than 1 and width other than 1 simply by multiplying by the actual $d^3_{\min}$ and B. This modified value of f'_{A1} is designated by k. Therefore the expression for stiffness is

$$\text{Abs } K_A = k_A \frac{EB\,d^3_{\min}}{12L}$$

or

$$\text{Rel } K_A = k_A \frac{B\,d^3_{\min}}{L}.$$

This factor k may be considered to be a dimensionless number which is related to the absolute and relative stiffnesses of a member in the manner indicated by the above two equations. Actually, it is the fiber stress in the analogous column, directly under the load of 1 rad, when the analogous column corresponds to a very particular beam. This particular beam will be called the *reference beam*. As an alternate to the dimensionless number concept suggested above, the factor k might be considered to be the relative stiffness of this reference beam. In this case the units of k will be either ft^3 or in^3, and B, d^3_{min}, and L must be considered to be dimensionless when substituted in the above equations for absolute or relative stiffness. This reference beam has a length, width, and minimum depth of 1 in. or 1 ft, depending upon whether the stiffness is desired in in·k or ft·k. If the stiffness is desired in in·k, then E must be in k/in^2; if it is desired in ft·k, then, of course, E must be in k/ft^2. In addition to the unit dimensions, the reference beam must have a proportional change in I along its length which is identical with the group of beams for which it is the reference. A discussion of this requirement follows.

Most haunched beams used in practice are of a few general types. Within each type, however, there can be many different combinations of dimensions. Consider the beam of Fig. 10–9. In this beam the coeffi-

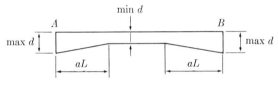

FIGURE 10–9

cient a (defining the length of the haunch) can have any value from 0 to 0.5. Minimum d and maximum d may be any practical dimension, and the ratio of minimum d to maximum d will usually vary from 0.35 to 1.00. Obviously, a large number of different beams will result from the combination of different values within the above limits. Consider one particular combination, as shown in Fig. 10–10. Here, $a = 0.2$ and the ratio of

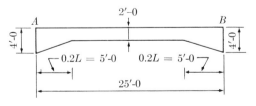

FIGURE 10–10

minimum d to maximum d is 0.5. But these two values may be held constant and we may still obtain many different beams by varying the values of the beam width B, the span L, and the value of the minimum d. The expression for absolute stiffness was

$$\text{Abs } K_A = k_A \frac{EB \, d_{min}^3}{12L},$$

and that for relative stiffness,

$$\text{Rel } K_A = k_A \frac{B \, d_{min}^3}{L}.$$

It is clear, therefore, that if the value of k is known for the group of beams as represented by Fig. 10–10, with $a = 0.2$ and the ratio of minimum d to maximum d of 0.5, it is very easy to find either Abs K_A or Rel K_A for any combination of values for B, L, and minimum d. In this case, then, the reference beam is completely defined as having values of unity for B, L, and the minimum d, a value of 0.2 for a, and a ratio of minimum d to maximum d of 0.5.

Carry-over factors, fixed-end moment coefficients and the values of k for many different groups of beams may be obtained from curves prepared by the Portland Cement Association. [These curves are reproduced in the Appendix, through the courtesy of the Association. Similar information, but more complete, and in some respects superior to that which may be obtained from the curves, has been prepared in tabular form by the Association under the title *Handbook of Frame Constants*. These tables are not reproduced in this book.]

Two examples to illustrate the application of the analogous column in the computation of fixed-end moments, stiffness, and carry-over factors now follow.

EXAMPLE 10–2. The haunched beam of Fig. 10–11 is 12 in. wide. E is 2000 k/in^2. Using the analogous column, and checking by the curves, find:

(a) Fixed-end moments at A and B

(b) Carry-over factor

(c) Relative stiffness

(d) Absolute stiffness

Use a reference depth of 20 in. and reduce the beam to a simple beam.

458 ANALYSIS OF FRAMES WITH NONPRISMATIC MEMBERS [CHAP. 10

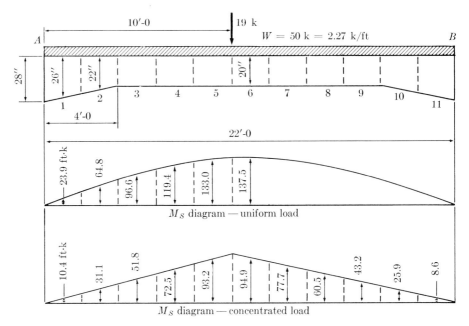

FIGURE 10–11

TABLE 10–3.

Segment number	Length (ft)	Relative width $\dfrac{d_{ref}^3}{d_{seg}^3}$	Relative elastic area A	x (ft)	$I_{y_{cg}}$	Ax^2	M_s (ft·k)	W	M_y
1	2	0.455	0.91	−10	0.3	91	34.3	31.2	−312
2		0.75	1.5	−8	0.5	96	95.9	143.8	−1150
3		1	2	−6	0.67	72	148.4	296.8	−1782
4		1	2	−4	0.67	32	191.9	383.8	−1537
5		1	2	−2	0.67	8	226.2	452.4	−905
6		1	2	0	0.67	0	232.4	464.8	0
7		1	2	+2	0.67	8	210.7	421.4	+843
8		1	2	+4	0.67	32	179.9	359.8	+1439
9		1	2	+6	0.67	72	139.8	279.6	+1678
10		0.75	1.5	+8	0.5	96	90.7	136.0	+1089
11		0.445	0.91	+10	0.3	91	32.5	29.6	+296
			18.82		6.29	598		2999.2	−341

$$M_A = \frac{2999.2}{18.82} + \frac{(-341)(-11)}{604.3} = 159.1 + 6.2 = 165.3 \text{ ft·k},$$

$$M_B = 159.1 - 6.2 = 152.9 \text{ ft·k}.$$

Checking values of fixed-end moments by curves [see Appendix], we find

$$a = \frac{4}{22} = 0.182, \quad b = \left(\frac{20}{28}\right)^3 = 0.364, \quad v = \frac{10}{22} = 0.454.$$

Uniform load:

fixed-end moment coefficient = 0.092,

fixed-end moment = fWL = 0.092 × 50 × 22 = 101.2 ft·k.

Concentrated load:

left end—

for b = 0.30, f = 0.155,

for b = 0.60, f = 0.145,

for b = 0.36, use f = 0.153;

fixed-end moment = 0.153 × 19 × 22 = 64.0 ft·k;

right end—

for b = 0.30, f = 0.12,

for b = 0.60, f = 0.12;

fixed-end moment = 0.12 × 19 × 22 = 50.2 ft·k.

Total fixed-end moment:

left end = 101.2 + 64.0 = 165.2,

right end = 101.2 + 50.2 = 151.4 ft·k.

460   ANALYSIS OF FRAMES WITH NONPRISMATIC MEMBERS   [CHAP. 10

Carry-over factors:

place unit load at left end of the analogous column—

$$f'_A = \frac{1}{18.82} + \frac{1(-11)(-11)}{604.3} = 0.053 + 0.201 = 0.254,$$

$$f'_B = 0.053 - 0.201 = -0.148.$$

Therefore,

$$C_A = \frac{0.148}{0.254} = 0.58.$$

Check by curves:

$$C = 0.58.$$

Relative stiffness:

$$\text{Rel } K_A = f'_A B \, d_{\text{ref}}^3$$

$$= 0.254 \times 1 \times 1.667^3 = 1.17 \text{ ft}^3$$

Check by curves:

$$k = 5.6,$$

$$\text{Rel } K_A = \frac{k}{L} \times B \, d_{\text{min}}^3 = \frac{5.6}{22} \times 1 \times 1.667^3 = 1.17 \text{ ft}^3.$$

Absolute stiffness:

$$\text{Abs } K_A = \text{Rel } K_A \times \frac{E}{12} = 1.17 \times \frac{288{,}000}{12} = 28{,}100 \text{ ft·k.}$$

In the above problem no solution is necessary for the location of the centroid of the analogous column, since the beam is symmetrical. In the case of an unsymmetrical beam the centroid would, of course, have to be located. The reference depth is taken as 20 in. in order to give an easier solution, but any depth could be so used, even though the depth chosen for reference does not actually exist in the beam.

EXAMPLE 10-3. Using the analogous column, find the carry-over factors, the relative stiffnesses, and the absolute stiffnesses for the steel member of Fig. 10-12. Express stiffnesses in inch kilopounds and, using the curves, check the values of the stiffnesses.

10-3] CURVES FOR STIFFNESS AND CARRY-OVER FACTORS 461

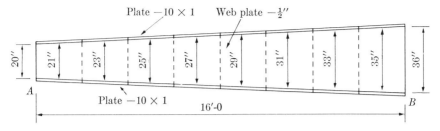

FIGURE 10-12

In Table 10-4, the fifth column shows values of the total I at the center of each segment. This is noted as also being the value of the equivalent d_{seg}^3. The explanation is that the expression substituted for I_{seg} in Eq. (10-2) to obtain Eq. (10-3) is for the moment of inertia of a rectangular cross section only. Consequently, whenever a member is nonrectangular in cross section, it is necessary to determine an equivalent member with a rectangular cross section and with the same variation in I as the given member, before the derived formulas will apply. That is, this equivalent member will have the same length and its I will vary along the span in the same manner as the given member. If the width of this equivalent member is arbitrarily made 12 in., then the value of I at any section of the given member is equal to d^3 for the corresponding section of the equivalent member. That is,

$$I = \frac{bd^3}{12} = \frac{12 \times d^3}{12}.$$

This arrangement obviously saves considerable work.

TABLE 10-4.

Segment number	Web plate depth (in.)	I Web plate (in⁴)	I Flange plates (in⁴)	Total I = equivalent d^3 segment	Relative width $\dfrac{d_{\text{ref}}^3}{d_{\text{seg}}^3}$	Segment length (ft)
1	21	386	2420	2806	1	2
2	23	507	2880	3387	0.828	
3	25	651	3380	4031	0.696	
4	27	820	3920	4740	0.592	
5	29	1016	4500	5516	0.509	
6	31	1241	5120	6361	0.441	
7	33	1497	5780	7277	0.385	
8	35	1787	6580	8267	0.339	

TABLE 10-5.

Segment number	Relative elastic area A	x (ft)	Ax	x' (ft)	x'^2	$I_{y_{cg}}$	Ax'^2
1	2.000	1	2.000	−5.395	29.106	0.7	58.2
2	1.656	3	4.968	−3.395	11.526	0.5	19.1
3	1.392	5	6.960	−1.395	1.946	0.5	2.7
4	1.184	7	8.288	0.605	0.366	0.4	0.4
5	1.018	9	9.162	2.605	6.786	0.3	6.9
6	0.882	11	9.702	4.605	21.206	0.3	18.7
7	0.770	13	10.010	6.605	43.626	0.3	33.6
8	0.678	15	10.170	8.605	74.046	0.3	50.2
	9.580		61.260			3.3	189.8
							3.3
							193.1

$$x_0 = \frac{61.260}{9.580} = 6.39 \text{ ft.}$$

Place a unit load of 1 rad at A:

$$f'_A = \frac{1}{9.58} + \frac{(-6.39)(-6.39)}{193.1} = 0.316,$$

$$f'_B = 0.104 + \frac{(-6.39)(+9.61)}{193.1} = -0.214,$$

$$C_A = \frac{0.214}{0.316} = 0.676.$$

Place a unit load of 1 rad at B:

$$f'_B = 0.104 + \frac{(+9.61)(+9.61)}{193.1} = 0.583,$$

$$f'_A = 0.104 + \frac{(+9.61)(-6.39)}{193.1} = -0.214,$$

$$C_B = \frac{0.214}{0.583} = 0.367.$$

Relative stiffness:

$$\text{Rel } K_A = f'_A \times \frac{B \, d^3_{\text{ref}}}{12} = 0.316 \times \frac{12}{12} \times 2806 = 886 \text{ in}^3,$$

$$\text{Rel } K_B = f'_B \times \frac{B \, d^3_{\text{ref}}}{12} = 0.583 \times \frac{12}{12} \times 2806 = 1635 \text{ in}^3.$$

Absolute stiffness:

$$\text{Abs } K_A = \text{Rel } K_A \times \frac{E}{12} = 886 \times \frac{30{,}000}{12} = 2{,}216{,}000 \text{ in·k},$$

$$\text{Abs } K_B = \text{Rel } K_B \times \frac{E}{12} = 1635 \times \frac{30{,}000}{12} = 4{,}088{,}000 \text{ in·k}.$$

Check by curves:

Equivalent d^3 at ends—

$$I_A = 333 + 2210 = 2540 = \text{Equiv } d^3_{\min},$$

$$I_B = 1944 + 6850 = 8794 = \text{Equiv } d^3_{\max},$$

$$b = \frac{2540}{8794} = 0.289, \quad a = 1.00.$$

If the curves for straight haunches are used, the results are

$$C_A = 0.69, \quad C_B = 0.37,$$

$k_A = 5.5,$ $\quad \text{Abs } K_A = \dfrac{5.5}{16} \times 12 \times \dfrac{2540}{12} \times \dfrac{30{,}000}{12} = 2{,}190{,}000 \text{ in·k},$

$k_B = 10.4,$ $\quad \text{Abs } K_B = \dfrac{10.4}{5.5} \times 2{,}190{,}000 = 4{,}135{,}000 \text{ in·k}.$

Although the values of C and k, as just obtained from the curves, are only approximate, they are usually close enough for purposes of design because the variation in depth of the equivalent rectangular member is neither linear nor parabolic, and hence the curves do not exactly apply

464   ANALYSIS OF FRAMES WITH NONPRISMATIC MEMBERS   [CHAP. 10

10–4 Fixed-end moments, stiffness, and carry-over factors by the conjugate beam. The conjugate beam provides an effective method for computing fixed-end moments, stiffness and carry-over factors for nonprismatic members. To demonstrate the method, Example 10–1 will be re-solved in Example 10–4.

EXAMPLE 10–4. Compute the fixed-end moments, the absolute and relative stiffnesses, and the carry-over factors for the beam of Fig. 10–13, which is of reinforced concrete and which has a uniform width of 1 ft. E is 288,000 k/ft^2.

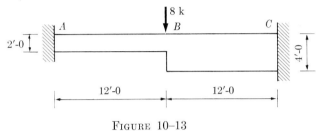

FIGURE 10–13

The conjugate beam for computing the fixed-end moments will have no real supports and will be held in equilibrium by the elastic loads. The width of the conjugate beam corresponding to the segment AB of the real beam will be taken as unity, and the width corresponding to the segment BC will be taken as one-eighth. The fixed-end moments M_A and M_C will be assumed to be positive, that is, to cause compression on the top side of the real beam.

The total intensity of the elastic load on the conjugate beam will be the summation of three separate diagrams, as shown in Fig. 10–14(a), (b), and (c). The total elastic load for each segment AB and BC resulting from the intensity of load represented by each diagram is shown, both in magnitude and line of action, on diagrams (a), (b), and (c). The total elastic load on each segment will act through the centroid of the corresponding part of the intensity diagram.

Two unknowns, M_A and M_C, exist and two condition equations must be written for their evaluation. The first of these is provided by the fact that $\Sigma V = 0$ for the conjugate beam. Physically this means that the algebraic sum of all flexural strains in the loaded real beam must be zero. If all loads are added in the above diagrams, the result is:

$$9M_A + 0.375M_A + 3M_C + 1.125M_C + 288 + 36 = 0,$$

which reduces to

$$9.375M_A + 4.125M_C + 324 = 0. \qquad (10\text{–}9)$$

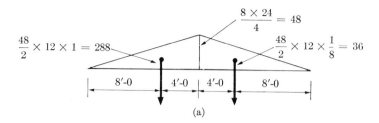

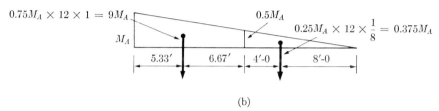

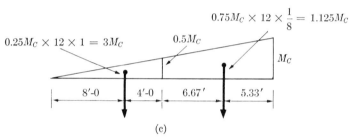

FIGURE 10-14

The second condition equation is provided by the requirement that $\Sigma M = 0$ for the loaded conjugate beam with respect to any desired point in the vertical plane which includes the longitudinal axis of the beam. The physical significance of this condition is somewhat more difficult to understand. It means that if the center of moments is to the right of the right end of the beam or to the left of the left end, and is connected to the near end of the unflexed beam with a horizontal straight line, and if the end of the beam most distant from the center of moments is fixed, then the net vertical deflection of the center of moments must be zero after the beam is flexed by the applied load. If the center of moments is between the two beam ends, the condition means simply that the vertical deflection of the section of the beam on a vertical line with the center of moments, as caused by (or consistent with) the flexural strains to the right of the center of moments, will be equal to the vertical deflection caused by (or consistent with) the flexural strains to the left of the center of moments.

If the beam center line is taken as the center of moments, the second condition equation is

$$-288 \times 4 + 36 \times 4 - 9M_A \times 6.67 + 0.375M_A \times 4 - 3M_C \times 4 + 1.125M_C \times 6.67 = 0,$$

which reduces to

$$+58.5M_A + 4.5M_C + 1008 = 0. \tag{10-10}$$

A simultaneous solution of Eqs. (10–9) and (10–10) will result in

$$M_A = -13.6 \text{ ft·k} \quad \text{and} \quad M_C = -47.5 \text{ ft·k}.$$

Stiffness and carry-over factors are computed by considering that the left support of the real beam is simple, with the right end fixed, and that a moment equal to K_A is applied at A. The moment induced at the fixed end C will be $C_A K_A$. The real beam is shown in Fig. 10–15 and the loaded

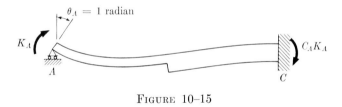

FIGURE 10–15

conjugate beam in Fig. 10–16.

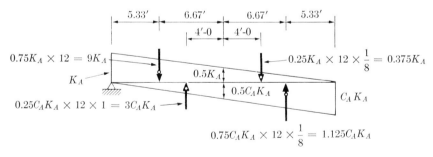

FIGURE 10–16

Since no vertical deflection of A in Fig. 10–15 has been permitted, the moment of all elastic loads on the conjugate beam must be zero about A. Writing $\Sigma M = 0$ about A, we obtain

$$+9K_A \times 5.33 + 0.375K_A \times 16 - 1.125C_A K_A$$
$$\times 18.67 - 3C_A K_A \times 8 = 0.$$

This reduces to
$$54.0K_A - 45.0C_A K_A = 0,$$
from which
$$C_A = 1.20.$$

It is also apparent from Fig. 10–15 that the algebraic summation of all flexural strains from A to C must be 1 rad; in other words, the shear at A in the conjugate beam must be 1 rad. It will be recalled that in computing the loads in Fig. 10–16 the relative width of the conjugate beam for the segment AB was taken as unity, while that for BC was one-eighth. Both widths, and therefore all loads, should be divided by EI_{AB} to obtain absolute values:

$$\frac{+9.0K_A + 0.375K_A - 1.125C_A K_A - 3C_A K_A}{288{,}000 \times 0.667} = 1,$$

from which
$$9.375K_A - 4.125C_A K_A = 192{,}000.$$

If the previously determined value of 1.20 is substituted for C_A, a solution will give $K_A = 43{,}400$ ft·k.

Problems

10–5. The beam of Fig. 10–17 is of reinforced concrete and is 10 in. wide. $E = 4000$ k/in². Using the curves in the Appendix, find (a) fixed-end moments, (b) carry-over factors, (c) relative stiffness, and (d) absolute stiffness. [Ans.: $M_A = 181$ ft·k, $M_B = 157$ ft·k, $C_A = C_B = 0.595$, Rel $K_A =$ Rel $K_B = 0.258$ ft³, Abs $K_A =$ Abs $K_B = 12{,}400$ ft·k.]

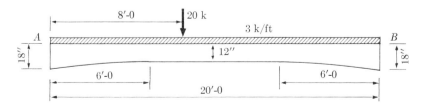

Figure 10–17

10-6. The reinforced concrete beam shown in Fig. 10–18 is 1 ft wide. $E = 4000$ k/in^2. Using the curves in the Appendix, determine values for (a) fixed-end moments, (b) carry-over factors, (c) relative stiffness, and (d) absolute stiffness. [*Ans.:* $M_A = 147$ ft·k, $M_B = 64$ ft·k, $C_A = 0.46$, $C_B = 0.74$, Rel $K_A = 3.0$ ft^3, Rel $K_B = 1.84$ ft^3, Abs $K_A = 144{,}000$ ft·k, Abs $K_B = 88{,}300$ ft·k.]

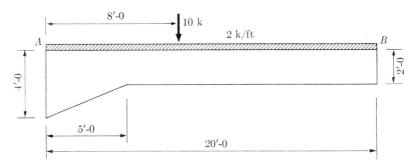

FIGURE 10–18

10-7. In Fig. 10–19, use the column analogy to compute (a) fixed-end moments, (b) carry-over factors, (c) relative stiffness, (d) absolute stiffness. The beam is 1 ft wide, with an E of 2000 k/in^2. Use segments as indicated and cut beam free at right end for fixed-end moment computations. Check values by curves. [*Ans.:* $M_A = 46.6$ ft·k, $M_B = 16.2$ ft·k, $C_A = 0.416$, $C_B = 0.874$, Rel $K_A = 4.08$ ft^3, Rel $K_B = 1.95$ ft^3, Abs $K_A = 98{,}000$ ft·k, Abs $K_B = 46{,}900$ ft·k.]

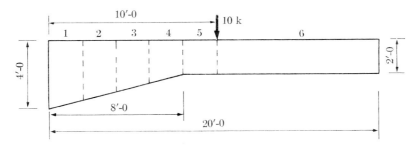

FIGURE 10–19

10-8. The beam of Fig. 10–20 is 1 ft wide, and $E = 3000$ k/in^2. Using segments as indicated, compute carry-over factors, relative stiffness, and absolute stiffness. Check values by curves. [*Ans.:* $C_A = 0.343$, $C_B = 0.729$, Rel $K_A = 6.36$ ft^3, Rel $K_B = 3.00$ ft^3, Abs $K_A = 153{,}000$ ft·k, Abs $K_B = 72{,}000$ ft·k.]

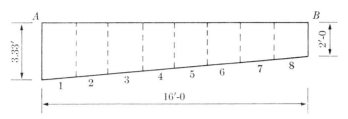

FIGURE 10–20

10–9. The beam of Fig. 10–21 is 1.5 ft wide. $E = 3000$ k/in². Using segments 2 ft long, compute fixed-end moments, carry-over factors, relative stiffness, and absolute stiffness. Cut the beam free at the right end in order to obtain the M_s diagram. Bottom curves are parabolic. [*Ans.:* $M_A = 27.7$ ft·k, $M_B = 36.4$ ft·k, $C_A = 0.953$, $C_B = 0.526$, Rel $K_A = 0.60$ ft³, Rel $K_B = 1.09$ ft³, Abs $K_A = 21,700$ ft·k, Abs $K_B = 39,400$ ft·k.]

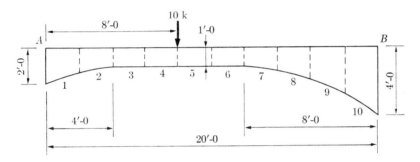

FIGURE 10–21

10–10. Using the analogous column, compute the carry-over factors, relative stiffness, and absolute stiffness of the steel member shown in Fig. 10–22. Use segments as indicated. $E = 30,000$ k/in². Check values by curves. [*Ans.:* $C_A = 0.694$, $C_B = 0.462$, Rel $K_A = 0.76$ ft³, Rel $K_B = 1.14$ ft³, Abs $K_A = 272,000$ ft·k, Abs $K_B = 408,000$ ft·k.]

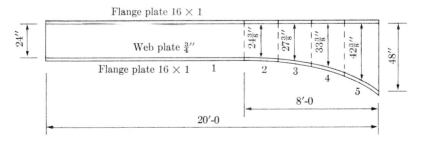

FIGURE 10–22

10-5 Stiffness of a nonprismatic member with far end pinned. The absolute stiffness of a prismatic member with the far end fixed has been shown to be $4EI/L$. It has also been shown that if the far end of the member is simply supported, the absolute stiffness is reduced to $3EI/L$. In other words, the stiffness at one end of a prismatic member is reduced to three-fourths of its former value when a simple support is substituted for a fixed support at the far end. The use of this reduced stiffness for members having terminal ends simply supported has been shown to decrease materially the labor necessary to analyze a structure by moment distribution. A similar advantage may be realized in the case of simply supported terminal reactions for nonprismatic members. The reduction factor, however, is a variable quantity. An expression for this reduction factor will now be derived.

Consider the nonprismatic member of Fig. 10-23. Assume temporarily

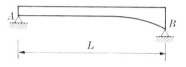

FIGURE 10-23

that the member has a fixed support at B, as shown in Fig. 10-24. A moment is applied at end A of a magnitude sufficient to cause the left end

FIGURE 10-24

of the beam to rotate through 1 rad. This moment is designated by, and will be equal to K_A, the absolute stiffness at A. Since the carry-over factor from A to B is C_A, the induced moment at B is $C_A K_A$. These moments are shown as external to the beam in Fig. 10-24. Actually, however, the support at B is not fixed, and thus the induced moment $C_A K_A$ at end B must be cancelled. This is accomplished by rearranging the supports

FIGURE 10-25

of the member as shown in Fig. 10–25, where end B is reduced to a simple support and end A is fixed in its previously rotated position of 1 rad.

The applied moment at B in Fig. 10–25 is $C_A K_A$, acting in a direction opposite to the induced $C_A K_A$ in Fig. 10–24. The induced moment at A is now $C_B C_A K_A$, C_B being the carry-over factor from B to A. If we add the moments of Figs. 10–24 and 10–25, the result is as shown in Fig. 10–26,

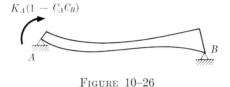

FIGURE 10–26

and the absolute stiffness at A for the member AB with a simple support at B is $K_A(1 - C_A C_B)$. Obviously the corresponding value for end B with a simple support at A would be $K_B(1 - C_A C_B)$.

A point to be noted is that although K_A has been considered to represent absolute stiffness in the above discussion, the correction factor $(1 - C_A C_B)$ may, if desired, be applied to relative stiffnesses.

10–6 Fixed-end moments induced in a nonprismatic member by relative displacement of the member ends. We have previously shown that when one end of a prismatic member is laterally displaced a distance Δ with respect to the other end, with rotation of the member ends being prevented, the fixed-end moments resulting from this displacement will be given by $6EI\Delta/L^2$. It is necessary to develop a corresponding expression for nonprismatic members.

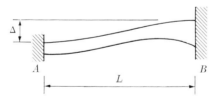

FIGURE 10–27

The distorted member of Fig. 10–27 results from superimposing the effects of three successive actions. In its original position the two ends of the member are at the same level. The first action is to drop end A through the distance Δ with respect to B, without any restraint at the ends. Figure 10–28 results. An external moment of a magnitude sufficient to bring end

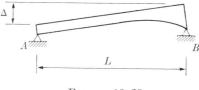

Figure 10-28

A back to zero slope is now applied at A. The angle through which end A must be rotated to accomplish this will be Δ/L radians, and the moment necessary to effect the rotation will be $K_A \Delta/L$, where K_A is the absolute stiffness of the member AB. While this moment is being applied at A, end B is held in the same rotated position it had in Fig. 10–28. This results in a moment being induced at B equal to $C_A K_A \Delta/L$, where C_A is the carry-over factor from A to B. (See Fig. 10–29.)

Figure 10-29

Now consider that the member of Fig. 10–29 is subjected to an additional external moment at end B of a magnitude sufficient to rotate this end through the angle Δ/L radians in order to bring it to the horizontal. End A is fixed in the horizontal position it attained in Fig. 10–29. The external moments developed and induced by this rotation of end B are

Figure 10-30

shown in Fig. 10–30. These moments, it should be noted, are in addition to those shown in Fig. 10–29.

If the moments of Figs. 10–29 and 10–30 are added, the sums will be the desired expressions for the fixed-end moments resulting from the displacement Δ. The final results are shown in Fig. 10–31. These expres-

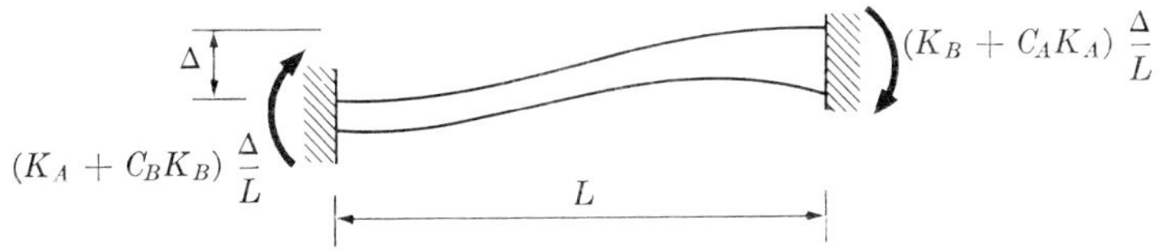

FIGURE 10–31

sions, however, may be simplified. By the Maxwell-Betti reciprocal theorem (see Example 3–2), the moment induced at A by a unit rotation of B is equal to the moment induced at B by a unit rotation of A. This is expressed as

$$C_A K_A = C_B K_B.$$

Consequently, the expression for the fixed-end moment at A reduces to

$$F_A = (K_A + C_B K_B)\frac{\Delta}{L} = (K_A + C_A K_A)\frac{\Delta}{L} = K_A(1 + C_A)\frac{\Delta}{L}.$$

Similarly, the expression for the fixed-end moment at B reduces to

$$F_B = (K_B + C_A K_A)\frac{\Delta}{L} = (K_B + C_B K_B)\frac{\Delta}{L} = K_B(1 + C_B)\frac{\Delta}{L}.$$

EXAMPLE 10–11. The first assumptions for the proportions of a continuous reinforced concrete bridge are indicated in Fig. 10–32. Find all moments resulting from the 2 k load shown. Consider a slice 1-ft wide, that is, 1-ft normal to the plane of the paper.

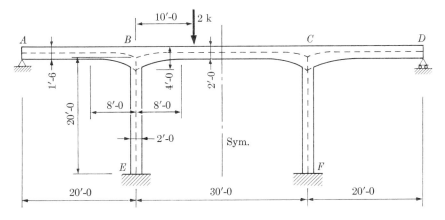

FIGURE 10–32

474   ANALYSIS OF FRAMES WITH NONPRISMATIC MEMBERS   [CHAP. 10

Fixed-end moments, stiffnesses, and carry-over factors are readily determined by using the curves in the Appendix.

Fixed-end moments:

$$v = \frac{10}{30} = 0.333, \quad a = \frac{8}{30} = 0.266, \quad \frac{\min d}{\max d} = 0.5,$$

$$F_{BC} = 0.19 \times 2 \times 30 = 11.40 \text{ ft·k},$$

$$F_{CB} = 0.07 \times 2 \times 30 = 4.20 \text{ ft·k}.$$

Stiffness and carry-over factors:

Members AB and CD:

$$a = \frac{8}{20} = 0.4, \quad \frac{\min d}{\max d} = \frac{1.5}{4.0} = 0.375.$$

Haunched end—

$$C_{BA} = C_{CD} = 0.43, \quad k = 9.7,$$

$$\text{Rel } K_{BA} = \text{Rel } K_{CD} = \frac{kB\,d_{\min}^3}{L} = 9.7 \times 1 \times \frac{1.5^3}{20} = 1.64.$$

Small end—

$$C_{AB} = C_{DC} = 0.87, \quad k = 4.8,$$

$$\text{Rel } K_{AB} = \text{Rel } K_{DC} = 4.8 \times 1 \times \frac{1.5^3}{20} = 0.81.$$

Check—

$$C_{AB}K_{AB} = C_{BA}K_{BA},$$

$$0.43 \times 1.64 = 0.87 \times 0.81,$$

or

$$0.705 = 0.705.$$

Member BC:

$$a = \frac{8}{30} = 0.266, \quad \frac{\min d}{\max d} = \frac{2.0}{4.0} = 0.5,$$

$$C_{BC} = C_{CB} = 0.645, \quad k = 7.5,$$

$$\text{Rel } K_{BC} = \text{Rel } K_{CB} = 7.5 \times 1 \times \frac{2^3}{30} = 2.00.$$

Members BE and CF:

$$C_{BE} = C_{CF} = 0.50, \quad k = 4.0,$$

$$\text{Rel } K_{BE} = \text{Rel } K_{CF} = 4 \times 1 \times \frac{2^3}{20} = 1.60.$$

As previously demonstrated for frames with prismatic members, the balancing process will be considerably shortened if the relative stiffnesses K_{BA} and K_{CD} are reduced in order to correct for the simple supports at A and D. The correction is easily made by means of the formula just derived, and the revised relative K_{BA} is given by

$$\text{Revised Rel } K_{BA} = \text{Rel } K_{BA} (1 - C_{BA}C_{AB})$$

$$= 1.64 (1 - 0.43 \times 0.87) = 1.03.$$

The distribution of moments is shown in Table 10–6.

TABLE 10–6.

Joint	A	B			E	C			F	D
Member	AB	BA	BE	BC	EB	CB	CF	CD	FC	DC
K	0.81	1.03	1.60	2.00	1.60	2.00	1.60	1.03	1.60	0.81
Distribution factor		0.222	0.346	0.432		0.432	0.346	0.222		
Carry-over factor			0.50	0.645		0.645	0.50			
				+11.40		−4.20				
		−2.53	−3.94	−4.93	−1.97	−3.18				
				+2.06		+3.19	+2.56	+1.63	+1.28	
		−0.46	−0.71	−0.89	−0.35	−0.57				
				+0.15		+0.24	+0.20	+0.13	+0.10	
		−0.03	−0.05	−0.07	−0.02	−0.04				
						+0.02	+0.01	+0.01		
Total		−3.02	−4.70	+7.72	−2.34	−4.54	+2.77	+1.77	+1.38	

Problem

10-12. The dimensions first assumed for a reinforced concrete bridge are shown in Fig. 10-33. It is required to find the maximum value of the moment M_{BC} for a strip of bridge 1-ft wide (normal to the plane of the paper), as caused by a uniform load of 0.1 k/ft and two concentrations of 2 k each, all loads being placed so as to give the required maximum moment. A previously constructed influence line for M_{BC} indicates that spans AB, BC, and DE should be completely covered with the uniform load for a maximum. In addition, it is indicated (approximately) that one concentration should be placed 15 ft from B in the span AB and 15 ft from B in the span BC. Find the maximum value of M_{BC}. Include correction for side lurch. [*Ans.:* $M_{BC} = +38.1$ ft·k.]

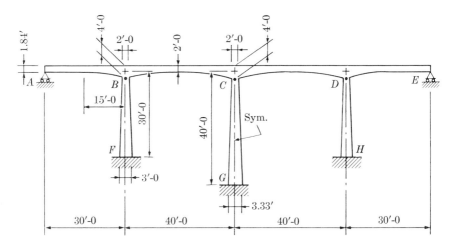

Figure 10-33

CHAPTER 11

THE SLOPE-DEFLECTION METHOD

11-1 General. The first satisfactory solution of the secondary stress problem in articulated structures was presented by Heinrich Manderla in a paper (2) published in 1879–80. His method was somewhat similar to the *slope-deflection* method, and quite possibly suggested its development. In a paper (4) published in 1892, Otto Mohr presented his improved method for handling the secondary stress problem, a method which was essentially what is now called the slope-deflection method. Axel Bendixen, in a book (1) published in Berlin in 1914, presented the method in greater detail, and a year later, in 1915, Professor G. A. Maney of the University of Minnesota published (3) his development of the method.

During the decade just prior to the introduction of moment distribution, nearly all continuous frames, for which rigorous analyses were required, were analyzed by the slope-deflection method. Many engineers still consider that slope-deflection is the better of the two methods and very seldom use moment distribution.

11-2 Development of the method. Two arbitrarily defined properties of flexural members previously discussed—the absolute stiffness and the carry-over factor—are now used again to develop the slope-deflection method. Since the reader already understands the significance of these terms, the method is readily developed.

Consider the nonprismatic member AB of Fig. 11–1. This is assumed

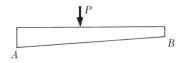

FIGURE 11–1

to be one of a number of members, both columns and beams, in a continuous frame. As the result of loads applied to this frame, one of which is shown acting directly on the member AB, the member is distorted as shown in Fig. 11–2. It is apparent that ends A and B have been caused to rotate through the angles θ_A and θ_B. In addition, because the support at B has settled a distance Δ relative to the support at A, the axis of the member has rotated through the angle $\rho = \Delta/L$ (with all rotations expressed in radians). Any clockwise rotation of a joint, represented by θ,

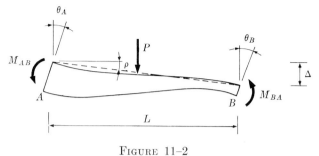

FIGURE 11-2

and any clockwise rotation of a member axis, represented by ρ, will be considered to be positive.

It is required to find expressions for the final member end moments M_{AB} and M_{BA}. As in moment distribution, these moments will always indicate the action of the member on the joint. A member end moment tending to rotate a joint in a clockwise direction will be considered to be positive. Each final member end moment will be the initial fixed-end moment (to be designated as F_{AB} or F_{BA}) plus corrections for end and axial rotations of the member. The configuration of the distorted member as shown in Fig. 11-2 may be considered to be the result of the superposition of several individual joint rotations.

Consider the member initially in an unflexed condition with end B a distance Δ below end A. (See Fig. 11-3.) End A is now caused to rotate

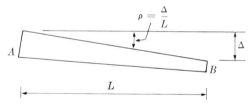

FIGURE 11-3

through ρ radians, in a counterclockwise direction, to bring it to the horizontal, with end B fixed in position during the rotation of end A. The distorted beam and resulting end moments are shown in Fig. 11-4.

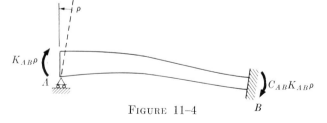

FIGURE 11-4

With end A fixed in the position shown in this figure, end B is now caused to rotate ρ radians in a counterclockwise direction. The resulting configuration of the beam and the additional end moments are shown in Fig. 11–5.

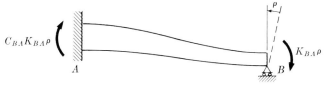

FIGURE 11–5

With end B held in position, a clockwise rotation of θ_A radians is impressed on end A of the distorted beam of Fig. 11–5. Additional end moments, as shown in Fig. 11–6, will result. Finally, with end A fixed,

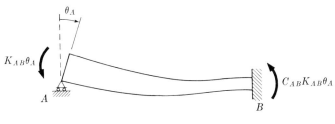

FIGURE 11–6

end B is caused to rotate θ_B radians in a clockwise direction, and the beam is now in the distorted position shown in Fig. 11–2. The additional end moments resulting from this last rotation of end B are shown in Fig. 11–7.

FIGURE 11–7

If, for each end of the beam, all the end moments of Figs. 11–4 through 11–7 are added to the initial fixed-end moments for the respective ends, the result will be the expressions for the final end moments of the member

AB in the loaded frame:

$$M_{AB} = F_{AB} - K_{AB}\theta_A - C_{BA}K_{BA}\theta_B + K_{AB}\rho + C_{BA}K_{BA}\rho,$$
(11-1)

$$M_{BA} = F_{BA} - K_{BA}\theta_B - C_{AB}K_{AB}\theta_A + K_{BA}\rho + C_{AB}K_{AB}\rho.$$
(11-2)

If the subscript N is taken to indicate the near end of a flexural member and F to represent the far end, a single equation will suffice:

$$M_N = F_N - K_N\theta_N - C_F K_F \theta_F + (K_N + C_F K_F)\rho.$$
(11-3)

This is the basic equation for the slope deflection method, and it applies to either prismatic or nonprismatic members. In the above equation, M_N represents the final moment at the near end (the end being considered) of the member. M_N is the action of the member on the joint, and if its sign is positive, it indicates that the member tends to rotate the joint in a clockwise direction. F_N is the initial fixed-end moment and is positive if the member tends to rotate the joint in a clockwise direction. θ_N and θ_F represent the rotations in radians of the ends of the member, as indicated by the subscripts, and are positive if clockwise. The rotation in radians of the member axis is designated by ρ, and it too is positive if clockwise. K_N and K_F represent the *absolute stiffness* of the member at the end indicated by the subscript. C_F is the carry-over factor for the far end.

If all the members in the frame being analyzed are prismatic and if both ends of the member are fixed to the frame (not pinned or restrained), then, since K_N and K_F are equal to $4EI/L$ and C_F is equal to $\frac{1}{2}$, Eq. (11-3) becomes

$$M_N = F_N - \frac{4EI\theta_N}{L} - \frac{2EI\theta_F}{L} + \frac{6EI\rho}{L}.$$
(11-4)

If, however, the far end of the prismatic member is pinned, the operations in the derivation indicated in Figs. 11-5 and 11-7 are unnecessary and are omitted. In addition, the value of K_N is reduced to $3EI/L$. Consequently, for a prismatic member pinned at the far end, Eq. (11-3) will reduce to

$$M_N = F_N - \frac{3EI\theta_N}{L} + \frac{3EI\rho}{L}.$$
(11-5)

If one or more of the members in the frame are nonprismatic, then Eq. (11-3) should be written as

11-3]   ANALYSIS OF CONTINUOUS BEAMS   481

$$M_N = F_N - \frac{k_N EB\, d_{min}^3 \theta_N}{12L} - \frac{C_F k_F EB\, d_{min}^3 \theta_F}{12L}$$

$$+ \left(\frac{k_N EB\, d_{min}^3}{12L} + \frac{C_F k_F EB\, d_{min}^3}{12L} \right) \rho. \quad (11\text{-}6)$$

In the above equation the symbols B, d_{min}, L, and k have the same significance as in the discussion of stiffness and carry-over factors as computed by the column analogy in Chapter 10. B represents the width of the member, d_{min} the minimum depth, L the span, and k the stiffness factor for the "reference beam." If the nonprismatic member is of a standard type for which information has been computed, the values of C_F, k_N, k_F, and F_N are most conveniently found from the curves in the Appendix or from the *Handbook of Frame Constants* [the Portland Cement Association].

11-3 Analysis of continuous beams. It is probable that most analysts who are familiar with both moment distribution and slope-deflection prefer the former method for the analysis of continuous beams. Nevertheless, the slope-deflection method is valuable for purposes of checking and every structural engineer should be thoroughly familiar with it. A single illustrative example should be sufficient to demonstrate its application.

EXAMPLE 11-1. In Fig. 11-8 find all moments by slope-deflection. The support at C settles 0.1 in. E is 30,000 k/in^2.

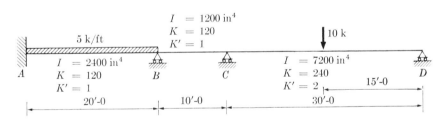

FIGURE 11-8

Fixed-end moments are first computed:

$$F_{AB} = \frac{5 \times 20^2}{12} = +166.5 \text{ ft·k}, \qquad F_{CD} = +\frac{10 \times 30}{8} = +37.5 \text{ ft·k},$$

$$F_{BA} = -166.5 \text{ ft·k}, \qquad\qquad F_{DC} = -37.5 \text{ ft·k}.$$

If we consider span CD as a propped cantilever with a simple support at D, the fixed-end moment F'_{CD} is easily computed by moment distribution from the above computed values for F_{CD} and F_{DC}:

$$F'_{CD} = +37.5 + \tfrac{1}{2}(37.5) = +56.25 \text{ ft·k}.$$

The values of ρ for spans BC and CD are computed as

$$\rho_{BC} = +\frac{0.1}{10 \times 12} = 0.000833 \text{ rad},$$

$$\rho_{CD} = -\frac{0.1}{30 \times 12} = -0.000278 \text{ rad}.$$

The expressions for the final moments at the ends of each span are written by means of Eqs. (11–4) and (11–5). Of special note here is the fact that although absolute stiffnesses have been used in deriving these equations, it is possible to simplify the computations by using relative values of I/L. The values of final end moments obtained by using these relative values are not affected and are correct. If, however, we wish to find the absolute value of any unknown θ or ρ, an adjustment must be made to correct for the use of relative values of I/L in writing the initial equations. (This will be demonstrated in subsequent examples.) The relative values of I/L to be used in the initial equations are shown as K' in Fig. 11–8.

By Eq. (11–4), and remembering that $\theta_A = 0$,

$$M_{AB} = F_{AB} - 2EK'\theta_B = +166.5 - 2E\theta_B, \tag{11-7}$$

$$M_{BA} = F_{BA} - 4EK'\theta_B = -166.5 - 4E\theta_B, \tag{11-8}$$

$$M_{BC} = -4EK'\theta_B - 2EK'\theta_C + \frac{6EI\rho}{L}$$

$$= -4E\theta_B - 2E\theta_C + \frac{6 \times 30{,}000 \times 1200 \times 0.000833}{10 \times 12 \times 12}$$

$$= -4E\theta_B - 2E\theta_C + 124.9, \tag{11-9}$$

$$M_{CB} = -4EK'\theta_C - 2EK'\theta_B + \frac{6EI\rho}{L}$$

$$= -4E\theta_C - 2E\theta_B + 124.9. \tag{11-10}$$

By Eq. (11-5),

$$M_{CD} = F'_{CD} - 3EK'\theta_C + \frac{3EI\rho}{L}$$

$$= +56.25 - 6E\theta_C + \frac{3 \times 30{,}000 \times 7200\,(-0.000278)}{30 \times 12 \times 12}$$

$$= -6E\theta_C + 14.5. \tag{11-11}$$

It is apparent that two independent unknowns, θ_B and θ_C, appear in the above expressions for the various end moments. These must be evaluated before the end moments can be determined, and two condition equations are required to effect a solution. These equations are obtained by writing expressions for the equilibrium of all or parts of the structure. In the given example it is most convenient to use expressions indicating that the sum of the internal moments at joints B and C must be zero. The following equations result:

$$M_{BA} + M_{BC} = 0, \tag{11-12}$$

$$M_{CB} + M_{CD} = 0. \tag{11-13}$$

By substituting in these equations the previously written values for the final end moments of the various spans, in terms of fixed-end moments and joint rotations, we obtain

$$-8.00E\theta_B - 2.00E\theta_C - 41.6 = 0, \tag{11-12a}$$

$$-2.00E\theta_B - 10.0E\theta_C + 139.4 = 0. \tag{11-13a}$$

A simultaneous solution will result in $E\theta_B = -9.14$ and $E\theta_C = +15.8$.

The foregoing values for $E\theta_B$ and $E\theta_C$ are *relative values*. In spite of this, however, they may be substituted back into the expressions for the various end moments to give the correct values for these moments. For example,

$$M_{AB} = +166.5 - 2(-9.14) = +184.8 \text{ ft·k},$$

$$M_{BA} = -166.5 - 4(-9.14) = -129.9 \text{ ft·k}.$$

Similarly, the remaining moments are found to be

$$M_{BC} = +129.9 \text{ ft·k}, \qquad M_{CB} = +80.1 \text{ ft·k}, \qquad M_{CD} = -80.1 \text{ ft·k}.$$

As previously noted, a positive sign indicates that the member tends to rotate the joint in a clockwise direction.

A point of importance here is the fact that if absolute values of θ_B and θ_C (or ρ, where this is unknown) are desired, certain factors must be introduced to compensate for using relative values for stiffness and to adjust units. An explanation of these factors follows.

When Eqs. (11–7) through (11–11) were written, all fixed-end moments were expressed in foot-kilopounds. Therefore, to be consistent the units of all other terms in these equations must be foot-kilopounds. These other terms are all in the form $QEI\beta/L$, where Q is a numerical coefficient and β is an angle of rotation (actually, either θ or ρ) expressed in radians. In this expression, if E is in k/in^2, I in in^4, and L in inches, the units will be inch-kilopounds. Consequently, an additional 12 should appear in the denominator to convert the units to foot-kilopounds. Actually, however, in writing Eqs. (11–7) through (11–11) the 12 was omitted from the denominator and L was expressed in feet. This will make the coefficient of each $E\theta$ term (and $E\rho$ term, if ρ is unknown) too large by the multiplier 144. Moreover, values of $K' = K/120$ were used in writing Eqs. (11–7) through (11–11) in order to permit the use of smaller numbers. This procedure will make the coefficient of each $E\theta$ term (and $E\rho$ term, if ρ is unknown) too small by the divisor 120. The net result, therefore, is that the coefficients of the $E\theta$ terms are 144/120 of their true values, and consequently, the values of $E\theta_B$ and $E\theta_C$ are 120/144 of their correct values. As a result, the absolute values of θ_B and θ_C are computed as follows:

$$\theta_B = \frac{-9.14 \times 144}{120 \times 30{,}000} = -110 \times 10^{-5} \text{ rad},$$

$$\theta_C = \frac{+15.8 \times 144}{120 \times 30{,}000} = +189 \times 10^{-5} \text{ rad}.$$

Problems

11-2. In Fig. 11–9 find all moments by slope-deflection. [*Ans.:* $M_{BA} = -36.9$ ft·k, $M_{CB} = -12.4$ ft·k.]

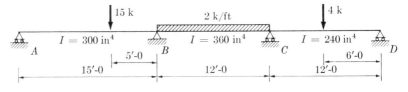

Figure 11–9

11-3. In Fig. 11-10 find all moments by slope-deflection. The support at A rotates clockwise 0.001 rad and the support at B settles 0.1 in. Find the rotation of the beam at B. $E = 30{,}000$ k/in^2. [Ans.: $M_{AB} = -12.0$ ft·k, $M_{BA} = -12.9$ ft·k, $\theta_B = +11.0 \times 10^{-4}$ rad.]

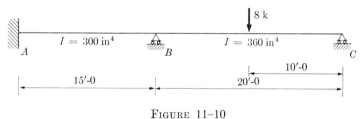

FIGURE 11-10

11-4. In Fig. 11-11 find all moments by slope-deflection. [Ans.: $M_{AB} = -5.7$ ft·k, $M_{BA} = -11.4$ ft·k, $M_{CB} = -50.2$ ft·k.]

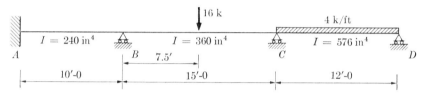

FIGURE 11-11

11-5. In Fig. 11-12 find all moments by slope-deflection. Also, find the rotation of joint C, draw the moment diagram, and sketch the deflected structure. $E = 30{,}000$ k/in^2. [Ans.: $M_{AB} = -7.1$ ft·k, $M_{BA} = -14.2$ ft·k, $M_{CB} = -36.0$ ft·k, $M_{CE} = +22.0$ ft·k, $M_{CD} = +14.0$ ft·k, $\theta_C = 2.24 \times 10^{-3}$ rad.]

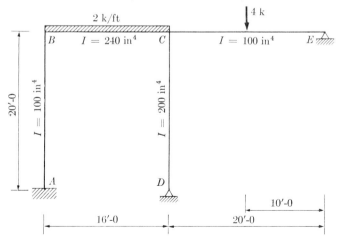

FIGURE 11-12

11–4 Frames with one degree of freedom. Frames having one degree of freedom to lurch are readily analyzed by the slope-deflection method. The independent unknowns consist of one θ for each joint which rotates as the loads are applied (unless a simple support at a terminal end permits an end span to be treated as a propped cantilever, as in Example 11–1) plus one or more unknown values for ρ due to the lurch of the frame.

EXAMPLE 11–6. In Fig. 11–13 find all moments by slope-deflection. Determine the rotation of joints B and C and the direction and magnitude of side lurch. $E = 30{,}000$ k/in^2. (This is the same structure previously analyzed in Example 8–17.)

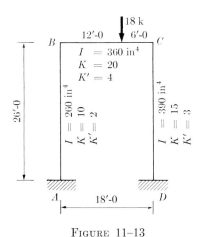

FIGURE 11–13

Fixed-end moments are computed as follows:

$$F_{BC} = \frac{18 \times 12 \times 6^2}{18^2} = +24 \text{ ft·k}, \quad F_{CB} = \frac{18 \times 6 \times 12^2}{18^2} = -48 \text{ ft·k}.$$

Equation (11–4) is used to write the expressions for the several final end moments. Values of K' are shown on Fig. 11–13.

$$M_{AB} = -2EK'\theta_B + 6EK'\rho = -4E\theta_B + 12E\rho, \tag{11-14}$$

$$M_{BA} = -4EK'\theta_B + 6EK'\rho = -8E\theta_B + 12E\rho, \tag{11-15}$$

$$M_{BC} = F_{BC} - 4EK'\theta_B - 2EK'\theta_C = +24 - 16E\theta_B - 8E\theta_C, \tag{11-16}$$

$$M_{CB} = F_{CB} - 2EK'\theta_B - 4EK'\theta_C = -48 - 8E\theta_B - 16E\theta_C, \tag{11-17}$$

$$M_{CD} = -4EK'\theta_C + 6EK'\rho = -12E\theta_C + 18E\rho, \tag{11-18}$$

$$M_{DC} = -2EK'\theta_C + 6EK'\rho = -6E\theta_C + 18E\rho. \tag{11-19}$$

Three independent unknowns (θ_B, θ_C, and ρ) appear in the above equations. These must be evaluated by writing three condition equations expressing equilibrium of all or parts of the frame, and these equations are then solved simultaneously for the three unknowns. The first two of these equations are

$$\Sigma M_B = M_{BA} + M_{BC} = 0,$$

from which

$$-6E\theta_B - 2E\theta_C + 3E\rho + 6 = 0; \tag{11-20}$$

$$\Sigma M_C = M_{CB} + M_{CD} = 0,$$

which reduces to

$$-4E\theta_B - 14E\theta_C + 9E\rho - 24 = 0. \tag{11-21}$$

The third condition equation is obtained by expressing $\Sigma H = 0$ for the entire frame. Since no horizontal loads are acting, this means that the sum of the column shears must be zero. The column shears in this case are found by adding the two moments at the ends of each column and dividing by the column length. Since both columns are the same length, the division by the length of each column may be omitted, and the condition equation is

$$M_{AB} + M_{BA} + M_{CD} + M_{DC} = 0,$$

from which

$$-2E\theta_B - 3E\theta_C + 10E\rho = 0. \tag{11-22}$$

A simultaneous solution of these three condition equations will result in

$$E\theta_B = +1.61, \quad E\theta_C = -2.44, \quad E\rho = -0.409,$$

and substitution of these values in Eqs. (11-14) through (11-19) will give the following values for the final end moments:

$$M_{AB} = -11.4 \text{ ft·k}, \quad M_{BA} = -17.8 \text{ ft·k},$$

$$M_{CD} = +21.9 \text{ ft·k}, \quad M_{DC} = +7.3 \text{ ft·k}.$$

Joint rotations are computed as follows:

$$\theta_B = \frac{+1.61 \times 144}{5 \times 30{,}000} = +1.55 \times 10^{-3} \text{ rad,}$$

$$\theta_C = \frac{-2.44 \times 144}{5 \times 30{,}000} = -2.34 \times 10^{-3} \text{ rad.}$$

The angle of side lurch is given by

$$\rho = \frac{-0.409 \times 144}{5 \times 30{,}000} = -3.9 \times 10^{-4} \text{ rad,}$$

and the lurch is

$$3.9 \times 10^{-4} \times 26 \times 12 = 0.12 \text{ in. to the left.}$$

EXAMPLE 11–7. In Fig. 11–14 find all moments by slope-deflection. (This frame was previously analyzed in Example 8–28.)

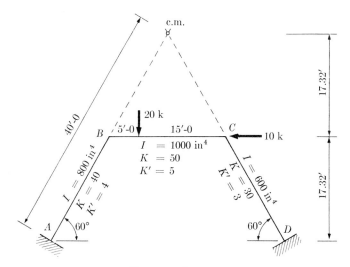

FIGURE 11–14

Fixed-end moments are:

$$F_{BC} = \frac{+20 \times 5 \times 15^2}{20^2} = +56.25 \text{ ft·k,}$$

$$F_{CB} = -\frac{20 \times 5^2 \times 15}{20^2} = -18.75 \text{ ft·k.}$$

In order to obtain the relationship between the values of ρ for the three members, a joint displacement diagram is drawn (see Fig. 11–15) with

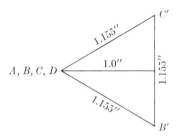

FIGURE 11–15

the assumption that joints B and C are so displaced that the horizontal projection of their displacements is 1 in. to the right. From this diagram it is apparent that if $\rho_{AB} = \rho_{CD} = +\rho$, then $\rho_{BC} = -\rho$.

The expressions for the various final end moments are now written by Eq. (11–4):

$$M_{AB} = -2EK'\theta_B + 6EK'\rho = -8E\theta_B + 24E\rho, \qquad (11\text{--}23)$$

$$M_{BA} = -4EK'\theta_B + 6EK'\rho = -16E\theta_B + 24E\rho, \qquad (11\text{--}24)$$

$$\begin{aligned} M_{BC} &= F_{BC} - 4EK'\theta_B - 2EK'\theta_C - 6EK'\rho \\ &= +56.25 - 20E\theta_B - 10E\theta_C - 30E\rho, \end{aligned} \qquad (11\text{--}25)$$

$$\begin{aligned} M_{CB} &= F_{CB} - 2EK'\theta_B - 4EK'\theta_C - 6EK'\rho \\ &= -18.75 - 10E\theta_B - 20E\theta_C - 30E\rho, \end{aligned} \qquad (11\text{--}26)$$

$$M_{CD} = -4EK'\theta_C + 6EK'\rho = -12E\theta_C + 18E\rho, \qquad (11\text{--}27)$$

$$M_{DC} = -2EK'\theta_C + 6EK'\rho = -6E\theta_C + 18E\rho. \qquad (11\text{--}28)$$

Two of the three condition equations necessary to evaluate the three independent unknowns (θ_B, θ_C, and ρ) express the fact that $\Sigma M = 0$ at joints B and C:

$$\Sigma M_B = M_{BA} + M_{BC} = 0,$$

from which
$$-36E\theta_B - 10E\theta_C - 6E\rho + 56.25 = 0; \quad (11\text{-}29)$$

$$\Sigma M_C = M_{CB} + M_{CD} = 0,$$

which results in
$$-10E\theta_B - 32E\theta_C - 12E\rho - 18.75 = 0. \quad (11\text{-}30)$$

The third condition equation is most easily obtained by expressing the fact that $\Sigma M = 0$ for the entire frame, with the center of moments as indicated in Fig. 11–14. Since the center of moments is located at the intersection of the two sloping legs, the axial stresses in these legs do not appear in the equation. Note that forces and moments external to the frame must be used in writing the equation, and that strict attention must be given to the signs of the various terms. A clockwise moment about the center of moments is considered to be positive. The third condition equation is

$$\Sigma M_{\text{c.m.}} = -M_{AB} + \frac{(M_{AB} + M_{BA})40}{20} - M_{DC} + \frac{(M_{CD} + M_{DC})40}{20}$$
$$+ 10 \times 17.32 - 20 \times 5 = 0. \quad (11\text{-}31)$$

The negative signs before the first and third terms are necessary because M_{AB} and M_{DC} are internal moments, and a positive (clockwise) internal moment will be opposed by a negative (counterclockwise) external moment. The second and fourth terms are the moments of the external shears acting at A and D, respectively. These are preceded by positive signs because positive internal end moments will induce external end shears at the reactions, which will tend to cause clockwise, or positive, rotation about the center of moments. Equation (11–31) reduces to

$$-40E\theta_B - 30E\theta_C + 126E\rho + 73.2 = 0. \quad (11\text{-}32)$$

A simultaneous solution of Eqs. (11–29), (11–30), and (11–32) results in

$$E\theta_B = +1.91, \quad E\theta_C = -1.09, \quad E\rho = -0.236.$$

Substitution of these values in Eqs. (11–23), (11–24), (11–27), and (11–28) will result in the following:

$$M_{AB} = -20.9 \text{ ft·k}, \quad M_{BA} = -36.2 \text{ ft·k},$$

$$M_{CD} = +8.9 \text{ ft·k}, \quad M_{DC} = +2.3 \text{ ft·k}.$$

EXAMPLE 11–8. In Fig. 11–16 find all moments by slope-deflection. (This frame was previously analyzed in Example 8–29.)

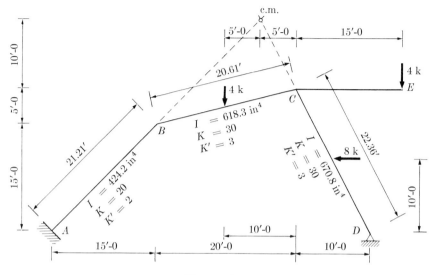

FIGURE 11–16

The fixed-end moments are:

$$F_{BC} = +10 \text{ ft·k}, \qquad F_{CB} = -10 \text{ ft·k}, \qquad F_{CD} = +20 \text{ ft·k},$$

$$F_{DC} = -20 \text{ ft·k}, \qquad F'_{CD} = +30 \text{ ft·k}.$$

To compute the relative values of ρ for the three members, it is assumed that joint B is displaced so that the horizontal component of this displacement is 1 in. to the right. (The joint displacement diagram is shown in Fig. 11–17.)

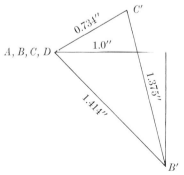

FIGURE 11–17

The relative values of ρ are computed as follows:

$$\text{Rel } \rho_{AB} = \frac{+1.414}{21.21} = +0.0667,$$

$$\text{Rel } \rho_{BC} = \frac{-1.375}{20.61} = -0.0667,$$

$$\text{Rel } \rho_{CD} = \frac{+0.734}{22.36} = +0.0328.$$

From these relative values it is apparent that if $\rho_{AB} = +\rho$, then $\rho_{BC} = -\rho$, and

$$\rho_{CD} = +\frac{0.0328}{0.0667}\rho = +0.49\rho.$$

By Eq. (11-4),

$$M_{AB} = -2EK'\theta_B + 6EK'\rho_{AB} = -4E\theta_B + 12E\rho, \quad (11\text{-}33)$$

$$M_{BA} = -4EK'\theta_B + 6EK'\rho_{BA} = -8E\theta_B + 12E\rho, \quad (11\text{-}34)$$

$$\begin{aligned} M_{BC} &= F_{BC} - 4EK'\theta_B - 2EK'\theta_C + 6EK'\rho_{BC} \\ &= +10 - 12E\theta_B - 6E\theta_C - 18E\rho, \end{aligned} \quad (11\text{-}35)$$

$$\begin{aligned} M_{CB} &= F_{CB} - 2EK'\theta_B - 4EK'\theta_C + 6EK'\rho_{CB} \\ &= -10 - 6E\theta_B - 12E\theta_C - 18E\rho. \end{aligned} \quad (11\text{-}36)$$

By Eq. (11-5),

$$\begin{aligned} M_{CD} &= F'_{CD} - 3EK'\theta_C + 3EK'\rho_{CD} \\ &= +30 - 9E\theta_C + 4.43E\rho. \end{aligned} \quad (11\text{-}37)$$

The necessary condition equations are written as follows:

$$\Sigma M_B = M_{BA} + M_{BC} = 0,$$

which reduces to

$$-20E\theta_B - 6E\theta_C - 6E\rho + 10 = 0; \quad (11\text{-}38)$$

$$\Sigma M_C = M_{CB} + M_{CD} + M_{CE} = 0,$$

which results in

$$-6E\theta_B - 21E\theta_C - 13.57E\rho + 80 = 0. \quad (11\text{-}39)$$

For the entire structure ΣM about the center of moments equals zero. Clockwise moments are positive.

$$\Sigma M_{\text{c.m.}} = -M_{AB} + \left(\frac{M_{AB} + M_{BA}}{21.21}\right) 42.42$$

$$+ \left(\frac{M_{CD}}{22.36}\right) 33.54 - \left(\frac{8 \times 10}{20}\right) 30$$

$$- 4 \times 5 + 8 \times 20 + 4 \times 20 = 0,$$

which reduces to

$$-20E\theta_B - 13.5E\theta_C + 42.64E\rho_{AB} + 145 = 0. \qquad (11\text{-}40)$$

A simultaneous solution of Eqs. (11–38) through (11–40) will give

$$E\theta_B = -0.473, \qquad E\theta_C = +5.22, \qquad E\rho = -1.97,$$

and if these values are substituted in Eqs. (11–33), (11–34), and (11–37), the following moments result:

$$M_{AB} = -21.8 \text{ ft·k}, \qquad M_{BA} = -19.9 \text{ ft·k}, \qquad M_{CD} = -25.7 \text{ ft·k}.$$

Problems

11–9. In Fig. 11–18 find all moments by slope-deflection.

[Ans.: $M_{BA} = -42.9$ ft·k,

$M_{BC} = -57.1$ ft·k,

$M_{CB} = -28.5$ ft·k.]

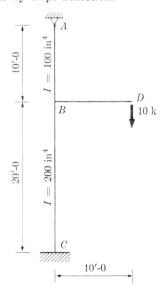

Figure 11–18

11-10. In Fig. 11-19 find all moments by slope-deflection. Determine the rotation of B, and compute the direction and magnitude of lurch at level of BDE. $E = 30,000$ k/in^2. [Ans.: $M_{AB} = +48.6$ ft·k, $M_{BC} = -30.3$ ft·k, $M_{BA} = +55.8$ ft·k, $M_{BE} = +54.9$ ft·k, lurch = 0.60 in. to right, $\theta_B = -0.17 \times 10^{-2}$ rad.]

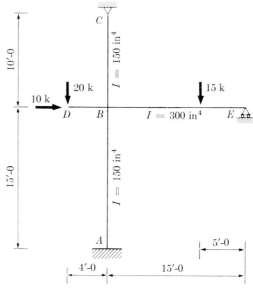

FIGURE 11-19

11-11. In Fig. 11-20 find all moments by slope-deflection. [Ans.: $M_B = 8.1$ ft·k, $M_C = 10.7$ ft·k.]

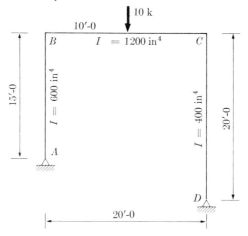

FIGURE 11-20

11-12. In Fig. 11-21 find all moments by slope-deflection. Determine rotation of joints B and C, as well as the direction and magnitude of side lurch. $E = 30{,}000$ k/in². [*Ans.:* $M_{AB} = +33.4$ ft·k, $M_{BA} = -2.5$ ft·k, $M_{CB} = -106.9$ ft·k, $M_{CD} = +66.9$ ft·k, $\theta_B = +4.31 \times 10^{-3}$ rad, $\theta_C = +1.79 \times 10^{-4}$ rad, $\rho_{AB} = +2.8 \times 10^{-3}$ rad, lurch = 0.67 in. to right.]

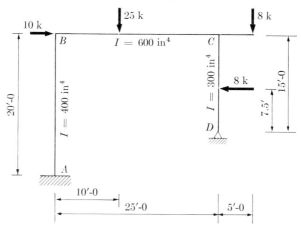

FIGURE 11-21

11-5 Analysis of gabled frames. The analysis of gabled frames will be simplified considerably if certain relationships which exist between the values of ρ for the various members are known. These relationships will now be derived.

Consider the frame of Fig. 11-22. Assume that D is held in position

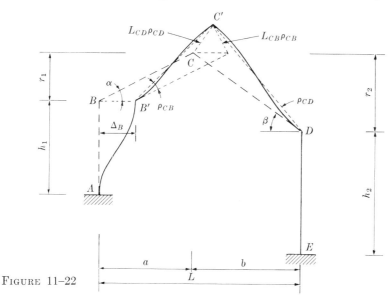

FIGURE 11-22

and that B is displaced a distance Δ_B to the right. In the joint displacement diagram shown in Fig. 11–23, it is obvious that

$$\Delta_{CB} \cos \alpha = \Delta_{CD} \cos \beta. \tag{11–41}$$

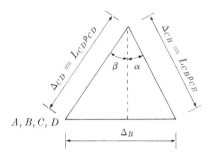

FIGURE 11–23

But, from Fig. 11–23,

$$\Delta_{CB} = L_{CB}\rho_{CB} \quad \text{and} \quad \Delta_{CD} = L_{CD}\rho_{CD}. \tag{11–42}$$

Substituting in Eq. (11–41) yields

$$a\rho_{CB} = b\rho_{CD}. \tag{11–43}$$

From Fig. 11–23,

$$\Delta_{CD} \sin \beta + \Delta_{CB} \sin \alpha = \Delta_B,$$

and substituting from Eq. (11–42),

$$L_{CD}\rho_{CD} \sin \beta + L_{CB}\rho_{CB} \sin \alpha = \Delta_B.$$

From Fig. 11–22, the above reduces to

$$r_2\rho_{CD} + r_1\rho_{CB} = h_1\rho_{AB}, \tag{11–44}$$

but, from Eq. (11–43),

$$\rho_{CD} = \frac{a}{b} \rho_{CB}.$$

Substitution in Eq. (11–44) yields

$$\frac{r_2 a \rho_{CB}}{b} + r_1\rho_{CB} = h_1\rho_{AB},$$

from which

$$\rho_{CB} = -\frac{bh_1}{ar_2 + br_1}\rho_{AB} \text{ (negative by inspection)}. \quad (11\text{--}45)$$

Similarly, from Eqs. (11–43) and (11–44),

$$\rho_{CD} = \frac{ah_1}{ar_2 + br_1}\rho_{AB} \text{ (positive by inspection)}. \quad (11\text{--}46)$$

Now if B is held fixed and D is moved to the right a distance Δ_D, then, proceeding as before, it can be shown that

$$\rho_{CB} = \frac{bh_2}{ar_2 + br_1}\rho_{DE} \text{ (positive by inspection)} \quad (11\text{--}47)$$

and

$$\rho_{CD} = -\frac{ah_2}{ar_2 + br_1}\rho_{DE} \text{ (negative by inspection)}. \quad (11\text{--}48)$$

If both B and D are displaced, the addition of Eqs. (11–45) and (11–47) will result in

$$\rho_{CB} = \frac{b}{ar_2 + br_1}(h_2\rho_{DE} - h_1\rho_{AB}), \quad (11\text{--}49)$$

and the addition of Eqs. (11–46) and (11–48) will give

$$\rho_{CD} = \frac{a}{ar_2 + br_1}(h_1\rho_{AB} - h_2\rho_{DE}). \quad (11\text{--}50)$$

If $h_1 = h_2 = h$ and $r_1 = r_2 = r$, then

$$\rho_{BC} = \frac{bh}{rL}(\rho_{DE} - \rho_{AB}), \quad (11\text{--}51)$$

$$\rho_{CD} = \frac{ah}{rL}(\rho_{AB} - \rho_{DE}). \quad (11\text{--}52)$$

If, in addition, $a = b = L/2$, as in a symmetrical bent, then

$$\rho_{BC} = \frac{h}{2r}(\rho_{DE} - \rho_{AB}), \quad (11\text{--}53)$$

$$\rho_{CD} = \frac{h}{2r}(\rho_{AB} - \rho_{DE}). \quad (11\text{--}54)$$

498   THE SLOPE-DEFLECTION METHOD   [CHAP. 11

If the symmetrical bent is loaded symmetrically, then $\rho_{DE} = -\rho_{AB}$ and

$$\rho_{BC} = -\frac{h}{r}\rho_{AB} = +\frac{h}{r}\rho_{DE}, \tag{11-55}$$

$$\rho_{CD} = +\frac{h}{r}\rho_{AB} = -\frac{h}{r}\rho_{DE}. \tag{11-56}$$

EXAMPLE 11-13. In Fig. 11-24 determine all moments by slope-deflection. The moment of inertia is constant.

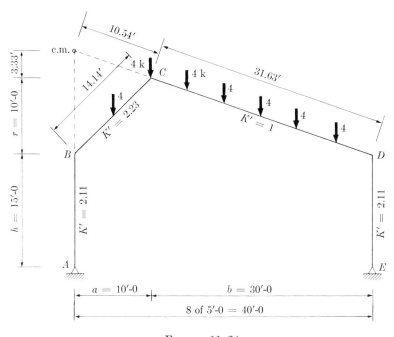

FIGURE 11-24

Fixed-end moments are first computed:

$$F_{BC} = +\frac{4 \times 10}{8} = +5.0 \text{ ft·k}, \qquad F_{CB} = -5.0 \text{ ft·k},$$

$$F_{CD} = \frac{(N^2 - 1)PL}{12N} = \frac{(6^2 - 1)4 \times 30}{12 \times 6} = +58.33 \text{ ft·k},$$

$$F_{DC} = -58.33 \text{ ft·k}.$$

11-5] ANALYSIS OF GABLED FRAMES

Using Eqs. (11–51) and (11–52),

$$\rho_{BC} = \frac{bh}{rL}(\rho_{DE} - \rho_{AB}) = \frac{30 \times 15}{10 \times 40}(\rho_{DE} - \rho_{AB}) = \frac{9}{8}(\rho_{DE} - \rho_{AB}),$$

$$\rho_{CD} = \frac{ah}{rL}(\rho_{AB} - \rho_{DE}) = \frac{3}{8}(\rho_{AB} - \rho_{DE}).$$

By Eq. (11–5),

$$M_{BA} = -3EK'\theta_B + 3EK'\rho_{AB} = -6.33E\theta_B + 6.33E\rho_{AB}, \quad (11\text{--}57)$$

$$M_{DE} = -3EK'\theta_D + 3EK'\rho_{DE} = -6.33E\theta_D + 6.33E\rho_{DE}. \quad (11\text{--}58)$$

By Eq. (11–4) and the relationships between the several values of ρ as computed above,

$$M_{BC} = F_{BC} - 4EK'\theta_B - 2EK'\theta_C + 6.75EK'(\rho_{DE} - \rho_{AB})$$
$$= +5.0 - 8.92E\theta_B - 4.46E\theta_C + 15.08E\rho_{DE} - 15.08E\rho_{AB}, \quad (11\text{--}59)$$

$$M_{CB} = F_{CB} - 2EK'\theta_B - 4EK'\theta_C + 6.75EK'(\rho_{DE} - \rho_{AB})$$
$$= -5.0 - 4.46E\theta_B - 8.92E\theta_C + 15.08E\rho_{DE} - 15.08E\rho_{AB}, \quad (11\text{--}60)$$

$$M_{CD} = F_{CD} - 4EK'\theta_C - 2EK'\theta_D + 2.25EK'(\rho_{AB} - \rho_{DE})$$
$$= +58.33 - 4E\theta_C - 2E\theta_D + 2.25E\rho_{AB} - 2.25E\rho_{DE}, \quad (11\text{--}61)$$

$$M_{DC} = F_{DC} - 2EK'\theta_C - 4EK'\theta_D + 2.25EK'(\rho_{AB} - \rho_{DE})$$
$$= -58.33 - 2E\theta_C - 4E\theta_D + 2.25E\rho_{AB} - 2.25E\rho_{DE}. \quad (11\text{--}62)$$

The necessary condition equations for evaluating the unknowns (θ_B, θ_C, θ_D, ρ_{AB}, and ρ_{DE}) are written as follows:

$$\Sigma M_B = M_{BA} + M_{BC} = 0,$$

from which

$$-15.25E\theta_B - 4.46E\theta_C - 8.75E\rho_{AB} + 15.08E\rho_{DE} + 5.0 = 0; \quad (11\text{--}63)$$

$$\Sigma M_C = M_{CB} + M_{CD} = 0,$$

which reduces to

$$-4.46E\theta_B - 12.92E\theta_C - 2E\theta_D - 12.83E\rho_{AB}$$
$$+ 12.83E\rho_{DE} - 5.0 = 0; \quad (11\text{-}64)$$

$$\Sigma M_D = M_{DC} + M_{DE} = 0,$$

which becomes

$$-2E\theta_C - 10.33E\theta_D + 2.25E\rho_{AB} + 4.08E\rho_{DE} - 58.33 = 0. \quad (11\text{-}65)$$

In Fig. 11–24, ΣM for $ABCD$ about the center of moments equals zero. Clockwise moments about the center of moments are considered to be positive.

$$\left(\frac{M_{BA}}{15}\right) 28.33 + \left(\frac{M_{CD} + M_{DC}}{31.63}\right) 42.17 - M_{DC}$$
$$+ 28 \times 20 - 10 \times 40 = 0,$$

which reduces to

$$-11.92E\theta_B - 6E\theta_C - 4E\theta_D + 15.67E\rho_{AB} - 3.75E\rho_{DE}$$
$$+ 218.2 = 0. \quad (11\text{-}66)$$

Finally,

$$\Sigma H = \frac{M_{BA} + M_{DE}}{15} = 0,$$

from which

$$E\theta_B + E\theta_D - E\rho_{AB} - E\rho_{DE} = 0. \quad (11\text{-}67)$$

A simultaneous solution of Eqs. (11–63) through (11–67) will result in the following:

$$E\theta_B = +2.13, \quad E\theta_C = +15.0, \quad E\theta_D = -10.1,$$
$$E\rho_{AB} = -8.98, \quad E\rho_{DE} = +1.02.$$

Substitution of the above values in the original expressions for end moments will result in

$$M_{BA} = -70.3 \text{ ft·k}, \quad M_{CB} = +3.0 \text{ ft·k}, \quad M_{DE} = +70.3 \text{ ft·k}.$$

11–6 Frames with several degrees of freedom. The slope-deflection method can be applied to frames with more than two degrees of freedom, but the solution of the necessary simultaneous equations usually involves so much labor that other methods are often preferred. For example, in the case of the three-story, two-bay building bent shown in Fig. 11–25

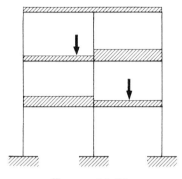

FIGURE 11–25

there would be twelve independent unknowns, nine joint rotations and three values of ρ, which would require the solution of twelve simultaneous equations for an analysis by slope-deflection. Although it would require four balancing operations, an analysis by moment distribution would necessitate the solution of only three simultaneous equations.

As another illustration consider the rigid frame of Fig. 11–26. An analysis

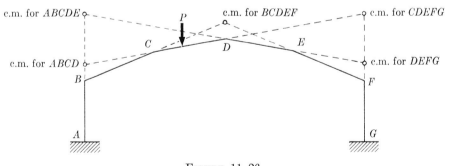

FIGURE 11–26

by slope-deflection would require the solution of eleven simultaneous equations. These equations would express that $\Sigma M = 0$ at joints B, C, D, E, and F; that $\Sigma H = 0$ for the entire structure; and that $\Sigma M = 0$ about the five moment centers shown in the sketch, for the parts of the structure indicated for each moment center. An analysis by moment distribution would require four AJR's, three balancing operations (if the frame is symmetrical), and the solution of four simultaneous equations. In this case the general method, using the conjugate structure to compute deflections, would be the easiest method to apply.

11-7 Secondary stresses. As previously stated, the slope-deflection method was developed by Otto Mohr as a solution for the secondary stress problem, and hence it is not surprising that it is particularly applicable here. The method is most easily explained with an illustrative example.

EXAMPLE 11-14. Determine the secondary moments for the truss of Fig. 11-27. All WF sections are oriented with their flanges in vertical planes. For the analysis, columns above and below the truss are assumed to be fixed at the far ends. $E = 30,000$ k/in^2.

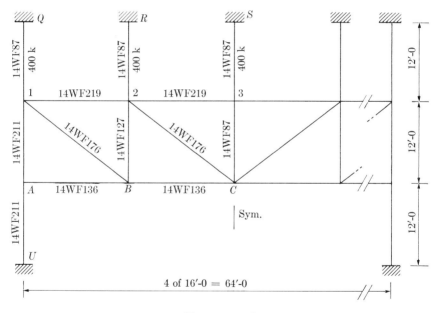

FIGURE 11-27

This structure was previously analyzed in Example 9-24; consequently, the axial strains, and the values of Δ resulting therefrom, are as determined in that example. The properties of the various members, the values of Δ as determined by the Williot diagram, and the fixed-end moments as computed for Example 9-24 are retabulated for convenience in Table 11-1.

TABLE 11–1.

Member	I (in^4)	Δ (in.)	L (in.)	Fixed-end moment = $\frac{6EI\Delta}{L^2}$ (ft·k)	$K = \frac{I}{L}$
1–2	1073	+0.765	192	+333	5.59
2–3	1073	+0.465	192	+204	5.59
A–B	568	+0.765	192	+177	2.95
B–C	568	+0.480	192	+111	2.95
1–A	1029	+0.310	144	+230	7.13
1–B	838	+0.735	240	+161	3.49
2–B	528	+0.230	144	+88	3.67
2–C	107	+0.385	240	+11	0.45
3–C	350	0	144	0	2.43
1–Q	350	−0.190	144	−48	2.43
2–R	350	−0.110	144	−28	2.43
3–S	350	0	144	0	2.43
A–U	1029	−0.120	144	−89	7.13

Since all members are prismatic and no loads are applied between joints, Eq. (11–4) becomes

$$M_N = \frac{-4EI\theta_N}{L} - \frac{2EI\theta_F}{L} + \frac{6EI\Delta}{L^2}$$

$$= -4EK\theta_N - 2EK\theta_F + \frac{6EI\Delta}{L^2}.$$

The above equation is used to write the expressions for the various final moments (secondary moments) which follow. (Note that, because of symmetry, joints 3 and C do not rotate.)

$$M_{1\text{-}Q} = -4 \times 2.43E\theta_1 - 48 = -9.72E\theta_1 - 48, \quad (11\text{–}68)$$

$$M_{1\text{-}2} = -4 \times 5.59E\theta_1 - 2 \times 5.59E\theta_2 + 333$$
$$= -22.36E\theta_1 - 11.18E\theta_2 + 333, \quad (11\text{–}69)$$

$$M_{1\text{-}B} = -4 \times 3.49E\theta_1 - 2 \times 3.49E\theta_B + 161$$
$$= -13.96E\theta_1 - 6.98E\theta_B + 161, \quad (11\text{–}70)$$

$$M_{1\text{-}A} = -4 \times 7.13E\theta_1 - 2 \times 7.13E\theta_A + 230$$
$$= -28.52E\theta_1 - 14.26E\theta_A + 230. \quad (11\text{–}71)$$

Since the sum of the final moments at joint 1 must be zero, then

$$M_{1\text{-}Q} + M_{1\text{-}2} + M_{1\text{-}B} + M_{1\text{-}A} = 0.$$

If Eqs. (11–68) through (11–71) are substituted, the resulting expression is

$$-74.56E\theta_1 - 11.18E\theta_2 - 14.26E\theta_A - 6.98E\theta_B + 676 = 0. \quad (11\text{–}72)$$

Similar equations are obtained by repeating the above process for joints 2, A, and B. The resulting four condition equations, which must be solved simultaneously for the four values of $E\theta$, follow:

$$-74.56E\theta_1 - 11.18E\theta_2 - 14.26E\theta_A - 6.98E\theta_B + 676 = 0, \quad (11\text{–}73)$$

$$-11.18E\theta_1 - 70.92E\theta_2 - 7.34E\theta_B + 608 = 0, \quad (11\text{–}74)$$

$$-14.26E\theta_1 - 68.84E\theta_A - 5.90E\theta_B + 318 = 0, \quad (11\text{–}75)$$

$$-6.98E\theta_1 - 7.34E\theta_2 - 5.90E\theta_A - 52.24E\theta_B + 537 = 0. \quad (11\text{–}76)$$

The above equations can be solved, of course, by the tabular method explained immediately following Example 5-14. Observe, however, that in each equation the coefficient of one of the unknowns is much larger than the coefficients of the other unknowns. Furthermore, the large coefficient applies to a different unknown in each equation. Consequently, these equations can be conveniently solved by iteration.

The procedure is to first obtain from each of the condition equations an approximate value for the unknown with the large coefficient. This is accomplished by neglecting the unknown terms with the small coefficients. The above four condition equations [(11–73) through (11–76)] then become

$$-74.56E\theta_1 + 676 = 0, \quad (11\text{–}73\text{a})$$

$$-70.92E\theta_2 + 608 = 0, \quad (11\text{–}74\text{a})$$

$$-68.84E\theta_A + 318 = 0, \quad (11\text{–}75\text{a})$$

$$-52.24E\theta_B + 537 = 0. \quad (11\text{–}76\text{a})$$

From the above equations the first approximate values for the unknowns are

$$E\theta_1 = +9.06, \qquad E\theta_2 = +8.59, \qquad E\theta_A = +4.63, \qquad E\theta_B = +10.3.$$

The next step is to substitute these first approximate values with the small coefficients into the original condition equations, and to solve again in each resulting equation for the revised and more exact value of the unknown with the large coefficient. When this is done in Eq. (11–73), the result is

$$-74.56E\theta_1 - 11.18(+8.59) - 14.26(+4.63) - 6.98(+10.3) + 676 = 0,$$

from which

$$E\theta_1 = +5.92.$$

Similarly, from the other three condition equations,

$$E\theta_2 = +5.53, \qquad E\theta_A = +1.87, \qquad E\theta_B = +7.88.$$

The above values are more nearly correct than the first approximate values. The process is repeated to obtain successive sets of values in which each set is more nearly correct than the last, until the increment between two consecutive sets is small enough to be neglected. In the present case, seven cycles were required to give the final values of

$$E\theta_1 = +6.82, \qquad E\theta_2 = +6.68, \qquad E\theta_A = +2.51, \qquad E\theta_B = +8.15.$$

These values are substituted in the expressions [as typified by Eqs. (11–68) through (11–71)] for the final (secondary) moments. For example,

$$M_{1-Q} = -9.72(+6.82) - 48 = -114 \text{ ft·k},$$

$$M_{1-2} = -22.36(+6.82) - 11.18(+6.68) + 333 = +106 \text{ ft·k},$$

$$M_{1-B} = -13.96(+6.82) - 6.98(+8.15) + 161 = +9 \text{ ft·k},$$

$$M_{1-A} = -28.52(+6.82) - 14.26(+2.51) + 230 = 0.$$

Comparison of these moments with the results of Example 9–24 will show that they agree almost exactly.

Specific References

1. Bendixen, A., *Die Methode der Alpha-Gleichungen zur Berechnung von Rahmenkonstruktionen*. Berlin: 1914.
2. Manderla, H., "Die Berechnung der Sekundärspannungen," *Allg. Bauztg.*, **45,** 34, 1880.
3. Maney, G. A., "Studies in Engineering, No. 1," University of Minnesota, 1915.
4. Mohr, O., *Zivilingineur*, **38,** 577, 1892.

General References

5. Amirikian, A., *Analysis of Rigid Frames*. Washington: Government Printing Office, 1942.
6. Andersen, P., *Statically Indeterminate Structures*. Chap. 5. New York: Ronald, 1953.
7. Johnson, J. B., Bryan, C. W., and Turneaure, F. E., *Modern Framed Structures*. 10th ed. Chap. 8. New York: Wiley, 1929.
8. Parcel, I. J., and Moorman, R. B. B., *Analysis of Statically Indeterminate Structures*. Chap. 5. New York: Wiley, 1955.
9. Wang, C. K., *Statically Indeterminate Structures*. Chap. 7. New York: McGraw-Hill, 1953.
10. Wilbur, J. B., and Norris, C. H., *Elementary Structural Analysis*. Chap. 14. New York: McGraw-Hill, 1948.
11. Williams, C. D., *Analysis of Statically Indeterminate Structures*. 2nd ed. Chap. 5. Scranton: International, 1946.

CHAPTER 12

INFLUENCE LINES

12-1 General. In 1867 the influence line was introduced by the German, E. Winkler. About twenty years later the important principle whereby influence lines for both determinate and indeterminate structures are quite easily determined was discovered by Professor Müller-Breslau.

It will be recalled that in 1886 Müller-Breslau published his improved version of the general method of Maxwell and Mohr. While developing this method, he became aware of the great value of Maxwell's theorem of reciprocal displacements, and also he discovered the principle now bearing his name. This principle is the basis for determining most influence lines for indeterminate structures, regardless of whether the method selected is mathematical or experimental.

12-2 The Müller-Breslau principle. This important principle may be stated as follows: *If an internal stress component, or a reaction component, is considered to act through some small distance and thereby to deflect or displace a structure, the curve of the deflected or displaced structure will be, to some scale, the influence line for the stress or reaction component.* This principle applies to beams, continuous frames, articulated structures, and to determinate as well as indeterminate structures. For indeterminate structures, however, it is limited to those for which the principle of superposition is valid. The significance of this statement will be made clear by several demonstrations of its validity.

Consider, as a beginning, that it is desired to compute the influence line for the horizontal reaction component at A for the frame shown in Fig. 12-1. Vertical loads may be applied to BC and horizontal loads to AB.

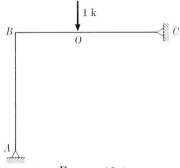

FIGURE 12-1

Assume that the influence line ordinate is to be computed for a 1 k load acting down at point O, where O may be any point in the span BC. If the value of the influence line ordinate (the horizontal reaction component at A) is to be computed by the general method, the procedure is to place rollers at A. The deflected structure will be as shown in Fig. 12–2, and the horizontal deflection of A is computed by any method desired.

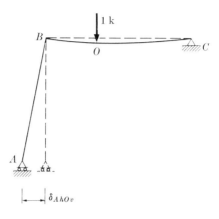

FIGURE 12–2

A unit horizontal force of 1 k is next applied at A, and the structure deflects as shown in Fig. 12–3. The resulting deflection of A is then computed.

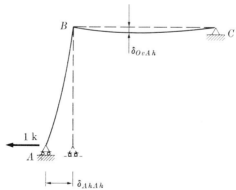

FIGURE 12–3

By the general method,

$$H_A = \frac{\delta_{AhOv}}{\delta_{AhAh}}.$$

But, by Maxwell's reciprocal deflection theorem,

$$\delta_{AhOv} = \delta_{OvAh}.$$

Therefore,

$$H_A = \frac{\delta_{OvAh}}{\delta_{AhAh}}. \tag{12-1}$$

From Eq. (12-1) it is apparent that the deflection curve of the member BC is, to the scale that δ_{AhAh} represents 1 k, the influence line for H_A for a vertical load on BC.

By a similar procedure, using a 1 k horizontal load acting at any point O on AB instead of the 1 k vertical load on BC, we can show that the deflected member AB in Fig. 12-3 is, to the same scale as above, the influence line for H_A for a horizontal load acting on AB.

As an additional demonstration, consider that an influence line for moment is desired for any point E between the supports of the continuous beam of Fig. 12-4. In accordance with the Müller-Breslau principle, the

FIGURE 12-4

internal stress component for which the influence line is desired is removed from the beam. In other words, the capability of the beam to resist moment at section E is removed. This is assumed to be accomplished by inserting a pin at E. Thus the shearing stress component and the thrust component (although this does not exist in the present case) are not removed. A unit load is applied at any point D along the beam, with the beam deflecting as shown in Fig. 12-5. The 1 k load is then removed

FIGURE 12-5

and a pair of unit couples are applied to the beam, one couple acting on each side of the pin. From this action the deflected beam of Fig. 12-6 will result.

FIGURE 12-6

Then, as before:

$$M_E = \frac{\alpha'_{ED}}{\alpha_{EE}} = \frac{\delta'_{DE}}{\alpha_{EE}}, \qquad (12\text{-}2)$$

and once again it is apparent that the curve of the deflected beam is, to some scale, the influence line for M_E.

As an additional example, consider that it is required to find an influence line for shear at point E of the beam of Fig. 12-4. In this case it must be assumed that the beam is cut at E and that a slide device is inserted which will permit a relative transverse deflection between the two beam ends at the cut, but which, at the same time, will always require these ends to have a common slope. In other words, the shearing resistance of the beam has been removed at E but the flexural resistance has not.

In Fig. 12-7 a 1 k load is applied at D, resulting in the relative linear

FIGURE 12-7

deflection δ_{ED} at E. With this load removed, a pair of 1 k loads are applied at E, and the beam deflects as shown in Fig. 12-8. As before, the shear at E is given by

$$S_E = \frac{\delta_{ED}}{\delta_{EE}} = \frac{\delta_{DE}}{\delta_{EE}}. \qquad (12\text{-}3)$$

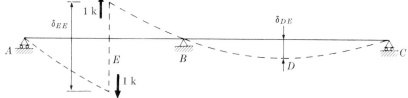

FIGURE 12-8

Obviously, then, the curve of the deflected beam of Fig. 12-8 is, to the scale that δ_{EE} represents 1 k, the influence line for the shear at E.

Consider now the two-hinged highway bridge arch of Fig. 12-9. In

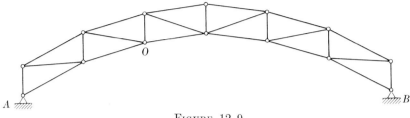

FIGURE 12-9

this case an influence line for H_A is desired. In order to compute the magnitude of H_A by the general method, assume that rollers are placed at A, and also that a 1 k load is applied at panel point O, where O might be any panel point along the arch. The left end of the arch will then move to the left, as shown in Fig. 12-10. Next, the 1 k vertical load having been

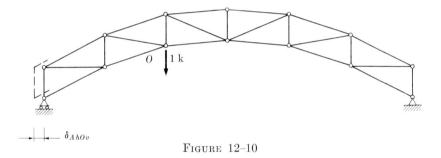

FIGURE 12-10

removed from panel point O, a 1 k horizontal force is caused to act to the left at A. The arch will then deflect as shown by the dashed lines in Fig. 12-11.

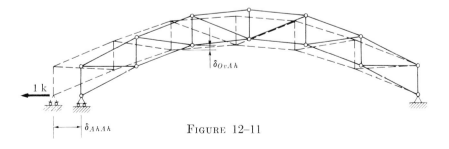

FIGURE 12-11

Again, by the general method and the reciprocal deflection theorem,

$$H_A = \frac{\delta_{AhOv}}{\delta_{AhAh}} = \frac{\delta_{OvAh}}{\delta_{AhAh}}. \qquad (12\text{-}4)$$

The above equation indicates once again that the vertical components of the deflections of the various panel points are, to the scale that δ_{AhAh} represents 1 k, ordinates for the influence line for H_A.

As a final example, consider the two-hinged arch for a highway bridge as shown in Fig. 12–12. An influence line is required for the stress in

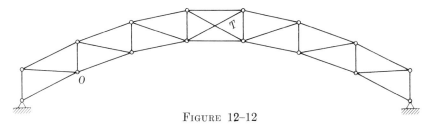

FIGURE 12–12

the redundant T. In order to find this stress T, as caused by a unit vertical load acting at any bottom chord panel point O, the redundant member is cut and a spread at the cut results, as shown in Fig. 12–13. After the value

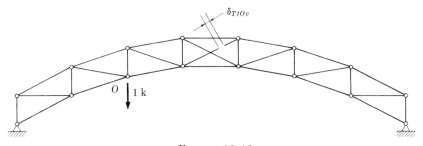

FIGURE 12–13

of δ_{TtOv} is determined, the 1 k vertical load is removed from O and a pair of 1 k forces are applied to the two cut ends of the redundant T. The arch will then deflect as shown in Fig. 12–14. As before,

$$T = \frac{\delta_{TtOv}}{\delta_{TtTt}} = \frac{\delta_{OvTt}}{\delta_{TtTt}}. \qquad (12\text{-}5)$$

The vertical components of the panel point deflections of Fig. 12–14 are thus seen to be, to the scale that δ_{TtTt} represents 1 k, the ordinates for the influence line for the redundant T.

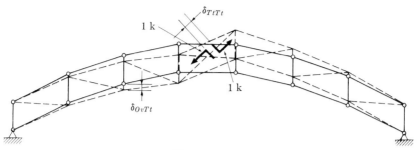

FIGURE 12-14

The preceding five examples have demonstrated that the Müller-Breslau principle applies to continuous beams and articulated structures. Actually, of course, it will apply to any structure for which the reciprocal deflection theorem is valid.

The Müller-Breslau principle suggests two different methods by which influence lines for redundant stresses and reactions may be obtained. The first of these is the experimental method requiring the use of models. The second method is, of course, by computation.

According to the Müller-Breslau principle, the procedure for obtaining the influence line (regardless of whether it be in fact, as with a model, or imaginary, as when computing the ordinates for the influence line) is to remove the redundant and to introduce a displacement in the structure at the point of action and in the direction of the redundant. When the influence line is being determined experimentally by means of a model, the magnitude of this impressed displacement is arbitrarily selected for convenience. The magnitude of the action causing the displacement is of no importance. When, however, the influence line ordinates are being computed, it is convenient to apply a unit action.

12–3 Influence lines for continuous beams with prismatic members. Probably the easiest method for computing influence line ordinates for continuous beams is by the conjugate beam. This method will be demonstrated with the several examples which follow.

EXAMPLE 12–1. Compute the ordinates, at intervals of 2.5 ft, of the influence line for R_A for the beam of Fig. 12–15. The moment of inertia is constant.

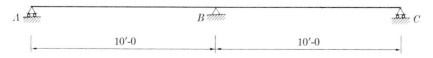

FIGURE 12-15

The redundant reaction at A is removed and a 1 k vertical force is applied instead, a force which may act either up or down. The beam will deflect as indicated by the dashed line in Fig. 12–16. The conjugate beam

FIGURE 12–16

for computing the deflections at the various sections of this beam is shown in Fig. 12–17. The ordinates of the elastic load on this conjugate beam, at

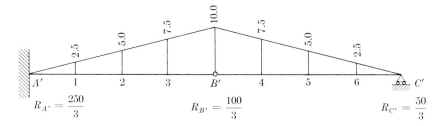

FIGURE 12–17

intervals of 2.5 ft, are shown in ft·k. (E and I are omitted for convenience.) The computations for the moments at the various sections are as follows:

$$M_6 = \frac{50}{3} \times 2.5 - 2.5 \times \frac{2.5}{2} \times \frac{2.5}{3} = 39.04,$$

$$M_5 = \frac{50}{3} \times 5 - 5 \times \frac{5}{2} \times \frac{5}{3} = 62.47,$$

$$M_4 = \frac{50}{3} \times 7.5 - 7.5 \times \frac{7.5}{2} \times \frac{7.5}{3} = 54.65,$$

$$M_B = 0,$$

$$M_3 = \frac{100}{3} \times 2.5 + 7.5 \times 2.5 \times \frac{2.5}{2} + 2.5 \times \frac{2.5}{2} \times 2 \times \frac{2.5}{3} = 111.99,$$

$$M_2 = \frac{100}{3} \times 5 + 5 \times 5 \times \frac{5}{2} + 5 \times \frac{5}{2} \times 2 \times \frac{5}{3} = 270.83,$$

$$M_1 = \frac{100}{3} \times 7.5 + 2.5 \times 7.5 \times \frac{7.5}{2} + 7.5 \times \frac{7.5}{2} \times 2 \times \frac{7.5}{3} = 461.09,$$

$$M_A = \frac{100}{3} \times 10 + 10 \times \frac{10}{2} \times 10 \times \frac{2}{3} = 666.66.$$

The moments computed above are proportional to the deflections at the corresponding sections of the beam of Fig. 12–16. Since a load of 1 k acting at A on the beam of Fig. 12–15 will cause a reaction at A of 1 k, the moment $M_A = 666.66$ represents an ordinate of unity. The deflection curve of the beam is, to this scale, the desired influence line. Therefore the influence line ordinates for other sections along the beam are obtained by dividing the conjugate beam moments by $M_A = 666.66$. The resulting influence line is shown in Fig. 12–18.

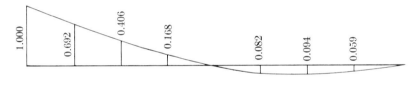

FIGURE 12–18

EXAMPLE 12–2. Compute the ordinates, at intervals of 2.5 ft, of the influence line for moment at the midpoint of span BC for the beam shown in Fig. 12–19. The moment of inertia is constant.

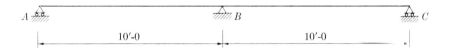

FIGURE 12–19

The capability of the beam to resist moment at the section for which the moment influence line is desired is removed by inserting a pin. Unit couples are applied to the beam on each side of the pin. The modified beam, which will deflect as indicated by the dashed line, is shown in

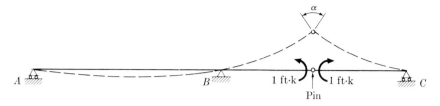

FIGURE 12–20

Fig. 12–20. The various influence line ordinates are computed, as indicated in Eq. (12–2), by evaluating α as well as the vertical deflections at the necessary sections of the modified beam of Fig. 12–20. The loaded conjugate beam is shown in Fig. 12–21, with the magnitudes of the ordinates at the required sections in ft·k.

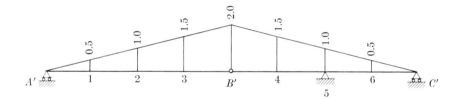

FIGURE 12–21

The shear on the pin at B' is

$$V_{B'} = \tfrac{2}{3} \times \tfrac{10}{2} \times 2 = 6.67.$$

The sum of the shears in the conjugate beam to the right and left of the support at 5 will be the reaction at this point of the conjugate beam, and it will also be the relative value of the angle α in Fig. 12–20. This is computed as

$$R_5 = 6.67 \times \frac{10}{5} + 2 \times \frac{10}{2} \times \frac{6.67}{5} = 26.67 \text{ up}.$$

The other two reactions are

$$R_{A'} = \tfrac{1}{3} \times \tfrac{10}{2} \times 2 = 33.3 \text{ up},$$

$$R_{C'} = 20 - 3.33 - 26.67 = 10.00 \text{ down}.$$

The various moments are computed as follows:

12-3] CONTINUOUS BEAMS WITH PRISMATIC MEMBERS 517

$$M_1 = 3.33 \times 2.5 - 0.5 \times \frac{2.5}{2} \times \frac{2.5}{3} = +7.81,$$

$$M_2 = 3.33 \times 5 - 1 \times \frac{5}{2} \times \frac{5}{3} = +12.48,$$

$$M_3 = 3.33 \times 7.5 - 1.5 \times \frac{7.5}{2} \times \frac{7.5}{3} = +10.94,$$

$$M_{B'} = 0,$$

$$M_4 = -6.67 \times 2.5 - 0.5 \times \frac{2.5}{2} \times \frac{2}{3} \times 2.5 - 1.5 \times 2.5 \times \frac{2.5}{2}$$
$$= -22.42,$$

$$M_5 = -6.67 \times 5.0 - 1 \times \frac{5}{2} \times \frac{2}{3} \times 5 - 1 \times 5 \times 2.5 = -54.18,$$

$$M_6 = -10 \times 2.5 - 0.5 \times \frac{2.5}{2} \times \frac{2.5}{3} = -25.52.$$

The value of the influence line ordinate at each of the above sections is computed by dividing each moment by relative $\alpha = 26.67$. The resulting influence line is shown in Fig. 12–22.

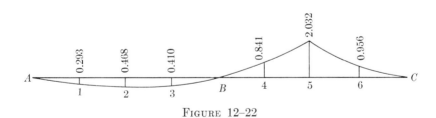

FIGURE 12–22

EXAMPLE 12–3. Compute the ordinates, at intervals of 2.5 ft, of the influence line for shear at the midpoint of span BC for the beam shown in Fig. 12–23. The moment of inertia is constant.

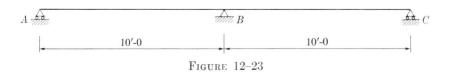

FIGURE 12–23

The shearing resistance of the beam at the midpoint of span BC must be removed. Moment resistance, however, must not be impaired. This can be accomplished by cutting the beam at this section and inserting a slide device which will permit a relative vertical displacement, between the two beam ends adjacent to the cut, but which will require that these two ends be tangent to a common slope in the distorted beam. When such a device is inserted at the midpoint of span BC, and opposing vertical forces of 1 k are applied to the two beam ends adjacent to the cut, the deflected beam and the induced reactions and couples will be as shown in Fig. 12–24.

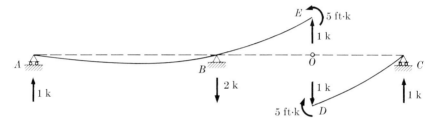

FIGURE 12–24

The loaded conjugate beam is shown in Fig. 12–25, and in this figure

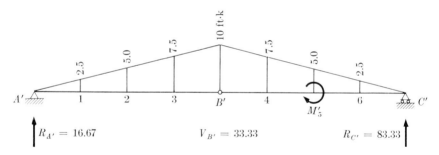

FIGURE 12–25

attention is called to the moment M'_5 at section 5. In Fig. 12–24, points D and E deflect relative to each other, D having an absolute deflection downward and E an absolute deflection upward. In the conjugate beam, therefore, the moment just to the right of the corresponding section must be different in sign, and probably different in magnitude, from the moment just to the left of the section. In addition, since the tangent to the deflected beam at D must be parallel to the tangent to the deflected beam at E, the shear just to the right of the corresponding section in the conjugate beam must be equal in magnitude to the shear just to the left. The action of the moment M'_5 will satisfy these requirements.

The values of the conjugate beam reactions $R_{A'}$ and $R_{C'}$, the shear $V_{B'}$ on the pin at B', and the moment M'_5 are computed below:

$$R_{A'} = \tfrac{1}{3} \times 10 \times \tfrac{10}{2} = 16.67 \text{ up},$$

$$R_{C'} = 10 \times \tfrac{20}{2} - 16.67 = 83.33 \text{ up},$$

$$V_{B'} = \tfrac{2}{3} \times 10 \times \tfrac{10}{2} = 33.33,$$

$$M'_5 = 33.33 \times 10 + 10 \times \tfrac{10}{2} \times \tfrac{2}{3} \times 10 = 666.6.$$

The correct sense for M'_5 is determined by inspection and indicated in Fig. 12-25. The relative deflection between points D and E in Fig. 12-24 is represented by M'_5.

The moments in the conjugate beam at intermediate sections are computed as follows:

$$M_1 = 16.67 \times 2.5 - 2.5 \times \frac{2.5}{2} \times \frac{2.5}{3} = +39.1,$$

$$M_2 = 16.67 \times 5.0 - 5 \times \frac{5}{2} \times \frac{5}{3} = +62.5,$$

$$M_3 = 16.67 \times 7.5 - 7.5 \times \frac{7.5}{2} \times \frac{7.5}{3} = +54.7,$$

$$M_4 = -33.33 \times 2.5 - 7.5 \times 2.5 \times \frac{2.5}{2} - 2.5 \times \frac{2.5}{2} \times \frac{5}{3}$$
$$= -111.9,$$

$$M_6 = 83.33 \times 2.5 - 2.5 \times \frac{2.5}{2} \times \frac{2.5}{3} = +205.7.$$

The moments in the conjugate beam just to the right and just to the left of section 5 are

$$M_{5R} = 83.33 \times 5 - 5 \times \tfrac{5}{2} \times \tfrac{5}{3} = 395.8,$$

$$M_{5L} = -666.6 + 395.8 = 270.8.$$

Each of the above moments must be divided by the relative deflection between D and E, as represented by M'_5, in order to obtain the required ordinates. These, when plotted, make it possible to draw the influence line as shown in Fig. 12-26.

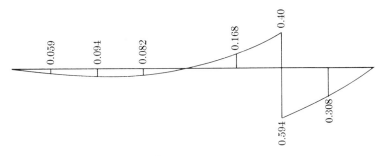

FIGURE 12-26

The foregoing illustrative problems demonstrate that any desired type of influence line for a continuous beam can be readily computed by use of the Müller-Breslau principle and the conjugate beam. Actually, when an influence line for shear is desired, it is usually easier to compute the required ordinates by statics after the influence lines for reactions have been computed. This probably will also hold true for moment influence lines.

Problems

12-4. Compute the ordinates, at intervals of 2.5 ft, of the influence line for the moment at A in Fig. 12-27. The moment of inertia is constant. [*Ans.* (in ft·k): $\text{Ord}_1 = 1.64$, $\text{Ord}_2 = 1.87$, $\text{Ord}_3 = 1.17$.]

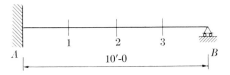

FIGURE 12-27

12-5. Compute the ordinates of the influence line for the reaction at C in Fig. 12-28. Use intervals of 2.5 ft in span AB and 4 ft in span BC. The moment of inertia is constant. [*Ans.*: $\text{Ord}_1 = -0.028$, $\text{Ord}_2 = -0.045$, $\text{Ord}_3 = -0.039$, $\text{Ord}_4 = +0.149$, $\text{Ord}_5 = +0.386$, $\text{Ord}_6 = +0.680$.]

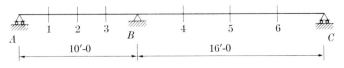

FIGURE 12-28

12-6. Compute ordinates of the influence line for the reaction at C in Fig. 12-29. The moment of inertia is constant. Compute ordinates at the centers of spans AB and BC, and at 10-ft intervals in span CD. [*Ans.:* $\text{Ord}_1 = -0.30$, $\text{Ord}_2 = +0.50$, $\text{Ord}_C = +1.00$, $\text{Ord}_3 = +1.31$, $\text{Ord}_4 = +0.85$.]

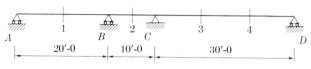

FIGURE 12-29

12-7. Compute ordinates of the influence line for the moment at B in Fig. 12-30. Use intervals of 2.5 ft for span AB and 4 ft for span BC. The moment of inertia is constant. [*Ans.* (in ft·k): $\text{Ord}_1 = +0.45$, $\text{Ord}_2 = +0.72$, $\text{Ord}_3 = +0.63$, $\text{Ord}_4 = +1.62$, $\text{Ord}_5 = +1.85$, $\text{Ord}_6 = +1.15$.]

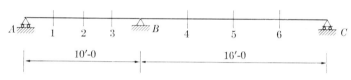

FIGURE 12-30

12-8. Compute ordinates of the influence line for the moment at B in Fig. 12-31. Use intervals of 2.5 ft. The moment of inertia is constant. [*Ans.* (in ft·k): $\text{Ord}_1 = +0.59$, $\text{Ord}_2 = +0.94$, $\text{Ord}_3 = +0.82$, $\text{Ord}_4 = +0.82$, $\text{Ord}_5 = +0.94$, $\text{Ord}_6 = +0.59$.]

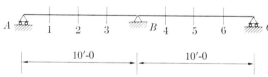

FIGURE 12-31

12-9. In Fig. 12-32 compute the ordinates, at the center of each span, of the influence line for the moment at A. The moment of inertia is constant. Use moment distribution to compute the loading for the conjugate beam. [*Ans.* (in ft·k): $\text{Ord}_1 = +6.35$, $\text{Ord}_2 = -1.76$, $\text{Ord}_3 = +0.57$.]

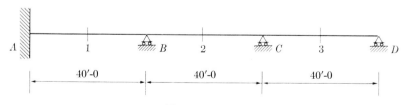

FIGURE 12-32

522   INFLUENCE LINES   [CHAP. 12

12–4 Continuous beams with nonprismatic members. Influence lines for continuous beams composed of nonprismatic members are, of course, computed in the same manner as previously demonstrated for prismatic members. The variation in the moment of inertia, however, must be taken into account, and this is most conveniently accomplished by the use of segments. The example which follows will illustrate the procedure.

EXAMPLE 12–10. The elevation of a two-span continuous steel girder is shown in Fig. 12–33. The bottom flange is parabolic. The web is $\frac{3}{4}$ in. thick and flange plates are 16 in. $\times$ 1 in. Out to out of flange plates is 36 in. at the ends and 72 in. over the center support. The moments of inertia at the centers of the various 5-ft segments are given in Table 12–1. Determine the influence line for the reaction at A, computing influence line ordinates at the centers of the 5-ft segments.

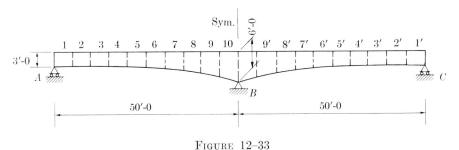

FIGURE 12–33

The support at A is removed and a 1 k force, acting up, is applied at this point, with the loaded simplified beam as shown in Fig. 12–34. It is neces-

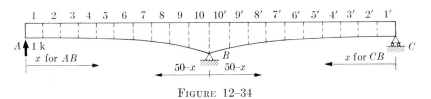

FIGURE 12–34

sary to compute the deflections at the centers of the various segments. The loaded conjugate beam is shown in Fig. 12–35, and the necessary computations are given in Table 12–2.

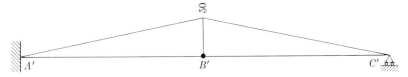

FIGURE 12–35

TABLE 12–1.

Segment	I (in^4)
1	12,300
2	13,000
3	14,000
4	15,900
5	18,700
6	22,600
7	27,600
8	34,300
9	43,300
10	54,700

TABLE 12–2.

(1) Segment number	(2) $M = x$	(3) Relative I	(4) $\dfrac{M\,\Delta S}{\text{Relative } I}$	(5) Shear $B'A'$	(6) Moment increment	(7) Moment at segment	(8) $50 - x$	(9) $(50 - x)R_{B'}$
10–10'	47.5	100.0	2.37	2.37	11.85	—	2.5	35
9–9'	42.5	79.1	2.68	5.05	25.25	11.85	7.5	106
8–8'	37.5	62.6	2.99	8.04	40.20	37.10	12.5	177
7–7'	32.5	50.5	3.22	11.26	56.30	77.30	17.5	247
6–6'	27.5	41.3	3.33	14.59	72.95	133.60	22.5	317
5–5'	22.5	34.2	3.29	17.88	89.40	206.55	27.5	388
4–4'	17.5	29.1	3.00	20.88	104.40	295.95	32.5	458
3–3'	12.5	25.6	2.44	23.32	116.60	400.35	37.5	529
2–2'	7.5	23.8	1.58	24.90	124.50	516.95	42.5	600
1–1'	2.5	22.4	0.56	25.46	63.60	641.45	47.5	670
A						705.05	50.0	705

Segment	(10) Moment at segment $+(50 - x)R_{B'}$	(11) Influence ordinate	Segment	(12) Moment at segment $-(50 - x)R_{B'}$	(13) Influence ordinate
10	+35	+0.025	10'	−35	−0.025
9	+118	+0.084	9'	−94	−0.067
8	+214	+0.152	8'	−140	−0.099
7	+324	+0.230	7'	−170	−0.121
6	+451	+0.320	6'	−183	−0.130
5	+594	+0.420	5'	−181	−0.129
4	+754	+0.535	4'	−162	−0.115
3	+929	+0.059	3'	−129	−0.091
2	+1117	+0.791	2'	−83	−0.059
1	+1311	+0.930	1'	−29	−0.020
A	+1410	+1.000			0.000

$$\text{Shear at } B' = \frac{705.05}{50} = 14.10.$$

In the above table, columns (2), (3), and (4) are used for computing the magnitudes of the elastic loads. The total elastic flexural strain for each segment is considered to act as a concentrated load on the conjugate beam at the center of that segment. Note that the values shown in column (5) are the shears in span $B'A'$ resulting from the elastic loads acting directly on span $B'A'$. The moments about the centers of the segments between B' and A', or between B' and C'', as caused by the concentrations acting at the centers of the segments (excluding the moment resulting from the shear at B'), are computed in columns (5), (6), and (7). The moments at the centers of the segments, resulting from the action of the shear at B', are shown in column (9). The total moments at the centers of segments 10 to 1 are obtained by adding the values of columns (7) and (9). The results of this addition are shown in column (10). The total moments at the centers of segments $10'$ to $1'$ are obtained by subtracting the values in (9) from (7), and are shown in column (12). The final influence line ordinates will result when the values in (10) and (12) are divided by the relative deflection of A in Fig. 12–34. This is found to be 1410, by adding the values in columns (7) and (9) for point A. The final ordinates are shown in (11) and (13). Note that ΔS could have been omitted in computing the values of the relative elastic weights in column (4) since segment length is constant.

Problems

12–11. Compute the ordinates of the influence line for moment at the center of span AB of the two-span continuous girder of the preceding example (Fig. 12–33). Compute ordinates at the centers of 5-ft segments and at center of AB.

	Segment number	Influence ordinate	Segment number	Influence ordinate
	2	+2.29 ft·k	2'	−1.46 ft·k
	4	+5.86 ft·k	4'	−2.88 ft·k
[Ans.:	℄ AB	+9.25 ft·k	6'	−3.26 ft·k
	7	+5.75 ft·k	8'	−2.46 ft·k
	9	+2.09 ft·k	10'	−0.62 ft·k

12–12. Assume that the support at A of the two-span continuous girder of Example 12–10 (see Fig. 12–33) is fixed. Compute the ordinates of the influence line for the moment at A at the centers of the 5-ft segments.

	Segment number	Influence ordinate	Segment number	Influence ordinate
	2	−5.55 ft·k	2'	+1.26 ft·k
[Ans.:	4	−7.35 ft·k	4'	+2.46 ft·k
	6	−5.78 ft·k	6'	+2.78 ft·k
	8	−3.12 ft·k	8'	+2.11 ft·k

12–13. Compute the ordinates of the influence line for the reaction at B of the three-span continuous bridge girder of Fig. 12–36. Compute ordinates at the centers of the 10-ft segments. The relative values of the moments of inertia at the centers of the segments are as follows:

Segment	Relative I
1, 10, 10', 1'	25.8
2, 9, 9', 2'	30.6
3, 8, 8', 3'	42.1
4, 7, 7', 4'	63.6
5, 6, 6', 5'	100.0

[Ans.:]

Segment number	Influence ordinate	Segment number	Influence ordinate
2	+0.425 k	2'	−0.089 k
4	+0.833 k	4'	−0.095 k
B	+1.000 k	C	0.000 k
7	+1.081 k	7'	+0.151 k
9	+0.990 k	9'	+0.474 k

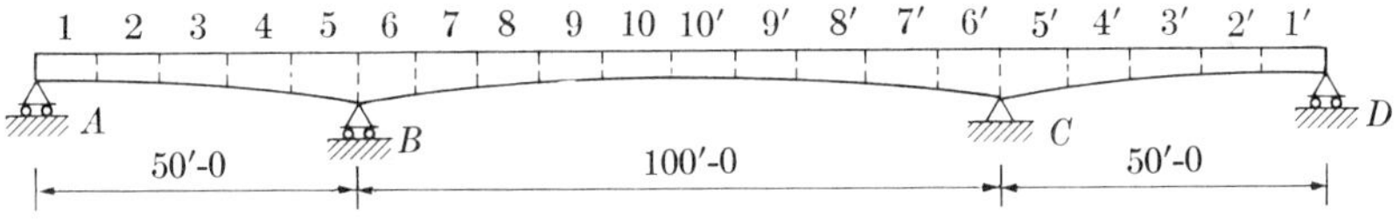

FIGURE 12-36

Suggested method of solution. Remove the support at B in Fig. 12-36 and apply a 1 k vertical force. The result is shown in Fig. 12-37, and this structure is

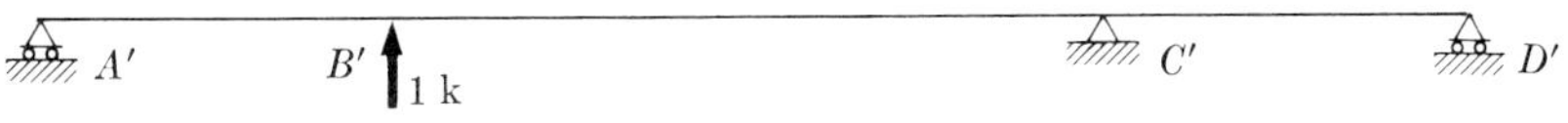

FIGURE 12-37

indeterminate. Remove the support at D' and apply an unknown upward force R which replaces the reaction. Figure 12-38 will result, and the loaded

FIGURE 12-38

conjugate beam will be as shown in Fig. 12-39. Since the girder at D' in Fig.

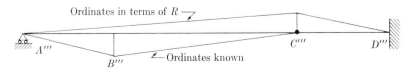

FIGURE 12-39

12-37 cannot deflect, the moment at D''' in the conjugate beam must be zero. This condition equation permits R to be determined. The value of R is used to obtain the magnitudes of the moment ordinates on the conjugate beam which were at first expressed in terms of R. Finally, the moments in the conjugate beam at the centers of the various segments are computed.

12-14. Compute the ordinates of the influence line for the moment at a section Q, which is 30 ft to the right of B, in the continuous girder of Problem 12-13 (see Fig. 12-36). Compute ordinates at the centers of the 10-ft segments and at the section Q.

	Segment number	Influence ordinate	Segment number	Influence ordinate
	2	−1.99 ft·k		
	4	−2.12 ft·k	4′	+0.14 ft·k
[Ans.:	7	+3.48 ft·k	7′	−0.28 ft·k
	Q	+9.42 ft·k	9′	−0.07 ft·k
	10	+3.15 ft·k	10′	+0.95 ft·k

12-5 Influence lines by moment distribution. The preceding sections have demonstrated how influence lines for continuous beams can be obtained by use of the conjugate beam. If the members of the continuous beam are prismatic, or if they are standard nonprismatic types for which the stiffness, carry-over factors, and fixed-end moments may be obtained from tables or curves, the various influence lines can be obtained more easily by *moment distribution*. If, however, this information is not available for the nonprismatic members, then the method previously demonstrated is to be preferred.

The method of moment distribution has one important additional advantage: it can be applied to more complicated structures, such as continuous bridge decks supported on, and continuous with, tall piers. Two examples are now presented to demonstrate the method.

EXAMPLE 12-15. Using the method of moment distribution, compute the ordinates at 5-ft intervals of the influence line for M_{BA} in Fig. 12-40.

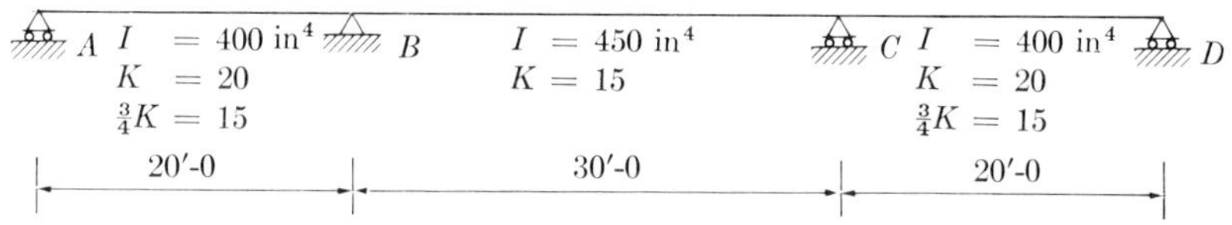

$A \quad I = 400 \text{ in}^4 \quad B \quad I = 450 \text{ in}^4 \quad C \quad I = 400 \text{ in}^4 \quad D$
$\quad K = 20 \quad\quad\quad K = 15 \quad\quad\quad K = 20$
$\tfrac{3}{4}K = 15 \quad\quad\quad\quad\quad\quad\quad\quad \tfrac{3}{4}K = 15$

$|\text{———} 20'\text{-}0 \text{———}|\text{———} 30'\text{-}0 \text{———}|\text{———} 20'\text{-}0 \text{———}|$

FIGURE 12-40

The method of solution is first to assume that a fixed-end moment of $+100$ ft·k exists in end B of the member BA. No other fixed-end moments are considered to exist. This 100 ft·k moment is balanced out in the usual way, and the moments obtained from the balancing operation are the member end moments which will result when a fixed-end moment of $+100$ ft·k is introduced at end B of member BA. A fixed-end moment of $+100$ ft·k is next introduced at end B of member BC. This, again, is considered to be the only fixed-end moment in the structure. The final moments are again found by moment distribution and, in the same way, the effects of $+100$ ft·k at C of CB and at C of CD are determined.

The balancing operations are shown in Table 12–3. Note that only one complete balance is necessary because of symmetry.

TABLE 12–3.

Joint	A	B		C		D
Member	AB	BA	BC	CB	CD	DC
K		15	15	15	15	
Distribution factor		0.5	0.5	0.5	0.5	
$F_{BA} = +100$ ft·k		$+100.0$ -50.0	-50.0	-25		
			$+6.2$	$+12.5$	$+12.5$	
		-3.1	-3.1	-1.5		
			$+0.3$	$+0.7$	$+0.7$	
		-0.1	-0.1			
		$+46.8$	-46.8	-13.3	$+13.2$	
$F_{BC} = +100$ ft·k			$+100.0$			
		-53.2	$+53.2$	-13.3	$+13.2$	
$F_{CB} = +100$ ft·k				$+100.0$		
		$+13.2$	-13.3	$+53.2$	-53.2	
$F_{CD} = +100$ ft·k					$+100.0$	
		$+13.2$	-13.3	-46.8	$+46.8$	

Having found the final moments resulting from the introduction of $+100$ ft·k at every point where fixed-end moments can exist in the structure, we can now write the equation for M_{BA} (and all other member end moments, if desired) in terms of the initial fixed-end moments:

$$M_{BA} = +0.468 F_{BA} - 0.532 F_{BC} + 0.132 F_{CB} + 0.132 F_{CD}.$$

The fixed-end moments, caused by a 1 k load placed successively at each of the points for which an influence line ordinate is desired, are computed in Table 12–4.

TABLE 12–4.

Span AB

$x = a$ (ft)	$b = 20 - a$ (ft)	a^2	b^2	L^2	$F_{AB} = \dfrac{ab^2}{L^2}$	$F_{BA} = \dfrac{a^2 b}{L^2}$	$C.O. = \dfrac{F_{AB}}{2}$	F_{BA}
5	15	25	225	400	+2.81 ft·k	−0.93 ft·k	−1.40 ft·k	−2.33 ft·k
10	10	100	100		+2.50 ft·k	−2.50 ft·k	−1.25 ft·k	−3.75 ft·k
15	5	225	25		+0.93 ft·k	−2.81 ft·k	−0.46 ft·k	−3.27 ft·k

Span BC

$x = a$ (ft)	$b = 30 - a$ (ft)	a^2	b^2	$F_{BC} = \dfrac{ab^2}{L^2}$	$F_{CB} = \dfrac{a^2 b}{L^2}$
5	25	25	625	+3.47 ft·k	−0.69 ft·k
10	20	100	400	+4.45 ft·k	−2.23 ft·k
15	15	225	225	+3.75 ft·k	−3.75 ft·k

The above values of the fixed-end moments are substituted in the equation above for M_{BA}, and the influence line ordinates are computed in Table 12–5.

TABLE 12–5.

Point	F_{BA}	F_{FC}	F_{CB}	F_{CD}	$+0.468F_{BA}$	$-0.532F_{BC}$	$+0.132F_{CB}$	$+0.132F_{CD}$	$=M_{BA}$ (ft·k)
A									
5	−2.33				−1.09				−1.09
10	−3.75				−1.75				−1.75
15	−3.27				−1.53				−1.53
B									
25		+3.47	−0.69			−1.85	−0.09		0.00 −1.94
30		+4.45	−2.23			−2.36	−0.29		−2.65
35		+3.75	−3.75			−2.00	−0.50		−2.50
40		+2.23	−4.45			−1.19	−0.59		−1.78
45		+0.69	−3.47			−0.37	−0.46		−0.83
C									
55				+3.27				+0.43	0.00 +0.43
60				+3.75				+0.49	+0.49
65				+2.33				+0.31	+0.31
D									

Problems

12-16. For Fig. 12-41 compute the ordinates, at 10-ft intervals, of the influence line for M_{BA}. The moment of inertia is constant. [*Ans.* (in ft·k): Ord$_1$ = -1.84, Ord$_2$ = -2.30, Ord$_3$ = -3.95, Ord$_4$ = -4.83, Ord$_5$ = -3.74, Ord$_6$ = -1.77, Ord$_7$ = $+0.71$, Ord$_8$ = $+0.57$.]

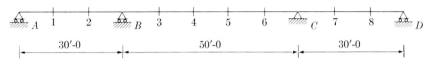

FIGURE 12-41

12-17. The first assumptions as to the proportions for a continuous reinforced concrete bridge are indicated in Fig. 12-42. Consider a strip 1-ft wide normal to the plane of the paper. Using moment distribution, write the equations for M_{BA} and M_{BC} in terms of the fixed-end moments. Compute influence line ordinates for M_{BA} at 5-ft intervals across the entire span.

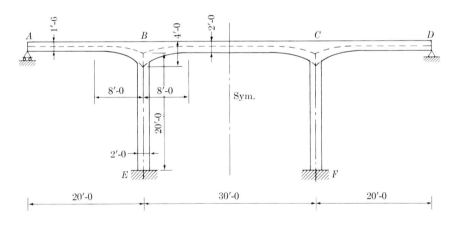

FIGURE 12-42

[*Ans.*:

$$M_{BA} = +0.760 F_{BA} - 0.240 F_{BC} + 0.066 F_{CB} + 0.066 F_{CD},$$

$$M_{BC} = -0.385 F_{BA} + 0.615 F_{BC} - 0.170 F_{CB} - 0.170 F_{CD}.$$

Distance from A (ft)	Influence ordinate (ft·k)	Distance from A (ft)	Influence ordinate (ft·k)
5	-3.03	40	-0.96
10	-4.73	45	-0.35
15	-3.54	55	$+0.31$
25	-1.03	60	$+0.41$
30	-1.56	65	$+0.26$
35	-1.38		

12–18. The dimensions first assumed for a reinforced concrete highway bridge are shown in Fig. 12–43. Continuity is provided between the bridge deck and piers. Using the method of moment distribution, write the equations for M_{BA}, M_{BC}, and M_{CB} in terms of the fixed-end moments of all deck spans. Compute the ordinates at the 0.1, 0.3, 0.5, 0.7, and 0.9 points of all spans for the influence line for M_{BC}.

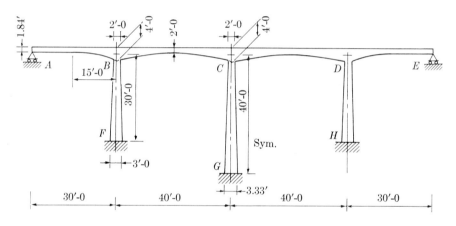

FIGURE 12–43

[Ans.:

$$M_{BA} = +0.720F_{BA} - 0.280F_{BC} + 0.085F_{CB}$$
$$+0.085F_{CD} - 0.028F_{DC} - 0.028F_{DE},$$

$$M_{BC} = -0.407F_{BA} + 0.593F_{BC} - 0.181F_{CB}$$
$$-0.181F_{CD} + 0.058F_{DC} + 0.058F_{DE},$$

$$M_{CB} = -0.200F_{BA} - 0.200F_{BC} + 0.622F_{CB}$$
$$-0.378F_{CD} + 0.122F_{DC} + 0.122F_{DE}.$$

Distance from A (ft)	Influence ordinate (ft·k)	Distance from A (ft)	Influence ordinate (ft·k)
3	+1.09	74	−0.67
9	+2.89	82	−1.58
15	+3.70	90	−1.57
21	+3.10	98	−0.91
27	+1.28	106	−0.25
34	+2.21	113	+0.18
42	+5.12	119	+0.44
50	+5.08	125	+0.53
58	+2.92	131	+0.41
66	+0.80	137	+0.15

12-6 Influence lines for articulated structures. In Section 12-2 it was demonstrated that the same methods which apply for computing influence lines for continuous structures also hold for articulated structures. Two problems relating to two-hinged arches were discussed in connection with Eqs. (12-4) and (12-5). It is believed unnecessary, therefore, to present any special illustrative problems dealing with influence lines for articulated structures. The reader should be particularly aware of the fact, however, that the Williot-Mohr diagram will usually be found to be the best method for determining the deflections of an articulated structure.

12-7 Qualitative influence lines by the Müller-Breslau principle. It has been demonstrated that the Müller-Breslau principle is of extreme importance in the determination of quantitative influence lines, that is, influence lines with definite arithmetical values for the ordinates. The principle also makes it possible to sketch *qualitative* influence lines. These are indispensable in determining the correct loading patterns for maximum values of moment, thrust, and shear in various types of structures.

For example, suppose it is necessary to determine which spans of a six-span continuous beam should be loaded with a uniform load to cause a maximum positive moment (compression on the top side) at the center of the span BC (see Fig. 12-44). A pin is inserted at this point in the span and couples are applied with a sense to cause compression on the top side. The beam will deflect as shown in Fig. 12-44. Any action which tends to restore the beam to its unstrained position will cause the same kind of stress due to flexure (compression on the top side) at the given section as the applied couples. Obviously, then, spans BC, DE, and FG should

FIGURE 12-44

be loaded if a maximum positive moment is desired at the center of span BC. Actually, of course, the effect of the load in span FG on the moment at the center of span BC is negligible.

If a maximum negative moment is desired at C, a pin is inserted at this point and couples (to cause tension on the top side) are applied on each side of the pin. The beam will deflect as shown in Fig. 12–45, and it is

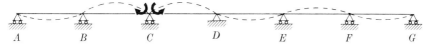

FIGURE 12–45

apparent that spans BC, CD, and EF should be loaded.

Again suppose that it is required to find the position of uniform load to give maximum positive shear (the part of the beam to the left of the section tends to move up with respect to the part on the right) at a section to the right of B. The beam is cut at the given section, a roller and slide device is inserted, and two vertical forces of equal magnitude are applied to the beam ends adjacent to the device. Each force acts with the sense required to give shear of the desired sign at the section. The deflected beam, and therefore the qualitative influence line, is shown in Fig. 12–46.

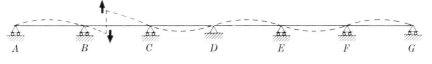

FIGURE 12–46

The qualitative influence line is particularly valuable in the analysis of building frames. For example, the influence line for positive moment at the center of span B4–C4 is shown in Fig. 12–47. The influence line for

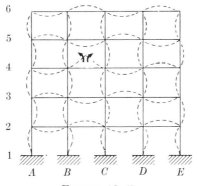

FIGURE 12–47

maximum negative moment at the left end of span $C4$–$D4$ is shown in Fig. 12–48. Note that in bay CD the indication is that a small portion at

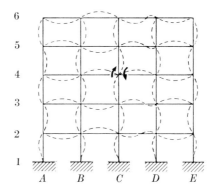

FIGURE 12–48

the right end of the beams at levels 3 and 5 should not be loaded. In addition, at levels 2 and 6, a short loaded length is theoretically required at the right end. For a practical analysis, levels 3 and 5 in bay CD would be entirely loaded, while levels 2 and 6 would be unloaded. The influence line for moment causing tension on the right side at the top of column $C3$–$C4$ is shown in Fig. 12–49.

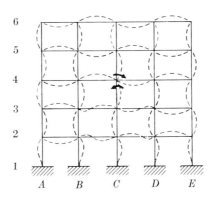

FIGURE 12–49

Of special note here is the fact that theoretical loading patterns are indicated in Figs. 12–47, 12–48, and 12–49. Practically, however, the effect of the load in a given bay on a stress component in a member two or three bays distant is negligible. Specifications, therefore, permit the analysis of large building frames by dividing them into sections. In analyzing a given run of girders and the adjacent columns immediately above and below these girders, it is permissible to consider that all columns are fixed at their far ends. The ends of girders two bays away from the stress component being evaluated are also considered to be fixed. This has the obvious effect of greatly simplifying the analysis, with only a slight loss in accuracy.

General References

1. Fife, W. M. and Wilbur, J. B., *Theory of Statically Indeterminate Structures*. 2nd ed. Chap. 4. New York: McGraw-Hill, 1937.
2. Grinter, L. E., *Theory of Modern Steel Structures*. Vol. II, revised ed. New York: Macmillan, 1949.
3. Johnson, J. B., Bryan, C. W., and Turneaure, F. E., *Modern Framed Structures*, Part II. 10th ed. New York: Wiley, 1929.
4. Maugh, L. C., *Statically Indeterminate Structures*. New York: Wiley, 1946.
5. Parcel, J. I. and Moorman, R. B. B., *Analysis of Statically Indeterminate Structures*. New York: Wiley, 1955.
6. Sutherland, H. and Bowman, H. L., *Structural Theory*. 4th ed. New York: Wiley, 1950.
7. Wilbur, J. B. and Norris, C. H., *Elementary Structural Analysis*. New York: McGraw-Hill, 1948.

CHAPTER 13

ELASTIC ARCHES

13-1 General. A *true arch* may be defined as a structure which depends (usually in considerable degree) for its ability to support applied vertical loads on the development of horizontal reaction components, acting toward the center of the arch span, at the two end supports. These horizontal reaction components are the means of differentiating between the true arch and the "corbeled arch," which consists of a series of cantilevers, each superimposed upon and projecting a little beyond the cantilever below, with a simple beam span closure at the top center.

As stated in Chapter 1, the corbeled arch was used extensively in ancient Egypt as an architectural unit. Only one true arch of ancient origin, discovered in a tomb at Thebes and dated about 1500 B.C., has been found in all that country. The Assyrians and Babylonians also used the corbeled arch quite commonly, with the latter using pointed brick arches in the construction of their sewers. These existed as early as 1500 B.C. and possibly many centuries before. It is said that a series of brick arches spanned the Euphrates in 2000 B.C. In general, however, very few of these early structures were true arches.

It appears that the true arch was first used to any great degree by the Etruscans, people of Asiatic origin who invaded northern Italy about 1300 B.C. Gradually, during the next thousand years, certain rules of thumb were evolved whereby true arches of a sort could be built. The engineers of the Roman Empire developed the true arch to the point where, for five centuries (from about 300 B.C. to perhaps 200 A.D.), many great arch bridges and aqueducts were built in all parts of the Empire. The outstanding examples of this period were of cut stone blocks placed without mortar.

The collapse of the Roman Empire in the West in 476 A.D. marked the beginning of the Dark Ages and the cessation of most road and bridge building throughout Europe. In 1176 A.D. the construction of the original London Bridge was begun, and this ugly structure, consisting of a series of short-span arches, lasted for six centuries. In the twelfth century the members of a religious order known as *Fratres Pontes* ("Brothers of the Bridge") dedicated themselves to the repair of existing bridges as well as the building of new ones. In the fourteenth century the first bridge with a masonry deck slab supported by masonry arch ribs was built. Throughout all these centuries, however, the engineering skill of the Romans was never equalled in Europe.

Beginning in Italy in the fourteenth century and spreading from there to the rest of Europe, the Renaissance was a period of general revival of economic life and new interest in art and in construction. Roads and bridges were needed and builders had to learn again how to construct arches. Like the arches of the Romans, these were built by rule of thumb.

The Frenchman, Lahire (1640–1718), first applied statics in an attempt to analyze the arch, and, in a memoir published in 1773, Coulomb considered the types of arch failures and proposed a theory. Both, however, considered the arch to be inelastic. In his work on the strength of materials published in 1826, Navier (1785–1836) seems to have made the first important contribution to the theory of bending of curved bars. Apparently including in his theory only the effects of flexural strain, Navier nevertheless explained how axial strains could be considered. He applied his theory to find the horizontal reaction components of circular and parabolic symmetrical arches.

No particular attention appears to have been given to the question of the most advantageous shape of the arch axis until Yvon Villarceau (1813–1883) presented a famous memoir on arches to the French Academy of Sciences in 1845. Villarceau understood that the complete solution required consideration of the elastic deformations of the rib. He reasoned, however, that insufficient information was available regarding the elastic properties of materials, and therefore considered the arch blocks to be absolutely rigid. It was his conclusion that the best arch axis to use in a given case would be that which coincided with the funicular polygon for the applied loads.

The Frenchman, J. V. Poncelet (1788–1867), apparently was the first to suggest (in an article published in 1852) that an arch must be considered as an elastic bar if a rational theory is to be applied.

Jacques Bresse (1822–1883), also a Frenchman, published a book in 1854 dealing with the deflection of curved bars, and in it he included the effects of both axial and flexural strains and demonstrated the application of his theory in the design of arches.

In 1866, C. Culmann (1821–1881), a German, published his famous book on graphic statics, a part of which is devoted to a consideration of the analysis of arches and retaining walls. Culmann introduced the very important concept of the *elastic center*, which permits the analysis of hingeless arches without the use of simultaneous equations.

E. Winkler (1855–1888), a German, published in 1867 his book on strength of materials. Whereas Navier and Bresse had previously considered the deflection of curved bars for which the ratio of the initial radius of curvature to the radial thickness of the bar is large, Winkler discussed the case where this ratio is small. When this ratio is less than ten, the errors involved in applying the ordinary theory of flexure usually become significant. This, of course, is not encountered in the design of arch bridges but applies to hooks, chain links, and similar designs. Winkler discussed the two-hinged arch problems previously considered by Bresse and extended the theory to include hingeless arches. In Winkler's book tables are included for different loading conditions of circular and parabolic arches with constant cross section.

Otto Mohr (1834–1918) made two important contributions to the elastic theory of arches. The first appeared in a paper published in 1870, in which Mohr presented the idea of computing the influence line for the horizontal reaction component of a two-hinged arch as the bending moment diagram for a conjugate beam loaded with the y/EI diagram for the arch. (The term y is the vertical distance from the plane of the hinges to the arch axis at any section.) Mohr obtained this moment diagram by graphical methods. His second contribution, appearing in a paper published in 1881, was an extension of the concept of the elastic center (neutral point) to include articulated arches.

Chiefly as the result of work by Winkler and Mohr, supplemented by carefully controlled tests conducted by the Society of Austrian Engineers and Architects, the elastic theory was finally accepted for the analysis of arches. More recent contributions to the theory and methods of arch analysis are those of A. Strassner in 1927, J. Melan and T. Gesteschi in 1931, and E. Morsch in 1935.

Down through the centuries, as methods of analysis and design and as materials and methods of construction have improved, the spans of new arch bridges have constantly increased. The longest steel highway arch bridge existing at the time of this writing is the Kill Van Kull Bridge, which connects Staten Island with Bayonne, New Jersey, and has a span of 1652 feet. The longest steel railway arch, with a span of 1650 feet, is over Sydney Harbor, Australia. Both bridges have braced ribs with two hinges. The longest reinforced concrete highway arch is the Sando Arch in Sweden, with a span of 866 feet. A reinforced concrete arch railway bridge at Esla, in Spain, holds the record for this type with a span of 645 feet.

For shorter spans it is possible to analyze the arch as an elastic structure and to neglect the effects of the change in shape of the arch axis resulting from the elastic strains; in other words, the deflection of the arch is neglected. For short spans the resulting errors are not significant. When the arch span is long, however, the neglect of the elastic deflection will usually result in significant, and perhaps serious, overstress. [A deflection theory for arches was published by A. Freudenthal in 1935. Additional information is available in the form of papers in the Transactions of the American Society of Civil Engineers.]

13–2 Types of arches. Arches may be classified, of course, on the basis of the materials of which they are built. The most common are steel, reinforced concrete, and timber. From the standpoint of structural behavior, arches are also classified as hingeless (sometimes designated as fixed), two-hinged, or three-hinged. (The three-hinged arch is determinate and will not be considered in this discussion.) Arches are also conveniently, and necessarily, classified as to the shape and structural arrangement of the rib, with several types shown in Fig. 13–1.

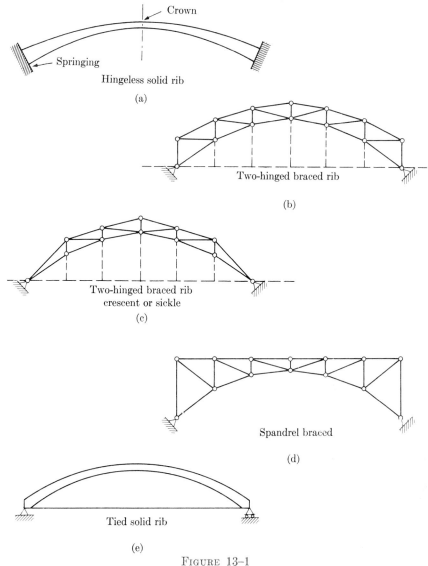

FIGURE 13–1

A solid rib arch, as shown in Fig. 13–1(a), may be used with two hinges at the ends or may be hingeless. Reinforced concrete ribs are almost always hingeless. Shorter span steel arches with solid ribs, say spans of 500 or 600 feet and under, are usually two-hinged, but solid steel ribs of longer spans are always hingeless. For example, the Henry Hudson steel arch in New York City, spanning 800 feet, and the Rainbow Arch at Niagara Falls, spanning 950 feet, are hingeless. In the case of a two-hinged trial arch designed as a preliminary study for the Rainbow Arch, it was found that deflection of the rib increased the quarter point stress by 29% over the values given by an analysis by the elastic theory. In the hingeless arch which was actually erected, the deflection of the rib increased the quarter point stresses by only 7%. Thus the greater stiffness of the hingeless arch is a distinct advantage in long spans.

The type of braced rib shown in Fig. 13–1(b) illustrates the essential features of both the Kill Van Kull and the Sydney Harbor arches. In both cases the deck is hung from the arch as shown by the dashed lines. Many other steel arches of this same type have been built.

The crescent arch of Fig. 13–1(c) is not commonly used. The spandrel-braced arch of 13–1(d) is essentially a deck truss with horizontal thrusts developed at the two ends.

Since horizontal thrust reaction components are essential for an arch, it is apparent that excellent foundation conditions must exist at the site. If such conditions are not available, then the horizontal reaction component may be provided by a tie, as shown in Fig. 13–1(e).

13–3 Curve of arch axis. The arch is structurally advantageous because the internal moments resulting from applied loads are very much smaller than those which would result if the same loads were applied to a truss or beam of the same span. This, of course, is due to the negative moments resulting from the horizontal thrusts at the ends. Figure 13–2(a) shows a simple beam and the moment diagram resulting from a single concentrated load. Figure 13–2(b) shows an arch of the same span subjected to the same

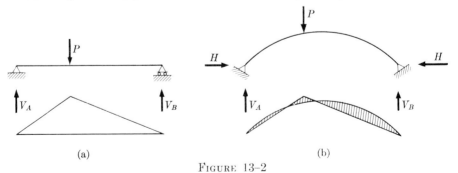

Figure 13–2

load, and the resulting moment diagram. It is obvious that the maximum positive moment is greatly reduced by the arch action. At the same time, of course, important axial stresses will exist in the arch rib.

The moments in the rib should be as small as possible. Consequently, the center line of the rib, whether it be solid or braced, should closely approximate the funicular polygon for dead load plus, perhaps, some portion of the live load. The funicular polygon will usually be fairly well approximated by a segment of a circle, a multi-centered circular curve, or a parabola.

The economy of any particular arch will, of course, be influenced by the rise-to-span ratio. In many cases this ratio will be determined by conditions at the site, if the structure is to be a bridge, or by headroom requirements, if the design is for a hangar, drill hall, or similar building. Where conditions permit the most economical rise-to-span ratio to be used, it is probable that this ratio will be from 0.25 to 0.30, although special considerations may void these limits.

13–4 Analysis of two-hinged articulated arch ribs. This classification includes both the braced rib and the braced spandrel arch. For the purpose of the discussion consider the rib of Fig. 13–3, which is indeterminate

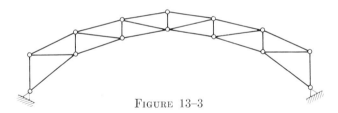

FIGURE 13–3

to the first degree. If this arch is subjected to static loads, it is most easily analyzed by the general method. One end is placed on rollers and the horizontal displacement of this end, as caused by the loads, is determined by virtual work. Stresses in the determinate truss, due to both the real and fictitious loads, are readily found by graphical methods. The horizontal displacement of the end on rollers, as caused by a horizontal load of 1 k acting at this end, is next computed by virtual work. The division of the first deflection by the second will give the magnitude of the horizontal reaction component. All bar stresses may then be computed by statics.

If the arch of Fig. 13–3 is to be analyzed for moving loads, an influence line for H_A is required. End A is placed on rollers, and a horizontal load of any convenient magnitude is applied at this end. The resulting horizontal displacement of end A, and the simultaneous vertical deflections of all panel points at which loads will be applied, are evaluated. These are most easily determined with a Williot or, if necessary, a Williot-Mohr diagram.

Then, by the Müller-Breslau principle, the several ordinates of the influence line for H_A are determined by dividing each vertical displacement of a panel point by the horizontal displacement of end A. After the influence line for H_A has been determined, the influence lines for stress in the various members of the arch can be computed by statics.

As is the case with all indeterminate structures, complete dimensions of all parts of the arch rib obviously must be known or assumed before an analysis is possible. If experience is lacking, the first analysis may be made by assuming that all members have the same cross-sectional area. All bar stresses are determined on the basis of this assumption and the members are designed for these stresses. Then the arch is again analyzed, using the new member cross-sectional areas. The resulting new bar stresses are used to revise the first design for the individual members of the rib, if this is necessary. A third analysis will usually be required to determine to what extent the various stresses may have been changed by the last revision of cross-sectional areas. The process is continued until further adjustment of members is unnecessary, with three or four analyses usually being sufficient.

Problems

13-1. Compute the ordinates, for loads at U_1 and U_2, of the influence line for the horizontal reaction component of the braced rib of Fig. 13-4. Assume that all members have the same cross-sectional area. Find the stresses in the various members as caused by the indicated loads. [*Ans.:* Influence ordinates—$U_1 = 0.38$, $U_2 = 0.44$; stresses—$L_0L_1 = -8$ k, $L_0U_1 = -11$ k, $U_1L_1 = -3$ k, $L_1L_2 = -6$ k, $U_1L_2 = -2$ k, $U_1U_2 = -5$ k, $U_2L_2 = -6$ k.]

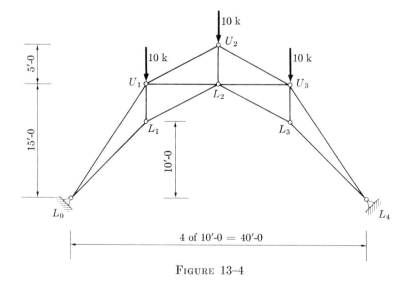

Figure 13-4

13-2. Compute the ordinates, for loads at U_1, U_2, U_3, and U_4, of the influence line for the horizontal reaction component of the braced rib shown in Fig. 13-5. All members are assumed to have the same cross-sectional areas. Find the stresses in the members as caused by the indicated loads. [*Ans.:* Influence ordinates— $U_1 = 0.14$, $U_2 = 0.24$, $U_3 = 0.31$, $U_4 = 0.33$; horizontal reaction component for loads = 16.8 k; stresses— $L_0U_0 = +8$ k, $U_1U_2 = -20$ k, $U_3U_4 = -25$ k, $L_1U_1 = -26$ k, $L_0L_1 = -46$ k, $L_1L_2 = -20$ k, $L_2L_3 = +8$ k, $U_3L_3 = 0$ k.]

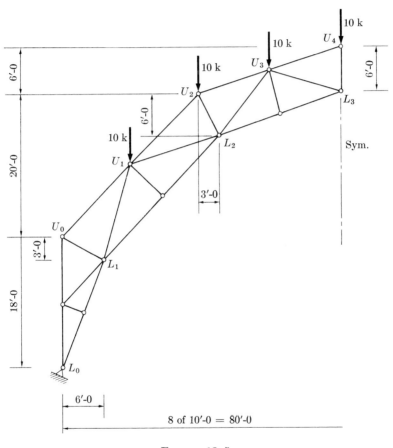

FIGURE 13-5

13-5 Two-hinged solid arch ribs.

This type of rib is analyzed by the general method, using a procedure similar to that described for an articulated rib. In the case of the articulated rib, the horizontal displacement of the end on rollers is the result of axial elastic strain in the component bars. In the solid rib, however, this horizontal displacement is the result of a combination of shearing, axial, and flexural elastic strains. Theoretically all these strains should be considered in the analysis. The following example will demonstrate how this may be accomplished.

EXAMPLE 13–3. Determine the value of the horizontal reaction component of the indicated two-hinged solid rib arch (see Fig. 13–6) as caused by a concentrated vertical load of 10 k at the center line of the span. Consider shearing, axial, and flexural strains. Assume that the rib is a 24WF130 with a total area of 38.21 in^2, that it has a web area of 13.70 in^2, a moment of inertia equal to 4000 in^4, E of 30,000 k/in^2, and a shearing modulus G of 12,000 k/in^2.

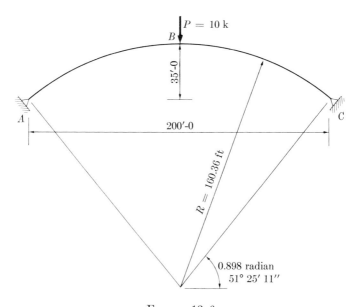

FIGURE 13–6

Consider that end C is placed on rollers, as shown in Fig. 13–7. A unit fictitious horizontal force is applied at C. The axial and shearing components of this fictitious force and of the vertical reaction at C, acting on any section θ in the right half of the rib, are shown at the right end of the rib in Fig. 13–7. The expression for the horizontal displacement of C is

$$\Delta_{Ch} = 2\int_C^B \frac{Mm\,dx}{EI} + 2\int_C^B \frac{Vv\,ds}{A_wG} + 2\int_C^B \frac{Nn\,ds}{AE}. \qquad (13\text{–}1)$$

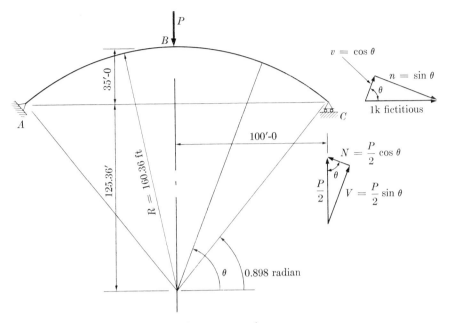

FIGURE 13-7

From Fig. 13-7, for the rib from C to B,

$$M = \frac{P}{2}(100 - R\cos\theta), \quad m = 1(R\sin\theta - 125.36),$$

$$V = \frac{P}{2}\sin\theta, \quad v = \cos\theta,$$

$$N = \frac{P}{2}\cos\theta, \quad n = -\sin\theta,$$

$$ds = R\,d\theta.$$

If the above values are substituted in Eq. (13–1) and integrated between the limits of 0.898 and $\pi/2$, the result will be

$$\Delta_{Ch} = 22.55 + 0.023 - 0.003 = 22.57. \quad (13\text{-}2)$$

The load P is now assumed to be removed from the rib, and a real horizontal force of 1 k is assumed to act toward the right at C in conjunction with the fictitious horizontal force of 1 k acting to the right at the same point. The horizontal displacement of C will be given by

$$\delta_{ChCh} = 2\int_C^B \frac{m^2\, ds}{EI} + 2\int_C^B \frac{v^2\, ds}{A_w G} + 2\int_C^B \frac{n^2\, ds}{AE} \qquad (13\text{–}3)$$

$$= 2.309 + 0.002 + 0.002 = 2.313 \text{ in.}$$

The value of the horizontal reaction component will be

$$H_C = \frac{\Delta_{Ch}}{\delta_{ChCh}} = \frac{22.57}{2.313} = 9.75 \text{ k.}$$

If only flexural strains are considered, the result is

$$H_C = \frac{22.55}{2.309} = 9.76 \text{ k.}$$

For the given rib and the single concentrated load at the center of the span it is obvious that the effects of shearing and axial strains are insignificant and can be disregarded. Erroneous conclusions as to the relative importance of shearing and axial strains in the usual solid rib may be drawn, however, from the values shown in Eq. (13–2). These indicate that the effects of the shearing strains are much more significant than those of the axial strains. This is actually the case for the single concentrated load chosen for the demonstration, but only because the rib does not approximate the funicular polygon for the single load. As a result, the shearing components on most sections of the rib are more important than would otherwise be the case. The usual arch encountered in practice, however, is subjected to a series of loads, and the axis of the rib will approximate the funicular polygon for these loads. In other words, the line of pressure is nearly perpendicular to the right section at all points along the rib. Consequently, the shearing components are so small that the shearing strains are insignificant and are neglected.

Axial strains, resulting in rib shortening, become increasingly important as the rise-to-span ratio of the arch decreases. It is advisable to determine the effects of rib shortening in the design of arches. The usual procedure is to first design the rib by considering flexural strains only, and then to check for the effects of rib shortening.

EXAMPLE 13–4. Design a two-hinged, solid welded-steel arch rib for a hangar. The moment of inertia of the rib is to vary as necessary. The span, center to center of hinges, is to be 200 ft. Ribs are to be placed 35 ft center to center, with a rise of 35 ft. Roof deck, purlins, and rib will be assumed to weigh 25 lb/ft² on roof surface, and snow will be assumed at 40 lb/ft² of this surface. Twenty purlins will be equally spaced around the rib.

13-5]　　　TWO-HINGED SOLID ARCH RIBS　　　547

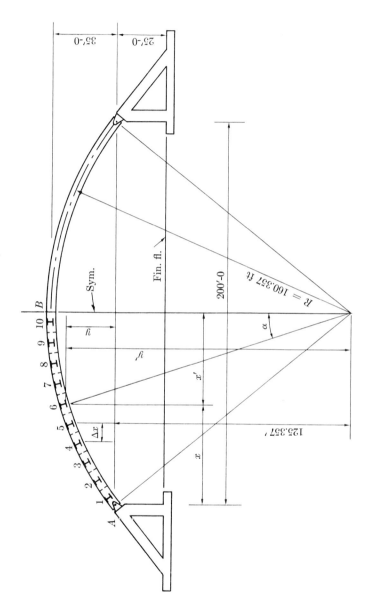

FIGURE 13-8

The center line of the rib will be taken as the segment of a circle. By computation the radius of this circle is found to be 160.357 ft, and the length of the arc AB to be 107.984 ft. For the analysis the arc AB will be considered to be divided into ten segments, each with a length of 10.798 ft. Thus a concentrated load is applied to the rib by the purlins framing at the center of each segment. (The numbered segments are indicated in Fig. 13-8.) Since the total dead and snow load is 65 lb/ft² of roof surface, the value of each concentration will be

$$10.798 \times 35 \times 65 = 24.565 \text{ k}.$$

The computations necessary to evaluate the coordinates of the centers of the various segments, referred to the hinge at A, are shown in Table 13-1. Also shown are the values of Δx, the horizontal projection of the distance between the centers of the several segments.

If experience is lacking and the designing engineer is therefore at a loss as to the initial assumptions regarding the sectional variation along the rib, it is recommended that the first analysis be based on the assumption of a constant moment of inertia. Even the experienced engineer will often find that this is the best procedure. The value of H having thus been determined, moments and axial thrusts are computed throughout the rib. Cross-sectional areas at the various sections are then designed for these moments and axial thrusts. A new value for H, as well as revised values for moments and axial thrusts throughout the rib, are determined for the new rib with the varying cross section. Stresses are checked in the rib, and the whole process is repeated if necessary. This procedure will be followed in the present case.

By the general method, assuming that end A is on rollers,

$$\Delta_{Ah} + H_A \delta_{AhAh} = 0.$$

Deflections will be computed by virtual work, and thus this expression becomes

$$\sum \frac{Mm \, \Delta S}{EI} + H_A \sum \frac{m^2 \, \Delta S}{EI} = 0.$$

Since E and ΔS are constant and, for the first analysis, the moment of inertia is constant, the final expression for H_A is

$$H_A = -\frac{\Sigma Mm}{\Sigma m^2}. \tag{13-4}$$

All the computations necessary for the first evaluation of H_A are shown in Table 13-2.

13–5]   TWO-HINGED SOLID ARCH RIBS   549

TABLE 13-1.

Segment	α	$\sin \alpha$	$\cos \alpha$	$x' = R \sin \alpha$ (ft)	$y' = R \cos \alpha$ (ft)	$x = 100 - x'$ (ft)	$y = y' - 125.36$ (ft)	Δx (ft)
A	38°34'49"	0.62361	0.78173	100.000	125.36	0	0	
1	36°39'05"	0.59695	0.80229	95.725	128.65	4.275	3.29	4.275
2	32°47'36"	0.54161	0.84063	86.851	134.80	13.149	9.44	8.874
3	28°56'07"	0.48382	0.87516	77.584	140.34	22.416	14.98	9.267
4	25°04'38"	0.42384	0.90574	67.966	145.24	32.034	19.88	9.618
5	21°13'09"	0.36194	0.93220	58.040	149.48	41.960	24.13	9.926
6	17°21'40"	0.29840	0.95444	47.851	153.05	52.149	27.69	10.189
7	13°30'11"	0.23350	0.97236	37.443	155.92	62.557	30.57	10.408
8	09°38'42"	0.16754	0.98586	26.866	158.09	73.134	32.73	10.577
9	05°47'13"	0.10083	0.99490	16.169	159.54	83.831	34.18	10.697
10	01°55'44"	0.03367	0.99943	5.399	160.27	94.601	34.91	10.770
Crown					160.36	100.000	35.00	5.399

TABLE 13–2.

Segment	(2) x (ft)	(3) $y = m$ (ft·k)	(4) Shear to right (k)	(5) Δx (ft)	(6) M increment (ft·k)	(7) M simple beam (ft·k)	(8) Mm	(9) m^2
A	0.0	0.0	245.650	4.275	1,050	1,050	3,500	10.9
1	4.275	3.29	221.085	8.874	1,962	3,010	28,400	89.2
2	13.149	9.44	196.520	9.267	1,821	4,830	72,400	224.4
3	22.416	14.98	171.955	9.618	1,654	6,490	129,100	395.4
4	32.034	19.88	147.390	9.926	1,463	7,950	191,800	582.2
5	41.960	24.13	122.825	10.189	1,251	9,200	254,800	767.0
6	52.149	27.69	98.260	10.408	1,023	10,220	312,400	934.4
7	62.557	30.57	73.695	10.577	779	11,000	360,000	1,071.4
8	73.134	32.73	49.130	10.697	526	11,530	394,100	1,168.4
9	83.831	34.18	24.565	10.770	265	11,790	411,600	1,218.6
10	94.601	34.91	0.000	5.399	0	11,790		
Crown	100.00	35.00	—					
Σ							2,158,100	6,461.9

$$H_A = -\frac{2\Sigma Mm}{2\Sigma m^2} = -\frac{2 \times 2{,}158{,}100}{2 \times 6461.9} = -333.97 \text{ k.}$$

The moments at the centers of the various segments of the rib, using the approximate value just determined for H_A, are computed in Table 13–3. A moment is considered to be positive if it tends to cause compression on the upper surface of the rib.

TABLE 13–3.

Segment	y (ft)	M simple beam (ft·k)	$H_A y$ (ft·k)	Total M at segment (ft·k)
A	0			
1	3.29	1,050	−1,100	−50
2	9.44	3,010	−3,150	−140
3	14.98	4,830	−5,000	−170
4	19.88	6,490	−6,640	−150
5	24.13	7,950	−8,060	−110
6	27.69	9,200	−9,250	−50
7	30.57	10,220	−10,210	+10
8	32.73	11,000	−10,930	+70
9	34.18	11,530	−11,420	+110
10	34.91	11,790	−11,660	+130
Crown	35.00	11,790	−11,690	+100

The thrust N, normal to the cross section at the center of each segment, must be computed before the section requirements can be determined. As a matter of interest the various values of S, the total shear on the cross section, will also be computed for the center of each segment. N and S are obtained by combining components of H and the vertical shear V at the center of each segment, a combination which is shown in Fig. 13–9.

From this figure it is apparent that

$$S = V \cos \alpha - H \sin \alpha,$$
$$N = V \sin \alpha + H \cos \alpha.$$

Values of S and N are computed in Table 13–4. The values of S are included in support of the statement previously made to the effect that shearing forces are small in the usual arch rib.

Inspection of the values for the moment M in Table 13–3 and for the thrust N in Table 13–4 will indicate that segment 3 is critical in the region of negative moment and segment 10 in the region of positive moment. The necessary sections for the rib, at the centers of these two segments, must be determined in accordance with the specifications of the American Institute of Steel Construction before Table 13–5 can be completed.

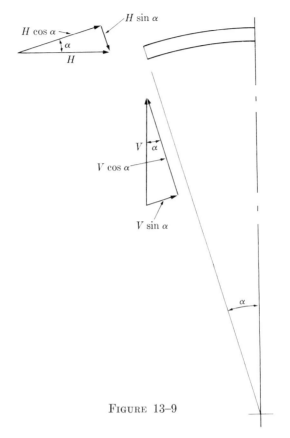

FIGURE 13-9.

TABLE 13-4.

Segment	V (k)	$V \cos \alpha - H \sin \alpha = S$ (k)			$V \sin \alpha + H \cos \alpha = N$ (k)		
A	245.6	192	-208	$= -16$	153	$+261$	$= 414$
1	221.1	177	-199	$= -22$	132	$+268$	$= 400$
2	196.5	165	-181	$= -16$	106	$+281$	$= 387$
3	172.0	150	-162	$= -12$	83	$+292$	$= 375$
4	147.4	133	-142	$= -9$	62	$+303$	$= 365$
5	122.8	114	-121	$= -7$	44	$+311$	$= 355$
6	98.3	94	-100	$= -6$	29	$+319$	$= 348$
7	73.7	72	-78	$= -6$	17	$+325$	$= 342$
8	49.1	48	-56	$= -8$	8	$+329$	$= 337$
9	24.6	24	-34	$= -10$	2	$+332$	$= 334$
10	0.0	0	-11	$= -11$	0	$+334$	$= 334$
Crown						$+334$	$= 334$

13–5] TWO-HINGED SOLID ARCH RIBS 553

TABLE 13-5.

Flanges: 12 in. × 1 in.   Allowable stress for bending = F_B = 20 k/in^2

Web thickness: $\tfrac{3}{4}$ in.   Allowable axial stress* = F_A = 17,000 − 0.485 L/r

Segment	(2) Total depth (in.)	(3) Area (in^2)	(4) I (in^4)	(5) M (ft·k)	(6) Axial thrust N (k)	(7) $f_a = \dfrac{N}{A}$ (k/in^2)	(8) $f_b = \dfrac{Mc}{I}$ (k/in^2)	(9) Allowable F_a (k/in^2)	(10) $\dfrac{f_a}{F_a}$	(11) $+ \dfrac{f_b}{F_b} =$	(12) Decimal of capacity
A	23.00	39.75	3483		414	10.4		15.88	0.66		0.66
1	23.00	39.75	3483	50	400	10.1	2.0	15.88	0.63	0.10	0.73
2	23.00	39.75	3483	140	387	9.7	5.9	15.88	0.61	0.28	0.89
3	23.00	39.75	3483	170	375	9.4	6.8	15.88	0.60	0.34	0.94
4	22.07	39.05	3172	150	365	9.3	6.3	15.90	0.59	0.32	0.91
5	21.13	38.35	2865	110	355	9.2	4.9	15.92	0.58	0.24	0.82
6	20.20	37.65	2585	50	348	9.2	2.3	15.94	0.58	0.11	0.70
7	19.27	36.95	2329	10	342	9.3	0.5	15.96	0.58	0.02	0.61
8	18.32	36.24	2071	70	337	9.3	3.7	15.97	0.58	0.19	0.77
9	17.40	35.55	1838	110	334	9.4	6.3	15.99	0.59	0.31	0.90
10	16.47	34.85	1631	130	334	9.6	7.9	16.01	0.60	0.40	1.00
Crown	16.00	34.50	1527	100	334	9.7	6.3	16.03	0.60	0.32	0.92

* In this equation the value of L is 10.798, since ribs are laterally braced at purlin framing points.

Thickness of flange and web plates and width of web plates are matters of judgment. The total depth determined for the center of segment 3 is used from the center of this segment to the end of the rib. The depth determined for the center of segment 10 is arbitrarily used for the rib depth at the crown. The total depth of the rib from the center of segment 3 to the crown is made to vary linearly. The adequacies of the sections thus determined for the centers of the several segments are checked in Table 13–5.

It is necessary to recompute the value of H_A because the rib now has a varying moment of inertia. Equation (13–4) must be altered to include the I of each segment and is now written as

$$H_A = -\frac{\Sigma Mm/I}{\Sigma m^2/I}.$$

The revised value for H_A is easily determined as shown in Table 13–6. Note that the values in column (2) of Table 13–6 are found by dividing the values of Mm for the corresponding segments in column (8) of Table 13–2 by the total I for each segment as shown in Table 13–5. The values in column (3) of Table 13–6 are found in a similar manner from the values in column (9) of Table 13–2. The simple beam moments in column (5) of Table 13–6 are taken directly from column (7) of Table 13–2.

TABLE 13–6.

(1) Segment	(2) $\dfrac{Mm}{I}$	(3) $\dfrac{m^2}{I}$	(4) Revised $H_A y$ (ft·k)	(5) M simple beam (ft·k)	(6) Revised moment (ft·k)
1	1.00	0.0031	1,100	1,050	−50
2	8.16	0.0256	3,160	3,010	−150
3	20.79	0.0644	5,020	4,830	−190
4	40.67	0.1247	6,660	6,490	−170
5	66.95	0.2032	8,080	7,950	−130
6	98.57	0.2967	9,270	9,200	−70
7	134.13	0.4012	10,240	10,220	−20
8	173.82	0.5174	10,960	11,000	+40
9	214.41	0.6357	11,450	11,530	+80
10	252.35	0.7472	11,690	11,790	+100
Crown			11,720	11,790	+70
Σ	1010.85	3.0192			

Revised $H_A = -\dfrac{2\Sigma Mm/I}{2\Sigma m^2/I} = -\dfrac{2 \times 1010.85}{2 \times 3.0192} = -334.81$ k.

The revised values for the axial thrust N at the centers of the various segments are computed in Table 13-7.

TABLE 13-7.

Segment	$V \sin \alpha$	$+ H \cos \alpha =$	Revised N (k)
A	153	262	415
1	132	269	401
2	106	282	388
3	83	293	376
4	62	303	365
5	44	312	356
6	29	320	349
7	17	326	343
8	8	330	338
9	2	333	335
10	0	335	335
Crown		335	335

The sections previously designed at the centers of the segments are checked for adequacy in Table 13-8. From this table it appears that all

TABLE 13-8.

Segment	$\dfrac{N}{A}$ (k/in²)	$\dfrac{Mc}{I}$ (k/in²)	$\dfrac{N/A}{F_A}$	$+\dfrac{Mc/I}{F_B}=$	Decimal of capacity
A	10.4		0.66		0.66
1	10.1	2.0	0.64	0.11	0.75
2	9.8	6.0	0.62	0.30	0.92
3	9.5	7.6	0.60	0.38	0.98
4	9.4	7.1	0.59	0.36	0.95
5	9.3	5.7	0.58	0.29	0.87
6	9.3	3.3	0.58	0.17	0.75
7	9.3	1.0	0.58	0.05	0.63
8	9.3	2.1	0.58	0.11	0.69
9	9.4	4.5	0.59	0.23	0.82
10	9.6	6.1	0.60	0.31	0.91
Crown	9.7	4.4	0.61	0.23	0.84

sections of the rib are satisfactory. This cannot be definitely concluded, however, until the secondary stresses caused by the deflection of the rib are investigated. These stresses will be considered in Section 13-8.

EXAMPLE 13-5. Determine the effects of rib shortening and temperature changes in the arch rib of the preceding example (see Fig. 13-8). Consider a temperature drop of 100°F.

In order to determine the effects of rib shortening it is first necessary to find the horizontal displacement of end A which will result if the elastic axial strains, which occur in the actual loaded arch, are introduced into the rib while end A is on rollers. It is then necessary to find the horizontal displacement of end A, caused by a horizontal force of 1 k acting at A, considering both flexural and axial strains in the rib. The horizontal displacement of A, resulting from the rib shortening caused by the real loads, is given by

$$\sum \frac{Nn \, \Delta L}{AE},$$

where N is the axial thrust taken from Table 13-7, n is the axial thrust caused by a 1 k fictitious horizontal force acting at A and equal to $\cos \alpha$ in Table 13-1, ΔL is the length of each segment, and A is the right sectional area. The horizontal displacement of A, resulting from the rib shortening caused by a 1 k horizontal force acting at A, is given by

$$\sum \frac{n^2 \, \Delta L}{AE}.$$

The necessary computations for evaluating these two deflections of A are shown in Table 13-9.

TABLE 13-9.

Segment	N (k)	$n = \cos \alpha$ (k)	A (in²)	$\dfrac{Nn}{A}$	$\dfrac{n^2}{A}$
1	401	0.8023	39.75	8.07	0.0162
2	388	0.8406	39.75	8.21	0.0178
3	376	0.8752	39.75	8.30	0.0193
4	365	0.9057	39.05	8.49	0.0210
5	356	0.9322	38.35	8.68	0.0227
6	349	0.9544	37.65	8.86	0.0242
7	343	0.9724	36.95	9.04	0.0256
8	338	0.9859	36.24	9.22	0.0268
9	335	0.9949	35.55	9.41	0.0286
10	335	0.9994	34.85	9.60	0.0278
Σ				87.88	0.2300

The horizontal displacement of A, resulting from the rib shortening caused by the real load, is

$$\sum \frac{Nn\,\Delta L}{AE} = \frac{2 \times 87.88 \times 10.798 \times 12}{30{,}000} = 0.760 \text{ in.,}$$

and that resulting from the rib shortening caused by a 1 k horizontal load at A is

$$\sum \frac{n^2\,\Delta L}{AE} = \frac{2 \times 0.2300 \times 10.798 \times 12}{30{,}000} = 0.002 \text{ in.}$$

The relative value of the horizontal displacement of A, resulting from the flexural strain caused by a 1 k horizontal force at A, has previously been determined in column (3) of Table 13–6. The absolute value for this displacement is

$$\frac{2 \times 3.019 \times 10.798 \times 1728}{30{,}000} = 3.755 \text{ in.}$$

The total value for the displacement of A due to both flexural and axial strains resulting from a 1 k horizontal force at A is therefore

$$3.755 + 0.002 = 3.757 \text{ in.}$$

The reduction in H_A due to rib shortening is given by

$$\frac{0.760}{3.757} = 0.20 \text{ k.}$$

An additional reduction in H_A will result from any drop in the temperature of the rib below that temperature at which the arch is erected. This occurs, of course, because some additional shortening of the rib will result. The decrease in the length of the horizontal projection of the rib resulting from a temperature drop of 100°F will be

$$\delta L = 0.0000065 \times 100 \times 200 \times 12 = 1.56 \text{ in.}$$

The resulting decrease in H_A will be

$$\frac{1.56}{3.757} = 0.42 \text{ k.}$$

The total decrease in H_A due to both rib shortening and temperature drop is

$$0.20 + 0.42 = 0.62 \text{ k.}$$

This will have the effect of increasing the moment near the center of the span. Inspection of column (6) in Table 13–8 indicates that segment 10 should be checked for possible overstress as the result of this increase. The increase in moment at segment 10 will be given by

$$\Delta M = \Delta H_A(y \text{ for segment } 10) = 0.62 \times 34.91 = 21.6 \text{ ft·k}.$$

The moment of 100 ft·k, as previously determined for segment 10, is accurate only in the first two places, and hence the revised total moment for segment 10 will be

$$100 + 20 = 120 \text{ ft·k}.$$

The thrust of 335 k, previously determined for segment 10, is accurate to three places. The change in H_A due to rib shortening and temperature change must therefore be rounded to 1 k in order to determine the corrected thrust in segment 10. This corrected thrust will be

$$335 - 1 = 334 \text{ k}.$$

The extreme fiber stress is

$$f = \frac{Mc}{I} = \frac{120 \times 8.23 \times 12}{1631} = 7.25 \text{ k/in}^2.$$

The section at segment 10 is checked, as indicated in Table 13–8, by

$$\frac{N/A}{F_A} + \frac{Mc/I}{F_B} = \frac{334/34.85}{16.01} + \frac{7.25}{20} = 0.60 + 0.36 = 0.96,$$

and obviously the section at segment 10 is adequate.

No consideration has been given to the effects of a temperature rise, since such a rise can occur only during the summer months when no snow loading is possible. Consequently, this phase of the problem has no practical significance.

The method of analysis just demonstrated will apply equally well to a reinforced concrete or a timber two-hinged arch. As previously stated, however, very few reinforced concrete arches are built with two hinges.

Of special note here is the fact that, in addition to axial strain and temperature drop, the shrinkage of the concrete in a reinforced concrete rib will also cause rib shortening. The shrinkage coefficient should be determined for the mix to be used and applied in the analysis in the same general way as demonstrated above for the temperature coefficient.

13-6 Two-hinged parabolic arch with a secant variation of the moment of inertia. A method for designing a two-hinged arch rib was demonstrated in Example 13-4. It was suggested therein that the first analysis for the redundant H can very well be based on the assumption of a constant moment of inertia for the rib. An alternate approach to the problem is available because of the fact that it is possible to devise a rather simple expression for the value of H provided two requirements are imposed upon the shape and proportions of the rib. These two requirements are, first, that the curve of the axis of the rib must be parabolic; and, second, that the moment of inertia of the rib at any particular section must be equal to the moment of inertia at the crown multiplied by the secant of the angle α, where α is the angle between the horizontal and the tangent to the arch axis at that particular section.

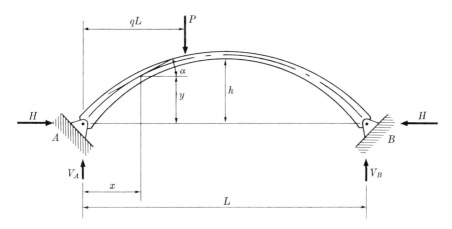

FIGURE 13-10

Consider the rib of Fig. 13-10 loaded with the single concentration P. Assume that the moment of inertia of the rib varies so that

$$I = I_c \sec \alpha,$$

where I_c is the moment of inertia at the crown and I the moment of inertia at any section. The arch axis is parabolic. The equation for the parabola, referred to A, is

$$y = 4h \left(\frac{x}{L} - \frac{x^2}{L^2} \right).$$

Assume that end A is on rollers. Then, by the general method,

$$H = -\frac{\Delta_{Ah}}{\delta_{AhAh}} = -\frac{\int Mm\,ds/EI}{\int m^2\,ds/EI},$$

where Δ_{Ah} is the horizontal deflection of A as caused by P, and δ_{AhAh} is the horizontal deflection of A resulting from the action of a 1 k horizontal force acting to the left at A. The moment in the rib at any section (with end A on rollers) as caused by P is represented by M, and m is the moment caused by the 1 k horizontal force at A.

In the equation above the following relations apply:

$$m = y, \qquad ds = \sec\alpha\,dx, \qquad I = I_c \sec\alpha.$$

When these substitutions are made, the above expression for H reduces to

$$H = -\frac{\int My\,dx}{\int y^2\,dx}.$$

For $x < qL$,

$$M = P(1-q)x,$$

and for $x > qL$,

$$M = P(1-q)x - P(x - qL) = Pq(L - x).$$

When the values for M and y are substituted in the above equation for H and the indicated integrations performed, the result is

$$H = \frac{5PL}{8h}(q - 2q^3 + q^4). \tag{13-5}$$

Although the above equation is derived for the particular type of rib as previously described, it is of some value in the preliminary analysis of two-hinged solid ribs having axial curves other than parabolic and having I variations other than proportional to the secant of α. The resulting values of H will, of course, be approximate but may be used in proportioning a first trial rib. If, for example, Eq. (13-5) is applied to the loaded rib of Example 13-4, the computed value of H will be 339.14 k. The correct value was found to be 334.81 k.

Note that if an influence line for H is required, then P is taken as unity and values of q are substituted that correspond to the various sections for which influence line ordinates are desired.

13-7 Hingeless arches. The hingeless arch has two advantages. First, the positive moments near the center of the span will be smaller than the positive moments near the center of a corresponding two-hinged arch. Second, the deflections of the hingeless arch will be somewhat less than those of the two-hinged type, and this advantage is important when spans are long. The hingeless type has the disadvantage, however, that it may be difficult to assure absolute fixity at the ends of the arch.

When an arch is to support a highway or railway, influence lines for the stress components (thrust, shear, and moment) will be required for various sections of the rib. The rib as finally designed in Example 13-4 will be considered to be fixed at the ends to illustrate the method of computing influence lines for a hingeless arch. Influence line ordinates will be computed for H_0, V_0, and M_0, the redundant reaction components acting at the elastic center. This procedure will not only demonstrate the method for computing influence lines but will also indicate the effects of fixing the ends of the rib.

Attention is called to the fact that when hingeless arches are subjected to static loads, they may be conveniently analyzed either by the method of the elastic center or by the column analogy. Neither method offers any real advantage over the other. However, when influence lines are required, the elastic center, combined with the Müller-Breslau principle, is definitely the better method. Note that the method of the elastic center may be applied in two ways, as demonstrated in Example 5-30. If the arch is unsymmetrical, the rigid bracket is attached to one end and conjugate axes of inertia may be used. If, however, the arch is symmetrical, it is best to cut it at the center and use two rigid brackets, as indicated in the second analysis of Example 5-30. This latter method will be used in the following example.

EXAMPLE 13-6. The loaded arch rib as finally designed in Example 13-4 is considered to be fixed at the ends (see Fig. 13-11). The flange plates of the rib are 12 in. $\times$ 1 in. and the web is $\frac{3}{4}$ in. thick. Depths, sectional areas, and moments of inertia are given in Table 13-10. Compute ordinates at segment centers and the crown for the influence lines for H_0, V_0, and M_0. Using these influence lines, determine the decimal of capacity load throughout the rib, at segment centers and the crown, when it is loaded as in Example 13-4.

The first step in the analysis is to locate the elastic center. This, of course, will be on the center line of the span. The elevation above the two ends of the arch will be given by

$$\bar{y} = \frac{\Sigma \Delta S y/EI}{\Sigma \Delta S/EI} = \frac{\Sigma y/I}{\Sigma 1/I}.$$

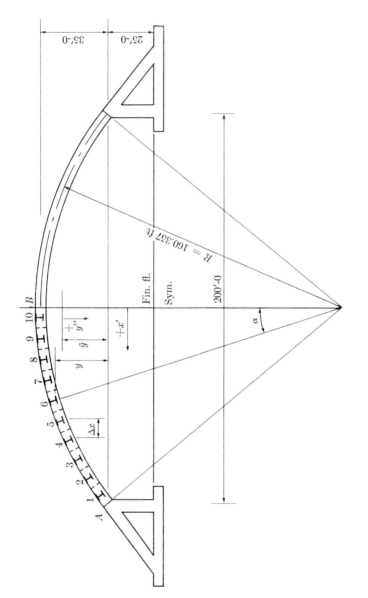

FIGURE 13-11

TABLE 13-10.

Section or segment	Total depth (in.)	Area (in²)	Moment of inertia (in⁴)
A	23.00	39.75	3483
1	23.00	39.75	3483
2	23.00	39.75	3483
3	23.00	39.75	3483
4	22.07	39.05	3172
5	21.13	38.35	2865
6	20.20	37.65	2585
7	19.27	36.95	2329
8	18.32	36.24	2071
9	17.40	35.55	1838
10	16.47	34.85	1631
Crown	16.00	34.50	1527

The necessary computations are shown in Table 13-11. The last column in this table gives the value of y'', the vertical distance from the elastic center to the center of each segment.

TABLE 13-11.

Segment	y (ft)	$\dfrac{I}{1000}$	$\dfrac{1000y}{I}$	$\dfrac{1000}{I}$	$y'' = \bar{y} - y$
1	3.29	3.483	0.94	0.2871	+22.40
2	9.44	3.483	2.71	0.2871	+16.25
3	14.98	3.483	4.30	0.2871	+10.71
4	19.88	3.172	6.27	0.3153	+5.81
5	24.13	2.865	8.42	0.3490	+1.56
6	27.69	2.585	10.71	0.3868	−2.00
7	30.57	2.329	13.12	0.4294	−4.88
8	32.73	2.071	15.80	0.4829	−7.04
9	34.18	1.838	18.60	0.5441	−8.49
10	34.91	1.631	21.40	0.6131	−9.22
			102.28	3.9819	

$$\bar{y} = \frac{102.28}{3.9819} = 25.69 \text{ ft.}$$

Rigid brackets are attached to the cut ends of the arch rib at the center and two 1 k horizontal forces are applied as shown in Fig. 13–12(a). By the Müller-Breslau principle the resulting vertical deflection component of any definite point on the arch rib, divided by the horizontal spread between the ends of the two rigid brackets, will be the H_0 influence line ordinate for a vertical load at the definite point on the rib.

The conjugate structure for the left half rib, for computing deflections, is shown in Fig. 13–12(b). Vertical deflections are computed for the centers of the segments and relative values for these deflections appear in column (8) of Table 13–12. One-half the spread between O_L and O_R is represented relatively by the sum of column (9). Ordinates for the influence line for H_0 are shown in column (10). The value for H_0 for the loaded rib is obtained by multiplying twice the sum of column (10) by the purlin concentration on the rib. Consequently, for the loaded rib,

$$H_0 = 2 \times 6.973 \times 24.565 = 342.6 \text{ k}.$$

In Fig. 13–13(a) two vertical forces of 1 k, one up and one down, are applied to the ends of the rigid brackets. Again, by the Müller-Breslau principle the resulting vertical deflection component of any definite point on the rib, divided by the resulting vertical spread between O_L and O_R, will be the V_0 influence line ordinate for a vertical load at the definite point on the rib. The conjugate structure for the left half of the rib is shown in Fig. 13–13(b), and all computations are shown in Table 13–13. Note that all V_0 influence line ordinates for points in the right half of the arch will have the same value as for the opposing points in the left half, but will have a negative sign. Consequently, V_0 for the loaded rib will be zero.

Finally, a pair of unit couples are applied to the ends of the rigid brackets in Fig. 13–14(a). As before, by the Müller-Breslau principle the vertical deflection component of any definite point on the rib, divided by the resulting rotational spread between O_L and O_R (in radians), will be the M_0 influence line ordinate for a vertical load at the definite point. The conjugate structure is shown in Fig. 13–14(b), and all computations are shown in Table 13–14. The value of M_0 for the loaded arch is determined by adding the influence line ordinates for segments 1 through 10, doubling this value to include the other half of the rib, and multiplying the total by the purlin concentration. For the loaded rib,

$$M_0 = 2 \times 65.55 \times 24.565 = 3220 \text{ ft·k}.$$

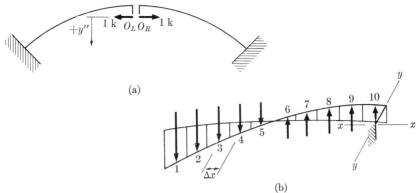

Figure 13-12

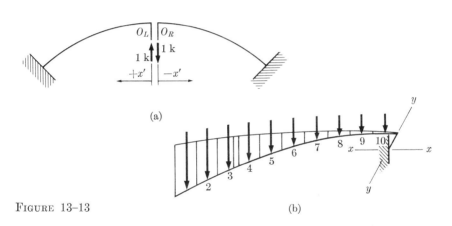

Figure 13-13

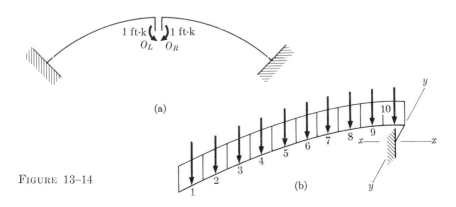

Figure 13-14

TABLE 13-12.

Segment	(2) $y'' = m$ (ft·k)	(3) $\dfrac{I}{1000}$	(4) Relative elastic load = $\dfrac{1000 y''}{I}$	(5) Shear to right	(6) Δx (ft)	(7) Moment increment	(8) Relative vertical deflection of segment center = total M	(9) $\Delta_{oh} =$ (elastic load) y''	(10) H_0 influence line ordinate = $(8) \div 2\Sigma(9)$
1	+22.40	3.483	+6.43	+6.43	8.874	57.1	57.1	+144.0	0.073
2	+16.25	3.483	+4.66	+11.09	9.267	102.8	159.9	+75.8	0.204
3	+10.71	3.483	+3.07	+14.16	9.618	136.3	296.2	+32.9	0.378
4	+5.81	3.172	+1.83	+15.99	9.926	158.8	455.0	+10.6	0.581
5	+1.56	2.865	+0.54	+16.53	10.189	168.6	623.6	+0.9	0.797
6	−2.00	2.585	−0.78	+15.75	10.408	164.1	787.7	+1.6	1.007
7	−4.88	2.329	−2.09	+13.66	10.577	144.6	932.3	+10.2	1.191
8	−7.04	2.071	−3.40	+10.26	10.697	109.9	1042.2	+23.9	1.332
9	−8.49	1.838	−4.62	+5.64	10.770	60.9	1102.1	+39.2	1.410
10	−9.22	1.631	−5.64	0				+52.1	
Σ								391.2	6.973

TABLE 13-13.

Segment	(2) $m = x'$ (ft·k)	(3) $\dfrac{I}{1000}$	(4) Relative elastic load = $\dfrac{1000\,x'}{I}$	(5) Shear to right	(6) Δx (ft)	(7) Moment increment	(8) Relative vertical deflection of segment center = total M	(9) Δ_{o_v} = (elastic load) x'	(10) V_0 influence line ordinate = $(8) \div 2\Sigma(9)$
1	95.72	3.483	27.48	27.48	8.874	243.9	243.9	2,631	0.011
2	86.85	3.483	24.94	52.42	9.267	485.8	729.7	2,166	0.033
3	77.58	3.483	22.28	74.70	9.618	718.4	1,448.1	1,728	0.065
4	67.97	3.172	21.43	96.13	9.926	954.1	2,402.2	1,456	0.108
5	58.04	2.865	20.26	116.39	10.189	1185.8	3,587.0	1,176	0.161
6	47.85	2.585	18.51	134.90	10.408	1403.9	4,991.9	886	0.224
7	37.44	2.329	16.08	150.98	10.577	1596.8	6,588.7	602	0.300
8	26.87	2.071	12.97	163.95	10.697	1753.7	8,342.4	349	0.374
9	16.17	1.838	8.80	172.75	10.770	1860.4	10,202.8	142	0.457
10	5.40	1.631	3.31	176.06	5.399	950.5	11,153.3	18	0.500
Crown									
Σ								11,154	

568   ELASTIC ARCHES   [CHAP. 13

TABLE 13-14.

Segment	(2) m (ft·k)	(3) $\dfrac{I}{1000}$	(4) Relative elastic load = $\dfrac{1000}{I}$	(5) Shear to right	(6) Δx (ft)	(7) Moment increment	(8) Relative vertical deflection of segment center = total M	(9) M_0 influence line ordinate = (8) ÷ 2Σ(4)
1	1	3.483	0.287	0.287	8.874	2.54	2.54	0.32
2	1	3.483	0.287	0.574	9.267	5.32	7.86	0.99
3	1	3.483	0.287	0.861	9.618	8.28	16.14	2.03
4	1	3.172	0.315	1.176	9.926	11.68	27.82	3.49
5	1	2.865	0.349	1.525	10.189	15.54	43.36	5.45
6	1	2.585	0.387	1.912	10.408	19.91	63.27	7.95
7	1	2.329	0.429	2.341	10.577	24.77	88.04	11.06
8	1	2.071	0.483	2.824	10.697	30.22	118.26	14.85
9	1	1.838	0.544	3.368	10.770	36.28	154.54	19.41
10 Crown	1	1.631	0.613	3.981				
Σ			3.981					65.55

Having determined the values of H_0, V_0, and M_0 for the loaded rib, we must now evaluate the thrust and moment at the centers of the various segments of the rib and at the crown. The moment at the crown is computed by statics (see Fig. 13–15) as follows:

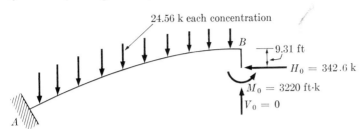

FIGURE 13–15

$$M_B = +3220 - 342.6 \times 9.31 = +30 \text{ ft·k (counterclockwise)}.$$

The moments at the centers of the various segments are most easily computed by first finding the value of the external moment at the left end of the arch. This is easily determined with the information shown in Fig. 13–16.

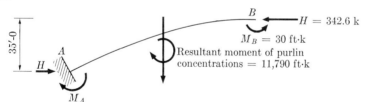

FIGURE 13–16

If M_A is assumed to be clockwise, and clockwise moments are considered to be positive, the equation expressing the fact that $\Sigma M_A = 0$ is

$$M_A + 11{,}790 - 30 - 342.6 \times 35.00 = 0,$$

from which

$$M_A = +230 \text{ ft·k}.$$

The moments at the centers of the various segments are computed in Table 13–15. Note that the simple beam moments shown in column (3) are taken from Table 13–2.

TABLE 13–15.

Segment	(2) y (ft)	(3) Simple beam M (ft·k)	(4) M_A (ft·k)	(5) Hy (ft·k)	(6) Total moment (ft·k)
A	0.00	0	+230		+230
1	3.29	+1,050	+230	−1,130	+150
2	9.44	+3,010	+230	−3,230	+10
3	14.98	+4,830	+230	−5,130	−70
4	19.88	+6,490	+230	−6,810	−90
5	24.13	+7,950	+230	−8,270	−90
6	27.69	+9,200	+230	−9,490	−60
7	30.57	+10,220	+230	−10,470	−20
8	32.73	+11,000	+230	−11,210	+20
9	34.18	+11,530	+230	−11,710	+50
10	34.91	+11,790	+230	−11,960	+60
Crown	35.00	+11,790	+230	−11,990	+30

The values for the axial thrust N at the centers of the various segments are computed in Table 13–16.

TABLE 13–16.

Segment	$V \sin \alpha$	$+ H \cos \alpha$	$= N$(k)
A	153	268	421
1	132	275	407
2	106	288	394
3	83	300	383
4	62	310	372
5	44	319	363
6	29	327	356
7	17	333	350
8	8	338	346
9	2	341	343
10	0	343	343
Crown	0	343	343

Finally, in Table 13–17, the adequacy of the rib is checked at the springing, the crown, and the centers of the various segments.

TABLE 13-17.

Segment	$\dfrac{N}{A}$ (k/in²)	$\dfrac{Mc}{I}$ (k/in²)	$\dfrac{N/A}{F_A} + \dfrac{Mc/I}{F_B} =$		Decimal of capacity
A	10.6	9.1	0.67	0.45	1.12
1	10.2	6.0	0.64	0.30	0.94
2	9.9	0.4	0.62	0.02	0.64
3	9.6	2.8	0.61	0.14	0.75
4	9.5	3.7	0.60	0.18	0.78
5	9.5	4.0	0:60	0.20	0.80
6	9.5	2.8	0.59	0.14	0.73
7	9.5	1.0	0.59	0.05	0.64
8	9.5	1.1	0.60	0.06	0.66
9	9.7	2.8	0.60	0.14	0.74
10	9.8	3.6	0.61	0.18	0.79
Crown	9.9	1.9	0.62	0.10	0.72

The values in the extreme right column of Table 13-17 indicate that a larger section is required at the springing, and that from segment 1 to the crown the rib may be somewhat over-designed. This cannot be stated with assurance, however, until the secondary stresses due to deflection of the rib are determined. These will be considered in Section 13-8.

EXAMPLE 13-7. An arch bridge with a span of 200 ft is to be built on a private road. The deck is to be wide enough for one lane of traffic. The anticipated loads are such that two ribs like those of Example 13-6 may be satisfactory. Construct the influence lines for thrust, shear, and moment at the crown, left quarter-point, and left springing so that the adequacy of the rib may be checked at these sections. The general arrangement of the bridge is shown in Fig. 13-17.

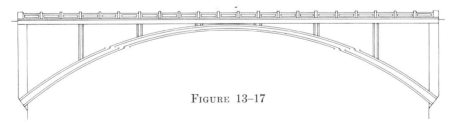

FIGURE 13-17.

When loads are transmitted from the bridge deck to the rib through columns, the resulting structure is known as an *open spandrel arch*. The spandrel of an arch is that portion of the structure which is located above the ribs or, in the case of a filled spandrel arch, above the ring. The open spandrel arch usually has two ribs supporting the deck. In the case of a filled spandrel arch only one rib (ring) is used, but the width of this ring is equal to the full width of the structure. Extending up to the roadway level from each side of the ring are masonry or reinforced concrete walls. The space between these walls is filled with gravel or with other suitable fill and the roadway slab built on top. Most (if not all) arch bridges now built are of the open spandrel type.

In the case of Example 13–6 the rib was analyzed for full load. The procedure was to compute the ordinates, at segment centers, for the influence lines for H_0, V_0, and M_0. The ordinates for each of these were then added and multiplied by 2 in order to obtain the total for the entire rib. In each case this total was then multiplied by the concentration to be applied at each segment center in order to obtain the values for H_0, V_0, and M_0 for the fully loaded rib. Then, by statics, stresses were obtained throughout the rib.

This procedure is based on the assumption that a partial loading of the rib, resulting in a more critical combination of stresses, is impossible. That is, it is considered very unlikely that a heavy snow load will accumulate on one-half of the roof while the other half remains unloaded. Occasionally it may be necessary or desirable to consider partial conditions of load, even for arch ribs supporting the roof of a structure. In some cases movable loads may be suspended from the inside. When the arch rib is to support a bridge deck, moving loads must, of course, be considered. Influence lines will be required, and these will now be discussed.

The extreme right colums of Tables 13–12, 13–13, and 13–14 show the ordinates of the influence lines for H_0, V_0, and M_0 for the rib under consideration. These act as shown in Fig. 13–18. It is apparent that H_c and

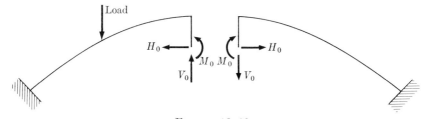

FIGURE 13–18

V_c, the thrust and shear, respectively, at the crown, are equal to H_0 and V_0. Therefore the ordinates for the influence lines for H_0 and V_0 may be plotted directly to obtain the influence lines for H_c and V_c. The influence line ordinates for crown moment, M_c, must be computed from the ordinates for H_0 and M_0 in accordance with

$$M_c = M_0 - 9.31 H_0.$$

These computations are given in Table 13-18.

TABLE 13-18.

Segment	M_0 (ft·k)	H_0 (k)	$9.31 H_0$ (ft·k)	M_c (ft·k)
1	—	—	—	—
2	0.32	0.07	0.65	−0.33
3	0.99	0.20	1.86	−0.87
4	2.03	0.38	3.54	−1.51
5	3.49	0.58	5.40	−1.91
6	5.45	0.80	7.45	−2.00
7	7.95	1.01	9.40	−1.45
8	11.06	1.19	11.08	−0.02
9	14.85	1.33	12.38	+2.47
10	19.41	1.41	13.13	+6.28
Crown	22.11	1.41	13.13	+8.98

The influence lines for H_c, V_c, and M_c are drawn as dashed curves in Fig. 13-19. These curves would be the correct influence lines for H_c, V_c, and M_c if loads could be applied directly to the rib throughout its entire length. It is proposed, however, to build the bridge with seven panels in the deck and six columns extending down to the rib. Therefore, loads actually can only be applied to the rib through these six columns. Consequently, the six points corresponding to these columns will be the only points of the required influence lines which will be on the dashed curves of Fig. 13-19. Each influence line will consist of a series of straight lines between these points, and these are shown on this figure. The values of the ordinates at the six column points, as read from the dashed curves, are shown in Table 13-19.

TABLE 13-19.

Load point	Thrust H_c (k)	Shear V_c (k)	Moment M_c (ft·k)
1	0.30	−0.052	−1.30
2	0.90	−0.193	−1.80
3	1.35	−0.390	+3.00
4	1.35	+0.390	+3.00
5	0.90	+0.193	−1.80
6	0.30	+0.052	−1.30

574 ELASTIC ARCHES [CHAP. 13

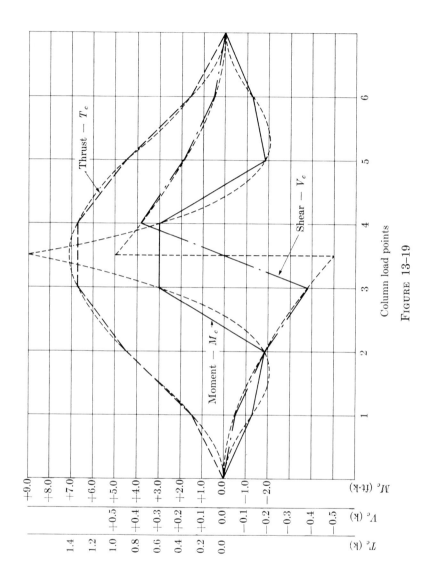

FIGURE 13-19

Having obtained the critical ordinates for the influence lines at the crown, we may in turn use these ordinates to compute the critical ordinates for the influence lines at the left springing and at the left quarter-point.

In Fig. 13–20, α is the angle between the horizontal and the tangent to the rib at any section for which the influence lines are desired. H_c, V_c, and M_c are all shown with positive senses. That is, M_c will cause compression on the top side of the rib, H_c will cause a thrust in the rib, and V_c indicates the sense of the shearing action when positive shear exists. In this case, positive shear is arbitrarily defined as the shearing action when the part of the rib to the right of a section tends to move down relative to the part of the rib to the left of the section. Note in Fig. 13–20 that the 1 k load is shown dotted at load point 2 merely to indicate the two components into which this unit load will be resolved when acting at load points 1, 2, or 3. When this 1 k load is actually acting at any one of these three points, V_c will have a sense opposite to that indicated in this figure.

From Fig. 13–20 it should be apparent that the thrust, shear, and moment at the left springing, respectively designated by T_s, S_s, and M_s, will be given by the following:

$$+H_c \cos \alpha_s + V_c \sin \alpha_s + 1 \cdot \sin \alpha_s = T_s,$$
$$-H_c \sin \alpha_s + V_c \cos \alpha_s + 1 \cdot \cos \alpha_s = S_s, \qquad (13\text{-}6)$$
$$+35.00 H_c - 100.00 V_c + M_c - 1(x \text{ of 1 k load}) = M_s.$$

Note that the last term in the expression for M_s will have a value only when the 1 k load is at column points 1, 2, and 3.

The equations for T_q, S_q, and M_q, the quarter-point thrust, shear, and moment, respectively, are

$$+H_c \cos \alpha_q + V_c \sin \alpha_q + 1 \cdot \sin \alpha_q = T_q,$$
$$-H_c \sin \alpha_q + V_c \cos \alpha_q + 1 \cdot \sin \alpha_q = S_q, \qquad (13\text{-}7)$$
$$+7.99 H_c - 50.00 V_v + M_c - 1(x \text{ of 1 k load}) = M_q.$$

The last term of the expression for M_q will have a value only when the 1 k load is at column points 2 and 3.

The coordinates for the required influence lines at the left springing and at the left quarter-point are computed in Tables 13–20 and 13–21. These influence lines are plotted in Figs. 13–21 and 13–22. The rib can be checked for adequacy at these sections by placing the live loads on the span in positions, as indicated by the influence lines, to give possible critical combinations of stresses.

576 ELASTIC ARCHES [CHAP. 13

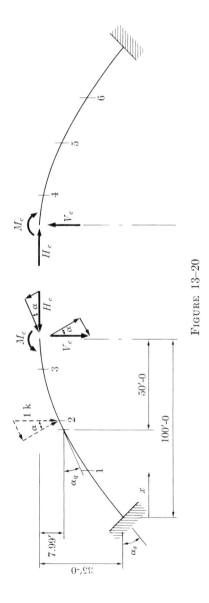

FIGURE 13-20

TABLE 13-20.

ORDINATES FOR INFLUENCE LINES FOR LEFT SPRINGING

Load point	Thrust				
	$+H_c \cos \alpha_s$ $+0.624 H_c$	$+V_c \sin \alpha_s$ $+0.782 V_c$	$+1 \cdot \sin \alpha_s$ $+0.782$	=	T_s (k)
1	+0.19	−0.04	+0.78		+0.93
2	+0.56	−0.15	+0.78		+1.19
3	+0.84	−0.30	+0.78		+1.32
4	+0.84	+0.30	0		+1.14
5	+0.56	+0.15	0		+0.71
6	+0.19	+0.04	0		+0.23

Load point	Shear				
	$-H_c \sin \alpha_s$ $-0.782 H_c$	$+V_c \cos \alpha_s$ $+0.624 V_c$	$+1 \cdot \cos \alpha_s$ $+0.624$	=	S_s (k)
1	−0.23	−0.03	+0.62		+0.36
2	−0.70	−0.12	+0.62		−0.20
3	−1.06	−0.24	+0.62		−0.68
4	−1.06	+0.24	0		−0.82
5	−0.70	+0.12	0		−0.58
6	−0.23	+0.03	0		−0.20

Load point	Moment					
	$+35.00 H_c$	$-100.00 V_c$	$+M_c$	$-1(x \text{ of } 1 \text{ k load})$	=	M_s (ft·k)
1	+10.5	+5.2	−1.3	−28.6		−14.2
2	+31.5	+19.3	−1.8	−57.1		−8.1
3	+47.2	+39.0	+3.0	−85.7		+3.5
4	+47.2	−39.0	+3.0	0		+11.2
5	+31.5	−19.3	−1.8	0		+10.4
6	+10.5	−5.2	−1.3	0		+4.0

TABLE 13–21.

ORDINATES FOR INFLUENCE LINES AT LEFT QUARTER-POINT

Load point	Thrust				
	$+H_c \cos \alpha_q$ $+0.950 H_c$	$+V_c \sin \alpha_q$ $+0.312 V_c$	$+1 \cdot \sin \alpha_q$ $+0.312$	$=$	T_q (k)
1	$+0.29$	-0.02	0		$+0.27$
2	$+0.86$	-0.06	$+0.31$		$+1.11$
3	$+1.28$	-0.12	$+0.31$		$+1.47$
4	$+1.28$	$+0.12$	0		$+1.40$
5	$+0.86$	$+0.06$	0		$+0.92$
6	$+0.29$	$+0.02$	0		$+0.31$

Load point	Shear				
	$-H_c \sin \alpha_q$ $-0.312 H_c$	$+V_c \cos \alpha_q$ $+0.950 V_c$	$+1 \cdot \cos \alpha_q$ $+0.950$	$=$	S_q (k)
1	-0.09	-0.05	0		-0.14
2	-0.28	-0.18	$+0.95$		$+0.49$
3	-0.42	-0.37	$+0.95$		$+0.16$
4	-0.42	$+0.37$	0		-0.05
5	-0.28	$+0.18$	0		-0.10
6	-0.09	$+0.05$	0		-0.04

Load point	Moment					
	$+7.99 H_c$	$-50.00 V_c$	$+M_c$	$-1(x \text{ of } 1 \text{ k load} - 25.00)$	$=$	M_q (ft·k)
1	$+2.4$	$+2.6$	-1.3	0		$+3.7$
2	$+7.2$	$+9.6$	-1.8	-7.1		$+7.9$
3	$+10.8$	$+19.5$	$+3.0$	-35.7		-2.4
4	$+10.8$	-19.5	$+3.0$	0		-5.7
5	$+7.2$	-9.6	-1.8	0		-4.2
6	$+2.4$	-2.6	-1.3	0		-1.5

13-7]   HINGELESS ARCHES   579

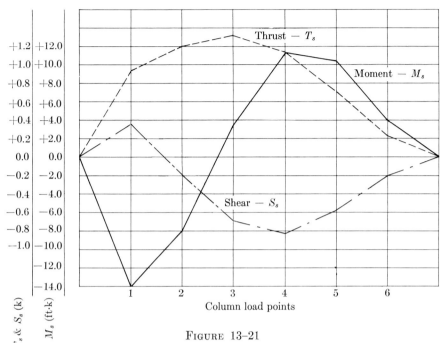

Figure 13-21

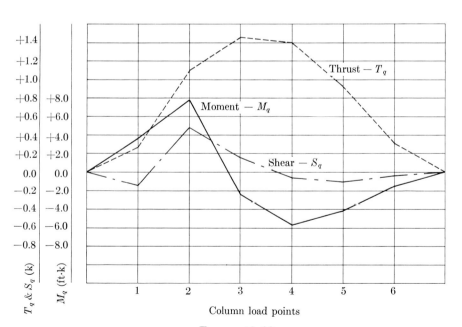

Figure 13-22

13-8 Secondary stresses in arches. As indicated in Section 3-7, the stresses in an arch rib may be materially affected by the deflection of the rib when loads are applied. If a rib is analyzed by the "elastic theory," the effects of deflection are ignored. When the effects of deflection are included in the analysis, the arch is considered to have been analyzed by the "deflection theory." In each case the rib is considered to be elastic.

When the elastic theory is applied, it is assumed that the lever arms of the horizontal reaction components will be equal to the vertical ordinates to the axis of the unloaded rib. Obviously this is incorrect because the horizontal reaction components cannot exist until the arch is loaded and in its deflected position. The deflected position cannot be determined until the moments are known throughout the loaded rib.

A satisfactory procedure for including the effects of deflection in the analysis of an arch rib is as follows:

(1) Having analyzed the rib by the elastic theory and having designed it, compute the vertical deflection components for all segment centers and the crown as caused by the moments obtained by the analysis. Since only vertical deflection components are required, these may be obtained by applying the elastic loads to a conjugate beam with a span equal to the span of the arch. If a hingeless arch is being analyzed, the conjugate beam will have no reactions, since the arch ends will not rotate or deflect. If the arch is two-hinged, the conjugate beam will be a simple beam with an external reaction at each end. Apply the deflections thus determined as corrections to the vertical ordinates to the undeflected arch axis in order to obtain the ordinates to the deflected rib.

(2) Make a second analysis by the elastic theory, using the vertical ordinates to the deflected rib as determined in (1). If the arch is two-hinged, the computations of Table 13-6 will have to be repeated using the new values of $m = y$. If the arch is hingeless, the computations of Tables 13-11 and 13-12 will have to be repeated and a new value for H_0 determined. In either case new moments must be computed for all segments.

(3) Compute a new set of deflection components using the revised moments obtained in (2), and then make another analysis as in step (2). The cycle is repeated until the differences between the values of the horizontal reaction components, moments, and deflections for two succeeding cycles are considered to be negligible.

It will be found that two-hinged arches are more seriously affected by deflections than are hingeless arches. It is important to note that unless the series of corrections as determined by the cycles of computations (as described above) converge, the rib will collapse.

In the case of the two-hinged rib of Example 13-4, the computations of steps (2) and (3) above were repeated five times. The differences between corresponding values, as given by the fourth and fifth cycles, were small enough to indicate that additional cycles were unnecessary. These computations were based on initial moments, obtained from a first cycle, which were accurate to the nearest ft·k. Moments and deflections for the first, second, and fifth cycles are shown in Table 13–22. Negative moments cause tension on top of the rib and positive deflections are up.

TABLE 13–22.

Segment or section	Moment (ft·k)			Deflection (in.)		
	Cycle 1	Cycle 2	Cycle 5	Cycle 1	Cycle 2	Cycle 5
1	−54	−65	−70	+0.31	+0.39	+0.43
2	−151	−182	−196	+0.87	+1.11	+1.21
3	−186	−229	−249	+1.21	+1.55	+1.70
4	−172	−219	−243	+1.24	+1.62	+1.78
5	−130	−175	−198	+0.94	+1.26	+1.39
6	−73	−105	−122	+0.34	+0.51	+0.55
7	−12	−25	−33	−0.45	−0.52	−0.61
8	+42	+50	+55	−1.29	−1.64	−1.89
9	+82	+110	+124	−2.01	−2.61	−3.00
10	+103	+141	+163	−2.43	−3.19	−3.67
Crown	+72	+111	+133	−2.43	−3.19	−3.67

The values of H, the horizontal reaction component, for the first through the fifth cycles are 334.8 k, 335.5 k, 335.9 k, 336.0 k, and 336.1 k. A check on the decimal capacity of the rib, as in Table 13–8, will indicate that its design capacity is exceeded at segments 3 and 10 by about 10%. The rib section should be enlarged and the entire analysis repeated.

In the case of the hingeless rib of Example 13–6, three cycles of computations were required, and the results of these are shown in Table 13–23. Moments are accurate to the nearest ft·k.

The value of H, the horizontal reaction component, increases from 342.6 k in the first cycle to 343.8 k in the second cycle, and finally to 344.1 k in the third cycle. A check of the decimal capacity of the rib indicates that it is satisfactory at the crown but that its design capacity is exceeded by about 20% at the springing.

It is interesting to note that while designing the Rainbow Arch for Niagara Falls, Hardesty, Garrelts, and Hedrick (1) found that values for the moments at various sections of a two-hinged rib, as obtained by the deflection theory, can be rather closely approximated by completing two cycles of computations and then applying the following equation:

$$M_d = M_e \left[\frac{1}{1 - \Delta m / M_e} \right]. \tag{13-8}$$

TABLE 13-23.

Segment or section	Moment (ft·k)			Deflection (in.)		
	Cycle 1	Cycle 2	Cycle 3	Cycle 1	Cycle 2	Cycle 3
A	+227	+253	+258			
1	+149	+170	+174			
2	+5	+12	+13	+0.24	+0.27	+0.28
3	−72	−79	−80	+0.49	+0.57	+0.58
4	−95	−115	−118	+0.63	+0.75	+0.77
5	−89	−109	−114	+0.58	+0.71	+0.73
6	−59	−77	−81	+0.33	+0.42	+0.44
7	−21	−30	−33	−0.07	−0.06	−0.07
8	+17	+17	+17	−0.53	−0.64	−0.67
9	+46	+57	+60	−0.95	−1.16	−1.23
10	+61	+78	+83	−1.20	−1.48	−1.57
Crown	+30	+47	+51	−1.20	−1.48	−1.57

In Eq. 13-8 M_d is approximately the value for the moment at a given section which would be obtained if the deflection theory were applied; M_e is the value obtained for the moment at the same section by the first cycle of the elastic theory; and Δm is the difference in the given moment as indicated by the first and second cycles of the elastic theory. If the appropriate values in Tables 13-22 and 13-23 are substituted in this equation, the results will be as shown in Table 13-24.

TABLE 13-24.

Segment or section	Moments (ft·k)	
	Two-hinged rib	Hingeless rib
A	0	+256
1	−68	+173
2	−190	0
3	−242	−80
4	−237	−120
5	−199	−115
6	−130	−85
7	0	−37
8	+52	+17
9	+124	+61
10	+163	+85

These results agree very well, particularly for the hingeless arch, with the final values for moments in Tables 13-22 and 13-23.

Specific Reference

1. "Rainbow Arch Bridge over Niagara Gorge—A Symposium," *Trans. Am. Soc. Civ. Engrs.*, **110,** 1–178, 1945.

General References

2. FREUDENTHAL, A., "Deflection Theory for Arches," *Publ. Int. Assn. Bridge Struct. Engrs.* (Zurich), Vol. 3, 1935.
3. FRITZ, B., *Theorie und Berechnung vollwändiger Bogenträger bei Berücksichtigung des Einflusses der Systemverformung.* Berlin: Springer, 1934.
4. GARRELTS, J. M., "Design of St. George's Tied Arch Span," *Trans. Am. Soc. Civ. Engrs.*, **108,** 543–54, 1943.
5. GRINTER, L. E., *Theory of Modern Steel Structures.* Revised ed. Chap. 8. New York: Macmillan, 1949.
6. HOOL, G. A. and KINNE, W. S., *Reinforced Concrete and Masonry Structures.* 2nd ed. Sec. 8. New York: McGraw-Hill, 1944.
7. JOHNSON, J. B., BRYAN, C. W., and TURNEAURE, F. E., *Modern Framed Structures.* 10th ed. Part II, chap. 4. New York: Wiley, 1929.
8. KASARNOWSKY, S., "Beitrag zur Theorie weitgespannter Brückenbogen mit Kämpfergelenken," *Stahlbau,* Copy 6, 1931.
9. MAUGH, L. C., *Statically Indeterminate Structures.* Chap. 8. New York: Wiley, 1946.
10. McCULLOUGH, C. B. and THAYER, E. S., *Elastic Arch Bridges.* New York: Wiley, 1931.
11. MELAN, J., "Genauere Theorie des Zweigelenkbogens mit Berücksichtigung der durch die Belastung erzeugten Formänderung," *Handbuch der Ingenieur-Wissenschaften,* Vol. 2, 1906.
12. O'ROURKE, C. E., URQUHART, L. C., and WINTER, G., *Design of Concrete Structures.* Sec. 10. New York: McGraw-Hill, 1954.
13. PIPPARD, A. J. S. and BAKER, J. F., *The Analysis of Engineering Structures.* 2nd ed. Chap. 12. London: Arnold, 1943.
14. SPOFFORD, C. M., *The Theory of Continuous Structures and Arches.* New York: McGraw-Hill, 1937.
15. WANG, C. K., *Statically Indeterminate Structures.* Chap. 10. New York: McGraw-Hill, 1953.
16. WILLIAMS, C. D., *Analysis of Statically Indeterminate Structures.* 2nd ed. Chap. 4. Scranton: International, 1946.

CHAPTER 14

MODEL ANALYSIS OF STRUCTURES

14-1 General. Structural model analysis has been increasingly recognized during the past two or three decades as an important tool of research and as a valuable supplement to the usual theoretical methods of structural analysis and design. The reasons for using a model to solve a structural problem are varied. A model analysis is indicated, for example, when an alternate mathematical analysis is either impossible or impractical. Occasionally a model analysis is required to verify the mathematical analysis of an unusual or complicated structure, as has been the case with several suspension bridges built in recent years. As another example, an engineer who works alone will often find that a simple model analysis is the best way to verify the final design of an indeterminate structure.

Unfortunately, however, it is not always possible to design and load a model so that its response will be similar to that of the structure it is supposed to simulate. For example, structural details such as welding or riveting do not scale up or down successfully. Reliable experimental information as to stress distribution in the immediate vicinity of these details must therefore be determined directly from the prototype. Another difficulty is due to the fact that the various conditions which must be satisfied to achieve complete similarity between the model and the prototype may be incompatible. In this case a model analysis is either impossible or of doubtful value. This difficulty is most often encountered with dynamic loads.

If the prototype is to be subjected to static loads, then the model will be loaded statically. In this case the principles involved in the design of the model, and the actual testing, are quite simple. Valuable experimental verification of a mathematical analysis can often be obtained with a simple model of cellulose acetate; hard cardboard can also be used with considerable success. If dynamic loads are to be considered, however, the principles involved are more complicated and expensive recording equipment is required. (Dynamic loading of models will not be considered in this discussion, although some pertinent references will be found in the bibliography at the end of this chapter.)

The various methods for the model analysis of structures subjected to static loads are conveniently divided into two classifications. These very logically have been designated (16) as the *indirect* and the *direct* types. In the indirect type of analysis the loading of the model is completely unrelated to the loading of the prototype. No readings of strain are taken on the model. A known displacement or distortion is impressed at the point of action and in the direction of a desired external reaction or internal stress component, and the resulting displacements of all load points are measured. Then, in accordance with the Müller-Breslau principle, an influence line is obtained for the reaction or stress component. In the case of a direct type of analysis, however, the model is usually loaded in exactly the same manner as the prototype. Strain measuring devices are often mounted directly on the model and the strains recorded. Model deflections are also observed, and corresponding stresses and deflections in the prototype are determined from these model strains and deflections.

Regardless of which type of analysis is used, the model must be designed in accordance with certain principles in order to establish definite relationships or similarities between the response of the loaded model and the response of the loaded prototype. The principles which establish these relationships are known as the *principles of similitude:* they govern both the design of the model and the extrapolation of the results to predict the prototype response. The principles are few and simple for any indirect type of model analysis and are slightly more numerous, but still simple, for a direct analysis of many common types of structures under static loads.

14-2 Structural similitude. Although this chapter will be restricted to consideration of the design and use of statically loaded structural models, it is nevertheless advisable to consider briefly the physical significance of the mechanics of similitude. In this discussion, frequent use will be made of the words *homologous* and *prototype*. The word *homologous* signifies the state or condition of having, or being in, the same relative position, proportion, or value. The word *prototype* means the original (in this case, the actual) structure. Thus homologous strains are strains in corresponding fibers of prototype and model for the same relative condition of load.

Similitude, or similarity, between two objects may exist with regard to any one of their physical characteristics. In reference to the required relationships between a prototype and a suitable model, however, three kinds of similitude will, in the general case, be necessary and sufficient: (1) *geometric* similitude, (2) *kinematic* similitude, and (3) *dynamic* and/or *mechanical* similitude.

Geometric similitude is similarity of form. This means that all homologous dimensions of prototype and model must be in some constant ratio. As this discussion develops, it will become evident that in some structural models certain dimensions, which do not affect the response of the model, may be exempted from this requirement.

Kinematic similitude is similarity of motion. This means that during any impressed movement or deflection all homologous particles of prototype and model must traverse geometrically similar paths and that the velocities of these homologous particles must always be in some constant ratio. When the loading is static this requirement means simply that the deflections of all homologous points of prototype and model must be in some constant ratio.

Dynamic and/or mechanical similitude is similarity of masses and/or forces. This means that kinematic similitude must, first of all, exist. In addition, the masses of all homologous parts of prototype and model, and all homologous forces which affect the motions of prototype and model, must be in some constant ratio. Absolute dynamic similitude is always difficult and usually impossible to achieve. In the case of static loadings, however, mechanical rather than dynamic similitude is required. Only gravitational and elastic forces are involved, and similitude can often be realized.

To summarize the last three paragraphs as they apply to statically loaded structural prototypes and models, the requirements for similitude are: (a) all homologous linear dimensions must be in some constant ratio, (b) all homologous linear deflections must be in some constant ratio, and (c) all homologous external loads, internal elastic forces, and dead loads must be in some constant ratio. These constant ratios are known as *scale factors*. When these scale factors are determined and applied to the several dimensions, properties, and loads of the prototype, the corresponding dimensions, properties, and loads for the model will result.

Scale factors for structural models may be derived by either of two methods:

(1) By the application of structural mechanics to express mathematically the conditions of similitude between the model and the prototype.

(2) By dimensional analysis.

Both methods will be discussed in this chapter.

14–3 Fundamentals of indirect model analysis. Consider the prototype beam shown in Fig. 14–1 on which the point O may be located anywhere

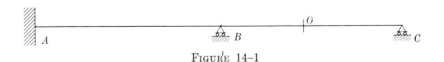

FIGURE 14–1

between A, B, and C. Assume that this beam is acted upon by two different systems of loads and reactions, as shown in Figs. 14–2(a) and 14–2(b). Applying the Maxwell-Betti reciprocal theorem to Systems 1 and 2, we obtain

$$R_B \cdot \Delta_{BB} - 1 \cdot \Delta_{OB} = 0,$$

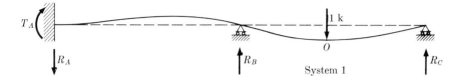

(a)

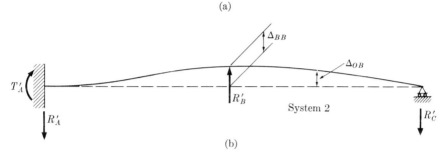

(b)

FIGURE 14-2

from which

$$R_B = 1 \cdot \frac{\Delta_{OB}}{\Delta_{BB}}. \tag{14-1}$$

Thus it is apparent, since O may be any point along the beam between reactions, that the deflected beam of Fig. 14-2(b) is the influence line for R_B.

As an alternate to System 2 in Fig. 14-2(b), consider System 3 as shown

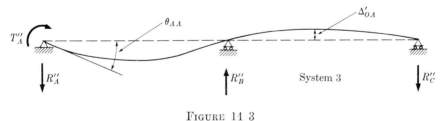

FIGURE 14-3

in Fig. 14-3. The Maxwell-Betti reciprocal theorem applied to Systems 1 and 3 results in

$$T_A \cdot \theta_{AA} - 1 \cdot \Delta'_{OA} = 0,$$

and therefore

$$T_A = 1 \cdot \frac{\Delta'_{OA}}{\theta_{AA}}. \tag{14-2}$$

As before, since O may be any point along the beam between reactions, the deflected beam of Fig. 14–3 is the influence line for T_A.

Equations (14–1) and (14–2) have been written for the prototype. The same procedure can be applied to a model. The corresponding equations (if a bar over a symbol represents a quantity on the model) would be

$$\overline{R}_B = 1 \cdot \frac{\overline{\Delta}_{OB}}{\overline{\Delta}_{BB}}, \tag{14-3}$$

$$\overline{T}_A = 1 \cdot \frac{\overline{\Delta}'_{OA}}{\overline{\theta}_{AA}}. \tag{14-4}$$

Before the model results can be extrapolated to obtain influence lines for the prototype, however, the relationships of R_B with $\overline{R}_B$ and of T_A with $\overline{T}_A$ must be established. Obviously these relationships depend on the relative deflections of the prototype and the model, and these can be determined by virtual work deflection equations. For the prototype of Fig. 14–2(b),

$$\Delta_{OB} = \int M_B \cdot m_O \cdot \frac{dx}{EI}. \tag{14-5}$$

In the above equation, M_B is the moment at any section of the prototype caused by R'_B, and m_O is the moment at any section due to a unit vertical fictitious load at O. The corresponding equation for a model of the prototype of Fig. 14–2(b) would be

$$\overline{\Delta}_{OB} = \int \overline{M}_B \cdot \overline{m}_O \cdot \frac{\overline{dx}}{\overline{EI}}. \tag{14-6}$$

In this case, $\overline{M}_B$ is the moment at any section of the model caused by $\overline{R}'_B$, and $\overline{m}_O$ is the moment at any section due to a unit vertical fictitious load at O on the model.

Assume that the following scale relationships exist between the model and the prototype:

$$k \cdot \overline{L} = L, \quad n \cdot \overline{I} = I, \quad q \cdot \overline{E} = E, \quad v \cdot \overline{F} = F,$$

where k, n, q, and v are scale factors and where L represents length, I is moment of inertia, E is the modulus of elasticity, and F represents force. It follows that

$$kv\overline{M}_B = M_B, \quad k\overline{m}_O = m_O, \quad k\,\overline{dx} = dx.$$

Substituting in Eq. (14–5),

$$\Delta_{OB} = \int (kv\overline{M}_B)(k\overline{m}_O) \frac{k\,\overline{dx}}{q\overline{E} \cdot n\overline{I}}$$

$$= \frac{k^3 v}{qn} \cdot \overline{\Delta}_{OB}. \tag{14-7}$$

This same relationship holds for all linear deflections of corresponding points of the prototype and the model. If Eq. (14–7) is substituted in Eq. (14–1) and compared with Eq. (14–3), the result is

$$R_B = 1 \cdot \frac{(k^3 v/qn) \cdot \overline{\Delta}_{OB}}{(k^3 v/qn) \cdot \overline{\Delta}_{BB}} = 1 \cdot \frac{\overline{\Delta}_{OB}}{\overline{\Delta}_{BB}} = \overline{R}_B. \tag{14-8}$$

It is thus evident that force reaction components for the prototype can be computed directly from an influence line obtained from a model and that they will be independent of the scale factors k, n, q, and v. In other words, the influence line for any force (external reaction component or internal shear or axial thrust) induced by loading of the prototype will be identical to the corresponding influence line obtained for the model.

Next, consider the prototype of Fig. 14–3. By virtual work the expression for θ_{AA} is

$$\theta_{AA} = \int M_A \cdot m_A \cdot \frac{dx}{EI}, \tag{14-9}$$

where M_A is the moment at any section of the prototype resulting from T_A'', and m_A is the moment at any section caused by a unit fictitious couple at A. In a model of the prototype of Fig. 14–3,

$$\overline{\theta}_{AA} = \int \overline{M}_A \cdot \overline{m}_A \cdot \frac{\overline{dx}}{\overline{EI}}, \tag{14-10}$$

where $\overline{M}_A$ is the moment at any section of the model resulting from the couple $\overline{T}_A''$, and $\overline{m}_A$ is the moment at any section caused by a unit fictitious couple at A. In this case,

$$v \cdot \overline{M}_A = M_A \quad \text{and} \quad \overline{m}_A = m_A.$$

Substituting in Eq. (14–9) and comparing with Eq. (14–10), we obtain

$$\theta_{AA} = \int (v\overline{M}_A)\overline{m}_A \cdot \frac{k\,\overline{dx}}{(q\overline{E})(n\overline{I})} = \frac{kv}{qn} \cdot \overline{\theta}_{AA}. \tag{14-11}$$

The expression for Δ'_{OA} for the prototype of Fig. 14–3 is

$$\Delta'_{OA} = \int M_A m_O \frac{dx}{EI}, \qquad (14\text{--}12)$$

where M_A is the moment at any section of the prototype as caused by T''_A, and m_O is the moment at any section caused by a unit vertical fictitious force at O. For the model,

$$\overline{\Delta'_{OA}} = \int \overline{M}_A \overline{m}_O \frac{\overline{dx}}{\overline{EI}}, \qquad (14\text{--}13)$$

where $\overline{M}_A$ is the moment at any section of the model as caused by $\overline{T}''_A$, and $\overline{m}_O$ is the moment at any section caused by a unit vertical fictitious force at O. Since the moments T''_A and $\overline{T}''_A$ are applied loads, then

$$v\overline{T}''_A = T''_A,$$

and consequently,

$$v\overline{M}_A = M_A.$$

Also,

$$k\overline{m}_O = m_O.$$

Substituting in Eq. (14–12) and comparing with Eq. (14–13), we find that

$$\Delta'_{OA} = \int (v\overline{M}_A)(k\overline{m}_O) \frac{k\,\overline{dx}}{(q\overline{E})(n\overline{I})} = \frac{k^2 v}{nq} \cdot \overline{\Delta'_{OA}}. \qquad (14\text{--}14)$$

If Eqs. (14–11) and (14–14) are substituted in Eq. (14–2), the result is

$$T_A = k \cdot \frac{\overline{\Delta'_{OA}}}{\overline{\theta}_{AA}}. \qquad (14\text{--}15)$$

Substitution of Eq. (14–4) in Eq. (14–15) yields

$$T_A = k \cdot \overline{T}_A. \qquad (14\text{--}16)$$

Thus it is apparent that influence line ordinates for a moment reaction component of the prototype may be obtained by multiplying by the linear dimension scale factor k the corresponding ordinates of the corresponding moment influence line obtained from the model. It can be demonstrated, in a manner similar to that used above, that this multiplication by k is also necessary when evaluating the ordinates of influence lines for internal moments in the prototype.

On the basis of the preceding discussion, it is apparent that a model which is to be used for an indirect analysis of a prototype in which only flexural strain is important must be designed in accordance with the following relationships:

$$k\overline{L} = L, \quad n\overline{I} = I, \quad q\overline{E} = E, \quad v\overline{F} = F.$$

The first of these relationships requires that some constant k times a linear dimension in the model must equal the corresponding dimension in the prototype. This is applied to axial dimensions only. When flexural strains alone are important, it is not necessary to apply the linear scale factor to cross-sectional dimensions.

The second relationship, however, does govern the cross-sectional dimensions to the extent that the moments of inertia of corresponding sections of homologous flexural members of model and prototype must be in some constant ratio, designated by n.

The third condition simply indicates that in all practical cases the modulus of elasticity of the construction material must be uniform throughout the model at any instant. It is permissible for this modulus to vary with time provided that the same variation occurs simultaneously for all the material in the model.

The fourth condition stipulates that homologous forces and loads, external and internal, must be in a constant ratio, designated by v. Although this requirement does apply, it is not important in indirect analysis since the magnitude of the applied force or moment necessary to produce a given distortion need not be known.

A noteworthy fact is that although the fundamentals of indirect model analysis have been developed in connection with a structure composed of flexural members, a similar development would have resulted if the prototype had been articulated. It is important to note that the second condition, $n\overline{I} = I$, applies only to flexural members. If an articulated prototype had been used for the demonstration, then this condition would have been replaced by $e\overline{A} = A$, where e is the scale factor for cross-sectional areas and A represents the cross-sectional area.

If several members are subjected to both axial and flexural loads, then both conditions theoretically should be satisfied. Although this may be possible, the conditions will often be found to be incompatible. Consequently, that condition should be satisfied which applies to the primary function of the member, that is, whether it be to resist flexural or axial loads, and the other condition should be satisfied as nearly as possible.

The properly designed model is subjected to an impressed displacement at the point of application and in the line of action of an external reaction component or internal stress component. The resulting deflection curve of the model, as measured at the points of application and in the direction of the loads to be applied to the prototype, will to some scale be an influence line for that external reaction component or internal stress component.

The impressed displacements necessary to obtain various influence lines may be explained by Maxwell's reciprocal deflection theorem. Suppose, for example, it is desired to compute the ordinates for an influence line for the reaction at B in the beam of Fig. 14-4(a). Let D be any point along

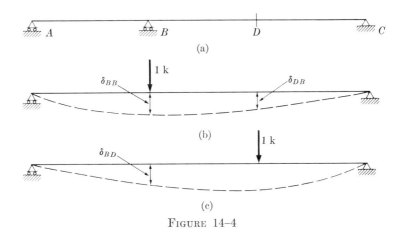

FIGURE 14-4

the beam between supports. The support at B is considered to be removed and a load of 1 k is assumed to act at B as shown in Fig. 14-4(b). The resulting deflections, δ_{BB} and δ_{DB}, are evaluated. The 1 k load is then moved to act at point D and the magnitude of the deflection δ_{BD} is determined. By the general method,

$$R_B = \frac{\delta_{BD}}{\delta_{BB}},$$

but by the reciprocal deflection theorem,

$$\delta_{BD} = \delta_{DB}.$$

Therefore,

$$R_B = \frac{\delta_{DB}}{\delta_{BB}}.$$

It is thus demonstrated that the deflection curve of the beam, with the reaction at B removed and a unit load applied at B, is an influence line for the reaction at B. Actually, any vertical force P could have been assumed instead of 1 k without changing the value of the influence line ordinate as given by the right side of the last equation above. In this case both numerator and denominator would have been multiplied by P.

If the influence line is to be determined experimentally by indirect model analysis, the magnitude of the force applied at B is of no importance. Consequently, the point B is caused to deflect a known amount. The influence line ordinate for the reaction at B, for any load point such as D, will then be the deflection of D divided by the impressed deflection at B.

If influence line ordinates are to be computed for the internal moment at B, the flexural continuity of the beam is assumed to be removed at this point. This is accomplished by considering a pin as inserted in the beam at B. A load of 1 k is applied at any point D and the beam deflects as shown in Fig. 14–5(a). The rotational deflection α'_{BD} having been

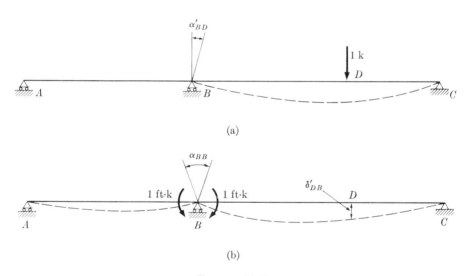

FIGURE 14–5

evaluated, the 1 k load is removed. Two unit couples are then applied to the ends of the two spans at B, and the values of the deflections α_{BB} and δ'_{DB} are determined. Then, by the general method and the reciprocal deflection theorem,

$$M_B = \frac{\alpha'_{BD}}{\alpha_{BB}} = \frac{\delta'_{DB}}{\alpha_{BB}}.$$

The manipulation of a model to determine the same influence line experimentally is indicated by Fig. 14–5(b). The model is cut through at B but each span is still pin-connected to the support. A known relative rotation α_{BB} is impressed between the ends of the two spans. The distorted model must be free to rotate about the pin support at B to a position of equilibrium. Any interference with this rotation will result in incorrect results. The influence line ordinates for M_B for a load at any point such as D will be equal numerically to the linear deflection δ'_{DB} divided by the rotational deflection α_{BB}, measured in radians. As previously demonstrated, each quotient must be multiplied by the linear scale factor k to obtain the proper value for the prototype. If δ'_{DB} is measured in inches, the influence line ordinate indicates the moment M_B in inch-kilopounds or in inch-pounds. If δ'_{DB} is measured in feet, the ordinate signifies the value of M_B in foot-kilopounds or foot-pounds.

Computation of the ordinates for the influence line for moment at any point E between supports is explained by means of Fig. 14–6(a) and (b). The beam at this section is assumed to be incapable of resisting the particular stress component for which the influence line is desired. Conse-

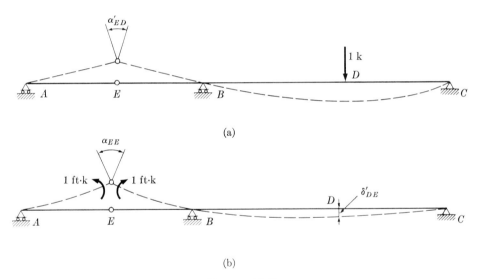

FIGURE 14–6

quently, a pin is considered to exist in the beam at E. In Fig. 14–6(a) a unit load is applied at D and the magnitude of the relative rotation α'_{ED} of the two portions of the beam on each side of the pin is computed. The unit load is removed and a pair of unit couples is applied as shown in 14–6(b). The resulting deflections, α_{EE} and δ'_{DE}, are evaluated. Then, as before,

$$M_E = \frac{\alpha'_{ED}}{\alpha_{EE}} = \frac{\delta'_{DE}}{\alpha_{EE}}.$$

Since D may be any point along the beam between supports, the curve of the deflected structure in Fig. 14–6(b) is an influence line for the internal moment at E.

When the influence line is to be determined directly from observed deflections of a model, the model is cut at section E. A device is clamped to the model at this cut which makes it possible to impress some small α_{EE} of known value. This device must not exert any external action other than to cause the relative rotation α_{EE}. Thus the model is free to assume a position of equilibrium consistent with the impressed distortion. The influence line ordinate for the moment at E for any load point such as D will then be equal to the observed linear deflection of D divided by the impressed relative rotation α_{EE}. The quotient must be multiplied by the linear scale factor k to obtain the correct value for the prototype.

Ordinates for an influence line for shear at any point such as E can also be computed. Consider that the beam of Fig. 14–7(a) is cut at E and that

(a)

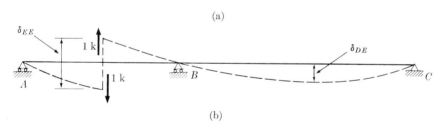

(b)

FIGURE 14–7

a device is inserted which permits a relative transverse deflection between the two beam ends at the cut, but which requires that these ends shall always have a common slope. In other words, the shearing resistance of the beam has been removed at E, but flexural resistance has not. In Fig. 14–7(a), a 1 k load is applied at D resulting in the relative linear deflection δ_{ED} at E. Application of a pair of 1 k loads at E, as in 14–7(b), results in δ_{EE} and δ_{DE}. As before,

$$S_E = \frac{\delta_{ED}}{\delta_{EE}} = \frac{\delta_{DE}}{\delta_{EE}}.$$

In the use of a model to obtain the influence line, a cut is made at E and a device as described above is actually clamped to the model on the two sides of the cut. A known δ_{EE} is applied and the deflected model, after assuming a position of equilibrium consistent with the impressed distortion, is to some scale the required influence line.

14-4 Model materials for indirect types of analysis. Certain characteristics are obviously desirable in a material to be used for constructing a model. The material should, of course, be inexpensive and easily obtainable. It should be easy to cut, shape, and join together. Its physical properties should not change with time, humidity, stress, or temperature. It should obey Hooke's law and it should be isotropic.

No one material will satisfy all these requirements. Steel, brass, aluminum, wood, and concrete have all been used as model construction materials with varying degrees of success. All things considered, however, it is probable that cellulose acetate will be found to be the most satisfactory material for the construction of models to be used in indirect analysis. It is easily machined and it is isotropic. It is readily joined by means of a cement made by dissolving scraps of the material in acetone. It is inexpensive. Time, humidity, and temperature will, however, cause the elastic properties to change. In addition, cellulose acetate will creep considerably under load. When a constant load is applied to a cellulose acetate model, about 85% of the total deflection will occur during the first few seconds. Another 13%, or perhaps 14%, will occur during the next fifteen to thirty minutes. Additional creeping will continue slowly after this. Thus, if a definite constant load is applied to a model made of this material and the resulting deflection is desired, there will be considerable question as to exactly when the final deflection has been reached.

As previously demonstrated, in indirect model analysis it is not necessary (fortunately) to apply a definite load and read the resulting deflections. Instead, the model is caused to deflect a known amount at a given point and held in this position while simultaneous deflections of other points are noted. The magnitude of the force causing the known deflection is of no importance, and consequently the creep of the model will, in this respect, cause no difficulty.

One other possible effect of creep has to be considered, however, before cellulose acetate can be judged acceptable as a model construction material. This is the possibility that the creep may cause the shape of the deflected structure to change with time. This, of course, would mean that a different influence line for a given stress function would be obtained for every different time of reading. For a demonstration that this does *not* occur, consider the beam of Fig. 14-8. A load P_t is applied at point C of span AB of sufficient magnitude to cause the deflection Δ_C. P_t is not a constant load; instead, it varies as necessary to maintain a constant Δ_C as the modulus of elasticity E_t of the cellulose acetate varies with time.

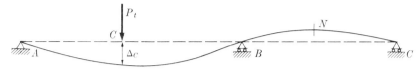

FIGURE 14–8

It has been established that the modulus of elasticity of this material does not vary with stress intensity, and therefore E_t is uniform at any instant throughout the loaded model. Consequently, the deflection of point C may be expressed as

$$\Delta_C = Q_B \cdot \frac{P_t}{E_t},$$

where Q_B is a constant depending on the dimensions of the beam. Thus,

$$\frac{P_t}{E_t} = \frac{\Delta_C}{Q_B} = K,$$

where K is a constant and therefore does not vary with time. The deflection of any other point N along the beam may be expressed as

$$\Delta_N = Q_N \cdot \frac{P_t}{E_t} = Q_N \cdot K.$$

We therefore conclude that the deflection curve of a cellulose acetate model which is subjected to a constant deflection at a given point will not change with time, and that as a consequence, this material is suitable for the construction of models for indirect analysis.

14–5 The spline method of indirect analysis. The simplest of the indirect methods utilizes a long flexible strip, or spline, of some material such as brass, steel, or wood for the model. The moment of inertia of homologous sections of the prototype and model must, of course, be in some fixed ratio. Consequently, if brass or steel strips or wires are to be used, the method is practically limited to continuous beams with constant moments of inertia within individual spans. Otherwise, strips or wires would be required with depths or diameters varying so as to agree with the prototype, and these would require special machining. The strip is mounted on a flat surface with small nails driven as close as possible against the top and bottom edges at all points of support, except the one for which the reaction influence line is desired. This reaction point of the model is then displaced a known amount and the resulting deflections, in the direc-

598   MODEL ANALYSIS OF STRUCTURES   [CHAP. 14

tion of applied loads, are recorded for all load points. The reaction influence line ordinate for each load point will be the recorded load point deflection divided by the impressed deflection of the reaction point.

The Gottschalk continostat (11) is a refinement of the above method. The simplest way of measuring deflections is to mount the model on cross-section paper and to read these deflections directly on this as a background. When this is done, a rather large impressed deflection must be used in order to minimize the effect of errors in reading. This introduces other errors, however, as will be discussed in the section which follows.

14-6 Errors resulting from changes in geometry. Large deflections introduce errors because of the change in geometry of the model. These may be minimized by impressing one-half of the desired total displacement alternately in each direction from the neutral position of the model. The total movement of any load point is then measured from one deflected position to the other. This has the advantage of reducing the possibility of overstraining the model. A demonstration similar to that given by Wilbur and Norris (16) will explain why this procedure will tend to eliminate errors caused by a change in geometry.

Suppose that in Fig. 14-9 an influence line is required for the vertical

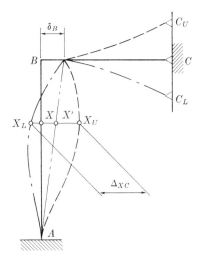

FIGURE 14-9

reaction component at C. Horizontal loads will be applied to AB and vertical loads on BC in the prototype. By the Müller-Breslau principle, if end C of the model is moved either up or down, the deflection curve is supposed to be, to some scale, the desired influence line. Actually, if the impressed displacement of C is relatively large, the curvature of the member BC will cause a horizontal displacement δ_B of B, which will introduce an error in all measured horizontal deflections of load points on AB. Vertical deflections of load points on BC will not be affected and will be correct. Considering a definite load point on AB, such as X, if C is moved to C_U and the horizontal deflection of X is measured from the neutral to the deflected position, the distance XX_U will be recorded. This will be too large for influence line computations by the distance XX'. If, on the other hand, C is moved to C_L, then the horizontal deflection of X from the neutral position to the deflected position will be XX_L, and this will be too small by the distance XX'. Obviously, if C is first moved to one deflected position, say C_U, and the location of X_U noted, and then C is moved to C_L, the distance $X_U X_L$ can be measured and will be correct for influence line ordinate computations. The correct ordinate value for the influence line would be the distance $X_U X_L$ divided by the distance $C_U C_L$.

The above procedure is satisfactory for eliminating errors due to changing geometry in models of the general type shown in Fig. 14–9. Errors due to changing geometry can be eliminated in all cases, however, if the impressed deflections are sufficiently small. This is a common procedure when accurate results are desired. In this case, the impressed deflection must be carefully controlled and the deflections thereby induced must be precisely measured. Various instruments have been designed to permit this exact control and precise measurement. Two of these are well known and give excellent results. A third, developed and built at Rensselaer Polytechnic Institute in 1953, also gives excellent results. These instruments will now be described.

14–7 The Beggs deformeter. The Beggs deformeter (1),(2),(3) was introduced in 1922 by the late Professor George E. Beggs of Princeton University. This instrument and the technique of its operation have been extremely well discussed by McCullough and Thayer (12) but a brief description will be given here. A typical assembly is shown in Fig. 14–10.

The model is usually cut from a sheet of some elastic material $\frac{1}{10}$ to $\frac{1}{8}$ in. thick. Hard cardboard has been used with some success, even though it is not isotropic, but cellulose acetate is usually preferred because of its greater strength and ease of fabrication. The deformeter consists essentially of several gages and one or more microscopes. The gages are placed at each reaction of the model and may be used either to impress a known displacement or to supply the reaction components. Each gage is constructed as shown in Fig. 14–11. The fixed bar is rigidly and permanently fastened to the clamp base plate at each end. A vertical hole for a wood screw is provided at each end of the fixed bar for fastening the gage to the mounting board. The movable bar is free to slide on the gage base plates.

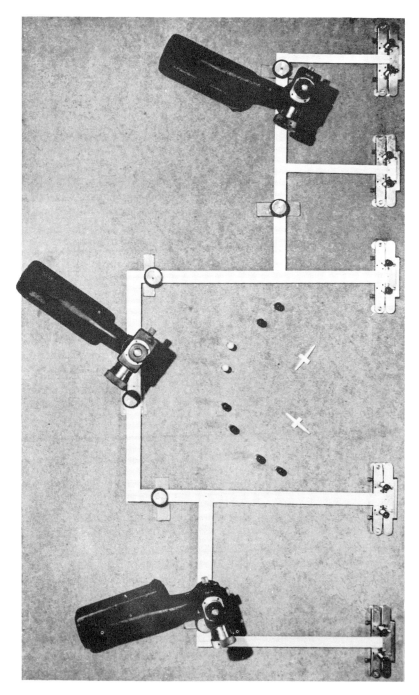

Fig. 14-10. The Beggs deformeter.

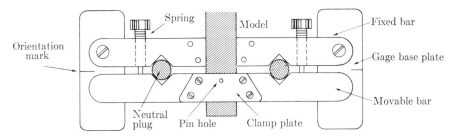

FIGURE 14-11

Its only connection with the rest of the gage is effected by two small rods which project through oversized holes in the fixed bar and are connected with a flexible joint in oversized holes in the movable bar. Between the outer edge of the fixed bar and the knurled nut on each rod are compression springs. These springs tend to pull the movable bar toward the fixed bar. The rods on which they operate do not, however, interfere with sidewise movement of the movable bar. The gages are placed in their neutral position by inserting a pair of "neutral" plugs in the plug slots.

Assume that a model for a two-span rigid frame with rigid column bases is to be mounted for analysis. A smooth table top or drawing board is placed in a horizontal position, and on this board the positions of the column bases are carefully located. The three gage clamps required, one for each column base, are carefully screwed to the board in the proper location with neutral plugs in each gage. Orientation marks are provided on the gages for accurate mounting. The model columns are then clamped to the movable bar of each gage with the clamp plate and screws; these columns are usually made 2 or 3 in. longer than scale length to permit mounting in the gages. The model must be free to move at all points between gages, and this is made possible by supporting it on small ball bearings, perhaps $\frac{1}{4}$ in. in diameter, which roll on small pieces of plate glass. Lead weights are usually placed on top to prevent buckling. The model is now ready for the analysis.

If an analysis is desired for column bases pinned, holes are drilled in the base of the model columns at points to correspond to the actual pin location in the prototype. The clamp plates are removed from the movable bars and pins are inserted in the pin holes. The holes in the model columns, which must be drilled for a smooth fit, permit the mounting of the model on these pins.

To obtain the influence lines for thrust, shear, and moment at the base of a given column, it is necessary to impress known axial, transverse, and rotational displacements of the column base of the model. To understand how the Beggs gage makes this possible, consider

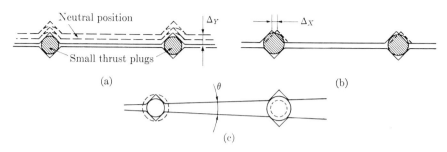

FIGURE 14–12

Fig. 14–12(a), (b), and (c). The known axial displacement is impressed by the use of two pairs of "thrust" plugs, each pair having a different diameter, as illustrated in 14–12(a). The neutral plugs are removed from the gage and the smaller thrust plugs are inserted. A reading is taken on one or more load points, depending on the number of reading instruments available. The small thrust plugs are then removed from the gage, the larger plugs are inserted, and another reading is taken on each load point previously observed. The impressed displacement Δ_Y is known by the observer from a previous calibration of the thrust plugs.

The transverse displacement of the column base is impressed by use of the "shear" plugs [see Fig. 14–12(b)], plugs of equal diameter but flattened on one side. They are inserted in order to produce a displacement, first in one transverse direction and then in the other, to give a total calibrated throw of Δ_X.

The rotational column base displacement is produced by the "moment" plugs, which are round and of different diameters. As indicated in Fig. 14–12(c), these plugs are inserted in the gage and will cause the movable bar to rotate a small amount in one direction from the neutral position. Reversal of these plugs will cause rotation in the opposite direction from this neutral position. The total rotational throw will be 2θ, and its value will have been determined by a previous calibration. It is important to note that only one type of displacement is impressed at one time; for example, the shear plugs will cause transverse displacement without any axial or rotational movement. The impressed displacements are very small. To illustrate, the axial throw Δ_Y of Fig. 14–12(a) is about four-hundredths of an inch.

The reading instruments are microscopes set in heavy frames for stability (see Fig. 14–10). A scale is engraved on glass within the instrument and in the field of view. Cross-hairs on another glass are caused to move along this scale by a micrometer movement. The system is so arranged that one division on the micrometer index represents a movement of an observed point on the model of about one six-thousandth of an inch.

In the event an influence line is required for internal thrust, shear, or moment at a given cross section of the model, one of the gages, with neutral plugs inserted, is located at that section. It is positioned so that the orientation marks coincide with the model cross section extended. The model is rigidly fastened to both bars of the gage with clamp plates. After the model is clamped, it is carefully cut between the two bars of the gage, and ball bearings running on pieces of plate glass are inserted beneath the gage base plates. Thus the gage is capable of impressing axial, transverse, or rotational relative displacements between the cut ends of the model without restraining the resultant model deflections. This arrangement provides what is called a "floating" gage.

The Beggs deformeter permits extremely accurate work in indirect model analysis when it is operated by a skilled analyst. However, considerable experience in the use of the equipment (which is relatively expensive) and great care in its manipulation are necessary to produce the excellent results of which it is capable. For the best results the Beggs deformeter should be used in a room with controlled temperature and humidity. Incandescent lamps should be kept away from the cellulose acetate model because the resultant differential heating will seriously disturb the deflections. Without the observance of these precautions the results are likely to be mediocre to poor. It should be noted that extended periods of use of this deformeter will result in considerable eye strain.

14-8 The Eney deformeter. The Eney deformeter was developed about 1935 by Professor W. J. Eney of Lehigh University (6),(7),(8). A typical assembly is shown in Fig. 14-13.

Gages are used in the Eney deformeter to perform the same functions as in the Beggs equipment. They are, however, very different in detail. The external reaction gages consist essentially of two plates. The lower plate is rectangular in shape and is screwed to the mounting board or table top. The upper plate is smaller and semi-circular in shape. It is provided with threaded holes and small clamping plates for rigidly connecting the model if a fixed support is desired. Upper and lower plates are not connected together but a series of pin holes are provided in both plates. Two removable pins inserted in matching holes prevent translation and rotation of the upper plate relative to the lower, unless removed. These holes are drilled so that axial or transverse displacements may be impressed on the model in $\frac{1}{4}$-in. increments in either direction from the neutral position simply by removing the pins, sliding the top plate to the desired position, and then replacing the pins in the new sets of matching holes.

If it is desired to impress a rotational deflection, both pins are removed and one is inserted in a pair of matching holes provided at points in the two plates corresponding to the center of the model reaction. The rota-

604   MODEL ANALYSIS OF STRUCTURES   [CHAP. 14

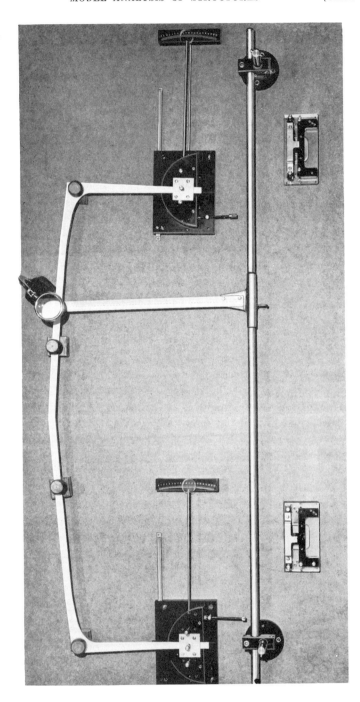

Fig. 14-13. The Eney deformeter.

tional deflection may be impressed in definite increments in either direction from the neutral position by rotating the top plate and inserting the other pin in matching holes provided on the arc of a circle centered on the hole in which the first pin was inserted.

The internal gage, used for determining influence lines for internal sections, is mounted on a rectangular plastic base. A bar is rigidly connected to this base and is provided with threaded holes to receive screws to hold a clamping plate for gripping the model. Another bar is provided which has threaded holes for clamping the model. This second bar is not permanently connected to the base but is provided with pin holes which serve the same purpose as already described in the case of the upper plate of the external gage. The entire gage is mounted on the model by carefully orienting it with respect to the section for which the influence line is desired and with two pins inserted in the pin holes of the movable bar to hold it in the neutral position. After the model is clamped to both bars, it is carefully cut between them. The entire gage is "floated" on ball bearings on glass. This gage permits an impressed relative axial, transverse, or rotational displacement between the two ends of the model at the cut. This is accomplished in a manner similar to that already described for the external reaction gage.

Deflections are read on a scale graduated to one-hundredth of an inch. The lower end of this scale is fastened to a fitting on a heavy steel rod screwed to the mounting surface. This rod is parallel to the main axis of the model and permits the sliding of the scale along the model to read deflections at any desired point.

The Eney deformeter is easier to use than the Beggs equipment, costs very much less, and gives excellent results. (It is not patented, since Professor Eney has preferred to make it available to the profession.)

14-9 The R.P.I. deformeter. The R.P.I. deformeter (see Fig. 14-14) was designed and built (14) in the structural model laboratories of Rensselaer Polytechnic Institute in 1953. In principle it is the same as the deformeters just described. In detail it is entirely different. The reaction gage is provided with three movements, as were the corresponding gages in the Beggs and Eney equipment. These movements are controlled, however, with three micrometers reading directly to one-thousandth of an inch. The internal or floating gage also is provided with three movements controlled with three micrometers.

Deflections are read by clamping targets on the model load points for which influence line ordinates are desired. These targets have a vertical face projecting up from the model. In the center of this face is mounted a needle projecting horizontally therefrom, and opposite this needle are set

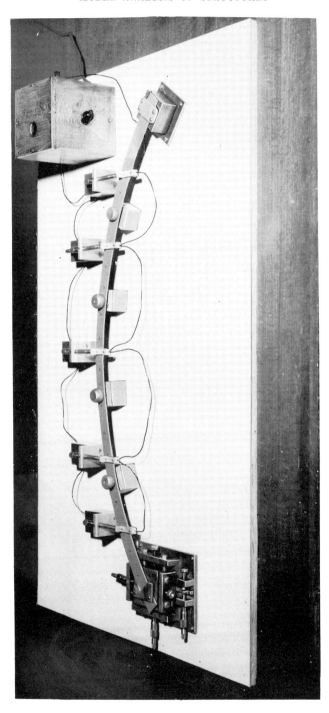

Fig. 14-14. The R.P.I. deformeter.

micrometers mounted in heavy steel bases. Needle and micrometer are oriented in the line of action of the prototype load. Each micrometer and its opposing target are connected in series with a contact indicator (to be described in the section which follows). When a reading is to be taken, the micrometer is carefully turned up until the indicator shows that the circuit has been closed.

When it is used with a contact indicator, the R.P.I. deformeter will give results comparable to the Beggs equipment (and no eye strain is involved in its use). The floating gage is more difficult to use, however, than that in the Eney deformeter. The R.P.I. deformeter is more expensive than the Eney equipment, but much less so than the Beggs deformeter.

14-10 The contact indicator. The contact indicator (13) is an extremely valuable piece of auxiliary equipment when it is desired to measure deflections accurately with micrometers. Simple and inexpensive to build and easy to use, it is extremely sensitive and will indicate contact of a micrometer and an opposing needle within three- or four-millionths of an inch. The small box shown in Fig. 14-14 contains one of these instruments. The wiring diagram is shown in Fig. 14-15.

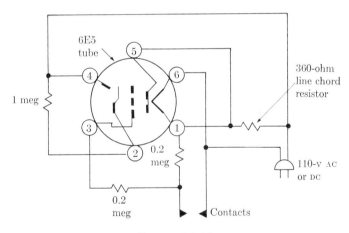

FIGURE 14-15

14-11 The moment deformeter. The moment deformeter was developed (16),(17) at the Massachusetts Institute of Technology. It is most ingenious and provides a means for determining the influence line for internal moment in a structural member. The instrument is clamped to the model member and centered on the section for which the moment influence line is desired. This deforms the model so that it is possible to obtain the influence line for internal moment at the desired section.

The moment deformeter is easy to use. It is unnecessary to cut the model in order to obtain the influence line. However, this deformeter has the disadvantage that it can only be used on members which have a constant section.

14–12 The brass spring model for articulated structures. The brass spring model was introduced (5) by Anders Bull in 1930. The members are made of #26 brass drill rod, and the model is made geometrically similar to the prototype so far as the length of all members are concerned. Mechanical similitude for static loading is achieved by assuring that homologous elastic forces in model and prototype are in some fixed ratio. In other words, the ratio of the total axial strain in a given member of the model resulting from unit loads applied at its ends, to the total axial strain in the corresponding member of the prototype as caused by unit loads applied at its ends, must always be constant. This is achieved by inserting a brass leaf spring in each member of the model.

The method was not used to any great extent until it was much improved by Professor Eney at Lehigh University. In 1940 he obtained excellent results with this method in an analysis of a large bridge (9).

Actually the brass spring method is something more than an indirect type of analysis. A brass spring model can, of course, be used to determine influence lines in the same way as a model made from cellulose acetate. In addition, however, it can be used to determine deflections of the prototype.

14–13 Model design for indirect analysis. It was demonstrated in Section 14–3 that usually only two scale factors have to be considered in the design of models for indirect analysis. The first of these, the linear scale factor k, is applied to axial dimensions only. The second scale factor may be either e or n, depending upon whether the prototype is an articulated structure or acts primarily in flexure. If the prototype is articulated, with the various members consequently being subjected chiefly to axial loads, the proper relationship between homologous cross-sectional areas of model and prototype will be $e\overline{A} = A$. If the stresses are primarily the result of flexure, the relationship between homologous moments of inertia is expressed by $n\overline{I} = I$. If both flexural and axial strains are important, then both relationships should be satisfied, and in this case, $e^2 = n$. An illustrative example will most clearly explain the method of design.

EXAMPLE 14–1. The prototype is a two-span continuous steel girder as shown in Fig. 14–16. The curve of the bottom flange is parabolic. The web is $\frac{3}{4}$ in. thick and flange plates are 12 in. $\times$ 1 in. Design a model for indirect analysis.

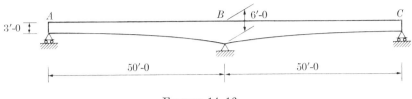

FIGURE 14-16

Two restrictions are assumed to be imposed by the equipment available. The first of these requires that the model must not exceed 4 ft in length. Consequently, k is taken as 25, although in the given problem this actual value is of no particular interest. In the case of a model design for an articulated structure, however, the factor k would be applied to the lengths of all members, as well as to the span length. The second restriction on the model dimensions is imposed by the dimensional capacity of the deformeter gages. In the given case it will be assumed that the gages will accommodate the model if the depth at the ends is 1 in.

As previously indicated, the model depth is designed so that at homologous sections

$$n\bar{I} = I.$$

Since the model is to be cut from a sheet of material of uniform thickness, the above equation becomes

$$n \frac{\bar{t}\,\bar{d}^3}{12} = I,$$

where $\bar{t}$ is the thickness of the sheet and $\bar{d}$ the required depth (both in inches) of the model at a given section. Consequently,

$$\bar{d} = C\sqrt[3]{I},$$

where C is a constant and equal to $\sqrt[3]{12/n\bar{t}}$ and can have any desired value, since n may be arbitrarily chosen. In Table 14-1, the $\sqrt[3]{I}$ at the end of the prototype is 23.1. Therefore the constant C is taken as $1/23.1$, the model depth at the end will be 1 in. as desired, and the depths of the model at the other sections are readily computed.

The depths shown in Table 14-1 should be carefully marked out on the model material at the corresponding sections. The points thus obtained are carefully joined with a smooth curve and the material cut and filed to this line. Observation points are scribed on the surface of the model at desired intervals and, after mounting the deformeter gages as previously described, the model is ready for use.

TABLE 14-1.

Distance from A on prototype (ft)	Distance from A on model (ft)	I (in^4)	$\sqrt[3]{I}$	Model depth d (in.)
0	0	12,300	23.1	1.00
5	0.2	12,550	23.3	1.01
10	0.4	13,400	23.8	1.03
15	0.6	14,900	24.6	1.07
20	0.8	17,300	25.9	1.13
25	1.0	20,500	27.4	1.19
30	1.2	25,000	29.3	1.27
35	1.4	31,000	31.4	1.36
40	1.6	38,600	33.8	1.46
45	1.8	48,700	26.5	1.58
50	2.0	61,900	39.6	1.72

EXAMPLE 14-2. Assume that a trial design has been made for a two-span continuous truss to have the dimensions indicated in Fig. 14-17. The heaviest truss member has been designed to have a cross-sectional area of 107.6 in^2, and the smallest an area of 23.6 in^2. It is desired to build a model by means of which influence lines may be obtained to check the first mathematical analysis. Design the model.

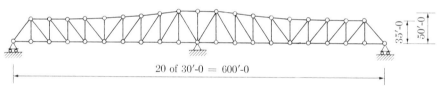

FIGURE 14-17

Assume that because of space limitations the model cannot exceed 6 ft in length. Consequently, the linear scale factor k is 100, and the depth of the model at the center will therefore be 0.5 ft. The lengths of all members of the model are determined by applying the same linear scale factor to the lengths of corresponding members of the prototype.

The cross-sectional areas of all truss members must be such that

$$e\overline{A} = A.$$

Since (as before) the members are to be cut from material of uniform thickness, the above equation becomes

$$e\overline{tb} = A,$$

where $\bar{t}$ is the material thickness and $\bar{b}$ the required width of the model member. Consequently,

$$\bar{b} = QA,$$

where Q is a constant and equal to $1/e\bar{t}$. Q may have any desired value (since e can be arbitrarily selected), and is chosen so that the widest model member will not be too wide and stiff and, at the same time, so that the narrowest member will not be too narrow. This is a matter of judgment based on experience, but this is experience easily obtained. In the present case, it seems that a maximum width of 0.75 in. might be suitable. Accordingly, based on the heaviest member,

$$Q = \frac{\max \bar{b}}{\max A} = \frac{0.7500}{107.6} = 0.00697.$$

The narrowest member, therefore, will have a width of

$$\min \bar{b} = Q(\min A) = 0.00697 \times 23.6 = 0.164 \text{ in.}$$

The widths of all other members in the model are determined by multiplying the areas of corresponding members in the prototype by Q. If the lighter model members are so located that they will not buckle when a displacement is impressed, the above value of Q is probably satisfactory.

If the model members are made of cellulose acetate or similar material, they are connected at the joints with cement made by dissolving scraps of the material in acetone. The prototype would be analyzed on the assumption of frictionless pins at the joints. Despite this, the quite rigid joints of the model often do not appear to have any significant effect on the results. However, heavy members in the prototype which may, depending on the value of Q, result in unusually wide and therefore stiff model members, can affect the results. In the example above, for instance, if it is decided that the narrowest members should be doubled in width to prevent buckling in the model, then the heaviest member would be 1.5 in. wide. For the length of the members involved, this would be too stiff. In this case it would be advisable to fabricate the initial member in the model 0.5 in. wide, and then to cement strips 0.5 in. wide on each face in order to build the member up to the required area.

14–14 Direct model analysis. It was indicated in Section 14–2 that in order for structural similitude to exist between prototype and model, all homologous linear dimensions must be in some fixed ratio, all homologous linear deflections must be in some fixed ratio, and all homologous loads and forces must be in some fixed ratio. In this discussion these three ratios will be called *primary scale factors*.

The first primary scale factor defines the length relationship between prototype and model, and the third defines the force relationship. The second primary scale factor defines the deflection relationship. Deflection, however, is a function of strain and is a linear dimension. Strain is a function of force per unit area and area is a linear dimension squared. Consequently, the deflection scale factor is a function of both the length and force scale factors. In other words, any two of the three primary scale factors may be assigned arbitrary values, but the third must then be computed from these two.

In addition to the primary scale factors, certain secondary scale factors must be evaluated before a model can be designed. These define the ratios, between prototype and model, of such measurable quantities as cross-sectional areas, density or mass, moment of inertia, or moment of a force. All of these, however, are easily determined from the primary scale factors.

When we design a model for direct analysis, the logical procedure is first to assign a value for the linear scale factor k. This is selected so that over-all dimensions of the model will be convenient and workable. One of the other two primary scale factors must be assigned a value, or otherwise defined. Since the most valuable information obtained from the model will probably be the strains as determined by SR-4 strain gages, it will be most convenient to design the model so that homologous unit strains in prototype and model will be equal. This immediately establishes three important relationships: (1) homologous rotational deformations will be equal, (2) the scale factor for homologous total strains will be k, and (3) as a consequence of (1) and (2), the scale factor for linear deflections will be k. The other required scale factors will now be derived.

In accordance with the above discussion,

$$k\overline{L} = L. \tag{14-17}$$

The various forces acting on and in any structure may be divided into external and internal. The external forces are those resulting from the action of gravity on the loads applied to the structure. These loads will be represented by W. Internal elastic forces are those which develop concurrently with the elastic straining of the various members. An internal elastic force will be designated by F. Therefore,

$$F = A\sigma = AE\epsilon. \tag{14-18}$$

In the above expression:

A = cross-sectional area of an axially loaded member,
σ = unit stress, E = modulus of elasticity,
ϵ = unit strain.

For mechanical similitude it is necessary that the ratio of all homologous forces in the prototype and the model be a constant value. Consequently,

$$\frac{W}{\overline{W}} = \frac{F}{\overline{F}} = \text{constant.} \tag{14-19}$$

(This constant is selected for greatest convenience.) Since $k = L/\overline{L}$ and since cross-sectional area is the product of two linear dimensions, it is possible to write

$$\frac{A}{\overline{A}} = \frac{L^2}{\overline{L}^2} = \frac{(k\overline{L})^2}{\overline{L}^2} = k^2. \tag{14-20}$$

Therefore, if desired, k^2 may be used as the scale factor for the cross-sectional areas of axially loaded members. In practice, however, since the value of k is primarily established to give an over-all workable model dimension, the scale factor k^2 will quite often result in cross-sectional areas for model members which will be difficult, if not impossible, to fabricate. Consequently, to give greater flexibility in the design of the model, an additional factor z, which will be called a *dimensional slicing factor*, is introduced and Eq. (14-20) becomes

$$\frac{A}{\overline{A}} = k^2 z. \tag{14-21}$$

Providing for the possibility that prototype and model may be of different materials, and remembering that homologous strains are to be equal, we now write Eq. (14-19) as

$$\frac{W}{\overline{W}} = \frac{F}{\overline{F}} = \frac{AE\epsilon}{\overline{A}\overline{E}\epsilon} = k^2 z \frac{E}{\overline{E}}. \tag{14-22}$$

This scale factor for forces, $k^2 z(E/\overline{E})$, is correct and can be used. However, the expressions for several other scale factors still have to be derived and these will be simplified if this force ratio is arbitrarily modified to $k^2 z$. If $k^2 z$ is substituted for the value of the constant in Eq. (14-19), and the internal elastic forces are expressed as before, the result is

$$\frac{W}{\overline{W}} = \frac{F}{\overline{F}} = \frac{AE\epsilon}{\overline{A}\overline{E}\epsilon} = k^2 z, \tag{14-23}$$

from which

$$\frac{A}{\overline{A}} = k^2 z \frac{\overline{E}}{E}. \tag{14-24}$$

Equation (14–24) defines the scale factor to be used in order to determine the cross-sectional areas of axially loaded members in models designed for direct analysis.

Since moment is the product of force times distance,

$$\frac{M}{\overline{M}} = \frac{FL}{\overline{FL}} = k^2 z \cdot k = k^3 z, \qquad (14\text{–}25)$$

and this is the scale factor for moments applied to a model, as well as for internal resisting moments.

The scale factor for moment of inertia for a flexural member may be obtained from two relationships of structural mechanics. The first of these is

$$M = \frac{EI}{R}, \qquad I = \frac{MR}{E},$$

where R is the radius of curvature. Thus

$$\frac{I}{\overline{I}} = \frac{MR/E}{\overline{MR}/\overline{E}} = \frac{MR\overline{E}}{\overline{M}\,\overline{R}\,E} = k^3 z \cdot k \cdot \frac{\overline{E}}{E} = k^4 z \frac{\overline{E}}{E}. \qquad (14\text{–}26)$$

The second is the ordinary flexure formula:

$$M = \frac{\sigma I}{c}, \qquad I = \frac{Mc}{\sigma},$$

from which

$$\frac{I}{\overline{I}} = \frac{Mc/\sigma}{\overline{M}\overline{c}/\overline{\sigma}} = \frac{Mc\overline{\sigma}}{\overline{M}\overline{c}\sigma}. \qquad (14\text{–}27)$$

If the model is geometrically similar to the prototype as to depth, as it should be, then $c/\overline{c} = k$. Also, since homologous unit strains will then be equal, $\epsilon = \overline{\epsilon}$. Consequently,

$$\frac{\overline{\sigma}}{\sigma} = \frac{\overline{E}\epsilon}{E\epsilon} = \frac{\overline{E}}{E}.$$

Substituting in Eq. (14–27), we obtain

$$\frac{I}{\overline{I}} = k^3 z \cdot k \cdot \frac{\overline{E}}{E} = k^4 z \frac{\overline{E}}{E}. \qquad (14\text{–}26\text{a})$$

It will be practically impossible to simulate the dead weight of the prototype with the weight of the model. The expression for weight is γAL, where γ is the specific weight of the material. For mechanical similitude,

$$\frac{W}{\overline{W}} = \frac{\gamma A L}{\overline{\gamma}\,\overline{A}\,\overline{L}} = k^2 z, \tag{14-28}$$

which, after substituting Eq. (14-24), results in

$$\frac{\gamma}{\overline{\gamma}} = \frac{1}{k}\,\frac{E}{\overline{E}}. \tag{14-29}$$

It is apparent from the above expression that in most cases the required density $\overline{\gamma}$ will far exceed that of any material obtainable. Consequently, in practice, the dead load of the prototype is simulated by hanging appropriate weights on the model.

It is possible to check the various scale factors which have just been derived. If all these factors have been correctly evaluated, homologous linear deflections should be in the same ratio as homologous linear dimensions; that is,

$$\frac{\Delta}{\overline{\Delta}} = k.$$

Assume that the prototype is a structure in which flexure is of primary importance. The formula for deflection by virtual work will have the form $\Delta = \int mM\,dx/EI$. Therefore,

$$\frac{\Delta}{\overline{\Delta}} = \frac{\int mM\,dx/EI}{\int \overline{m}\,\overline{M}\,\overline{dx}/\overline{EI}}, \tag{14-30}$$

but

$$M = k^3 z \overline{M},$$

$$m = k\overline{m},$$

$$dx = k\,\overline{dx},$$

$$I = k^4 z \overline{I}\,\frac{\overline{E}}{E}.$$

If these values are substituted in Eq. (14-30), the result is

$$\frac{\Delta}{\overline{\Delta}} = k.$$

If the prototype is articulated, the virtual work deflection expression is in the form

$$\Delta = \sum \frac{uSL}{AE}.$$

Therefore

$$\frac{\Delta}{\bar{\Delta}} = \frac{\Sigma uSL/AE}{\Sigma \overline{uSL/AE}}, \qquad (14\text{-}31)$$

but

$$S = k^2 z \bar{S},$$
$$u = \bar{u},$$
$$L = k\bar{L},$$
$$A = k^2 z \bar{A}\, \frac{\bar{E}}{E}.$$

Substituting in Eq. (14-31), we obtain

$$\frac{\Delta}{\bar{\Delta}} = k.$$

14–15 Dimensional analysis. Dimensional analysis provides one of the most effective methods available for determining the requirements for similitude between prototype and model and for interpreting experimental data. It is based on the fact that the various physical properties of, and quantities acting upon, a body, such as length, mass, temperature, velocity, force, etc., can be expressed in terms of one or more of the fundamental physical dimensions. The fundamental physical dimensions are length, time, and force or mass.

Two different systems of dimensions are used. In the F-L-T system, known as the *engineer's system*, the dimensions are force, length, and time. The *physicist's system*, in which the dimensions are mass, length, and time, is designated as the M-L-T system. In this discussion the F-L-T system will be used.

The object of dimensional analysis is to find some means of directly relating prototype and model, not only as to size but also as to response. For example, the length of a beam model is not equal to the length of the prototype, nor is the depth of the model equal to the depth of the prototype. However, for geometric similitude, the length-to-depth ratio of the model will be equal to the length-to-depth ratio of the prototype. These ratios can be equated because they are dimensionless.

Various combinations of the several physical properties and quantities which define model and prototype can be found which will give dimensionless terms. In each case, if complete similitude has been realized in the model, corresponding dimensionless terms for model and prototype will be equal. Consequently, if enough of these dimensionless terms can be found to define similitude, they can be used to compute scale factors and to interpret test data.

In 1915, E. Buckingham published an excellent paper (4) in the appendix of which he derived his π theorem. This theorem makes it possible to obtain in a systematic manner the dimensionless terms discussed above. One possible statement of the theorem may be phrased as follows: *If a physical phenomenon can be defined in terms of n variables, and if each of these n variables can be expressed in terms of no more than m dimensions, then the general equation for the phenomenon can be expressed as a function of $n - m$ dimensionless π terms. Each dimensionless term will be composed of $m + 1$ variables, two of which will be common for all π terms.* The significance of this theorem can best be explained with an example.

EXAMPLE 14–3. Given a flexural member as indicated in Fig. 14–18, supporting any kind of load. Using dimensional analysis and the expression for strain, find the π terms necessary to define similitude.

For a flexural member

$$\epsilon = \frac{\sigma}{E} = \frac{Md}{2EI} = \frac{KWSd}{2EI}, \qquad (14\text{–}32)$$

where

ϵ = unit strain,
M = bending moment,
K = a constant for the beam and type of load,
W = total applied load,
S = length of span,
d = depth of member,
E = Young's modulus,
I = moment of inertia.

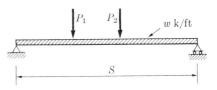

FIGURE 14–18

In general,

$$\epsilon = f(W, S, d, E, I)$$

or

$$g(\epsilon, W, S, d, E, I) = 0.$$

When the structure is statically loaded the fundamental dimension of time does not appear. All the above variables can be expressed in terms of units of force and length. In accordance with the π theorem, two of the above variables may be common to all π terms. These common variables may be selected as desired, except that the dimension of force must be used at least once, and the dimension of length must be used at least once, in the dimensional expressions for these common variables. In this case, W and S will be used.

The general expression for the π terms is

$$\pi_i = W^a S^b Q_i^{-1} = F^a L^b Q_i^{-1}. \tag{14-33}$$

In the last term above, the F indicates that the load is expressed in units of force, the L that the span is expressed in units of length. The term Q_i will be replaced by the dimensional expressions for the other variables of the general g function, one at a time. If each π term is to be dimensionless, then, as each successive Q_i term is substituted, the exponents a and b must have a value such that the summation of all exponents for F and all exponents for L will be zero. Having determined a and b for a given Q, we write the π term by substituting these values for the exponents in the general expression for the π terms [Eq. (14-33)] and inserting the corresponding variable for Q_i. This procedure will now be demonstrated.

For π_1, $Q_1 = d$:

$$\pi_1 = W^a S^b d^{-1} = F^a L^b L^{-1},$$

$$F: a = 0,$$

$$L: b - 1 = 0, \therefore b = 1;$$

$$\therefore \pi_1 = W^0 S^1 d^{-1} = \frac{S}{d}.$$

For π_2, $Q_2 = E$:

$$\pi_2 = W^a S^b E^{-1} = F^a L^b (FL^{-2})^{-1},$$

$$F: a - 1 = 0, \therefore a = 1;$$

$$L: b + 2 = 0, \therefore b = -2;$$

$$\therefore \pi_2 = W^1 S^{-2} E^{-1} = \frac{W}{S^2 E}.$$

For $\pi_3, Q_3 = I$:

$$\pi_3 = W^a S^b I^{-1} = F^a L^b L^{-4},$$

$$F: a = 0,$$
$$L: b - 4 = 0, \therefore b = 4;$$

$$\therefore \pi_3 = W^0 S^4 I^{-1} = \frac{S^4}{I}.$$

For $\pi_4, Q_4 = \epsilon$:

$$\pi_4 = W^a S^b \epsilon^{-1} = F^a S^b,$$

$$F: a = 0,$$
$$L: b = 0;$$

$$\therefore \pi_4 = W^0 S^0 \epsilon^{-1} = \frac{1}{\epsilon}.$$

For complete similitude the four π terms obtained above must be equal for model and prototype, and the necessary scale factors may therefore be determined from these π terms. As before, the linear scale factor will be designated by k. Therefore,

$$\frac{S}{\overline{S}} = k.$$

From π_1: $\quad \dfrac{S}{d} = \dfrac{\overline{S}}{\overline{d}}, \qquad \therefore \dfrac{d}{\overline{d}} = \dfrac{S}{\overline{S}} = k.$

From π_2: $\quad \dfrac{W}{S^2 E} = \dfrac{\overline{W}}{\overline{S}^2 \overline{E}}, \qquad \therefore \dfrac{W}{\overline{W}} = \dfrac{S^2 E}{\overline{S}^2 \overline{E}} = k^2 \dfrac{E}{\overline{E}}.$

From π_3: $\quad \dfrac{S^4}{I} = \dfrac{\overline{S}^4}{\overline{I}}, \qquad \therefore \dfrac{I}{\overline{I}} = \dfrac{S^4}{\overline{S}^4} = k^4.$

From π_4: $\quad \dfrac{1}{\epsilon} = \dfrac{1}{\overline{\epsilon}}, \qquad \therefore \dfrac{\epsilon}{\overline{\epsilon}} = 1.$

The above scale factors agree with those previously derived by structural mechanics, with the exception of the dimensional slicing factor. Note that π_4 could have been predicted, since strain is a dimensionless quantity, and thus for complete similitude must be equal in model and prototype.

This section represents no more than a superficial introduction to the subject of dimensional analysis, a method used extensively in all branches of science and engineering. However, reference material at the end of this chapter will give the interested reader considerable information on the subject.

14-16 Design of models for direct analysis. The proper use of the scale factors just developed can best be explained by demonstration. Accordingly, three examples are presented.

EXAMPLE 14-4. The prototype is a steel cantilever beam 30 ft long. It is a 30WF116 with an actual depth of 30 in. The moment of inertia of the cross section with respect to the major axis is 4919 in^4, and the modulus of elasticity is 30,000 k/in^2. A concentrated load of 10 k acts down on the free end of the cantilever. It is required to design a model for direct analysis using aluminum plate $\frac{1}{8}$ in. thick with a modulus of elasticity of 10,000 k/in^2.

FIGURE 14-19

As previously indicated, the scale factor k is usually chosen to give a model of a convenient length. This, of course, will depend on the space available for fabricating and testing the model, the sizes of the model construction material available, the accuracy with which it is desired to reproduce details of the prototype, the dimensional capacity of any special testing equipment, and other similar considerations.

Suppose in the problem under discussion that it has been decided that the model is to be 3 ft long. Then $k = 10$. The depth of the model for geometric similitude will be 3 in. It is arbitrarily decided to make the flange plates 1 in. wide. The model cross-section is sketched in Fig. 14-20.

The moment of inertia of the model is:

$$I \text{ of web} = \frac{0.125 \times 2.75^3}{12} = 0.217 \text{ in}^4$$

$$I \text{ of flanges} = 2(0.125 \times 1.437^2) = \underline{0.516 \text{ in}^4}$$

$$\text{Total } I = 0.733 \text{ in}^4.$$

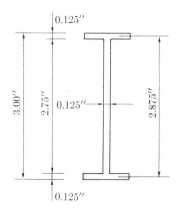

FIGURE 14-20

From Eq. (14-26),

$$z = \frac{EI}{\overline{EI}k^4} = \frac{30 \times 4919}{10 \times 0.733 \times 10^4} = 2.02.$$

The load to be applied to the model, by Eq. (14-23), will be

$$\overline{W} = \frac{W}{k^2 z} = \frac{10{,}000}{10^2 \times 2.02} = 49.6 \text{ lb.}$$

Actually it is not absolutely necessary that the depth of the model be $1/k$ times the depth of the prototype. To illustrate, assume that it is desired to make the model 2 instead of 3 in. deep. By maintaining the same I as before (0.733 in^4), homologous linear deflections will be kept in the ratio k. Since corresponding values of $\overline{M}\,\overline{ds}/\overline{EI}$ (flexural deformations) will be equal in the two models, the extreme fiber strains will vary directly as the depths of the models. This means that the strains in the second model will be $\frac{2}{3}$ times the strains in the first model. Therefore, if extreme fiber strain readings taken on the second model are to be used to predict stresses in the prototype, these readings must be converted to what they would have been on the first model. Therefore,

$$\bar{\epsilon}_1 = \frac{\bar{d}_1}{\bar{d}_2} \cdot \bar{\epsilon}_2,$$

and the extreme fiber stress in the prototype will be

$$\sigma = E\bar{\epsilon}_2 \cdot \frac{\bar{d}_1}{\bar{d}_2}.$$

EXAMPLE 14–5. Design a model for direct analysis of the truss of Fig. 14–21. The cross-sectional areas of the members (in square inches) are marked on the sketch. The prototype is of steel, with $E = 30{,}000$ k/in². The model is to be fabricated of aluminum, with $E = 10{,}000$ k/in².

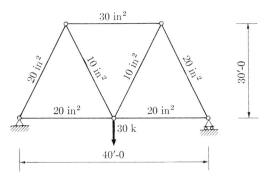

FIGURE 14–21

Assume that it has been decided to make the model 4 ft long. Then $k = 10$. From Eq. (14–24),

$$\overline{A} = \frac{E}{\overline{E}} \cdot \frac{A}{k^2 z} = \frac{3A}{100z} = 0.03\,\frac{A}{z}.$$

The value of z may be selected to ensure model members which will be stable (no buckling of compression members) and yet convenient to fabricate. If z is taken as unity, the area of the top chord member, for example, will be

$$0.03 \times 30 = 0.9 \text{ in}^2.$$

If it is felt that this cross-sectional area is too large for convenient fabrication, then z may be taken as 2, or any other desired value. If z is taken as 2, then from Eq. (14–23),

$$\overline{W} = \frac{W}{k^2 z} = \frac{30{,}000}{10^2 \times 2} = 150 \text{ lb}.$$

If this load is too large, one way of reducing it would be to use a larger value for z. If z is too large, however, the cross-sectional areas of the model members will be too small for stability when subjected to compression. In the event that the computed scaled load for the model is too large and z cannot be made any larger because of fabrication difficulties, an additional load slicing factor q may be introduced to reduce the load. For example, if in the above problem the indicated load of 150 lb is too large for the testing equipment available, and z cannot be increased in value, then q may be taken as 3. This simply means that a load of 50 lb may be applied. The necessary readings are then taken on the model and multiplied by 3 before using them to predict the response of the prototype.

EXAMPLE 14–6. Design a model, to be used in direct analysis, of the eccentrically loaded column shown in Fig. 14–22. The prototype is a 24WF100 with cross-sectional area = 29.43 in^2, moment of inertia = 2987 in^4, out-to-out of flanges = 24.00 in., and $E = 29{,}000$ k/in^2. The model is to be built of aluminum plates 0.100 in. thick, with $E = 10{,}000$ k/in^2.

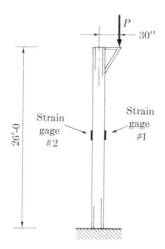

FIGURE 14–22

In this example, flexure and axial stress are both important. Theoretically the cross-sectional area, moment of inertia, and depth of the model should all be designed in accordance with the scaling factors previously derived. If this is attempted, however, it will be found that one of the three requirements will be incompatible with the other two. Nevertheless, it is possible to design a model which will be suitable for predicting stresses in the prototype.

It is decided to make the model 2 ft high. Therefore, k is 13.00. The theoretical out-to-out of model flanges would be $24.00/13.00 = 1.846$ in. Aluminum strip in the required thickness of 0.100 in. is at hand in widths of 1.000 in., which will be used for the flanges, and 2.000 in., which will be used for the web. The eccentricity of the load on the model will be $30.00/13.00 = 2.31$ in.

The moment of inertia of the model section will be

$$I \text{ of web} = \frac{0.100 \times 2^3}{12} = 0.0667 \text{ in}^4$$

$$I \text{ of flanges} = 2 \times 0.100 \times 1.05^2 = 0.2205 \text{ in}^4$$

$$\text{Total } I = 0.2872 \text{ in}^4.$$

From Eq. (14–26),

$$z = \frac{EI}{\overline{EI}k^4} = \frac{29.00 \times 2987}{10.00 \times 0.2872 \times 13.00^4} = 1.056.$$

The load to be applied to the model will be, from Eq. (14–23),

$$\overline{P} = \frac{P}{k^2 z} = \frac{P}{13.00^2 \times 1.056} = 0.00560 P.$$

With the value of $\overline{P}$ as computed above, homologous flexural deformations will be equal. Homologous extreme fiber strains associated with flexure will not, however, be equal. The theoretical out-to-out of the model flanges, as previously computed for similitude, is 1.846 in. The actual out-to-out of model flanges is 2.200 in. The correct extreme fiber strain ϵ associated with bending in a fully scaled model may be computed by multiplying the extreme fiber strain ϵ_b associated with bending in the actual model by a correction factor C_b. This may be written as

$$\epsilon_B = C_b \epsilon_b = \frac{1.846}{2.200} \epsilon_b = 0.839 \epsilon_b.$$

Similitude with respect to the cross-sectional area must now be investigated. In accordance with Eq. (14–24), the area of the model should have been

$$\overline{A} = \frac{AE}{k^2 z \overline{E}} = \frac{29.43 \times 29.00}{13.00^2 \times 1.056 \times 10.00} = 0.478 \text{ in}^2.$$

The actual area of the model is 0.400 in^2. It is apparent that the model strains associated with axial stress will be too large. The correct strain ϵ_A associated with axial stress in a fully scaled model may be computed by multiplying the strain ϵ_a associated with axial stress in the actual model by a correction factor C_a.

This is written as

$$\epsilon_A = C_a \epsilon_a = \frac{0.400}{0.478}\epsilon_a = 0.837\epsilon_a.$$

The two strains ϵ_a and ϵ_b, associated with axial stress and bending, respectively, in the actual model, can be determined if two strain gages are mounted directly opposite each other on the model flanges in positions corresponding to those gages shown on the prototype in Fig. 14–22. If we represent the observed strain in gage number 1 by ϵ_1, and that in gage number 2 by ϵ_2, it is apparent that

$$\epsilon_a + \epsilon_b = \epsilon_1 \quad \text{and} \quad \epsilon_a - \epsilon_b = \epsilon_2.$$

If these equations are solved simultaneously, the result is

$$\epsilon_a = \frac{\epsilon_1 + \epsilon_2}{2} \quad \text{and} \quad \epsilon_b = \frac{\epsilon_1 - \epsilon_2}{2}.$$

The above two strains may now be corrected to what they would have been in a fully scaled model and then combined to give the correct strains in the prototype.

Note that the three examples of model design just presented are not meant to imply that a model of such simple structures would ever be built, except as an exercise. Simple structures were used as examples in order that full attention could be given to the few principles involved in designing the models.

A model of the two-hinged arch rib which was designed by the elastic theory in Example 13–4, and checked by the deflection theory in Section 13–8, is shown in Fig. 14–23. The model was designed and the experimental results interpreted as illustrated in Example 14–6.

PROBLEMS

14–7. A 24WF100 beam, with a depth of 24.00 in. and an I_x of 2990 in^4, supports a concentrated load of 20 k on a span of 30 ft. Design a model for direct analysis in accordance with the following data: $k = 10$; aluminum alloy plates to be 0.100 in. thick; width of flange plates to be 1 in. The over-all depth of the model is arbitrarily selected as 2 in. Strains are to be measured with SR-4 strain gages. $E = 29{,}000$ k/in^2 and $\overline{E} = 10{,}000$ k/in^2. Determine the value of the concentrated load which should be applied to the model for similitude. What is the value of the factor by which the observed strains should be multiplied to correct for the fact that the requirements for similitude have not been satisfied relative to the depth of the model? [*Ans.:* $\overline{P} = 52.9$ lb, $\epsilon = 1.2\bar{\epsilon}$.]

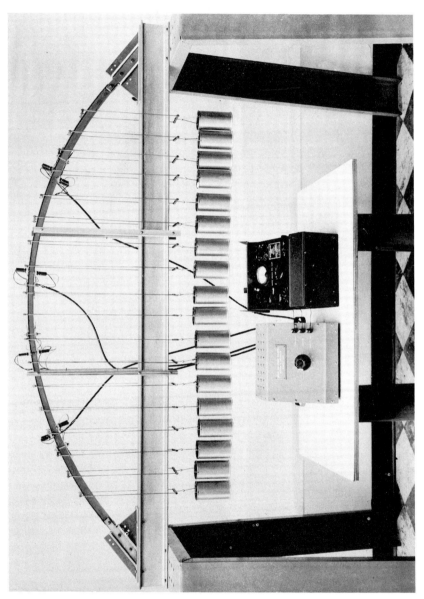

Fig. 14-23. Model of two-hinged arch rib.

14-8. The prototype is a column loaded eccentrically with respect to the major (x–x) axis. The column shaft is a 27WF145 with a height of 20 ft, a cross-sectional area of 42.68 in^2, a moment of inertia with respect to the x–x axis of 5414 in^4, and an out-to-out of flanges of 26.88 in. E of steel = 29,000 k/in^2. The load is 210 k applied on a bracket with an eccentricity of 2 ft. The top of the prototype is braced laterally. A model for direct analysis is to be designed using a linear scale factor of $k = 10$ applied to the height of the model and the eccentricity of the load. The model is to be fabricated of aluminum alloy using a web plate 2.0 in. × 0.10 in. and flange plates 1.0 in. × 0.10 in. E for the aluminum alloy is 10,000 k/in^2. Using a load slicing factor of 10, determine what load should be applied to the model. Having determined the magnitude of the load, assume that the model is fabricated and tested. The strains, as determined by two SR-4 strain gages mounted directly opposite each other on the center lines of the two model flanges at mid-height, are: strain gage "x," on the flange further from the eccentric load, a strain of $+25.5\mu$; strain gage "y," on the flange nearer to the eccentric load, a strain of -44.7μ. In this case, the + sign indicates a tensile strain. [The Greek μ is often used as a symbol for microunits (one microunit is a strain of 1×10^{-6}).] Determine the stresses in the prototype at points corresponding to the positions of the strain gages "x" and "y" on the model, as indicated by the above strain readings on the model. [*Ans.*: $\overline{P} = 38.5$ lb. The prototype stresses, as computed from the observed strain readings on the model, are: for the point corresponding to the position of the strain gage "x," 7510 lb/in^2; for the point corresponding to the position of strain gage "y," 17,400 lb/in^2.]

14-17 Fabrication and loading of models for direct analysis. The fabrication of models for direct analysis is, for several different reasons, somewhat more involved than for indirect analysis. It has been demonstrated that more attention must be given to scaling factors in designing a model for direct analysis. This means that the model must be more exact in detail, with more pieces to cut, fit, and join together. Very often small I-shaped sections must be fabricated. The system of loads applied must be similar to the loads on the prototype and thus the model may have to receive a complicated load pattern. No such problem exists in indirect analysis.

The material to be used in constructing the model should have all those qualities previously listed in the discussion of models for indirect analysis. It should be inexpensive, sufficiently strong, isotropic, and easily worked and joined. It will be recalled that for indirect model analysis, cellulose acetate, or a similar material, most nearly satisfies all these requirements. It was demonstrated that creep under load causes no difficulty in indirect analysis. For direct analysis, however, the variation in the modulus of elasticity of the cellulose acetate, which causes creep, renders this material unsuitable for models unless certain compensating devices are used. These devices will be subsequently discussed.

Steel can be used for constructing models but it is difficult to fabricate. The same thing is true, to a lesser degree, for brass. Good results have been obtained with aluminum alloy; the various pieces of aluminum may be joined with epoxy resin adhesive. One particular commercial adhesive requires no clamping of the pieces to be joined and no special preparation of the joints beyond wiping with carbon tetrachloride.

Curing temperatures vary from room temperatures to 200°F, depending on the time of cure. One and one-half to two hours is sufficient time for temperatures of 200°F. A small thermostat-controlled oven is satisfactory for small models. Joints of larger assemblies may be cured overnight with one or two 100-watt incandescent bulbs in reflectors placed close to them. Tensile and shearing strengths in excess of 3000 lb/in^2 in the joints of aluminum models can regularly be obtained.

Models under test may, of course, be mounted in a position corresponding to the prototype under load. Many times, however, this introduces a tendency for lateral buckling which does not exist in the prototype. A model analysis of a rigid frame, for example, involves the loading of a single frame. The prototype, however, would be loaded as one of several similar frames adequately braced as a group to prevent lateral buckling. Accordingly, it will be found advisable in many cases to mount the model in a horizontal position supported on ball bearings running between two glass plates. Loads may be applied in a horizontal direction through aircraft control cables extending over aircraft control pulleys mounted at the edge of a supporting table. Weights of the desired magnitude may be suspended on these cables. Lead shot in small buckets will provide a flexible system. A loading frame designed and built in the model laboratories at Rensselaer Polytechnic Institute is shown in Fig. 14–24. This frame will accomodate models 9 ft long and 4 or more ft high mounted horizontally, as described above. In addition to direct loading of the model, several levers are provided to apply up to 5000 lb each.

14–18 Auxiliary equipment for direct model analysis. Strains and deflections are usually observed in a direct model analysis of a structure. SR–4 strain gages, for which there is no adequate substitute, are indispensable for reading strains. They are cemented to the surface of the model at the point for which the strain is desired and will indicate strain to 1×10^{-6}. When considerable accuracy is desired in reading deflections, it may be achieved with deflection gages reading directly to 0.001 in. or 0.0001 in. Scales graduated to 0.01 in. will in many cases give satisfactory results.

14–19 The moment indicator. The moment indicator, developed by Ruge and Schmidt (15) at the Massachusetts Institute of Technology

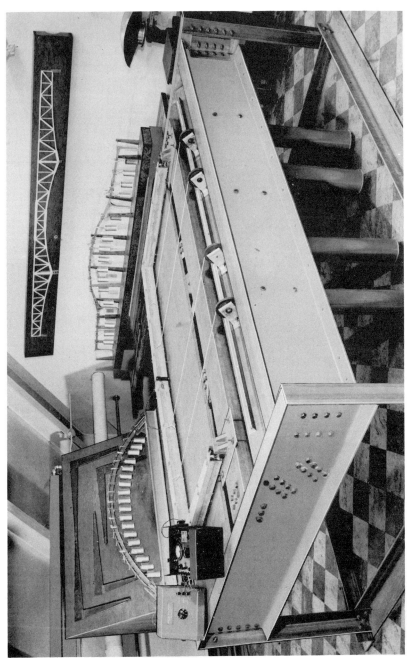

Fig. 14-24. Loading frame.

about 1936, is a most ingenious and useful instrument for direct model analysis. It provides a means for directly determining the absolute values of the internal moments in model flexural members. The model is designed and loaded in accordance with the principles previously developed for direct model analysis. The moment indicator, clamped to the model member at two sections where the moments are desired, provides information from which the desired internal moments may be easily computed. It must be fastened, however, to a length of model member which is initially straight and has a constant modulus of elasticity and moment of inertia between the two sections of attachment. No loads may be applied to the member between these two sections.

Consider the free body LR, of length D, shown in Fig. 14–25(a). This

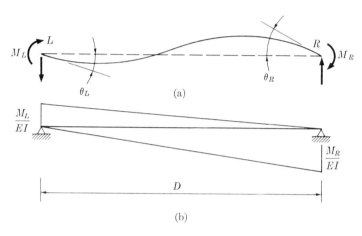

FIGURE 14–25

represents a segment of a model member, the segment having a constant E and I. The internal moments and shears acting at L and R in the uncut member are shown as external loads on the ends of the free body. From the conjugate beam shown in Fig. 14–25(b),

$$\theta_L = \frac{2M_L}{3EI} \cdot \frac{D}{2} - \frac{M_R}{3EI} \cdot \frac{D}{2}$$

and

$$\theta_R = -\frac{M_L}{3EI} \cdot \frac{D}{2} + \frac{2M_R}{3EI} \cdot \frac{D}{2}.$$

A simultaneous solution of these equations yields

$$M_L = \frac{2EI}{D}(2\theta_L + \theta_R), \qquad (14\text{–}34)$$

and
$$M_R = \frac{2EI}{D}(2\theta_R + \theta_L). \tag{14-35}$$

Note that in the above equations, M_L and M_R are positive when acting clockwise on the ends of LR and that θ_L and θ_R are positive when the tangent rotates clockwise relative to the chord LR.

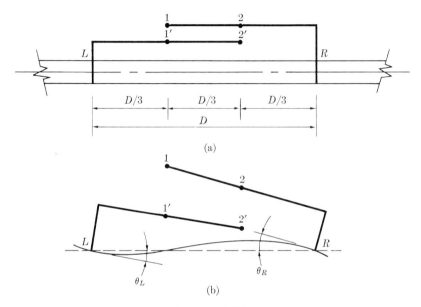

FIGURE 14-26

Figure 14-26 indicates the essential features of the moment indicator. It consists of two rigid arms which are clamped to the unstrained model member at sections L and R. Each arm extends parallel to the axis of the member to a point two-thirds of the distance to the clamping section of the other arm. In Fig. 14-26(b) the positions of these two arms are indicated after the model is loaded. The flexural strain in the model member causes displacement of point 1 relative to 1' and of point 2 relative to 2'. The change in the distance between points 1 and 1' will be designated as Δl, and the simultaneous change between points 2 and 2' will be Δr. Since θ_L and θ_R are actually very small angles,

$$\Delta r = \frac{2D\theta_L}{3} + \frac{D\theta_R}{3} = \frac{D}{3}(2\theta_L + \theta_R)$$

and

$$\Delta l = \frac{D\theta_L}{3} + \frac{2D\theta_R}{3} = \frac{D}{3}(2\theta_R + \theta_L).$$

Substituting in Eqs. (14–34) and (14–35), we obtain

$$M_L = \frac{6EI\Delta r}{D^2} \quad \text{and} \quad M_R = \frac{6EI\Delta l}{D^2}. \tag{14-36}$$

It is important to note in the above equations that Δl and Δr are positive when points 1 and 1', or 2 and 2', move farther apart. The signs of M_L and M_R depend on the signs of Δr and Δl, respectively. A positive M_L or M_R indicates a clockwise external moment acting on the corresponding end of the segment LR.

Excellent results can be obtained with the moment indicator if the values of Δr and Δl are carefully measured. It is suggested that the arms be made of aluminum about $\frac{1}{8}$ in. thick and perhaps $\frac{3}{4}$ in. wide. If two micrometers reading to 0.001 in. are attached at points 1 and 2, with opposing needles at points 1' and 2', the required displacements can be accurately determined. The micrometers and opposing needles should be placed in series with a contact indicator.

In an actual analysis the internal moments at sections L and R of the member are first determined. For a given moment indicator these two sections must always be the same distance apart along the axis of the member. For D, $4\frac{1}{2}$ or 6 in. will be satisfactory, although larger values may be used if desired. Having the moments at L and R, we can compute the moment at any other section along the member.

The moment indicator is particularly suited for model analyses of Vierendeel trusses, secondary stresses in articulated structures, and all types of rigid frames. It is suggested that sheet aluminum be used as the construction material. The required moments of inertia are obtained by varying the widths of the members. Individual members may be joined very easily with epoxy resin adhesive. The resulting model looks similar to those used for indirect analysis; in this case, however, the various moments of inertia must be obtained by application of the scaling factors for direct model analysis. In addition, the model must be loaded in the same way as the prototype and the loads must be scaled.

14–20 The cellulose acetate spring balance. It has been stated that models of cellulose acetate will creep considerably under load because of a change in the modulus of elasticity of the material. This cannot be disregarded in direct model analysis. It has been established (10), however, that by use of a cellulose acetate spring balance it is possible to predict deflections and obtain deflection influence lines for the prototype even though a cellulose acetate model is used. It has also been demonstrated (16) that with a different technique and a similar spring balance the observed strains in a cellulose acetate model may be extrapolated to predict prototype stresses. This latter use of the spring balance will now be described.

Suppose that it is desired to determine the stresses at various points of a rigid frame prototype, as caused by a horizontal concentrated load applied at one knee of the frame. Strains are to be measured at selected points on a cellulose acetate model, and stresses computed therefrom are to be extrapolated to give stresses in the prototype. The frame and load are shown in Fig. 14–27. In accordance with the principles previously de-

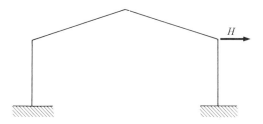

FIGURE 14–27

veloped the applied load H should have a constant scaled value. This constant load could be applied directly to the model, but the measured strains could not be interpreted in terms of stress because of the variation of E. The problem can be solved, however, with the arrangement shown in Fig. 14–28. The rectangle 1, 2, 3, 4 is the cellulose acetate spring balance. The straps 1–2 and 3–4 are each made of two pieces of steel, brass, or

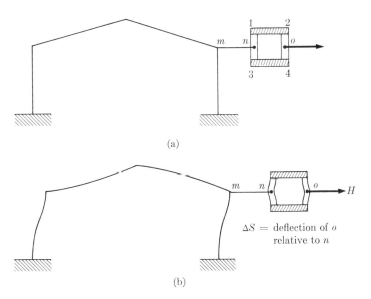

FIGURE 14–28

aluminum, between which the cellulose acetate straps 1–3 and 2–4 are inserted. The axial strain in the metal straps must be negligible. The straps 1–3 and 2–4 are cut from the same piece of cellulose acetate as the model and are rigidly clamped between the metal straps at their ends. The value of ΔS, the deflection of o relative to n, must be carefully measured. It is suggested that a micrometer mounted for clamping on point o with an opposing needle clamped on point n, connected in series with a contact indicator, will provide the best means for this purpose. The connecting member mn can be any convenient material.

The scaled load H is applied at point o of the spring balance. The deflection of the spring balance can be expressed as

$$\Delta S_t = K_s \cdot \frac{H}{E_t}, \tag{14-37}$$

where ΔS_t and E_t are the values of ΔS and E at any time t. K_s is a constant for the balance. Its value, which depends upon the dimensions of the cellulose acetate straps 1–3 and 2–4, may be determined by computation or by calibration.

If the strain ϵ_t is observed at a point on the model, the expression for the stress at the point is

$$\sigma_t = E_t \cdot \epsilon_t;$$

but by Eq. (14-37)

$$E_t = \frac{K_s}{\Delta S_t} \cdot H,$$

and therefore

$$\sigma_t = \frac{K_s}{\Delta S_t} H \cdot \epsilon_t. \tag{14-38}$$

Actually, for any given point on the model, the ratio $\epsilon_t/\Delta S_t$ will have a constant value, since the variation of the modulus of elasticity of cellulose acetate is a function of time after loading and is not affected by stress intensity in the usual range. Consequently, Eq. (14-38) can be written as

$$\sigma = \frac{K_s}{\Delta S_t} H \cdot \epsilon_t. \tag{14-39}$$

In other words, the computed value of stress at a given point of the model should always be the same, regardless of the time at which simultaneous readings of model strain and spring balance deflections are noted.

The assembly shown in Fig. 14–29 is used to determine the value of K_s by calibration. The cantilever beam is cut from the same piece of cellulose acetate as the model and the straps of the spring balance. To it is clamped

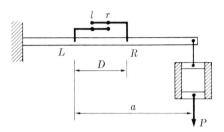

FIGURE 14–29

a moment indicator. The spring balance is connected to the end of the cantilever beam and through it the load P is applied to the cantilever. The moment at L in the beam, in terms of Δr obtained from the moment indicator, is given by Eq. (14–36) as

$$M_L = \frac{6E_t I \, \Delta r_t}{D^2}.$$

E_1 can be expressed by the equation of the spring balance as

$$E_t = \frac{K_s}{\Delta S_t} \cdot P. \qquad (14\text{--}40)$$

Substitution of this value for E_t in the equation above yields

$$M_L = \frac{6I K_s}{D^2} \frac{\Delta r_t}{\Delta S_t} P.$$

But $M_L = Pa$. If we make this substitution and solve for K_s, the result is

$$K_s = \frac{D^2 a \, \Delta S_t}{6I \, \Delta r_t}. \qquad (14\text{--}41)$$

It is suggested that the entire assembly should be mounted on ball bearings on a flat surface. The load should be applied by a cord passing over an aircraft control pulley at the edge of the surface to the proper weight hanging below.

One other difficulty, relative to the design of the model, must be eliminated.

From Eq. (14-26) for a flexural member,

$$\bar{I} = \frac{EI}{\bar{E}k^4z}. \tag{14-42}$$

From Eq. (14-24) for an axially loaded member,

$$\bar{A} = \frac{EA}{\bar{E}k^2z}. \tag{14-43}$$

Obviously, therefore, before a model can be properly designed for direct analysis, the value of the modulus of elasticity of the model construction material must be known. In the case of cellulose acetate this is impossible. It is suggested, therefore, that the model be designed on the basis of what is believed to be the most reasonable value for E. Assume that in a given case this value is 400 k/in^2 and that the model is designed accordingly. The model is fabricated and tested. Strains are measured and stresses are computed as indicated by Eq. (14-39). From observed values of ΔS_t in the spring balance, however, the value of E_t is found, by Eq. (14-40), to be 360 k/in^2. Consequently, the stresses computed for the model must be adjusted. If the model member is flexural,

$$\bar{\sigma} = \frac{\overline{Mc}}{\bar{I}}.$$

Substituting from Eq. (14-42), we obtain

$$\bar{\sigma} = \frac{\overline{Mc}\bar{E}k^4z}{EI}. \tag{14-44}$$

If the model member is axially loaded,

$$\bar{\sigma} = \frac{\bar{P}}{\bar{A}}.$$

If Eq. (14-43) is substituted in the above equation, the result is

$$\bar{\sigma} = \frac{\overline{PE}k^2z}{EA}. \tag{14-45}$$

It is apparent from Eqs. (14-44) and (14-45) that the observed stress in the model is directly proportional to the modulus of elasticity of the model material. Therefore the computed model stresses should be multiplied by 36/40 to obtain the correct values.

Specific References

1. Beggs, G. E., "An Accurate Mechanical Solution of Statically Indeterminate Structures by Use of Paper Models and Special Gages," *Proc. Am. Conc. Inst.*, Vol. 18, 1922.

2. Beggs, G. E., *Discussion of* "Design of a Multiple Arch System," *Trans. Am. Soc. Civ. Engrs.*, **88**, 1208–30, 1925.

3. Beggs, G. E., "The Use of Models in the Solution of Indeterminate Structures," *J. Franklin Inst.*, March, 1927.

4. Buckingham, E., "Model Experiments and the Forms of Empirical Equations," *Trans. Am. Soc. Mech. Engrs.*, Vol. 37, June, 1915.

5. Bull, A., "A New Method for the Mechanical Analysis of Trusses," *Civ. Eng.*, December, 1930.

6. Eney, W. J., "New Deformeter Apparatus," *Eng. News Rec.*, February 16, 1939.

7. Eney, W. J., "Model Analysis of Continuous Girders," *Civ. Eng.*, September, 1941.

8. Eney, W. J., "A Large Displacement Deformeter Apparatus for Stress Analysis with Elastic Models," *Proc. Soc. Exptl. Stress Anal.*, Vol. 6, No. 2.

9. Eney, W. J., "Studies of Continuous Bridge Trusses with Models," *Proc. Soc. Exptl. Stress Anal.*, Vol. 6, No. 2.

10. Eney, W. J., "Determining the Deflection of Structures with Models," *Civ. Eng.*, March, 1942.

11. Gottschalk, O., "Mechanical Calculation of Elastic Systems," *J. Franklin Inst.*, July, 1926.

12. McCullough, C. B. and Thayer, E. S., *Elastic Arch Bridges*, pp. 282–306. New York: Wiley, 1931.

13. Mills, B., "A Sensitive Contact Indicator," *Rev. Sci. Instr.*, February, 1941.

14. Moakler, M. W. and Hatfield, L. P., "The Design and Construction of a Detormeter for Use in Model Analysis." Thesis for degree of Master of Civil Engineering, Rensselaer Polytechnic Institute, June, 1953.

15. Ruge, A. C. and Schmidt, E. O., "Mechanical Structural Analysis by the Moment Indicator," *Trans. Am. Soc. Civ. Engrs.*, Vol. 104, 1939.

16. Wilbur, J. B. and Norris, C. H., *Elementary Structural Analysis*. New York: McGraw-Hill, 1948.

17. Wilbur, J. B., "Structural Analysis Laboratory Research," *Mass. Inst. Tech. Dept. Civ. San. Eng. Publ.*, **65**, December, 1938; **68**, December, 1939; **73**, December, 1940; **80**, December, 1941.

General References

18. Beggs, G. E., "Design of Elastic Structures from Paper Models," *Proc. Am. Conc. Inst.*, **19**, 53–66, 1923.

19. Beggs, G. E., "Skew Arch Reactions Measured by Reciprocal Method," *Eng. News Rec.*, **99**, 106, 1927.

20. Beggs, G. E., "Deformeter Analysis for Rigid Frame Bridges of High Indeterminacy." From *The Rigid Frame Bridge*, by A. G. Hayden. New York: Wiley, 1931.

21. Beggs, G. E., Birdsall, B., and Timby, E. K., "Suspension Bridge Stresses Determined by Model," *Eng. News Rec.*, **108**, 828–32, 1932.

22. Beggs, G. E., "Tests of a Celluloid Model of the Stevenson Creek Dam," *Report of Arch Dam Investigation of Engineering Foundation*, Vol. 1, *Proc. Am. Soc. Civ. Engrs.*, Part III, pp. 219–30.

23. Bertrand, M. J., "Notes on the Principle of Similitude," *Journal de l'Ecole Polytechnique*, Cahier 32, **19**, 189–97. 1848.

24. Bridgman. P. W., "Tolman's Principle of Similitude," *Phys. Rev.* Ser. 2, **8**, 423–31, 1916.

25. Bridgman, P. W., *Dimensional Analysis*. 2nd ed. Yale University Press, 1931.

26. Buckingham, E., "On Physically Similar Systems; Illustrations of the Use of Dimensional Equations," *Phys. Rev.*, Ser. 2, **4**, 345–76, 1914.

27. Buckingham, E., "Notes on the Method of Dimensions," *Phil. Mag.*, Ser. 6, **42**, 696–719, November, 1921.

28. Campbell, N., "Dimensional Analysis," *Phil. Mag.*, Ser. 6, **47**, 481–94, March, 1924.

29. Chick, A. C., "Dimensional Analysis and the Principle of Similitude as Applied to Hydraulic Experiments with Models." In *Hydraulic Laboratory Practice*, by J. R. Freeman, App. 15, pp. 775–827. *Am. Soc. Mech. Engrs.*, 1929.

30. Dawley, E. R., "Wind Pressure Tests Made on Large Model Building," *Eng. News Rec.*, 1928.

31. DenHartog, J. P., "Models in Vibration Research," *Trans. Am. Soc. Civ. Engrs.*, **54**, 153, 1932.

32. Ehrenfest-Afanassjewa, Mrs. T., "Dimensional Analysis Viewed from the Standpoint of the Theory of Similitudes," *Phil. Mag.* Ser. 7, **1**, 257–72, January, 1926.

33. Fleming, R., "Mechanical Method for Determining Reactions of a Continuous Girder," *Eng. News Rec.*, **83**, 428, 1919.

34. Fogg, R. J., "Load Test of Large Model of Cellular Concrete Arch," *Eng. News Rec.*, **102**, 418, 1929.

35. GIBSON, A. H., "The Principle of Dynamic Similarity, with Special Reference to Model Experiments," *Eng.* (London), **117**, 325–27, 357–59, 391–93, 422–23. 1924.

36. GIBSON, A. H., "The Use of Models in Engineering," *Proc. Inst. Mech. Engrs.*, pp. 53–64, 1924.

37. GLICK, G. W., "Rigid Rectangular Frame Foundation for Albany Telephone Building," *Eng. News Rec.*, **105**, 836–38, 1930.

38. GOTTSCHALK, O., "Structural Analysis Based upon Principles Pertaining to Unloaded Models," *Trans. Am. Soc. Civ. Engrs.*, **103**, 1019, 1938.

39. GREEN, N. B., "Flexible 'First Story' Construction for Earthquake Resistance," *Trans. Am. Soc. Civ. Engrs.*, **100**, 645, 1935.

40. GROAT, B. F., "Models, Properly Designed, Show Correctly Performance of Dams and Turbines," *Eng. Rec.*, **72**, 377, 1915.

41. GROAT, B. F., "Ice Diversion, Hydraulic Models, and Hydraulic Similarity," *Trans. Am. Soc. Civ. Engrs.*, **82**, 1139–90, 1918. Also *Can. Eng.*, **34**, 55–58, January 17, 1918.

42. GROAT, B. F., "Theory of Similarity and Models," *Trans. Am. Soc. Civ. Engrs.*, **96**, 273–386, 1932.

43. HERRMANN, W., "The Conditions for Dynamic Similarity," *Verein Deutscher Ingenieure*, **75**, 611–16, 1931.

44. HOUGH, B. K., "Stability of Embankment Foundations," *Trans. Am. Soc. Civ. Engrs.*, **103**, 1414, 1938.

45. JOHANSEN, F. C., "Research in Mechanical Engineering by Small Scale Apparatus," *Proc. Inst. Mech. Engrs.*, March 14, 1929. Also *Eng.* (London), **127** (1927): 371–73, 407–9, 469–70, 567–68, 655–58.

46. KARPOV, A. V. and TEMPLIN, R. L., "Building and Testing an Arch Dam Model," *Civ. Eng.*, **2**, 11–16, January, 1932.

47. LAMBERT, B. J., "Gravity Dams Arched Down Stream," *Trans. Am. Soc. Civ. Engrs.*, **96**, 1178, 1932.

48. LEVY, H., "The Principles of Dynamic Similarity." In Glazebrook, *Dictionary of Applied Physics*. New York: Macmillan, 1922.

49. LYSE, I. and MADSEN, I. E., "Structural Behavior of Battle Deck Floor Systems," *Trans. Am. Soc. Civ. Engrs.*, **104**, 244, 1939.

50. LYSE, I. and BLACK, W. E., "An Investigation of Steel Rigid Frames," *Trans. Am. Soc. Civ. Engrs.*, **107**, 127, 1942.

51. McCULLOUGH, C. B., "Derivation of Theories Underlying Mechanical Methods of Stress Analysis," *Eng. News Rec.*, **105**, 489–90, 1930.

52. MOISSEIFF, L. S., "Designing the Towers of the Hudson River Bridge," *Eng. News Rec.*, **100**, 819, 1928.

53. MOISSEIFF, L. S., "George Washington Bridge: Design of the Towers," *Trans. Am. Soc. Civ. Engrs.*, **97**, 179, 1933.

54. Nelson, M. E., "Laboratory Tests on Hydraulic Models of the Hastings Dam," *Univ. Iowa Stud. Eng., Bull.* 2, 1932.

55. Newton, Sir Isaac, *Principia*, Book II, Theorem XXVI, Propositions XXXII and XXXIII and Corollaries 1 and 2. 1687.

56. Norris, C. H., "Model Analysis of Structures," *Proc. Soc. Exptl. Stress Res.* Vol. 1, No. 2.

57. O'Brien, M. P., "Analyzing Hydraulic Models for Effects of Distortion," *Eng. News Rec.*, **109**, 313–15, 1932.

58. Ottley, J. W. and Brightmore, A. W., "Experimental Investigations of the Stresses in Masonry Dams Subjected to Water Pressure," *Proc. Inst. Civ. Eng.* (London), **172**, 89–106, 1908.

59. Pippard, A. J. S. and Baker, J. F., "An Experimental Investigation into the Properties of Certain Framed Structures Having Redundant Bracing Members," *Aeronaut. Res. Comm. Tech. Rep.* (Great Britain), **2**: 611–35, 1924–1925.

60. Porter, A. W., "Units, dimensions of (including the principles of dynamic similarity)," *Encyclopedia Britannica*, **22**, 853–86, 1929.

61. Rathbun, J. C., "Crown Stresses in a Skew Arch," *Trans. Am. Soc. Civ. Engrs.*, **94**, 135–60, 1930.

62. Rathbun, J. C., "Simple Model Checks Indeterminate Structures," *Civ. Eng.*, **1**: 131–32, 1930.

63. Rathbun, J. C., "An Analysis of Multiple Skew Arches on Elastic Piers," *Trans. Am. Soc. Civ. Engrs.*, **98**, 1–45, 1933.

64. Rathbun, J. C., "Wind Forces on a Tall Building," *Trans. Am. Soc. Civ. Engrs.*, **105**, 1, 1940.

65. Rayleigh, J. W. S., "The Principle of Similitude," *Nature*, **95**, 66–68, or *Sci. Pap.*, Vol. 6, pp. 300–305, 1915.

66. Reynolds, K. C., "Notes on the Laws of Hydraulic Similitude as Applied to Experiments with Models." In *Hydraulic Laboratory Practice*, by J. R. Freeman, App. 14, pp. 759–73. *Am. Soc. Mech. Engrs.*, 1929.

67. Routh, E. J., "The Principle of Similitude," "On Models," "Theory of Dimensions," *Dynamics of Rigid Bodies*. Part I, Arts. 367–74, pp. 292–96. 6th ed. New York: Macmillan, 1897.

68. Ruge, A. C., "Earthquake Resistance of Elevated Water-Tanks," *Trans. Am. Soc. Civ. Engrs.*, **103**, 889, 1938.

69. Savage, J. L. and Houk, I. E., "Checking Arch Dam Design with Models," *Civ. Eng.*, **1**, 695–99, 1931.

70. Savage, J. L. and Houk, I. E., "Model Tests of Hoover Dam," *Eng. News Rec.*, **108**, 494–99, 1932.

71. Seely, F. B. and James, R. V., "The Plaster-Model Method of Determining Stresses Applied to Curved Beams," *Bull.* No. 195, *Eng. Exptl. Sta., Univ. Ill.*, 1929.

72. Smith, B. A., "Experimental Deformations of a Cylindrical Arched Dam," *Trans. Am. Soc. Civ. Engrs.*, **91,** 705–20, 1927.

73. Steinman, D. B., "The St. John's Suspension Bridge," *Eng. News Rec.*, **104,** 272–77, 1930.

74. Steinman, D. B., "Rigidity and Aerodynamic Stability of Suspension Bridges," *Trans. Am. Soc. Civ. Engrs.*, **110,** 439, 1945.

75. Templin, R. L., "Tests of Engineering Structures and Their Models," *Trans. Am. Soc. Civ. Engrs.*, **102,** 1211, 1937.

76. Thomson, J. T., "Model Analysis of a Reinforced Concrete Arch," *Publ. Rds.* Vol. 9. No. 11.

77. Tietjeus, O. G., "Use of Models in Aerodynamics and Hydrodynamics," *Trans. Am. Soc. Mech. Engrs.*, **54,** 225–33, September, 1932.

78. Timby, E. K., Mead, L. M., and McEldowney, R., "Moments in Flexible Two-Hinged Arches," *Trans. Am. Soc. Civ. Engrs.*, **110,** 35, 1945.

79. Tolman, R. C., "The Principle of Similitude," *Phys. Rev.* Ser. 2, **3,** 244, 1914.

80. Tolman, R. C., "The Specific Heat of Solids and the Principle of Similitude," *Phys. Rev.* Ser. 2, **4,** 145–53, 1914.

81. Tolman, R. C., "The Principle of Similitude and the Principle of Dimensional Homogeneity," *Phys. Rev.* Ser. 2, **6,** 219–33, 1915.

82. Vogel, H. D. and Dean, J. P., "Geometric versus Hydraulic Similitude," *Civ. Eng.*, **2,** 467–71, 708–09, August, 1932.

83. Weber, W., "The Fundamentals of the Mechanics of Similitude and Its Use in Model Experiments," *Jahrbuch der Schiffsbautechnischen Gesellschaft*, **20,** 355. 1919.

84. Weber, W., "The General Principles of Similitude in Physics and Their Relation to Dimensional Theory and the Science of Model Testing," *Jahrbuch der Schiffsbautechnischen Gesellschaft*. Berlin: J. Springer, Bd. 31, 1930.

85. Weiner, B. L., *Discussion of Paper No.* 1827, "An Analysis of Multiple Skew Arches on Elastic Piers," *Trans. Am. Soc. Civ. Engrs.*, **98,** 46, 1933.

86. Williams, H. A., "Dynamic Distortions in Structures Subjected to Sudden Earth Shock," *Trans. Am. Soc. Civ. Engrs.*, **102,** 838, 1937.

87. Wilson, J. S. and Gore, W., "Stresses in Dams; An Experimental Investigation by Means of India Rubber Models," *Proc. Inst. Civ. Eng.* (London), **172,** 107–33, 1908.

88. Wilson, W. M., "Laboratory Tests of Multiple-Span Reinforced Concrete Arch Bridges," *Trans. Am. Soc. Civ. Engrs.*, **100,** 424, 1935.

89. Witmer, F. P. and Bonner, H. H., "Tall Building Frames Studied by Means of Mechanical Models," *Trans. Am. Soc. Civ. Engrs.*, **102,** 244, 1939.

APPENDIX

Coefficients for the determination of fixed-end moment, stiffness, and carry-over factors for nonprismatic members. (Courtesy of the Portland Cement Association.)

APPENDIX

SYMMETRICAL MEMBERS WITH STRAIGHT HAUNCHES

1. Stiffness Coefficient, k

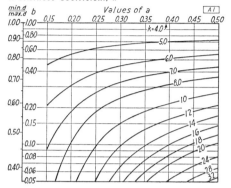

2. Carry-over Factor, C

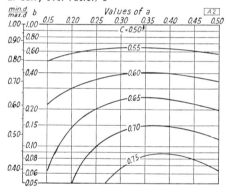

3. Uniform Load f.e.m. Coefficient, f

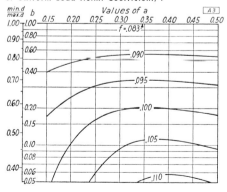

4. Concentrated Load f.e.m. Coefficient, f

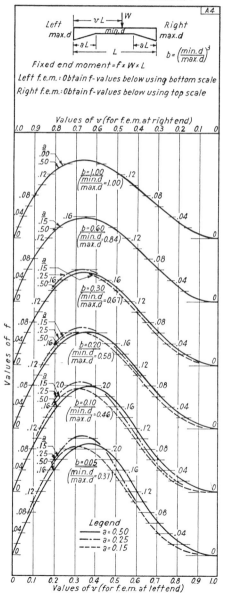

SYMMETRICAL MEMBERS WITH PARABOLIC HAUNCHES

1. Stiffness Coefficient, k

2. Carry-over Factor, C

3. Uniform Load f.e.m. Coefficient, f

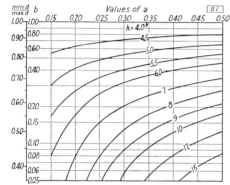
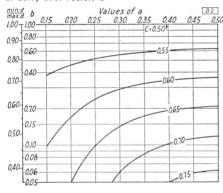
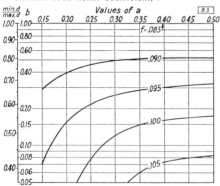

4. Concentrated Load f.e.m. Coefficient, f

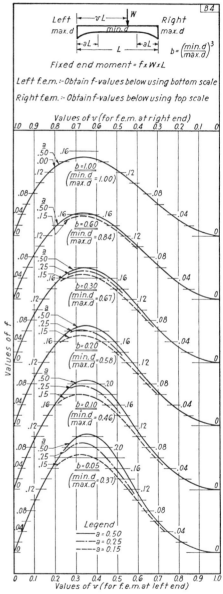

APPENDIX

UNSYMMETRICAL MEMBERS WITH STRAIGHT HAUNCH AT ONE END
Coefficients at Haunched End

1. Stiffness Coefficient, k

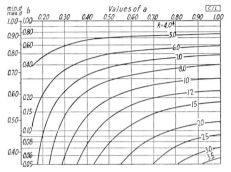

2. Carry-over Factor, C

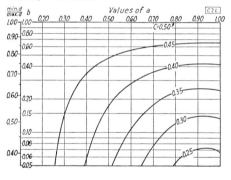

3. Uniform Load f.e.m. Coefficient, f

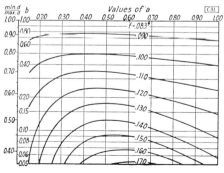

4. Concentrated Load f.e.m. Coefficient, f

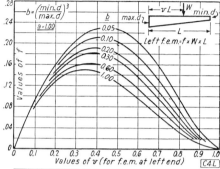

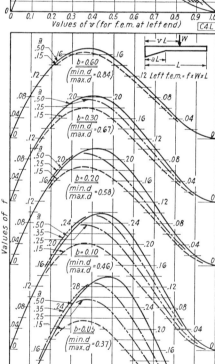

UNSYMMETRICAL MEMBERS WITH PARABOLIC HAUNCH AT ONE END
Coefficients at Small End

1. Stiffness Coefficient, k

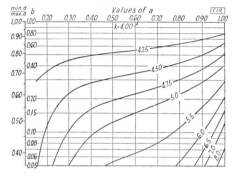

2. Carry-over Factor, C

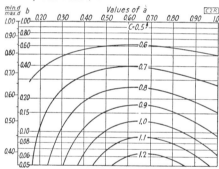

3. Uniform Load f.e.m. Coefficient, f

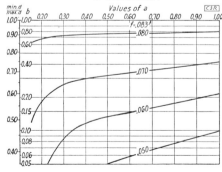

4. Concentrated Load f.e.m. Coefficient, f

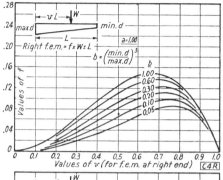

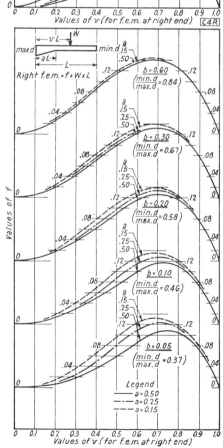

UNSYMMETRICAL MEMBERS WITH STRAIGHT HAUNCH AT ONE END
Coefficients at Haunched End

1. Stiffness Coefficient, k

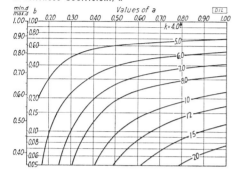

2. Carry-over Factor, C

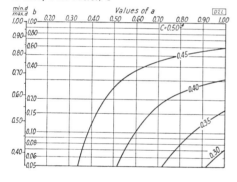

3. Uniform Load f.e.m. Coefficient, f

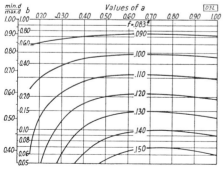

4. Concentrated Load f.e.m. Coefficient, f

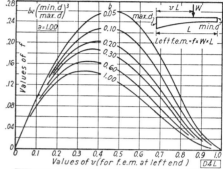

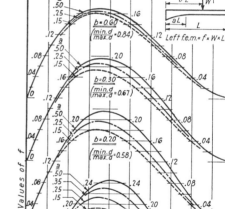

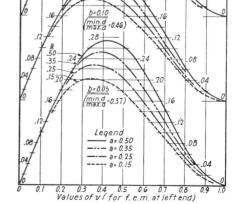

UNSYMMETRICAL MEMBERS WITH PARABOLIC HAUNCH AT ONE END
Coefficients at Small End

1. Stiffness Coefficient, k

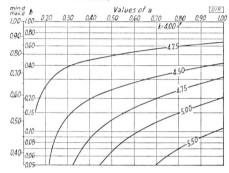

2. Carry-over Factor, C

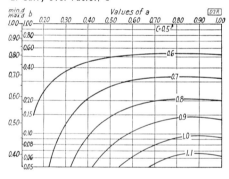

3. Uniform Load f.e.m. Coefficient, f

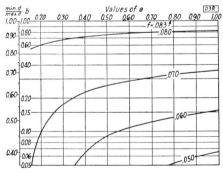

4. Concentrated Load f.e.m. Coefficient, f

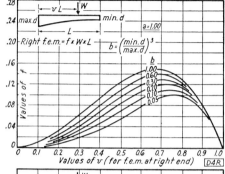

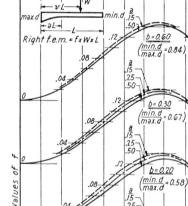

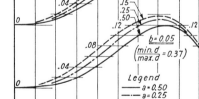

INDEX

Absolute stiffness, 303
Alexander the Great, 5
Alexandria, 5
Analogous column, 278
Analysis-design procedure for indeterminate structures, 43
Anaximander, 4
Angle changes in articulated structures, 180
Arabic numbers, 7
Arc welding, effect of introduction of, 20
Arches, curve of axis of, 540
   early analysis of, 10, 537
   hingeless ribs, 561
   influence lines, 571
   rib shortening and temperature stresses, 556
   secondary stresses, 580
   two-hinged articulated ribs, 541
   two-hinged parabolic ribs, 559
   two-hinged solid ribs, 543
Archimedes, 5
Aristotle, 1, 5
Articulated structure, definition of, 21
Astronomy, 4
Atomic theory of Democritus, 5

Babylon, 4
Bacon, Roger, 7
Beams, analysis of, 8, 10, 11
   curved, 12
   elastic curves of, 9
   vibration of, 9
Beggs deformeter, 599
Bendixen, Axel, 477
Bernoulli, Daniel, 9, 254
Bernoulli, James, 9
Bernoulli, Johann, 9, 55, 96
Betti, E., 12
Brass spring models, 608
Bresse, Jacques, 12, 537

Brewster, Sir David, 10
Brothers of the Bridge, 536
Buckingham, E., 617
Bull, Anders, 608

Carry-over factors, curves for, 454
Carus, Lucretius, 6
Castigliano, Alberto, 13, 254
Castigliano's first theorem, 13, 84
Castigliano's second theorem, 13, 254
   analysis of continuous beams and frames by, 258
   analysis of articulated structures by, 263
Cellulose acetate spring balance, 632
Characteristic points, 14
Clapeyron, 11, 184
Columns, eccentrically loaded, 10
   elastic curves of, 9
   Gordin-Rankine formula, 12
   flexure formula, 278
Column analogy, 278
   carry-over factors for nonprismatic members by, 445
   continuous frame analysis by, 291
   development of the method, 281
   fixed-end moments for nonprismatic members by, 445
   stiffness of nonprismatic members by, 445
   units in, 290
Conjugate beam method, 13
   continuous beam analysis by, 139
   deflections by, 129
   fixed-end moments, stiffness, and carry-over factors for nonprismatic members by, 464
Conjugate structure, 143, 157
Constantinople, 7
Contact indicator, 607
Continuous frame, definition of, 21
Coulomb, Charles Augustine, 10, 537

Cremona, 12
Cross, Hardy, 15, 20, 278, 302
Crotona, 4
Culmann, Carl, 12, 240, 537
Curve of arch axis, 540
Cyclic method of design, 45
Cylinders, stresses in hollow, 11

daVinci, Leonardo, 7
Deflections: Castigliano's first theorem, 84
  conjugate beam, 134
  conjugate structure, 143
  elastic weights:
    beams, 134
    articulated structures, 179
  moment area, 127
  real work:
    flexure, 81
    axial loads, 83
  virtual work:
    axial strains, 96
    flexural strains, prismatic members, 104
    flexural strains, nonprismatic members, 114
    shearing strains, 118
    torsional strains, 119
  Williot-Mohr diagram, 164
  condition equations, 48
Deflection theory, 54
Democritus, 5
Saint-Venant, 11
Determinate vs. indeterminate structures, 42
Degree of indeterminateness, 22
Design of models:
  direct analysis, 620
  indirect analysis, 608
Dimensional analysis, 616
Direct model analysis, 611
  design of models, 620
  fabrication and loading of models, 627
Distribution factor, 303

Earth pressure, theory of, 10

Eastern Roman Empire, 7
Elastic arches, 536
Elastic center, 240
Elastic theory, 54
Elastic weights, method of, 13
Elasticity, first book on theory of, 11
Eney deformeter, 603
Eney, W. J., 603, 608
Equilibrium, unstable, 21
Esla arch, 538
Etruscans, 536
Euclid, 5
Euler, Leonhard, 9, 254
External stability, 22

Fidler, Claxton, 14
Fixed-end moments, nonprismatic members:
  by the analogous column, 449
  by the conjugate beam, 464
  caused by relative displacement of member ends, 470
Fixed-end moments, prismatic members:
  caused by displaced supports, 308
  caused by various loads, 313
Fletcher, Sir Banister, 4
Fratres, Pontes, 536
Freudenthal, A., 538
Fundamentals of indirect model analysis, 586

Galilei, Galileo, 8
Galileo's problem, 8
General method, 52
  articulated structures, 198
  beams, 185
  continuous frames, 213
Geometric instability, 27
Geometry, 4
Gesteschi, T., 538
Graphic statics, 12
Greene, Charles E., 13
Grubenmann, Ulric, 9

Henry Hudson arch, 540
Herodotus, 4
Hingeless arch ribs, 561

Hooke's law, 8
Hooke, Robert, 8

Imhotep, 3
Indeterminate structural analysis, first theory for, 12
Indeterminate structure, definition of, 22
Indeterminateness, degree of, 22
Indirect model analysis:
  Beggs deformeter, 599
  brass spring models, 608
  contact indicator, 607
  Eney deformeter, 603
  errors from changes in geometry, 598
  fundamentals, 586
  model design, 608
  model materials, 596
  moment deformeter, 607
  R.P.I. deformeter, 605
Infinitesimal calculus, 9
Influence lines, 12, 507
  articulated structures, 532
  by moment distribution, 526
  continuous beams, prismatic members, 513
  continuous beams, nonprismatic members, 522
  qualitative by Müller-Breslau principle, 532
Internal determinateness, definition of, 27
  condition equation for, 28
Internal work, axial loads, 74

Kill Van Kull arch, 538

Lahire, 537
Lamé, G., 11
Land surveying, 4
Law of the lever, 7
Least work, method of, 9, 13, 254
  articulated structures, 263
  continuous beams and frames, 258
Leibnitz, 5
Leucippus, 5
Loaded chord, problem of the, 8

London bridge, 536
Lavoisier, 6
Lyceum, 5

Manderla, Heinrich, 14, 477
Maney, G. A., 14, 20, 477
Marcellus, 6
Mariotte, E., 9
Maxwell, James Clerk, 12
Maxwell-Betti reciprocal theorem, 12, 65, 67
Maxwell-Cremona diagram, 12
Maxwell's reciprocal deflection theorem, 14, 124
Mechanics of materials, founding of science of, 10
Melan, J., 538
Menabrea, 254
Method of sections, 12
Method of virtual work, 96
Model analysis of structures, 14, 584
Model design, 608
Model materials for indirect types of analysis, 596
Mohr, Otto, 13, 14, 15, 164, 184, 477, 538
Mohr rotation diagram, 172
Molecular action, nature of, 11
Moment distribution, 15
  absolute stiffness, 303
  absolute stiffness of prismatic members, 307
  carry-over factor, 306
  continuous beams with prismatic members, 314
  crane loads on gable frames, 404, 408
  dead and snow loads on gable frame, 387
  distribution factor, 303
  effect of introduction of, 20
  fixed-end moments induced by displaced supports, of prismatic members, 308
  of nonprismatic members, 471
  frames with one degree of freedom of joint translation, 338
  frames with restrained joints, 431

frames with several degrees of freedom, 420
frames with two degrees of freedom, 368
girder shortening in gable frames, 398
influence lines by, 526
irregular frames, 382
method for checking results, 324
multispan irregular gable frames, 420
relative stiffness, 305
secondary stresses, 426
sign convention, 302
symmetry and antisymmetry, 322
temperature stresses in gable frames, 398
tie elongation of gable frames, 398
two-story bents, 369
viaduct bents with sloping legs, 377
Vierendeel truss, 423
wind loads on gable frame, 393
Moment deformeter, 607
Moment-area method, 13
Moment indicator, 628
Moment of a force, concept of, 7
Morsch, E., 538
Müller-Breslau, Heinrich, 14, 61, 184
Müller-Breslau principle, 14, 70, 507, 509, 532

Navier, Louis Marie Henri, 10, 537
Newton, Sir Isaac, 1, 5, 9
Nonprismatic members:
 carry-over factors, 445
 curves for stiffness and carry-over factors, 454
 fixed-end moments, stiffness, and carry-over factors, by column analogy, 445
 by conjugate beam, 464
 influence lines for continous beams, 522
 stiffness, 445
Notation for deflections, 45

Ostenfeld, A., 14, 15

Palladio, Andrea, 8
Pascal, 3
Photoelasticity, 10, 14
Pi theorem, 617
Plastic theory, 54
Plasticity, theory of, 11
Poncelet, J. V., 537
Primary stresses, 21
Principia, 1, 9
Principle of superposition, 52
Principle of virtual displacements, 55
Principle of virtual work, 96
Ptolemy I, 5
Pythagoras, 4

Rainbow arch, 540
Rankine, William John M., 12
Reciprocal deflection theorem, Maxwell's, 12, 61, 63, 65, 509, 592
Reciprocal theorem, Maxwell-Betti, 12, 65, 587
Reinforced concrete, effect of introduction of, 20
Relative stiffness, 305
Relaxation method, 15
Renaissance, 537
Ritter, August, 12
Roman Empire in the west, collapse of, 536
R.P.I. deformeter, 605

Sando arch, 538
Secondary stresses, 13, 14, 21
 by moment distribution, 426
 by slope deflection, 502
 in arch ribs, 580
Second theorem of Castigliano, 254
Significant figures and approximate numbers, 209
Simultaneous equations, solution of, 208, 504
Slope-deflection method, 13, 14
 continuous beams, 481
 development of the method, 477
 frames with one degree of freedom, 486

frames with several degrees of freedom, 501
gable frames, 495
secondary stresses, 502
Society of Austrian Engineers and Architects, 538
Southwell, R. C., 11, 15
Spheres, stresses in hollow, 11
Spline method of indirect model analysis, 597
Stability, 21
Stable structure, 21
Stevin, Simon, 8
Stiffness of nonprismatic members, by the column analogy, 445
curves for, 454
by the conjugate beam, 464
with far end pinned, 470
Strassner, A., 538
String polygon, 13
Structural similitude, 585
Superposition, principle of, 52
Supports, types of, 22
Suspension bridges, early analysis of, 10
Sydney Harbor arch, 538
Syracuse, 6

Tension coefficients, 14
Thales, 4
Thebes, 536
Theorem of reciprocal deflections, 184
Theorem of three moments, 11
Torsion of shafts, 10
Tower of Babel, 4
Trajan's bridge, 6
Tredgold, 12
Trusses, first use of, 8, 9,
Two-hinged articulated arch ribs, 541
Two-hinged parabolic arch ribs, 559
Two-hinged solid arch ribs, 543

Universal gravitation, law of, 9
Unstable equilibrium, 21

Varignon, 12
Vierendeel truss, 423
Villarceau, Yvon, 537
Virtual displacement, definition of, 55
Virtual displacements, principle of, 55
Virtual velocity, 9
definition of, 55
Virtual work, definition of, 55
principle of, 57, 59

Western Roman Empire, 6
Whipple, Squire, 11, 20
Why moment distribution works, 309
Williot, 13, 164
Williot-Mohr diagram, 13, 164
Wilson, C. A. Carus, 14
Winkler, E., 12, 507, 537